Statistical Mechanics

Proceedings of the Sixth IUPAP Conference on Statistical Mechanics

STATISTICAL MECHANICS
NEW CONCEPTS, NEW PROBLEMS, NEW APPLICATIONS

Edited by
Stuart A. Rice, Karl F. Freed, and John C. Light

The University of Chicago Press
Chicago and London

The University of Chicago Press, Chicago 60637
The University of Chicago Press, Ltd., London

Published 1972
Printed in the United States of America
International Standard Book Number: 0-226-71115-3
Library of Congress Catalog Card Number: 72-85434

Contents

PREFACE

The sixth in a series of conferences on statistical mechanics sponsored by the International Union of Pure and Applied Physics was held from March 29 to April 2, 1971, at The University of Chicago. This volume contains the Proceedings of that conference.

I was asked in March 1970 to organize the IUPAP meeting in Chicago. Consultations with my colleagues quickly led to the definition of a set of guidelines which were used to plan the conference. Given the contents and themes of previous meetings it was decided to devote this one to the trinity: New Concepts, New Problems, New Applications. IUPAP meetings on statistical mechanics have traditionally been important points of information transfer. Through formal lectures and informal contacts summaries of achievements are transmitted, new problems outlined, and hints as to new avenues for investigation tentatively advanced. This has been possible because the attendance at the meetings has been limited. In the recent past the community of workers in the field of statistical mechanics was small, and each practitioner was acquainted with a sizable cross-section of the entire community. Then, even with a limited attendance, a wide and representative variety of views was presented at a meeting. Today the community interested in statistical mechanics has grown quite large. Despite this growth I and my colleagues felt that preservation of the function of the meeting was a desideratum of the planning. To achieve this it was necessary to restrict the attendance to approximately 250 and to organize the scientific sessions carefully.

Few scientists today feel that listening to a barrage of short papers is a satisfactory learning experience. When such papers are presented in sessions that are run in parallel, there results an almost universal feeling of frustration among those attending the meeting. Accordingly, it was decided that there should be only one session at a time, that the lectures given should be long enough to permit in-depth exploration of the subject, and that lectures should be grouped around common themes. To help expose the audience to new mathematical techniques and new problems, one session was organized around ergodic theory and another around theoretical models relevant to biological applications. Provision was made for the informal presentation of contributed papers. Lengthy abstracts of all these were given to all attendees at registration, and the informal talks were distributed so as to interfere as little as possible with the main sessions. Whether or not this organization was successful, and whether the inclusion of mathematicians and biologists as lecturers stimulated the audience, are not for me to judge. The attendance at all sessions was good, so the audience seemed willing to work at making the experiment a success.

The order of the papers in this volume is the same as presented at the conference. Some thought was given to rearranging them, especially when this would better group papers dealing with similar topics, but in the end this was not done. The entire set of papers does not form a linked text, although several subsets, taken in sequence, do provide good coverage of a subject. It is unlikely that anyone will read these entire proceedings in order, and it was thought better to retain the order of papers as presented.

The IUPAP meeting on statistical mechanics could not have been organized in the short time between March 1970 and November 1970, when final notices went out, without the help of my colleagues and the staff of the University of Chicago. Professors John Light and Karl Freed put forth prodigious efforts, and Professor Dieter Forster and Dr. William Gelbart were of very great assistance. I am grateful to Professor J. Lebowitz (Yeshiva University), Professor A. J. F. Siegert (Northwestern University), and Dr. F. Stillinger (Bell Laboratories) for mailing lists that helped ensure wide circulation of the meeting notices. Mrs. Rosalind O'Gallagher deftly handled the secretarial burden and the complications of negotiating with overseas and domestic visitors during the entire organization, planning, and execution periods. Mrs. Wendy Mobasser and the staff of the Center for Continuing Education, University of Chicago, managed the hotel accommodations, set up of meeting rooms, helped with the staffing, and made many other valuable contributions to making the meeting run smoothly. I am grateful to Mr. Rudolf Banovitch, Mr. Edward Wolowiec, and their associates for efficient processing of the enormous volume of meeting notices and contributed paper abstracts.

The IUPAP meeting on statistical mechanics could not have been held without the financial assistance of IUPAP, the Alfred P. Sloan Foundation, the National Science Foundation, and the Army Research Office (Durham). I, my colleagues, and--I am sure--all those who attended, are grateful for their contributions.

Stuart A. Rice

I. Fundamental Principles

PERSONAL PERSPECTIVES ON MATHEMATICS AND MECHANICS

Steve Smale

1.

These days especially, provocative questions confront a socially conscious scientist when he begins to contemplate where applications of his work might lead. As one whose main work has been in pure mathematics, and who is beginning to concern himself with areas of applied mathematics such as electrical circuit theory, I wonder to what extent I should explicitly direct my work toward socially positive goals. Many issues on the relations of the profession of a scientist to the social crises of this time deserve a format, for example, at various professional meetings and conferences, we well as in the departments of universities. There are strong pressures on the scientist, from the sources of funding research, scientific tradition, and a basic conservatism of maintaining the status quo, which act to prevent discussion of these issues. Although I am not going to pursue such a discussion here, I feel that mathematicians and scientists in general must face these questions in a much more serious way than they have done to date (myself included).

2.

A fairly general procedure for the mathematical study of a physical system starts with explication of the space of states of that system. Now this space of states could reasonably be one of a number of mathematical objects. However, in my mind, a principal candidate for the state space should be a differentiable manifold; and in case the system has a finite number of degrees of freedom, then this will be a finite dimensional manifold. Usually associated with the physical system is the notion of how a state progresses in time. The corresponding mathematical object is a dynamical system or a first order ordinary differential equation on the manifold of states.

Too often in the physical sciences, the space of states is postulated to be a linear space, when the basic problem is essentially nonlinear; this confuses the mathematical development.

For the International Union of Pure and Applied Physics Conference on Statistical Mechanics, Chicago, March 1971.

On the other hand, recent decades have seen big developments in parts of mathematics related to differentiable manifolds. Thus there should be no reason for reticence on the part of applied mathematicians to use these developments.

We have tried to provide a little background for what follows. I will describe briefly aspects of three areas in which I have been working. These are mechanics, differentiable dynamical systems, and electrical circuits.

3.

Recent years have seen several textbook accounts of the development of classical mechanics on manifolds. Two of these are Abraham and Marsden, Foundations of Mechanics [1], and Loomis and Sternberg, Advanced Calculus [2]. A number of journal articles also have appeared on aspects of this subject, including my "Topology and Mechanics" [3].

A basic motivation of this literature is to make the discipline of classical mechanics more elegant, more geometric (and so perhaps less analytic), more global. Furthermore, this development has done a lot to unify mechanics with large parts of geometry and part of topology. The globalization has removed the need for a linear space framework, such as one finds in the traditional classical mechanics text.

I will give a quick indication how this goes. Configuration space is a smooth manifold M. For the space of states one takes the tangent bundle T of M or the cotangent bundle T^* of M. The manifold T^* has a natural symplectic structure,i.e., a canonical nondegenerate 2-form Ω defined on it. This form defines an isomorphism between vector fields and 1-forms. Thus if $H:T^* \to R$ is a function on T^* (a "Hamiltonian"), dH is a 1-form which corresponds to a unique vector field X_H via Ω. A vector field is the same thing as a first order ordinary differential equation; in fact, X_H is one way of representing Hamilton's equations. It is immediate then that $X_H(H) = 0$ or that differentiation of H along X_H is zero. This is conservation of energy. Symmetry and other conservation laws are related similarly, simply and conceptually. Furthermore, X_H preserves Ω and hence the volume element Ω^n, where n = dim M. This is Liouville's theorem.

Suppose one starts with a kinetic and potential energy as well as M. How does that relate to this context? Potential energy is a smooth function $V:M \to R$, and kinetic energy can be thought of as a Riemannian metric on M with $K:T \to R$ given by norm squared of a tangent vector with respect to this Riemannian metric. Total energy $E:T \to R$ is the function $K + V \circ \Pi$, where $\Pi:T \to M$ is the projection. The Riemannian metric induces a bundle isomorphism (the Legendre transformation $T \to T^*$ and thus pulls Ω back to T to give a symplectic structure on T. Thus there is a vector field X_E naturally associated to E via this symplectic structure. The vector field X_E generalizes Newton's equations, and of course the energy E is conserved. If one has a Lie group $\mathfrak{G}$ of symmetries acting on M leaving E invariant, one has a natural induced map α of the Lie algebra $\mathcal{Y}$ of $\mathfrak{G}$ into vector fields on M. Let $\alpha_x : \mathcal{Y} \to T_x$ be the linear map obtained by evaluating the vector field $\alpha(X)$ at $X \in M$ where $X \in \mathcal{Y}$. One then has a map $J_1 : T^* \to \mathcal{Y}^*$ obtained by letting J_1 restricted to a fiber T_x^* be the dual linear map of α_x. The composition J, $T \to T^* \to \mathcal{Y}^*$ generalizes angular momentum. This map $J : T \to \mathcal{Y}^*$ is constant on orbits, and furthermore it can be shown that J is equivariant with respect to the adjoint action on G on $\mathcal{Y}^*$. This description of angular momentum helps clarify Jacobi's "elimination of the node" in celestial mechanics.

In the above if $V \equiv 0$, one has simply the case of Riemannian geometry, and the projections into M of the integral curves of X_E are geodesics. More generally the projections of these integral curves are the trajectories in configuration space.

We have given here the flavor of how this modernization of mechanics goes. The assertions above can usually be proved with very little effort, if one has accepted the basic elements of differential topology and geometry. The above-mentioned three references do cover and expand on this material.

For many modern mathematicians, the elegance of this treatment of mechanics would be justification for its development. Beyond this, some new insights have been obtained already by this globalization with, I believe, much more to come.

4.

A lot of research has been done in recent years in the area of mathematics now called differentiable dynamical systems. Let me describe briefly, roughly, what it is about and after that give an assessment. For more substantial and precise information on this subject one can see my "survey," "Differentiable Dynamical Systems" [4], and its references, as well as the more recent Global Analysis [5].

The discipline of differentiable dynamical systems attempts to study systematically dynamical systems on a manifold. So one forms the topological space Dyn(M) (with a topology respecting differentiable properties) of all dynamical systems on a manifold M.

One has a reasonable notion of "almost all" in Dyn (M) meaning a Baire set (a countable intersection of open dense sets), and when convenient one restricts attention to such a set. A theorem is regarded as useful if, for example, it gives a property valid for all systems of some Baire set of Dyn(M).

Related is the idea of structural stability and certain variations. This kind of stability is a property of a dynamical system itself (not of a state or orbit) and asserts that nearby dynamical systems have the same structure. The "same structure" can be defined in several interesting ways, but the basic idea is that two dynamical systems have "the same structure" if they have the same gross behavior, or the same qualitative behavior. For example, the original definition of "same structure" of two dynamical systems was that there was an orbit preserving continuous transformation between them. This yields the definition of structural stability proper. It is a recent theorem that every compact manifold admits structurally stable systems, and almost all gradient dynamical systems are structurally stable. But while there exists a rich set of structurally stable systems, there are also important examples which are not structurally stable, and have good but weaker stability properties.

One series of theorems has been in the direction of characterizing various stability properties of dynamical systems in terms of more tangible properties of the system. One cannot expect this development per se to help in the study of mechanical systems as in section 3 above, because small perturbations will destroy the Hamiltonian behavior of a mechanical system: in particular, dissipation accomplishes this. On the other hand, engineering, and problems from the biological sciences, could profit from this approach through stability questions. What happens near an attractor or sink would be of special interest in applications. Most simple is the steady-state case (fixed point of the dynamical system), then the oscillation (or periodic behavior); but recent studies have shown how easy it is to have far more complicated attractors appear. In particular, even the structurally stable attractors in rather simply given examples contain a vast mixture of periodic, almost periodic, homoclinic, and other kinds of phenomena.

So differentiable dynamical systems have developed as a purely mathematical subject: the attempt to study all ordinary differential equations on a manifold. As such, a wealth of theorems and knowledge has been accumulated. On the other hand, the subject has not achieved any definitive state.

One of the main recent achievements has been to remove the fear of making a global study of differential equations on manifolds of dimension greater than 2. Many examples, and in fact large classes of systems, with good stability properties are now known and well understood on manifolds of all dimensions. On the other hand, we have concrete examples not understood as to their stability properties; in fact, there remain good questions to be found. The subject is in a state of activity and flux at this time. As I have said, up to this point the study of differentiable dynamical systems has developed as a purely mathematical discipline with only minimal direct relations to physics and engineering. I suspect there is a good potential for interaction. For example, the above-described work might be used to extend the horizons of applied mathematics by expanding the realm of possibilities for mathematical models.

5.

I would like to give here the flavor of some recent work of mine on electric circuits--from "Mathematical Foundations of Electrical Circuits," preprints in existence. Our development proceeds in a natural manner, working with what is given by the circuit and making no arbitrary choices. We consider circuits--for example, as in C. Desoer and E. Kuh, Basic Circuit Theory [6] --with nonlinear elements, either resistors, inductors, or capacitors. Associated with a circuit is a linear graph, say connected and oriented, consisting of nodes or 0-cells and elements, branches, or 1-cells.

An unrestricted state is the set of currents and voltages in each branch and is thus a 2b-tuple of real numbers, where b is the number of branches. The space $\mathfrak{S}$ of all states then is real

Cartesian space of 2b dimensions. Kirchoff's and Ohm's laws will impose conditions so that a physical state must lie in a certain subset $\Sigma \subset \mathfrak{S}$ which we proceed to define.

First, Kirchoff's current and voltage laws define linear relations on the currents and voltages in the branches, and thus restrict states to lie in a linear subspace K of $\mathfrak{S}$.

Let $i_R \times \upsilon_R : \mathfrak{S} \to R \times R'$ be the projection defined by taking the resistance currents and resistance voltages. Thus R consists of currents through the resistors and R' is the linear space of voltages across the resistors. We take Ohm's law to say that for each resistor ρ, the characteristic is a closed one-dimensional submanifold Λ_ρ of the current voltage plane of that resistor. Then these Λ_ρ define a closed submanifold (their product) Λ in $R \times R'$. Let $\Pi' : K \to R \times R'$ be the restriction of $i_R \times \upsilon_R$. Then consider. Hypothesis: Π' is transversal to Λ.

Although this is a key hypothesis, rather than recalling its rather technical definition we note the main consequences that are of concern to us. Under this hypothesis, $\Pi^{-1}(\Lambda) = \Sigma$ is a smooth manifold of $\mathfrak{S}$ and dim Σ = the number of inductors plus the number of capacitors. Furthermore, this hypothesis is generic in that it will be satisfied with almost all choices of resistor characteristics.

The space of physical states is naturally defined as Σ (those satisfying both Kirchoff's and Ohm's laws). We see the above hypothesis as key, because in giving a manifold structure to Σ, it permits analysis of the space of states. In fact, via Faraday's laws, we can obtain an ordinary differential equation, perhaps singular, on Σ.

In brief, this goes as follows (but see the above reference for details). The inductances and capacitances define a symmetric form I over Σ; this form cannot, in general, be expected to be positive definite at each point of Σ. In fact, besides being indefinite, I may also be degenerate on part of Σ. The resistances define a closed 1-form w on Σ. Then the differential equations are the gradient of w with respect to I. These are going to be much more complicated than the usual gradients because of the general nature of I. These equations contain in increasing generality those of Van der Pol, Lienard, and Brayton-Moser. On the other hand, they contain examples which are of a different type corresponding to "relaxation oscillations" and thus contribute to achieving a certain unity in the differential equations of circuit theory.

6.

There is a remaining aspect of this global analysis approach to applied mathematics that I would like to mention. The approach taken in this paper may indeed be a little hard to accept for the applied mathematician trained in traditional methods. However, the approach here tends to make applied mathematics and also ordinary differential equations accessible and attractive to the modern mathematician, one who has been brought up in the purist, Bourbaki style of education.

For several decades, the most important mathematical tendencies in the United States and Western Europe have been very separated from applications and there has even been a certain disdain for applied mathematics by many leading mathematicians. I believe that there have been some healthy sides to this separation. Many fields in pure mathematics have flourished in this time, and perhaps because of the division a certain clarity has been achieved.

On the other hand, at least since before the time of Newton and Leibniz, never have the main schools of mathematics, the main mathematicians, been so sharply separated from other disciplines. I believe it worthwhile to try to change this course of events. The modernizations I have described in mechanics, ordinary differential equations, and circuits will hopefully help to break this isolation by bringing modern mathematics more substantially into communication with other fields.

REFERENCES

1. Abraham and Marsden, Foundations of Mechanics (New York, 1967).

2. Loomis and Sternberg, Advanced Calculus (Reading, Mass., 1968), chap. 13.

3. Steve Smale, "Topology and Mechanics," I and II, Inventiones Math. 1970.

4. Steve Smale, "Differentiable Dynamical Systems," Bull. Amer. Math. Soc. 1967.

5. S. S. Chern and Steve Smale, Global Analysis, vol. 1 (Providence, R.I.: American Mathematical Society, 1970).

6. C. Desoer and E. Kuh, Basic Circuit Theory (New York, 1969).

DISCUSSION

J. M. Deutch: Have the techniques you discussed for structural stability and electrical networks been applied to systems of coupled, nonlinear chemical reactions?

M. E. Fisher: Would you make some comments on the "realizability" of a given circuit. By this I mean the question of some sort of existence theorem that shows that the model is "complete" and does not require the specification of further conditions, such as jump conditions, or regularization devices to avoid singularities, etc.

P. T. Landsberg: Returning to the first part of your talk, one could, of course, introduce Riemann geometry into classical mechanics and exhibit it in the guise of the mathematical framework of general relativity (see Frankman, for example). Is this the kind of generalization of classical mechanics which you have in mind? Indeed, is it general enough for your purposes?

R. Mazo: You have shown us how many familiar results of classical mechanics arise in your "modern" reformulation. I assume that the purpose of reformulating the theory is not only to express known results in new language but also to provide tools for obtaining new results. I wonder if you could give us an example of the sort of new results to which your formulation leads.

I. Prigogine: As this session is devoted to perspectives in mechanics, it seems not out of place to mention some recent work which shows that classical and quantum dynamics may be viewed in a new,rather unusual way as linear eigenvalue problems in eigenprobabilities.

Let us first consider the case of classical dynamics (for more detail see I. Prigogine, Cl. George, and J. Rae, Physica [1971, in press]). As is well known, a basic method for solving the integration problem in classical dynamics is through canonical transformation to a "cyclic" Hamiltonian. Then all momenta are constants. This is generally done through the solution of the Hamilton-Jacobi equation. Let us indicate an alternative way which leads to a linear eigenvalue problem. This is at first very surprising, but it is, in fact a direct consequence of the introduction of the "subdynamics formalism" in our recent papers (I. Prigogine, Cl. George, and F. Henin, Physica 45, 418 [1969], Proc. Nat. Acad. Sci. 65, 489 [1971]). Let us consider the Hamiltonian

$$H(J, \alpha) = H_0(J) + \lambda\, H_1(J, \alpha), \tag{1}$$

where

$$J = (J_1, \ldots, J_N), \quad \alpha = (\alpha_1, \ldots, \alpha_N) \tag{2}$$

are action-angle variables (the generalization to other sets of canonical variables present no problem). To equation (1) corresponds the Liouville operator

$$L = L_0 + \lambda\, L_1 \tag{3}$$

defined in terms of the Poisson bracket $\{\ ,\ \}$

$$L_f = i\,\{H, f\}\ . \tag{4}$$

We introduce the projection operator P on the null space of L_0 together with its complement $\phi = 1 - \phi$ and the analogous projector π for the null space of L. Both P and π are self-adjoint operators. By definition each invariant I of the motion governed by H belongs to the null space of L so that

$$\pi I = I\ . \tag{5}$$

The operator π has remarkable properties implying a well-defined relation between the component (PI) in the null space of the unperturbed Hamiltonian H_0 and the complementary component ϕI:

$$\phi\, I = \mathbb{C}\,(PI) \tag{6}$$

where $\mathbb{C}$ is what we call a reaction operator (for its definition, see Prigogine et. al. 1969). Its exact form is of no importance here. It may be calculated explicitly in terms of J and derivatives $\partial^n/\partial J^n$. (The main point is that we now introduce a statistical ensemble characterized by the probability distribution $\tilde{\rho}$ and take the average of an invariant, say H, over the ensemble.)

Let us call $\tilde{J}$ the action variables corresponding to the complete Hamiltonian H such that

$$H = \tilde{H}(\tilde{J})\ . \tag{7}$$

We then have for the average value

$$\langle H \rangle = \frac{1}{(2\pi)^N} \quad \alpha\tilde{J}\ d\tilde{\alpha}\ \tilde{H}(\tilde{J})\,\tilde{\rho}(\tilde{J}, \alpha) = \int d\,\tilde{J}\,\tilde{H}\,(\tilde{J})\tilde{\rho}_o\,(\tilde{J})\ . \tag{8}$$

Let us now calculate $\langle H \rangle$ in the original variables J, α. We then have by using equation (5) and the fact that Π is self-adjoint,

$$\begin{aligned}\langle H \rangle &= \frac{1}{(2\pi)^N} \int dJ \int d\alpha\ H(J, \alpha)\ \rho\ (J, \alpha) \\ &= \frac{1}{(2\pi)^N} \int dJ \int d\alpha\ H(J, \alpha)\ \Pi\,\rho\,(J, \alpha)\ .\end{aligned} \tag{9}$$

This shows that we need only the projection $\rho = \Pi_\rho$ of ρ into the null space of H. By using equation (6), this leads to further simplification; relation (6) may be written more explicitly $I_n = C_n I_o$, where a Fourier expansion in angle variables is introduced:

$$\langle H \rangle = \frac{1}{(2\pi)^N} \int dJ \int d\alpha \left[H_o(J)\, \rho_o(J) + \sum_{n \neq 0} V_n \exp(in\alpha) \sum_{m \neq 0} \rho_m(J) \exp(in\alpha) \right]$$

$$= \int dJ \left[H_o\, \rho_o(J) + \sum_{n \neq 0} V_{-n} \rho_n \right]$$

$$= \int dJ\, (H_o + \sum V_{-n}\, C_n)\, \rho_o(J) \;. \tag{10}$$

Once we know the distribution $\rho_o(J)$ of the unperturbed action variables, we may therefore calculate the average value $\langle H \rangle$, but we have to associate with H the operator on the action variables

$$\mathcal{H} = H_o + \sum_{n \neq 0} V_{-n}\, C_n . \tag{11}$$

To this operator corresponds the eigenvalue problem

$$\mathcal{H} \varphi_{\tilde{J}}(J) = \tilde{H}(\tilde{J}) \varphi_{\tilde{J}}(J) . \tag{12}$$

The eigenvalues are precisely the values of the Hamiltonian as expressed in the "correct" action variables (see eq. [7]). Therefore, the solution of this problem is equivalent to the solution of the Hamilton-Jacobi equation. Moreover, the eigenfunctions are eigenprobabilities which give the distribution of J once J is given. The analogy to quantum mechanics is striking.

It is quite remarkable that the concepts of classical mechanics when combined with the uses of an ensemble ρ lead to a new formulation of dynamics. This method can be taken over practically without change to quantum mechanics (see I. Prigogine and Cl. George, Physica [1971, in press]). Of course, the Liouville operator $L\rho$ is now associated to the commutator $[H, f]_-$. The usual way of quantization consists in looking for special forms of the density ρ (pure states) such as

$$\rho = | \Psi > < \Psi | \;. \tag{13}$$

One obtains then the Schrodinger equation for $| \Psi >$ whence the usual H-eigenvalue problem which leads to the quantization with, as eigenfunctions, probability amplitudes. This procedure corresponds therefore to the factorization of the Hilbert-Schmidt space $\mathfrak{h}^{(2)}$ of density matrices ρ into an exterior product of two Hilbert spaces $\mathfrak{h}$ for the probability amplitudes:

$$\mathfrak{h}^{(2)} = \mathfrak{h} \otimes \mathfrak{h} \tag{14}$$

and to the setting up of an eigenvalue problem in $\mathfrak{h}$. On the contrary, in our new method $\mathfrak{h}^{(2)}$ is written as the sum of two subspaces, symbolically

$$\mathfrak{h}^{(2)} = \Pi + \hat{\Pi}\, \tfrac{1}{2} \tag{15}$$

and the eigenvalue problem is formulated in the Π space also. This leads then again to an eigenvalue problem of the form (12) with eigenprobabilities as eigenfunctions. Let us make two final remarks. First, the solution of the eigenvalue problem (12) corresponds to the elimination of "correlations" as induced by the creation operator $\mathcal{C}$ in equation (11). Therefore, both the Hamilton-Jacobi method and the Schrodinger-Heisenberg method of diagonalization of H acquire a new meaning as methods for the elimination of correlations. Second, one method may be extended to more complicated situations (corresponding roughly to nonseparable systems in classical dynamics).

In conclusion, one method seems to us to have a special methodological interest, for it bridges concepts of classical or quantum dynamics with the ensemble concept of statistical mechanics.

R. Balescu: I should like briefly to point out a very remarkable property of the time evolution process in a many-body system. (This remark is based on recent work of I. Prigogine, Cl. George, F. Henin, J. Wallenborn, L. Brenig, and myself.)

The distribution function f(t) can be separated into two terms by means of operators π and $\hat{\pi} \equiv 1 - \pi$:

$$f(t) = \pi f(t) + \hat{\pi} f(t) \quad . \tag{1}$$

It can be shown, under fairly general conditions, that (a) π (and hence $\hat{\pi}$) is a projection operator:

$$\pi^2 = \pi; \tag{2}$$

(b) π commutes with the Liouville operator $\mathcal{L}$:

$$\pi\mathcal{L} = \mathcal{L}\pi \; , \tag{3}$$

The property (3) is quite remarkable. If $u(t) \equiv \exp(t\mathcal{L})$, we see that

$$\pi f(t) = \pi u(t) f(o) = u(t) \pi f(o). \tag{4}$$

Hence $\pi f(t)$ [and also $\hat{\pi} f(t)$] obeys the Liouville equation. This implies that the set of distribution functions f(t) has been split into two subsets which are invariant under the motion. In other words, the representation u(t) of the group of translations is reducible (even in presence of interactions!).

The formalism can be extended into a relativistic theory. In our formulation we begin with the construction of ten generalized Liouvillians, $\mathcal{L}_i$, which generate the infinitesimal transformations of the Poincaré group. The generator of translations is just the ordinary Liouvillian $\mathcal{L}_H \equiv \mathcal{L}$. The pairwise commutators of these operators satisfy the characteristic Lie algebra relations of the Poincaré group.

In a recent paper (R. Balescu and L. Brenig, to be published in Physica, 1970) we have been able to prove that the separation (1) is invariant under all the transformations of the Poincare group. This is expressed by the commutation relations $\mathcal{L}_i \pi = \pi \mathcal{L}_i$, $i = 1, \ldots, 10$, valid for all the Liouvillian generators. In particular, this theorem implies that under a Lorentz transformation, the πf component transforms into a πf component, independently of $\hat{\pi} f$. This invariance property shows that the separation (1) expresses a deep intrinsic feature of the time evolution process.

We cannot discuss here the important connections of these concepts with the theory of kinetic equations. These questions have been treated in detail in the literature (see, e.g., R. Balescu and J. Wallenborn, to be published in Physica, 1970).

SOME NEW RESULTS IN ERGODIC THEORY

Donald S. Ornstein

Certain mechanical systems, such as gases, exhibit a certain randomness, and this randomness can be used to make predictions about what the systems will do. The motions of these systems are in general too complicated to analyze exactly, and we try to make the problem more tractable by ignoring those motions that occur with zero probability. This approach, started by Birkhoff and von Neumann, has led to a part of mathematics called ergodic theory. (For example, topological considerations tell us that we can balance a pencil on its point for as long as we want to, even on a moving train. The ergodic theory approach ignores such possibilities.)

As you all know, we can represent the state of the mechanical system as a point in a high-dimensional Euclidean phase space. As time progresses, the point moves on a surface of constant energy, giving us a flow; or, if we look at the system at discrete times--take a movie of it--we get a transformation of the surface of constant energy. Such flows or transformations turn out to be measure-preserving. We are thus led to the study of measure-preserving transformations. In keeping with our principle of ignoring motions of zero probability, we will say that the transformation T, acting on the space X, is equivalent or isomorphic to T_1 acting on X_1 if we can find invariant subsets $\bar{X}$ of X and $\bar{X}_1$ of X_1 such that the measures of $X - \bar{X}$ and $X_1 - \bar{X}_1$ are zero, and there exists an invertible measure-preserving transformation, d, mapping $\bar{X}$ onto $\bar{X}_1$ such that $T_1(d(x)) = d(T(x))$ for all x in $\bar{X}$.

So far there is only one kind of mechanical system that can be completely analyzed from this point of view (that is, we know exactly--up to isomorphism--what the transformation is). These systems are the so-called geodesic flows on surfaces or manifolds of negative curvature. I call these mechanical systems because[1] in three-dimensional Euclidean space one can specify a closed surface V and place near it a finite number of centers of attraction or repulsion, creating a force

---

This research was supported in part by NSF grant GP 21509.

[1]This was pointed out by Kolmogorov in reference [19].

potential in V such that the motion of a material point constrained to move in V under the above potential will be mathematically equivalent to geodesic flow on a surface of negative curvature.

The transformation associated with the above kind of mechanical system turns out to be a "Bernoulli shift." Bernoulli shifts are in a sense the simplest example of measure-preserving transformations. They are also in a sense the most random transformations possible. This means that it is essentially impossible to distinguish a sequence of observations made on our mechanical systems from a sequence of observations made on the successive spins of a roulette wheel.

Geodesic flow on manifolds of negative curvature are very special examples. There is, however, reason to believe that they are just mathematically simpler cases of a fairly general phenomenon, which includes the case of an ideal gas enclosed in a box. I should mention in this connection that Sinai has succeeded in proving that the ideal gas in a rectangular box is ergodic on manifolds of constant energy. He actually proved more, namely that the above transformations are "K-automorphisms." I will return to this later and explain what a "K-automorphism" is.

BERNOULLI SHIFTS

Let us now turn to a description of "Bernoulli shifts."

Take our measure space to be the unit square in the plane with ordinary (Lebesgue) measure. Then to each integer k and real numbers $0 \le p_i$, $1 \le i \le k$, and $\sum_{1}^{k} p_i = 1$, there corresponds a Bernoulli shift $(p_1, \ldots, p_k)$.

The Bernoulli shift $(\frac{1}{2}, \frac{1}{2})$ can be described as follows.

T will send the point (x, y) into $(2x, \frac{1}{2}y)$ if $0 \le x < \frac{1}{2}$ and $(2x - 1, \frac{1}{2}y + 1)$ if $\frac{1}{2} \le x < 1$. We can picture T in the following way: We first squeeze down the unit square and we then translate the part that has left the square back onto the part of the square that is now empty (see fig. 1).[2]

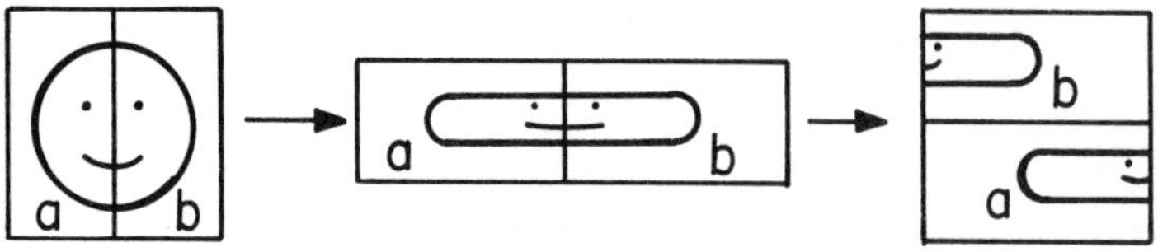

Fig. 1

This transformation is also called the Baker's transformation.

The Bernoulli shift $(\frac{1}{3}, \frac{2}{3})$ can be described follows.

---

[2]This nice type of picture was suggested in Ergodic Problems of Classical Mechanics by V. I. Arnold and A. Avez (New York: Benjamin, 1968).

T will send (x,y) into $(\frac{3}{2}x, \frac{2}{3}y)$ if $0 \le x < \frac{2}{3}$ and $(3x - 2, \frac{1}{3}y + \frac{2}{3})$ if $\frac{2}{3} \le x < 1$. We can picture T in the following way. We take the part of the square $0 \le x < \frac{2}{3}$, $0 \le y < 1$, and squeeze its height by $\frac{2}{3}$ and expand its width by $\frac{3}{2}$ (see fig. 2). We then take the part of the

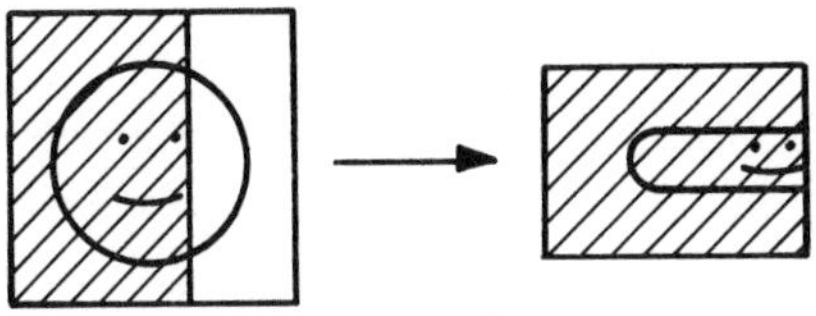

Fig. 2

square $\frac{2}{3} \le x < 1$ and squeeze its height by $\frac{1}{3}$ and expand its width by 3 (see fig. 3). We then reassemble these two pieces to get the configuration shown in figure 4.

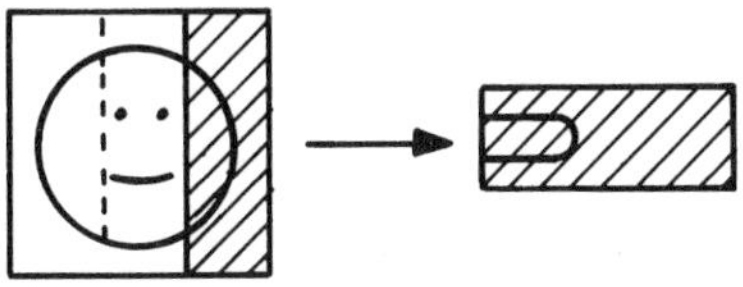

Fig. 3

Fig. 4

The Bernoulli shift $(p_1, \ldots, p_k)$ can be described as follows: Divide X into k rectangles whose height is 1 and whose base is an interval of length p_i $(1 \le i \le k)$. Squeeze the height of the i(th) rectangle by p_i and expand its width by $\frac{1}{p_i}$. Now reassemble these pieces by putting the first on the bottom, the second on top of it, the third on top of that, etc.

There is another way to describe Bernoulli shifts. Let Y be a set of k elements, and let us give the i(th) element measure p_i. Let Y_i, $-\infty < i < +\infty$, be copies of Y; and let X be the product of Y_i with the product measure. Thus each point in X is a doubly infinite sequence of points in Y. T will act by shifting each of the above sequences. That is, T will take the sequence $\{y_i\}$ into the sequence $\{y_i'\}$, where $y_i' = y_{i+1}$. We leave it as an exercise for the reader to verify that the above definition of Bernoulli shift is the same as (isomorphic to) the first one we gave.

This form of the definition shows that a Bernoulli shift is the realization of the process of spinning a roulette wheel with k slots of width p_i. The points in X are all the possible outcome of spinning the wheel forever. Our process gives us a probability of any finite sequence of spins, and this extends uniquely to the measure we put on X. T corresponds to looking at the process one unit of time later.

Another reason for giving the above form of the definition is that it shows that in a certain sense Bernoulli shifts are the simplest examples of ergodic measure-preserving transformations (T is ergodic if the only sets that are invariant under T have measure 0 or 1). There is a theorem that says that any invertible, ergodic, measure-preserving transformation can be obtained if we modify the above construction by taking Y to be countable and by taking some other measure invariant under the shift, instead of the product measure. Thus Bernoulli shifts are simplest in the sense that the product measure is the simplest measure invariant under the shift.

There is a third way to describe Bernoulli shifts. T is isomorphic to the Bernoulli shift $(p_1, \ldots, p_k)$ if and only if there is a partition P into sets $P_1, \ldots, P_k$ such that the measure of P_i is p_i and such that (1) the T^iP are independent and (2) the T^iP generate.

Condition 1 means that if we take a finite number of sets of the form $T^iP_{j(i)}$ (only one for each i), then the measure of their intersection is the product of their measure. Condition 2 means that if B is measurable, then given ϵ, we find n and B' in the algebra of sets generated by the $T^iP_{j(i)}$ $(-n \le i \le n)$ $(1 \le j(i) \le k)$ such that the symmetric difference between B and B' is ϵ. (The equivalence of this third description is a routine exercise in measure theory.)

When Halmos wrote his book on ergodic theory, one of the main problems was the following: Are all Bernoulli shifts isomorphic? In particular, is the Bernoulli shift $(\frac{1}{2}, \frac{1}{2})$ isomorphic to the Bernoulli shift $(\frac{1}{3}, \frac{1}{3}, \frac{1}{3})$? Up to this point there were various properties of transformations that were known, and because of these a large number of transformations could be distinguished from one another (properties such as ergodicity, mixing, weak mixing). None of these properties, however, distinguished any two Bernoulli shifts (two Bernoulli shifts even induced isomorphic unitary operators).

The breakthrough in this area came when Kolmogorov introduced a new invariant called entropy, which was motivated by Shannon's work on information theory. This invariant, which I will describe below, was easy to compute for the Bernoulli shift $(p_1, \ldots, p_k)$ and is simply $-\sum_{i=1}^{k} p_i \log p_i$. [Thus the Bernoulli shift $(\frac{1}{2}, \frac{1}{2})$ is not isomorphic to the Bernoulli shift $(\frac{1}{3}, \frac{1}{3}, \frac{1}{3})$].

The entropy of a transformation is defined as follows: We first define the entropy, H(P), of a partition P, whose i(th) set has measure p_i as $-\sum p_i \log p_i$. [If the T^iP were independent, then for all n large enough the size of most of the atoms in the partition $\bigvee_{i=1}^{n} T^iP$ is approximately $(\frac{1}{2})^{H(P)n}$. This follows from the law of large numbers.] We now define the entropy of P relative to T, H(P,T) as $\lim_{n \to \infty} n^{-1} H(\bigvee_{1}^{n} T^iP)$. Actually the limit exists. [There is a theorem due to Shannon and McMillan that says that for all n large enough the size of most of the atoms in $\bigvee_{i=1}^{n} T^iP$ is roughly the same and approximately equal to $(\frac{1}{2})^{H(P)n}$.]

We now define H(T) as sup H(P,T), where sup is taken over all finite partitions. At first glance one would expect that H(P,T) would be very large if P had a large number of atoms and that H(T) would always equal ∞. To get a feeling why this is not so, note that H(P,T) = $\lim_{n \to \infty} n^{-1} H(\bigvee_{1}^{n} T^iP)$ should equal $H(\bigvee_{1}^{\ell} T^iP, T) = \lim_{n \to \infty} n^{-1} H(\bigvee_{1}^{n+\ell} T^iP)$. This same argument suggests that sup H(P,T) should be attained if the T^iP generate. This is actually true (it would be easy if each B were actually in some $\bigvee_{-n}^{n} T^i P$). It is this last fact that enables us to calculate the entropy for Bernoulli shifts.

Let us now return to the isomorphism problem for Bernoulli shifts. There are still quite a lot of Bernoulli shifts with a given entropy. Are these all the same? Is there another property yet to be discovered that will distinguish some of these? Are they all different? Mesalkin showed that the answer to the last question was no by showing that the Bernoulli shifts $(\frac{1}{4}, \frac{1}{4}, \frac{1}{4}, \frac{1}{4})$ and $(\frac{1}{2}, \frac{1}{8}, \frac{1}{8}, \frac{1}{8}, \frac{1}{8})$ were isomorphic.

Sinai proved a beautiful and deep theorem along these lines, namely: If $H(T) = -\sum p_i \log p_i$, then we can find a partition P whose i(th) set has measure p_i such that the $T^i P$ are independent. (The $T^i P$ may not generate. If this could be shown, under the additional hypothesis that T is a Bernoulli shift, then all Bernoulli shifts of the same entropy would be isomorphic.)

We now know [7]:

Theorem. All Bernoulli shifts with the same entropy are isomorphic.

If one wants to show that a certain transformation is isomorphic to a Bernoulli shift, it is in general very hard to find an independent generator. However, a careful analysis of the proof of the above theorem shows that if T has a generator satisfying a condition much weaker than independence [8, 13], then the transformation is a Bernoulli shift. Sinai had already shown that for geodesic flow on manifolds of negative curvature there were "Markov partitions" that generated [6]. This, together with the above mentioned criteria, gave the following:

Theorem. Let T be the transformation on a manifold of constant energy in the phase space of a geodesic flow on a manifold of negative curvature, obtained by observing the system at multiples of a fixed time. Then T is isomorphic to a Bernoulli shift.

It is not essential here that time be discretized. We say that a flow is a Bernoulli flow if for each fixed time the transformation is Bernoulli. It can be shown (by elaborating the argument for the isomorphism of Bernoulli shifts) that any two Bernoulli flows are isomorphic (except for a normalization of the time scale), and in particular isomorphic to the following flow: Let T acting on X be the Bernoulli shift $(\frac{1}{2}, \frac{1}{2})$. Let $P = \{P_1, P_2\}$ be its independent generator. Let f(x) equal α for x in P_1 and equal β for x in P_2 where α/β is irrational. Our flow is defined on the area under the graph of f(x). Each point moves up at unit speed until it reaches the top of the graph, then (x, f(x)) goes to (Tx, 0).

I might mention here that by using the above-mentioned criteria one can show that Bernoulli shifts arise in other seemingly unrelated contexts. Katznelson has shown that any ergodic automorphism of the n-dimensional torus is a Bernoulli shift; and S. Ito, H. Murata, and H. Totoki have shown that the natural extension of the continued fraction transformation is a Bernoulli shift. Also a mixing Markov shift on a finite state space is a Bernoulli shift [8].

K-AUTOMORPHISMS

Kolmogorov introduced a class of transformations, now called K-automorphisms (K-transformations would be more consistent with our terminology), and conjectured that every K-automorphism was a Bernoulli shift. These transformations were of interest partly because they had certain randomness properties that made it reasonable to guess that they were the most random possible--i.e., Bernoulli shifts--and partly because it was possible to show that a fairly large class of transformations were K-automorphisms. As I mentioned before, Sinai showed that the transformation arising from an ideal gas in a box is a K-automorphism.

We say that T is a K-automorphism if there is a finite partition P such that the T^iP generate and $\bigcap_{n=1}^{\infty} \bigvee_{i=n}^{\infty} T^iP$ is trivial. (That is, $\bigvee_{i=n}^{\infty} T^iP$ is the class of measurable sets generated by the T^iP, $n \leq i < \infty$. The only sets which are contained in the above classes for all n have either measure 0 or 1.) A special case is when the T^iP generate and are independent.

There is a beautiful theorem due to Sinai and Rokhlin which says that T is a K-automorphism if and only if for every finite partition Q we have that $E(Q,T) > 0$. (This is the same as saying that Q is not contained in $\bigvee_{-\infty}^{-1} T^iQ$.)

We now know that Kolmogorov's conjecture is false, [9], i.e.,

Theorem. There is a K-automorphism that is not a Bernoulli shift.

The proof of this above theorem is closely related to the proof of the isomorphism of Bernoulli shifts. We need a usable criterion for when the T^iP generate a Bernoulli shift, and such a criterion comes out of the methods used in the proof of that theorem.

RANDOM PROCESSES

Suppose that we perform an experiment or take some measurement of our mechanical system and that this experiment or measurement has only a finite number of possible outcomes. Suppose also that we repeat this measurement at regularly spaced intervals of time (all multiples of some fixed unit of time). The sequence of outcomes gives us a process in the sense of probability theory. In some sense all the information we can get about a mechanical system is some process derived from it.

In general a process can be thought of as a box that prints out one letter each unit of time, where the probability of a given letter being printed may depend on the letters printed in the past but is independent of time (that is, the mechanism in the box does not change).

The mathematical model for a process is a transformation T acting on X and a partition P of X. The reason for this is the following: We know all about the process if we know the probability of its printing out any given finite sequence. Now for each point $x \in X$ corresponds a sequence whose i(th) term is the atom of P containing T^ix. If we think of the measure on X as a probability measure, then the probabilities of finite sequences determine the pair P, T. (The measure of the atoms of $\bigvee_{-n}^{n} T^i P$ is determined.)

The Birkhoff ergodic theorem says that if T is ergodic, then for almost all x the frequency of each finite sequence will tend to its probability. Thus we can tell all about such a process by watching it long enough.

If the process P, T has a zero entropy $[H(P,T) = 0]$, then P, T is deterministic in the following sense: $P \subset \bigvee_{-\infty}^{-1} T^i P$, which means that by knowing the past we can predict the next letter with probability 1, or by knowing enough of the past we can predict the next letter with arbitrarily high probability. We can even make a stronger statement. For all n, $P \subset \bigvee_{-\infty}^{-n} T^i P$, which means that we can predict a particular letter by knowing the distant past, however distant.

Now suppose T is a K-automorphism and P any finite partition, such that the $T^i P$ generate. Let us call such a process a K-process. The K-processes are exactly those processes that contain no deterministic part or are completely nondeterministic in the following sense: If P, T is not a K-process and the $T^i P$ generate (and T is ergodic[3]), then there is a set S (whose measure is as close to $\frac{1}{2}$ as we want) and an N such that $S \subset \bigvee_{0}^{N} T^i P$ and for all n, S can be approximated to within 99/100 by a set in $\bigvee_{-\infty}^{-n} T^i P$. This means that we can divide the sequences of length N into two classes (each having measure close to $\frac{1}{2}$) and we predict with probability 99/100 which class the next N symbols will belong to by knowing the distant past, however distant. (All we really need to know is a finite number of terms in the distant past. We will in general need more terms in the more distant past.) Thus Sinai's theorem says that any measurement made on an ideal gas is completely nondeterministic.

Now suppose T is a Bernoulli shift and B an independent generator. If P is a partition in $\bigvee_{-K}^{K} T^i P$ for some K, then the P, T can be physically constructed from a roulette wheel (with the same distribution as B). If P is arbitrary, then P, T is a limit of processes physically constructible from a fixed roulette wheel.

In these terms Kolmogorov's conjecture says that if P, T is completely nondeterministic, then T is a Bernoulli shift (and the process is essentially constructible from a roulette wheel). The fact that Kolmogorov's conjecture is false means that we have a completely nondeterministic process, $\bar{P}, \bar{T}$, such that if T is a Bernoulli shift (and P a finite partition), then for all n large enough we can find two disjoint collections of sequences of length n, C and $\bar{C}$, such that the measure of $\bar{C}$ under $\bar{P}, \bar{T}$ is greater than 99/100 and the measure of C under P, T is greater than 99/100. (It turns out that even more is true. There is a fixed $\sigma > 0$, associated with $\bar{P}, \bar{T}$, such that C and $\bar{C}$ can be chosen with the additional property that any sequence in C differs from any sequence in $\bar{C}$ in more than σn places. Thus $\bar{P}, \bar{T}$ and P, T are a definite distance apart (see [10]).

---

[3]The assumption is there only for simplicity. In fact, if T is not ergodic, we have an even stronger deterministic property.

These processes are somewhat mysterious, and I have no idea if they can arise from an actual mechanical system.

CODES

Now suppose T is a Bernoulli shift and P an arbitrary finite partition with k atoms. I would like to describe the process P, T in a somewhat more finitistic or physical way. For the sake of notation let us assume that T is the Bernoulli shift $(\frac{1}{2}, \frac{1}{2})$. Thus T is the shift based on flipping a coin. To each sequence of 0, 1's we will associate another sequence as follows: Divide all possible finite sequences of 0, 1's of length 2n + 1 into k classes 1, . . . , k. For each term in our infinite sequence of 0, 1's look at the block of 2n + 1 consecutive terms with it as center and print out the corresponding 1, . . . , k. Call the resulting mapping a "code" of length n. If we code our sequence of 0, 1's, we get another process. (The coding is physically realizable if we allow a time delay.) We say that a sequence of longer and longer codes converges if for each term all sufficiently long codes give the same result (except for a collection of sequences of 0, 1's of probability 0). This implies that any two sufficiently long codes agree with very high probability. It is easy to see that any process P, T (T the Bernoulli shift $(\frac{1}{2}, \frac{1}{2})$) is obtained by coding the sequence of 0, 1's or as the limit of a convergent sequence of longer and longer codes. In general, if two processes generate isomorphic transformations, then we can obtain one from the other by means of a convergent sequence of codes (and then get back the original process exactly by applying another convergent sequence of codes).

SOME HOPES FOR THE FUTURE

I think that one of the main goals for ergodic theory is to find (up to isomorphism) the measure-preserving transformation associated with a mechanical system. I expect that this will be accomplished for a large class of transformations and in particular for an ideal gas. I also expect that these transformations (for an ideal gas and "reasonably" random systems) will be Bernoulli shifts.

After determining what the measure-preserving transformation is, it will be important to get more information in the case of specific systems. That is, knowing the transformations tells us about all the possible measurements of the system, but it does not give us much information about one particular measurement. (In this direction Sinai has shown that the central limit theorem holds for a large class of topologically reasonable function under geodesic flow on a surface of constant negative curvature.)

REFERENCES

1. P. R. Halmos, Lectures on Ergodic Theory (Tokyo: Math. Soc. of Japan, 1956), 54-55; (New York: Chelsea).

2. P. Billingsley, Ergodic Theory and Theory of Information (New York: Wiley, 1965).

3. A. N. Kolmogorov, A new metric invariant of transitive dynamical systems and

automorphism of Lebesgue spaces, Dokl. Akad. Nauk SSSR, 128 (1958), 861-864.

4. I. G. Sinai, A weak isomorphism of transformations with an invariant measure, Dokl. Akad. Nauk SSSR, 147 (1962), 797-800, and Soviet Math. Dokl. 3 (1962), 1725-1729.

5. R. L. Adler and B. Weiss, Entropy, a complete metric invariant for automorphisms of the torus, Proc. Nat. Acad. Sci. U.S.A. 57 (1967), 1573-1576.

6. I. G. Sinai, Markov partitions and U-diffeomorphisms, Funkcional. Anal. i Prilozen, 2 (1968), 61-82.

7. D. S. Ornstein, Bernoulli shifts with the same entropy are isomorphic, Adv. in Math. 4 (1970), 337-352.

8. N. A. Friedman and D. S. Ornstein, On isomorphism of weak Bernoulli transformations, Adv. in Math. 5 (1970), 365-394.

9. D. S. Ornstein, A Kolmogorov automorphism that is not a Bernoulli shift, Adv. in Math. (in press).

10. D. S. Ornstein, An application of ergodic theory to probability theory, Adv. in Math. (in press).

11. D. S. Ornstein, Two Bernoulli shifts with infinite entropy are isomorphic, Adv. in Math. 5 (1970), 339-348.

12. D. S. Ornstein, Factors of Bernoulli shifts are Bernoulli shifts, Adv. in Math. 5 (1970), 349-364.

13. D. S. Ornstein, Imbedding Bernoulli shifts in flows, in "Contributions to Ergodic Theory and Probability," Lecture Notes in Mathematics Series (Berlin: Springer-Verlag, 1970), 178-218.

14. M. Smorodinsky, An exposition of Ornstein's isomorphism theorem, Adv. in Math. (in press).

15. Y. Katznelson, Ergodic automorphisms of T^n are Bernoulli shifts, Israel J. Math. (in press).

16. S. Ito, H. Murata, and H. Totoki, A remark on the isomorphism theorem for weak Bernoulli transformations (in press).

17. A. N. Kolmogorov, On dynamical systems with an integral invariant in the torus, Dokl. Akad. Nauk SSSR, 93 (1953), 763-766.

18. A. N. Kolmogorov, Entropy per unit time as a metric invariant of automorphisms, Dokl. Akad. Nauk SSSR, 124 (1959), 754-755.

19. A. N. Kolmogorov, A general theory of dynamic systems and classical mechanics, Proc. Internat. Congress Math. 1 (1954), 315-333.

20. V. A. Rokhlin and J. G. Sinai, Construction and properties of invariant measurable partitions, Soviet Math. Dokl. 2 (1961), 1611-1614.

21. J. G. Sinai, Theory of Dynamical Systems, Part I: Ergodic Theory, Lecture Notes Series No. 23, March 1970, Aarhus University.

22. V. A. Rokhlin, Lectures on the entropy theory of measure-preserving transformations, Russian Math. Surveys, 22 (1967), 1-52.

23. J. G. Sinai, Probabilistic ideas in ergodic theory, Internat. Math. Congr., 1962, 540-559; AMS Transl. 31 (1963), 62-81.

24. D. V. Anosov and Y. G. Sinai, Some smooth ergodic systems, Russian Math. Surveys, 22 (1967), 103-167.

25. L. D. Meshalkin, A case of isomorphism of Bernoulli schemes, Dokl. Akad. Nauk SSSR, 128 (1959), 41-44.

26. Y. G. Sinai, On weak isomorphism of transformations with invariant measure, Mat. Sb. 63 (1963), 23-42.

27. Y. G. Sinai, On the foundations of the ergodic hypothesis for a dynamical system of statistical mechanics, Dokl. Akad. Nauk SSSR, 153 (1963), 1261-1264 = Soviet Math. Dokl. 4 (1963), 1818-1822.

28. Y. G. Sinai, The central limit theorem for geodesic flows on manifolds of constant negative curvature, Dokl. Akad. Nauk SSSR, 133 (1960), 1303-1306 = Soviet Math. Dokl. 1 (1960), 983-986.

DISCUSSION

M. Green: Does Sinai's proof of the ergodicity of a finite hard-sphere gas answer the following question: Does a sum of variables associated with the system obey the central limit theorem?

D. Ornstein: No. Even if one could show that the transformation associated with the hard-sphere gas were a Bernoulli shift, one would only know that a dense class of random variables obeyed the central limit theorem.

O. De Pazzis: The ergodic conjecture has been proved for the infinite perfect gas in equilibrium in recent papers of Sinai et. al. The proof can be extended to the case of infinite systems of hard rods.

L. Galgani: I would like to make a comment and point out some conclusions that can be drawn from the recent developments in ergodic theory. Fifty years ago the general opinion was that, in general, systems are ergodic (Fermi, 1923). Now the position seems to have been reversed. At high energies, systems ought to be ergodic (Sinai, 1964), but at low energies in general they are not (Kolmogorov, Arnold, Moser, 1964).

This has some consequences in connection with the foundations of quantum mechanics. Indeed, for example, Dirac gives three arguments for saying that classical mechanics is not adequate. Two of them at least (blackbody radiation, specific heats of solids) are based on the assumption of ergodicity. It seems then that one is no longer allowed to rely on these arguments.

M. Green: One might speculate that the Kolmogoroff entropy of a transformation might be as useful in the macroscopic theory of nonequilibrium phenomena as the ordinary entropy is in the macroscopic theory of equilibrium phenomena.

D. W. Robinson: A well-known physical model which is an illustration of the concepts discussed by Dr. Ornstein is the one-dimensional classical spin system. The configurations of such a system are a string of + 1's and - 1's denoting whether the spins at each lattice site are up or down. The transformations are space translations of shifts in the one-dimensional lattice. If one considers an invariant measure on the space of configurations, then this determines a translationally invariant state of the system to which one can assign a Gibbsian entropy per unit volume. This entropy is identical to the Kolmogorov entropy.

EXACT RESULTS IN EQUILIBRIUM STATISTICAL MECHANICS

Robert B. Griffiths

I. INTRODUCTION

My topic is "exact results in equilibrium statistical mechanics." Now this is an enormous subject, and in order to say anything useful in even the very generous space of time allotted me by the organizers of this conference, I have had to pick only a few among the possible topics. It will not surprise you to learn that I intend to speak on the subjects closest to my professional interests, which is to say the application of exact or rigorous results to the theory of phase transitions [1].

Many physicists have an adverse reaction to theoretical work which emphasizes mathematical rigor. One has the picture of some researcher carefully brushing the dust out of some obscure corner in Hilbert space while devoting himself to problems abandoned by most theoretical physicists as quite uninteresting 50 years ago.

I would like to think that this unfortunate image does not apply to much of the work done in applying rigorous results to phase transitions. There are a variety of reasons why rigorous results in this field tend to be close to the front line of development; let me mention just one. Almost by definition, a phase transition is a point where one variable does <u>not</u> depend smoothly on another, a point where perturbation theory breaks down, where limits cannot be interchanged with impunity, and where the theoretician who does not pay at least minimal attention to his epsilons and deltas is likely to be doing not only bad mathematics but also bad physics.

I do not mean at all to imply that the only work of value in connection with phase transitions is that done with the highest standards of mathematical precision. This is clearly not the case. But I do think that both in the development of the theory and in our present understanding of the subject, exact or rigorous results play an important role.

Exact results on phase transitions are of two kinds. First, there are the exactly soluble models, as exemplified in Onsager's [2] solution to the two-dimensional Ising problem in 1944.

Research supported by the National Science Foundation, Grant GP-11454.

In recent years exact solutions for several other models have appeared--for example, the work of Lieb [3] and others on two-dimensional ferroelectrics, and Eisher's [4] one-dimensional fluids with many-particle interactions. While one cannot underestimate the importance of these results for our understanding of the subject, I want to restrict my remarks today to the second kind of exact results in which one establishes certain mathematical properties of a class of models without actually being able to solve them exactly. An example is the theorem of Lee and Yang [5] (about which I shall have more to say a little later) on the location of zeros of the partition function for an Ising ferromagnet. This theorem tells us that for the models in question all the thermodynamic functions are analytic functions of the magnetic field H except, possibly, at H = 0. No one has ever obtained the solution of the two-din ensional Ising model in a finite magnetic field (Onsager [2] obtained it for zero field), and yet this very elegant theorem of Lee and Yang enables us to say a lot about the functions on which we have yet to lay our hands.

It is perhaps worth pointing out that while most of the developments I will discuss have come about in the past decade, a large fraction of them are found in seminal form in, or were at least motivated by, the two papers of Yang and Lee [5,6] on phase transitions which appeared in 1952 and are rightly regarded as "classics." The combination of physical insight and very careful mathematics in these papers has set a high standard for others working in the field.

II. PHASE TRANSITIONS

To indicate what I mean by a phase transition, let me show you two examples (fig. 1). The

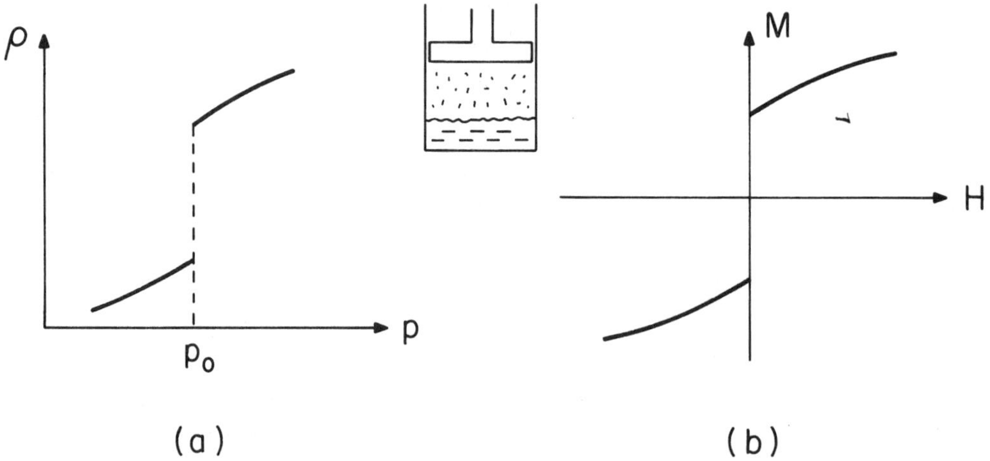

Fig. 1. A (first-order) phase transition in (a) a fluid and (b) a magnet. At the pressure p_0 two phases may coexist, as indicated in the inset.

first example is the vapor-to-liquid transition in a fluid as one increases the pressure p (holding temperature constant). It is well-known that the density ρ is discontinuous at that pressure p_0 where the liquid and vapor phases coexist. To the right and left of this transition the density depends smoothly on the pressure (though there may be additional phase transitions at higher pressures). The second example is a ferromagnet in which the magnetization M changes discontinuously as a function of the applied field H (corrected for demagnetizing effects), also at constant temperature. At H = 0 the coexisting "phases" are the domains in which the magnetization points in different directions. In both of these examples one has a discontinuity in a first (partial) derivative of a Gibbs free energy, and thus they are first-order transitions.

There are several interesting questions in mathematical physics associated with such phase transitions:

1. What are the characteristics of model Hamiltonians which give rise to such transitions? Can one say that some interactions will give rise to a phase transition and others will not?

2. What are the peculiar statistical properties of a system in which two phases coexist? How can you characterize by means of the particle distribution functions a system with more than one phase?

3. Why do phase transitions, as observed in nature, only occur on isolated lines in the p, T plane? Is this a general feature one would expect of almost any system, or does it reflect some rather special properties of the systems investigated experimentally?

4. Are the thermodynamic and other properties really smooth (hopefully analytic) functions of the pressure away from the phase transition?

5. How does one characterize "metastable states"?

We are a long way from answering these questions in any detail, but I hope my remarks may serve to indicate where some progress has been made in recent years.

III. THE THERMODYNAMIC LIMIT

In order even to begin to study the questions just posed, we must first ask how it is that one calculates density as a function of pressure, magnetization as a function of field, or, in general, the statistical properties of a particular model. I shall assume that we can use the canonical or grand-canonical prescription described in all the elementary textbooks. (The question of justifying these formulae on the basis of ergodic theory or the like belongs in the domain of nonequilibrium statistical mechanics.) But especially when discussing phase transitions it is essential to consider the so-called thermodynamic limit, or infinite-volume limit, a subject not usually treated, or not treated adequately, in the elementary texts. That is, you must not simply calculate the average density as a function of pressure for some finite box, you must then consider this function in the limit when the linear dimensions of the box become infinite.

There are both mathematical and physical reasons for taking this limit. Among the former is the fact that only in this limit do you get results which are independent of which ensemble you employ and independent of the size of the box and the boundary conditions at its edge. And in the grand ensemble it is only in this limit that phase transitions, in the form of mathematically sharp discontinuities, can appear. Thus the thermodynamic limit provides a "clean" mathematical problem from which certain complications have been removed. But there is also a physical justification in that in experiments, despite the fact that they are carried out on finite systems, the results may often, with reasonable confidence, be considered to represent the "bulk" properties of some material, to a good approximation independent of size and shape. As long as the experimental results are claimed to be independent of size and shape, it is most reasonable to compare them with corresponding theoretical quantities obtained in the thermodynamic limit.

During the last decade there has been a considerable advance in proving that a well-defined thermodynamic limit exists for a wide variety of nonrelativistic systems, and these proofs provide a necessary foundation for a precise discussion of phase transitions. You will find a detailed summary in Ruelle's book [7], and I will only discuss, briefly, the case where the potential energy is a sum of pair potentials, the pair potential depending on the relative location of the two particles. The existence of a well-defined limit for the thermodynamic functions can be demonstrated if two conditions are satisfied.

$$U_N = \sum_{i<j}^{N} \Phi\,(\underline{r}_i - \underline{r}_j) \geq -BN\,, \tag{1}$$

$$\text{for } |\underline{r}_i - \underline{r}_j| > R_o,\ \Phi\,(\underline{r}_i - \underline{r}_j) \leq C/|\underline{r}_i - \underline{r}_j|^{d+\epsilon}\,, \tag{2}$$

where B, C, and R_o are constants, N is the number of particles, d the dimensionality of the system (d = 3 in the real world, but not in every interesting model!) and ϵ a number which is strictly positive.

Most potentials of interest to statistical mechanicians satisfy these requirements--e.g., potentials with a hard core plus a finite-range attraction, Lennard-Jones 6-12 potentials, and the like. A rather embarrassing exception is the Coulomb potential, since it violates both conditions. However, Dyson and Lenard [8] have shown how quantum mechanics plus the Pauli exclusion principle may be used to place a lower bound on the total (kinetic plus potential) energy which is proportional to N. And Lebowitz and Lieb [9] have shown how to use the effect of shielding by opposite charges in a neutral system to get around the violation of condition 2. Now that proofs exist for Coulomb systems, there seem to be few important systems for which the existence of a thermodynamic limit for the thermodynamic functions remains in doubt. One system for which proofs are still lacking is the case of dipole-dipole interactions, e.g., in models of magnetism. When an external magnetic field is present, there is a shape-dependent contribution to the free energy, and one has yet to prove that this is of the expected form [10] even though one knows that a limit exists for zero field [11].

IV. THE LEE-YANG CIRCLE THEOREM AND ANALYTICITY

In 1952 Lee and Yang [5] published what must be considered one of the most remarkable results in the history of the theory of phase transitions, their celebrated "circle theorem." To be specific, they showed that, given the Hamiltonian

$$\mathcal{H} = -\sum_{i<j} J_{ij}\,\sigma_i\,\sigma_j - H\sum_i(\sigma_i - 1) \tag{3}$$

for an Ising model ($\sigma_i = \pm 1$), the zeros of the partition function

$$\Xi = \mathrm{Tr}\,[e^{-\beta\mathcal{H}}] \tag{4}$$

as a function of $z = \exp(-2\beta H)$ all lie on the unit circle $|z| = 1$ provided all the J_{ij} are nonnegative.

The importance of this result is that (combined with another theorem of Yang and Lee [6]) it shows that in the thermodynamic limit all the thermodynamic quantities for this model are analytic functions of z for z either inside or outside the unit circle. In particular, for physical values of z--real and positive--no nonanalytic behavior, in particular no phase transition, can occur, except perhaps at $z = 1$. This means that at $H = 0$ in the ferromagnet, or at a certain specific value of the chemical potential in the corresponding lattice gas, it is possible to have a phase transition. But elsewhere the thermodynamic functions are all smooth in the strongest possible sense; that is, they are analytic functions.

Some additional implications of the Lee-Yang theorem have been worked out by Lebowitz and Penrose [12]. They combined it with some arguments by Ruelle [13] (which I shall discuss presently) to prove that the thermodynamic functions are analytic functions of temperature as well as magnetic field (or pressure). They also showed that the correlation functions depend analytically on temperature and field away from $H = 0$.

Quite recently Asano [14] has succeeded in extending the Lee-Yang theorem to quantum lattice systems, in particular to the Heisenberg-Ising model with Hamiltonian:

$$\mathcal{H} = -\sum_{i<j} J_{ij}\,[\sigma_i^z\sigma_j^z + \gamma_{ij}(\sigma_i^x\sigma_j^x + \sigma_i^y\sigma_j^y)] - H\sum_i(\sigma_i^z - 1), \qquad J_{ij} \geq 0,\ -1 \leq \gamma_{ij} \leq 1. \tag{5}$$

Asano's work is of interest not simply because it considerably extends the set of models to which the very powerful results of Lee and Yang on phase transitions apply, but because it has provided a new approach to the whole problem of establishing regions of analyticity. Already Ruelle [15] has made use of Asano's ideas to provide a very interesting theorem of zeros which can be used to find other regions in the complex plane, besides the interior of the unit circle, where there are no zeros, and which can be applied to antiferromagnets as well as ferromagnets. And my own feeling is that we can expect still more interesting results in this area in the next few years.

A very different approach to the problem of analyticity is the use of integral equations for the correlation functions. Ruelle [13] showed that the Kirkwood-Salzburg equations may be used to establish a region at low density where the virial series for a classical gas with pair interactions converges and the pressure and correlation functions are all analytic functions of the activity or of the density, and also of the temperature. (Analogous results for quantum systems have been derived by Ginibre [16].) These results are applicable also to "lattice gases," that is, to the Ising model, though here one can obtain in some circumstances even stronger results by using the equations of Gallavotti and Miracle-Sole [17].

At the present time we are very far away from showing what we might hope to show, that for a large class of models the thermodynamic properties and correlation functions are analytic throughout large regions in the p,T or H,T plane, and that phase transitions (in the sense of nonanalyticity) only occur along a few isolated lines. But at least some progress has been made in this direction.

V. THE PEIERLS ARGUMENT FOR THE EXISTENCE OF PHASE TRANSITIONS.

In 1952 when Yang and Lee published their papers [5,6], essentially the only way by which one could prove rigorously that a model had a phase transition was to calculate explicitly its properties in the thermodynamic limit and exhibit the singular behavior. While this technique remains a valuable one, it has been supplemented in recent years by certain arguments which show that a transition is present even though they do not indicate the precise behavior of the limiting functions. Probably the most useful argument of this type goes back to some work of Peierls [18] in 1935, and I shall call it the Peierls argument. Rigorous versions were developed independently a few years ago by Dobrushin [19] and by me [20].

Without actually giving the mathematical details, let me point out the type of argument which is employed. Consider the square lattice shown in figure 2 with ferromagnetic interaction ($J > 0$) between nearest neighbors and zero magnetic field:

$$\mathcal{H} = -J \sum_{\langle ij \rangle} \sigma_i \sigma_j \quad . \tag{6}$$

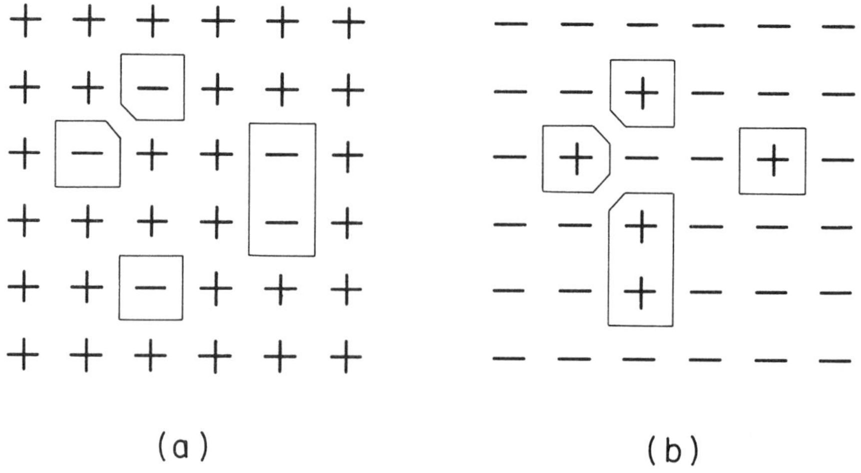

Fig. 2. Two configurations of a small piece of a square lattice with (a) the boundary spins required to be +1 and (b) the boundary spins required to be - 1.

At low temperatures this type of interaction tends to line up the spins parallel on all the sites--that is, they tend to be all +1 or all −1. If the spins on the boundary of a large square are all set equal to +1, one can show that at low temperature the fraction of sites where $\sigma_i = -1$ in the interior of even a very large square is relatively small. On the other hand, if all the spins on the boundary are −1, then the opposite situation occurs, which is to say the fraction of sites with $\sigma_i = +1$ in the interior is relatively small. We conclude that if $H = 0$, the net magnetization or, in lattice-gas language, the density, is very sensitive to the boundary conditions. From here it is a rather short step to show that the magnetization or density must be discontinuous functions of the field or pressure, respectively, in the thermodynamic limit.

There have been various extensions of the Peierls argument to other models besides the two-dimensional square lattice with nearest-neighbor interactions, where we know the value of the discontinuity from explicit calculation [21]. The argument works as well in three dimensions as in two, and it is possible to include interactions extending beyond the nearest-neighbors, provided their magnitudes are suitably restricted. Dobrushin [22] has used the Peierls argument to prove the existence of a phase transition in certain antiferromagnets and "hard square" lattice gases, and Ginibre [23] has extended the argument to include certain quantum lattice systems.

It is perhaps worth emphasizing that what arises most naturally in the Peierls argument is not the discontinuity of thermodynamic functions, but rather the fact that the properties in the interior of a large system can be influenced by what is happening at the boundaries, even when the boundaries become infinitely far away. This provides an alternative definition for what one means by a phase transition, an alternative which has been discussed in particular by Dobrushin [24] and by Lanford and Ruelle [25].

The paper by Lanford and Ruelle discusses the problem in terms of states on a C* algebra, and this seems as good a point as any to insert a remark on the C* algebra approach to things (see the last two chapters in Ruelle's book [7]). My knowledge of the subject is quite limited, but what little I understand suggests that there are some valuable physical insights to be gained if one can penetrate beneath the layers of rather formal and (to me) unfamiliar mathematics which characterize papers written from this point of view.

In the case of an Ising model of a lattice gas, the "state" of the infinite system is simply the set of probabilities of finding various distributions of particles--or spins up and down--in all regions of finite size in the lattice. One wants states which satisfy "equilibrium equations," which is to say that the probabilities of different situations in finite regions are related by appropriate Boltzmann factors. The natural question is as follows: Is there a unique state of the infinite system which satisfies the equilibrium equations? The Peierls argument shows us a specific situation in which the state is not unique. To be specific, there are at least two states possible: one with positive magnetization and one with negative magnetization.

One might expect that the existence of a nonunique state is closely associated with the presence of a phase transition of some sort in the thermodynamic sense; nonanalytic dependence of thermodynamic functions on a suitable parameter. However, the connection is as yet not entirely clear (to me at least!) and one may hope that the future will bring additional insights.

There are some rather severe limitations to the Peierls argument, and a discussion of these will focus attention on some important unsolved problems in the rigorous theory of phase transitions.

The first limitation is that the Peierls argument has only been successfully applied thus far to lattice systems. Nobody believes, of course, that the lattice, which is artificially imposed on these models, is essential for a phase transition, but thus far nobody has been able to prove that a phase transition occurs in a reasonable model of a continuum fluid with short-range forces. Some recent work of Lebowitz and Gallavotti [26] comes tantalizingly close, but still falls just short of establishing a phase transition for a particular continuum model.

A second limitation of the Peierls argument is that it is applicable only to cases where the two (or more) coexisting phases are related to each other by a particular symmetry. In the Ising model this is the "up-down" symmetry of $\sigma_i \to -\sigma_i$, which means time-reversal symmetry in magnetic language and particle-hole symmetry in lattice-gas language. Now phase transitions of this sort are fairly common in solid-state physics, but there are also many other examples--for instance, boiling, melting, and transformations between solid phases with a different crystalline structure--where the two phases are quite unrelated by any known symmetry. One finds experimentally that phase transitions exist, and most of us have the intuitive feeling that they could also be proved to exist in model systems if only we had the needed physical insight and mathematical finesse to push through a proof. One does not have to go to continuum systems, by the way, to find examples of phase transitions in which symmetry between the phases is lacking; they also occur (or rather, we <u>assume</u> they occur!) in certain lattice models.

A third situation in which the Peierls argument, at least in its present form, seems inapplicable is that in which the different "phases" are related by a continuous symmetry operation unlike the discrete symmetry (time reversal) in the Ising model. There are several examples of such transitions. The Heisenberg model of ferromagnetism, for example, can have the spontaneous magnetization pointing in any direction in space. Superfluidity involves a complex order parameter whose phase can vary continuously. And in a solid the symmetries of interest are translation and rotation.

For such models the rigorous arguments available at present are all negative: if the forces are of short range, such a system cannot exhibit a phase transition in one or two dimensions. An exact proof for the Heisenberg model has been published by Mermin and Wagner [27], and analogous arguments have been applied to the case of superfluidity [28] and solids [29].

The arguments just discussed <u>do</u> permit the existence of a phase transition of the ordinary sort in three dimensions. No one, so far as I know, has succeeded in proving the presence of such a transition in an appropriate model with short-range forces. Certainly the ordinary Peierls argument is not sufficient for the task, and an alternative approach is needed.

VI. CORRELATION INEQUALITIES

In the case of Ising ferromagnets one can extend considerably the number of models for which one knows there is a phase transition by using correlation inequalities. The most useful appear to be those put forward by Kelly and Sherman [30], generalizing some of my earlier work [31]. Rather than discuss the most general case, let us consider a particular example of these "GKS" inequalities and its application to proving the existence of phase transitions. With the pair Hamiltonian (3) for a finite system and provided H and all the J_{ij} are nonnegative, one can show that

$$\partial < \sigma_k > / \partial J_{ij} \geq 0 \ . \tag{7}$$

This in particular implies that the magnetization for such a model increases (or at least does not decrease) if one adds additional ferromagnetic interactions. This monotonic behavior persists in the thermodynamic limit, so that if a model exhibits a phase transition, the transition will persist if one adds additional ferromagnetic interactions. This is illustrated in figure 3.

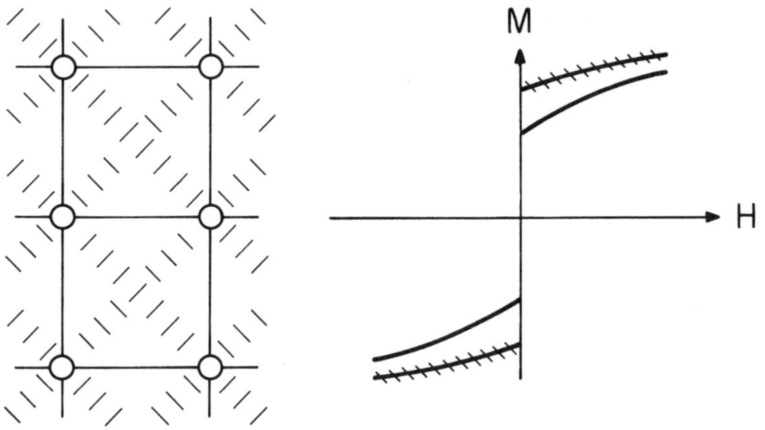

Fig. 3. The solid line is the magnetization curve in the thermodynamic limit for a square lattice with interactions between nearest-neighbor pairs. The addition of ferromagnetic interactions between next-nearest-neighbor pairs (indicated by the stripes) leads to a magnetization curve (indicated by solid line crossed with stripes) which lies above, or at least not below, the original magnetization curve for H > 0.

As well as enabling one to prove, with a minimum amount of effort, the existence of phase transitions in a wide variety of models, the GKS inequalities are useful for various other purposes, such as showing that correlation functions have well-defined thermodynamic limits in the presence of suitable boundary conditions.

The extension of GKS inequalities to other, non-Ising, lattice systems has met with some successes and some failures. Among the former we may mention Ginibre's [32] extension to a system of "plane rotators" and Gallavotti's [33] work on the "X-Y" model. However, Hurst and Sherman [34] showed by means of a counterexample that the most obvious extension of GKS to the quantum Heisenberg model breaks down. The establishment of GKS inequalities for certain systems with continuous symmetries may, hopefully, be a first step in establishing the existence of phase transitions in such systems, a problem which, as I mentioned before, is a challenge to us at the present time.

In addition to the GKS inequalities, certain other inequalities have been proved for correlation functions in Ising ferromagnets [1]. Some of these can be used to show the absence of a phase transition under suitable conditions.

VII. SUMMARY OF PHASE TRANSITIONS IN ISING FERROMAGNETS

Perhaps at this point I can use a diagram, figure 4, to summarize our current knowledge

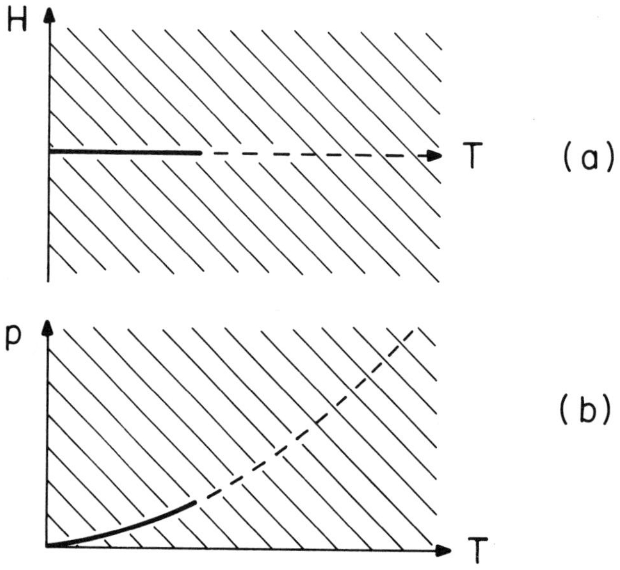

Fig. 4. (a) The cross-hatching indicates schematically the region where analytic behavior of the thermodynamic and correlation functions in the H, T plane has been proved for an Ising ferromagnet in two or more dimensions. At all temperatures this region includes $H > 0$ and $H < 0$, and at high temperatures it includes $H = 0$. At sufficiently low temperatures a phase transition occurs at $H = 0$ (indicated by the solid line). (b) The region of analyticity in the p, T plane for the corresponding lattice gas with negative pair interactions.

about phase transitions in Ising ferromagnets with pair interactions. This is the type of phase transition about which we have the most information, but even here our knowledge is still incomplete. Note, first of all, that in the regions $H > 0$ and $H < 0$, the thermodynamic functions and correlation functions are analytic as a function of field and temperature. At sufficiently high temperatures analyticity can be proved also at $H = 0$ [35] and at sufficiently low temperatures we can prove the existence of phase transitions for most models of interest by using the Peierls argument or the GKS inequalities. Note that this implies that there is a continuous path in the H, T plane from one side of the phase transition to the other (crossing the $H = 0$ line at sufficiently high temperatures) along which all functions are analytic. Proposals have occasionally been made to the effect that the first-order phase transition which one finds at low temperatures should never disappear entirely, but some trace of it should persist, in the form of a "higher-order transition" (some nonanalytic behavior) at arbitrarily high temperatures. It is therefore of some interest that we can rule out this possibility in a particular class of models. We have not yet, however, succeeded in showing that the range of temperatures where things are analytic at $H = 0$ extends right down to the critical temperature (where the first-order transition begins), though everyone believes this is the case for Ising models with Hamiltonian invariant under translations. In a certain class of "random" ferromagnets [36] one can show that nonanalytic behavior at $H = 0$ indeed commences at a temperature above that at which a first-order transition begins.

There is another unsolved problem associated with analytic behavior at the place where a first-order phase transition occurs, which is perhaps of some interest in connection with metastable states. With reference to figure 1, the question is: Can the function M(H) for $H > 0$ be smoothly, in particular analytically, continued into the region $H < 0$? Equivalently, can $\rho(p)$ for $p > p_0$ be analytically continued into the region $p < p_0$? Such an analytic continuation, if it exists, would presumably represent the magnetization (density) of a metastable state. Experts are divided into two camps on this question, though the actual difference between them is for all practical purposes quite small. Those who believe that you cannot analytically continue the functions are generally willing to concede that they are at least infinitely differentiable and hence very smooth at $H = 0$ ($p = p_0$). Thus the argument hinges on a very fine point indeed! Of course, some exact answer one way or another for a model with short-range forces would be most welcome; in particular, let me encourage anyone with a bright idea on how to solve the two-dimensional Ising model in a finite field to get to work!

VIII. CONCLUSION

I have not had time to go into a number of interesting exact results in connection with phase transitions, so let me at least mention a few of the subjects neglected. There is Dyson's [37] work proving the presence of phase transitions in certain one-dimensional spin systems with interactions decreasing (not too rapidly) with distance between pairs of spins. This nicely complements some other arguments [38] which show the absence of phase transitions if the interactions decrease rapidly enough. There is the rather extensive work by Minlos and Sinai [39], who used the tools developed in the Peierls argument to demonstrate certain properties of coexisting phases in an Ising ferromagnet. And then there are the inequalities at critical points [1] which have sometimes been useful in clamping down on the traffic in illegal exponents!

Since this is a conference devoted to "new problems" and the like, I think I should say something in closing about future prospects for exact results as applied to phase transitions.

Where are the interesting problems and what will be the significant advances in the future? I feel it is quite safe to give you my opinions, my guesses, because if they turn out to be true I shall be honored as a prophet and if, as is much more likely, they turn out to be all wrong, my colleagues can thank me for not exposing to the public the interesting ideas before they have had a chance to work on them.

First, I expect we shall see much more attention paid to phase transitions in systems for which the Hamiltonian lacks translational invariance. There are at least two cases of interest: a system with a surface in which one is interested in properties near the surface, and a system with impurities distributed more or less at random throughout the bulk. Intuitively one might expect the random elements to change a sharp phase transition into a continuous one, but at present we know very little about such systems.

Second, I expect that the infinite-system approach, equilibrium equations, and the like will be a fruitful source of physical insight, especially if the practitioners manage to educate the rest of us on what it is they are really up to.

Third, I think that we haven't seen the last of the interesting results involving zeros of the partition function, following the route laid out by Yang and Lee. Some of the work will no doubt be minor extensions of what is already known, but I think this can also be a fruitful source of new insight into phase transitions.

Fourth, I look forward to seeing someone prove the presence of phase transitions in systems with continuous symmetries. We all, I suppose, believe that solids really exist, but as long as proofs are lacking, it gives one the suspicion that something is defective in our understanding of statistical mechanics, the solid state, or perhaps both. The same remarks apply to superfluids and Heisenberg ferromagnets.

Fifth, there are lots of interesting and unanswered questions in the field of critical phenomena and lambda transitions. Exactly soluble models [2,3] have been of great importance in this field, but otherwise there are very few exact results. At the present time there is no general way of showing, apart from explicit calculation, that a model system exhibits a lambda transition; that is, we lack the proper analogue of the Peierls argument.

Finally, I predict a bright future for the study of time-dependent phenomena in connection with phase transitions--but this topic belongs to the next speaker.

REFERENCES

1. R. B. Griffiths, "Rigorous Results and Theorems," in Phase Transitions and Critical Phenomena, ed. C. Domb and M. S. Green (New York: Academic Press, to be published).

2. L. Onsager, Phys. Rev. **65**, 117 (1944).

3. E. H. Lieb, Phys. Rev. Letters, **18**, 1046 (1967).

4. M. E. Fisher, Physics, **3**, 255 (1967).

5. T. D. Lee and C. N. Yang, Phys. Rev. **87**, 410 (1952).

6. C. N. Yang and T. D. Lee, Phys. Rev. 87, 404 (1952).

7. D. Ruelle, Statistical Mechanics: Rigorous Results (New York: Benjamin, 1969).

8. F. J. Dyson and A. Lenard, J. Math. Phys. 8, 423 (1967).

9. J. L. Lebowitz and E. H. Lieb, Phys. Rev. Letters, 22, 631 (1969).

10. W. F. Brown, Jr., Magnetostatic Principles in Ferromagnetism (Amsterdam: North-Holland, 1962), chaps. 2, 3.

11. R. B. Griffiths, Phys. Rev. 176, 655 (1968).

12. J. L. Lebowitz and O. Penrose, J. Math. Phys. 5, 841 (1964).

13. D. Ruelle, Ann. Phys. (N.Y.), 25, 109 (1963); Rev. Mod. Phys. 36, 580 (1964).

14. T. Asano, Phys. Rev. Letters, 24, 1409 (1970); J. Phys. Soc. Japan, 29, 350 (1970).

15. D. Ruelle, Phys. Rev. Letters, 26, 303 (1971).

16. J. Ginibre in Statistical Mechanics, ed. T. A. Bak (New York: Benjamin, 1967), p. 148.

17. G. Gallavotti and S. Miracle-Sole, Commun. Math. Phys. 7, 274 (1968).

18. R. Peierls, Proc. Cambridge Phil. Soc. 32, 477 (1936).

19. R. L. Dobrushin, Theory Prob. Applic. 10, 193 (1965).

20. R. B. Griffiths, Phys. Rev. 136, A437 (1964).

21. C. N. Yang, Phys. Rev. 85, 808 (1952).

22. R. L. Dobrushin, Functional Analysis Applic. 2, 302 (1968).

23. J. Ginibre, Commun. Math. Phys. 14, 205 (1969).

24. R. L. Dobrushin, Functional Analysis Applic. 2, 292 (1968).

25. O. E. Lanford III and D. Ruelle, Commun. Math. Phys. 13, 194 (1969).

26. J. L. Lebowitz and G. Gallavotti, J. Math. Phys. 12, 1129 (1971).

27. N. D. Mermin and H. Wagner, Phys. Rev. Letters, 17, 1133, 1307 (1966).

28. P. C. Hohenberg, Phys. Rev. 158, 383 (1967).

29. N. D. Mermin, Phys. Rev. 176, 250 (1968).

30. D. G. Kelly and S. Sherman, J. Math. Phys. 9, 466 (1968).

31. R. B. Griffiths, J. Math. Phys. 8, 478, 484 (1967).

32. J. Ginibre, Commun. Math. Phys. 16, 310 (1970).

33. G. Gallavotti, Stud. Appl. Math. 50, 89 (1971).

34. C. A. Hurst and S. Sherman, Phys. Rev. Letters, 22, 1357 (1969).

35. G. Gallavotti, S. Miracle-Sole, and D. W. Robinson, Phys. Letters 25A, 493 (1967).

36. R. B. Griffiths, Phys. Rev. Letters, 23, 17 (1969).

37. F. J. Dyson, Commun. Math. Phys. 12, 91 (1969).

38. Reference 7, p. 134 and D. Ruelle, Commun. Math. Phys. 9, 267 (1968).

39. R. A. Minlos and Ya. G. Sinai, Sov. Phys.-Doklady, 12, 688 (1968); Math. USSR-Sbornik, 2, 335 (1967); Trans. Moscow Math. Soc. 17, 237 (1969); ibid., 19, 113 (1969).

DISCUSSION

R. Kubo: Could you comment how much of exact results are known about the behavior of matter near absolute zero? It is already about 60 years since the Nernst-Planck theorem was discovered, but now we realize that very little is known about how this theorem can be established for realistic systems. We know that this theorem is easily proved for ideal systems, but a more general proof seems rather hard. One has to know the distribution of low-lying states above the ground state. Knowledge of the ground state is not sufficient. Still I believe there should be some way to study something exact about the nature of singularities at absolute zero.

R. B. Griffiths: There is very little known exactly about the properties of systems near $T = 0$. On the negative side, it is in general not true that the properties of a system as $T \to 0$ after the thermodynamic limit are the same as the properties of the ground state (first $T \to 0$, then the thermodynamic limit). Most discussions of the "statistical basis" of the Nernst theorem which one finds in textbooks are misleading precisely because of carelessness in interchanging limits.[1] Quite recently H. Leff[2] has succeeded in showing that the entropy of various ferromagnetic Ising models goes to zero as $T \to 0$; he uses the GKS inequalities for this purpose.

M. S. Green: What rigorous results are foreseen for critical phenomena?

R. Griffiths: It is hard to say, but there might be a possibility of proving that the order parameter goes continuously to zero in a suitable class of models.

R. Brout: There is hope for a rigorous proof of thermodynamic scaling a là Widom.

---

1. R. B. Griffiths, J. Math. Phys. 6, 1447 (1965); article in A Critical Review of Thermodynamics, ed. E. B. Stuart et al. (Baltimore, Md.: Mono Book Corp., 1970), p. 101.

2. H. Leff, Phys. Rev. A2, 2368 (1970).

Anonymous: It is to be regretted that approximate calculations on second-order phase transitions are becoming less fashionable.

G. Wannier: I can answer that. Methods of the above type have been used for decades with very poor results.

HAMILTONIAN FLOWS AND RIGOROUS RESULTS IN NONEQUILIBRIUM STATISTICAL MECHANICS

J. L. Lebowitz

I. INTRODUCTION

I will begin my talk by calling your attention to Art Wightman's lecture, Statistical Mechanics and Ergodic Theory: An Expository Lecture, given about two years ago not far from here at the symposium in honor of Professor George Uhlenbeck at Northwestern in 1969 [2]. Wightman's lecture has now been published [1], and I urge you strongly to read it. It describes in a physicist's language many of the ideas developed by mathematicians in the last 40, and more particularly in the last 20, years for the study of the qualitative features of the time evolution of isolated dynamical systems [3].

These concepts on the one hand are of central importance to the understanding of nonequilibrium phenomena in physical systems and on the other hand are almost entirely unknown in the statistical mechanics fraternity, a goodly many of whom are gathered here today. (At least they were unknown before you all heard the lectures by Professors Smale and Ornstein this morning.) Even the nomenclature of this work, such as mixing flow, K-system, are rarely or not at all to be found in any of the many books on statistical mechanics and kinetic theory written for physicists in recent years or, for that matter, in the physics journals. Even when time evolution, irreversibility, and approach to equilibrium are discussed in the standard books [4], all that is usually mentioned is something about ergodic systems and the Birkhoff theorem that a system is ergodic if and only if the Hamiltonian flow in phase space, restricted to a surface of constant energy, is metrically transitive. The latter means that the energy surface is not decomposable, in a "sensitive way," into separate parts which are left invariant by the flow.

Now I believe that almost all real physical systems are "essentially" ergodic. Indeed, this is necessary for understanding why equilibrium statistical mechanics, which includes a description

---

Lecture given at IUPAP Conference on Statistical Mechanics at the University of Chicago, March 1971. Research supported by the U.S.A.F.O.S.R. under contract F44620-71-0013.

of fluctuations in thermal equilibrium, works so well in the real world. I think, however, that just ergodicity alone is not sufficient for either the understanding of irreversible phenomena or--and this is perhaps even more important to those of us who make a living working in this field--for the existence of the integrals which we use in the computation of transport coefficients, such as the diffusion constant, viscosity, etc. [5]. What is necessary is something stronger which is more like Hopf's mixing flow or still stronger.

I want to devote my lecture to (1) defining these concepts in ways which are useful to statistical mechanics, (2) discussing the role of the thermodynamic limit, where systems become very large or infinite, in the study of irreversible phenomena, (3) relating these concepts to some simple model systems which have been worked out explicitly, or almost explicitly, by various authors, some of whom are in this audience, (4) raising some questions about the time evolution of quantum systems which do not fit in a simple way into the structure of ergodic theory, mixing flows, etc., the thermodynamic limit being essential for mixing in quantum systems, (5) questions about the existence of a heat conductivity in random harmonic crystals, and finally (6) to a few remarks about some recent work on the metastable state of systems with very long range interactions.

I would like to make it clear at this point that (1) this talk does not contain even all the rigorous results known to me in this field, much less all those known to others, and (2) nonrigorous does not mean incorrect any more than rigorous means relevant or interesting.

I have in particular left out all discussion of kinetic equations. These equations associated with the names of Boltzmann, Pauli, Kirkwood, Born and Green, Prigogine, van Hove, Bogolubov, and Uhlenbeck, to mention just a few, have played and continue to play a central role in our understanding, interpretation and prediction of nonequilibrium phenomena. (Boltzmann's great H-theorem will have its centenary next year.) Some of these equations, like the Boltzmann equation, may be exact at all times in special limiting situations [such as the Grad limit for hard spheres; particle density $n \to \infty$, particle diameter $a \to 0$, mean free path $\ell = (na^2)^{-1}$ remaining finite]. Other equations may be valid only asymptotically (in the Prigogine-van Hove type of equations these conditions are clearly stated), or may hold only approximately. In any case their physical validity is a question to be settled first by experiment and only later, if we ever get clever enough, by rigorous mathematics.

I could continue with other things I have left out--Green-Kubo formulas, master equations, Brownian motion, hydrodynamics, etc.,--but I think I had better start on the few things which I do want to discuss.

II. ERGODICITY, MIXING, AND DECAY OF CORRELATIONS

Since most of these concepts, and many more, are presented and defined in a clear and precise way in Wightman's lecture [1] and in the book by Arnold and Avez [3], I shall only sketch here the definitions which I need for this talk. I shall use the physicist's language, which sometimes sacrifices precision and generality for the sake of simplicity and familiarity.

Consider a conservative classical system of particles with canonical variables q_i, p_i, i = 1, . . . , 3N. The state of the system is represented in the N-dimensional phase space Γ by a point $x = \{q_i, p_i\}$. The time evolution of the system or its representative point in Γ, x_t [or x(t)], is governed by the canonical equations of motion coming from a time independent Hamiltonian H(x),

$$H = \sum_{i=1}^{N} p_i^2/2m + V(q_1, \ldots, q_{3N}); \quad V(q_1, \ldots, q_{3N}) > -K, \; K < \infty. \qquad (2.1)$$

We assume in addition that the coordinates of the system are confined to a finite region of space Λ by rigid walls (or by some external potential which can be put into V), or that the system moves on some N-dimensional torus (periodic boundary conditions). (When the particles have hard cores, these can also be thought of as internal walls.) Under these conditions the motion of the system, i.e., the flow in Γ space, will be confined to a (6N - 1) = dimensional energy S_E which is piecewise smooth and has finite "area," given by the equation

$$H(x) \equiv E = H(x_t), \quad -\infty \leq t \leq \infty . \qquad (2.2)$$

The time evolution of any function f(x) is given by

$$\begin{aligned} &df(t)/dt = (f(t), H) = iLf, \\ &f(t) \equiv f(x_t) = f(\exp[itL]x) \equiv \exp[itL]f(x) \equiv U_t f(x), \end{aligned} \qquad (2.3)$$

where L is the Liouville operator corresponding to the usual Poisson bracket. Thus if $\rho(x,0)$ describes some Gibbsian ensemble density in Γ at t = 0, then the ensemble density at time t is given by the solution of the Liouville equation,

$$\frac{\partial\rho(x,t)}{\partial t} = (H, \rho) = -iL\rho \qquad (2.4)$$

or

$$\rho(x,t) = \exp[-itL]\ \rho(x,0) = U_{-t}\ \rho(x,0) = \rho(x_{-t}, 0). \qquad (2.5)$$

U_t is a unitary operator, and i is introduced so that L is self-adjoint.

Every $\rho(x,t)$ (which may be singular), which is nonnegative, defines a measure μ on Γ. Thus if A is any set in Γ,

$$\mu(A,t) = \int_A \rho(x,t)dx \equiv \int_A \rho(x_{-t},0)dx = \int_A \rho(x_{-t},0)dx_{-t} = \int_{A_{-t}} \rho(x,0)dx = \mu(A_{-t}0). \qquad (2.6)$$

The third equality following from the Liouville theorem that the volume element dx (Lebesque measure) is invariant in time. We shall always assume that μ is normalized:

$$\mu(\Gamma,t) = \int \rho(x,t)dx = \int \rho(x,0)dx = 1. \qquad (2.7)$$

Of particular importance in statistical mechanics are time independent measures (or ensemble densities),

$$\rho(x,t) = \rho(x,0) = \rho_o(x), \qquad \mu(A,t) = \mu_o(A). \qquad (2.8)$$

Particular examples of such time invariant ensembles are the micro-canonical and canonical ensembles. The first of these is

$$d\mu_o = \begin{cases} [\Omega(E)]^{-1} \dfrac{d\sigma}{|\mathrm{grad}\ H(x)|}, & x \in S_E, \\ 0, & \text{otherwise} \end{cases} \tag{2.9a}$$

where $d\sigma$ is an element of the "surface area" of an element of the (2N - 1) dimensional energy surface S_E and

$$\Omega(E) = \int_{S_E} \frac{d\sigma}{|\mathrm{grad}\ H|}, \tag{2.9b}$$

which is finite by virtue of our assumption that the system is confined to a bounded region of physical space, Λ. The canonical ensemble measure is even more familiar:

$$d\mu_o = Z^{-1} \exp[-\beta H(x)]\ dx, \qquad \beta > 0, \tag{2.10a}$$

where $\beta > 0$ is the reciprocal temperature and

$$Z = \int \exp[-\beta E]\ \Omega(E)\ dE = Z(\beta, N) \tag{2.10b}$$

is a normalizing constant. Another useful ensemble is the generalized microcanonical ensemble used by Griffith [6] in considering the thermodynamic limit problem in equilibrium statistical mechanics

$$d\mu_o = \begin{cases} \text{const. } dx, & \text{for } \{x;\ H(x) \le E\}, \\ 0, & \text{otherwise} \end{cases} \tag{2.11}$$

The grand canonical ensemble has as its domain the disjoint union of the phase spaces Γ_N of systems in Λ having different numbers of particles N, N = 0, 1, . . . , $\underset{\sim}{x}_N \in \Gamma_N$, $\underset{\sim}{x}_N = (\underset{\sim}{q}_1, \ldots, \underset{\sim}{q}_N, \underset{\sim}{p}_1, \ldots, \underset{\sim}{p}_N)$. The lower bound on the potential energy -K in equation (2.1) has now to be written as K = BN, $B < \infty$ independent of N. Its invariant measure is

$$d\mu_o = \Xi^{-1} (N!)^{-1} \exp\{-\beta[H(x_N) - \mu N]\}\ dx_N, \tag{2.12a}$$

with

$$\Xi = \sum_{N=0}^{\infty} (N!)^{-1} \exp[\beta\mu N] Z(\beta, N). \tag{2.12b}$$

We shall now restrict our attention to the microcanonical measure, given in equations (2.9), for some particular S_E.

Definition A. Ergodic System (Birkhoff). A system is said to be ergodic on an energy surface S_E if and only if

$$\bar{f}(x) \equiv \lim_{T\to\infty} \frac{1}{T}\int_0^T f(x_t)dt = \int f(x)d\mu \equiv \langle f\rangle \tag{2.13}$$

for all f(x) in L_1,

$$\int |f(x)|\ d\mu_o < \infty, \tag{2.14}$$

and for almost all x. The latter means that the set A consisting of all points x for which equation (2.13) fails has zero measure, $\mu_o(A) = 0$.

Definition B. Mixing Flow (Hopf). The Hamiltonian flow on S is called mixing if and only if

$$\lim_{t\to\infty} \mu_o(A_t \cap B) = \mu_o(A)\mu_o(B), \quad \mu_o(S_E) = 1 \qquad (2.15)$$

for all sets $A \subset S_E$ and $B \subset S_E$ of positive measure $\mu_o(A) > 0$. $A_t = U_t A$ is the set of points into which A is carried in a time t by the action of the time evolution operator U_t; i.e., $A_t = \{x : x_{-t} \in A\}$.

The significance of mixing is thus that the fraction of systems of the ensemble originally in the set A that are located in some set B at time t approaches, as $t \to \infty$, the ratio of the volume of B to the volume of S_E,

$$\frac{\mu_o(A_t \cap B)}{\mu_o(A)} \xrightarrow[t\to\infty]{} \mu_o(B); \quad \mu_o(S_E) = 1. \qquad (2.16)$$

This property of mixing is stronger than and implies ergodicity [3]. It represents the kind of irreversibility which one always hoped that "coarse graining" in some way or another would bring about. The most remarkable thing though is that Sinai [7] was actually able to prove that a system consisting of a finite number N $(N \geq 2)$ hard spheres (or hard disks in two dimensions) confined to a box is a mixing system.

To make the meaning and significance of mixing even clearer and more directly connected with the formalism commonly used in statistical mechanics, we define, as usual, the correlation functions

$$\langle f(t)\rangle \equiv \int f(x_t)\, g(x)\, d\mu_o, \quad \langle fg\rangle \equiv \langle f(0)g\rangle . \qquad (2.17)$$

It can then be shown, ([3]: Theorem 9.8) that

B'. A system is mixing if and only if

$$\langle f(t)g\rangle \xrightarrow[t\to\infty]{} \langle f\rangle\ \langle g\rangle \qquad (2.18)$$

for all square integrable functions f and g; i.e.,

$$\int |f|^2\, d\mu_o < \infty, \qquad \int |g|^2\, d\mu_o < \infty. \qquad (2.19)$$

It follows therefore in particular from Sinai's work that for a finite system of two or more hard spheres in a box, in two or three dimensions, the velocity autocorrelation of any particle, say particle one, approaches zero as $t \to \infty$,

$$\langle v_1(t)v_1\rangle \xrightarrow[t\to\infty]{} \langle v_1\rangle\ \langle v_1\rangle = 0. \qquad (2.20)$$

This is indeed a remarkable result and contrary to some folklore which holds that it is necessary to go to an infinite size system in order to obtain a decay of the correlation functions. Here, on the other hand, this is shown to be true for a system consisting of N $(N \geq 2)$ particles, as long as N is finite. (It is presumably true also for an infinite system, but this is far from proven.) The usual folklore reason for the belief in the necessity of going to an infinite system is that for a finite system one always has a finite Poincaré recurrence time. Note, however, that the mixing definition is meaningful only for sets of positive measure and that the equivalent decay of correlation functions (B') applies only to square integrable functions (with respect to $d\mu_o$). What mixing therefore implies is that the times when different systems which were initially together in the same set A [which can be as small as desired as long as $\mu_o(A) > 0$] return to the "neighborhood of A" are so different from each other that eventually the set A_t spreads out uniformly over all S_E. This is the important property of mixing flows which does not follow at all from ergodicity alone, and is in striking contrast to what occurs in assemblies of oscillators such as harmonic crystals where all the phase points have essentially the same time dependence. We shall discuss these later.

Having debunked one piece of folklore, we shall now show that another piece of folklore does survive and becomes a rigorous result for mixing systems. To this end consider a function f(x), which is equal to (F(x), H(x)), the Poisson bracket of F(x) with the Hamiltonian H(x), with f and F square integrable as in inequalities (2.19). (We are always dealing here with a finite system.) Let $\varphi(x)$ also be a square integrable function. Then

$$\int_0^T \langle f(t)\varphi \rangle \, dt = \int_0^T \langle \frac{d}{dt} F(x_t)\, \varphi(x) \rangle \, dt = \langle F(T)\, \varphi \rangle - \langle F\varphi \rangle \quad . \tag{2.21}$$

In a similar way if g(x) = (G(x), H(x)), with g and G square integrable, then

$$\int_0^T \langle f(t) g \rangle \, dt = \int_0^T \langle \varphi(t+t') \frac{dG(t')}{dt'} \rangle \, dt = -[\langle \varphi(T) G \rangle - \langle \varphi G \rangle] , \tag{2.22}$$

where we have used the time invariance of μ_o,

$$\langle \psi(t+t')\, \chi(t') \rangle = \langle \psi(t) \chi \rangle \quad . \tag{2.23}$$

Finally if f = (F, H) and g = (G, H), then we can also write

$$\int_0^T \langle f(t) g \rangle \, dt = \int_0^T \frac{d}{dt} F(t+t') \frac{d}{dt'} G(t') \quad dt = - \frac{d}{dT} \langle F(T) G \rangle + \langle (F, H) G \rangle . \tag{2.24}$$

If we now set f = g, F = G, then since $\langle (F,H)F \rangle = \frac{1}{2} \langle dF^2/dt \rangle = 0$, we find, for the time integral of the autocorrelation function,

$$\int_0^T \langle f(t) f \rangle \, dt = - \frac{d}{dT} \langle F(T) F \rangle = - \langle f(T) F \rangle \quad . \tag{2.25}$$

Now if our system is mixing, we can use condition (2.18) to take the limit $T \to \infty$ in the above equations and obtain the following theorem.

<u>Theorem</u>. Let f, g, φ, F, G be square integrable with f = (F, H), g = (G, H). For a <u>mixing</u> system

$$\lim_{T \to \infty} \int_0^T \langle f(t)\, \varphi \rangle \, dt = \langle F \rangle \langle \varphi \rangle - \langle F\varphi \rangle , \tag{2.26}$$

$$\lim_{T \to \infty} \int_0^T \langle \varphi(t) g \rangle \, dt = \langle \varphi G \rangle - \langle \varphi \rangle \langle G \rangle , \tag{2.27}$$

$$\lim_{T \to \infty} \int_0^T \langle f(t) g \rangle \, dt = \langle (F, H) G \rangle , \tag{2.28}$$

$$\lim_{T\to\infty} \int_0^T \langle f(t)f\rangle\, dt = 0. \tag{2.29}$$

Example: For a finite mixing system confined by a wall,

$$\lim_{T\to\infty} \int_0^T \langle v_1(t)v_1\rangle\, dt = \lim_{T\to\infty} \langle q_1(T)v_1\rangle = 0 \ . \tag{2.30}$$

Proof: $v_1 = (q_1, H)$ and $\int v_1^2\, d\mu_o < \infty$, $\int q_1^2\, d\mu_o < \infty$.

Note that when q_1 is an angle variable, e.g., in the case of periodic boundary conditions, then $v_1 \neq (q_1, H)$ and equation (2.30) need not hold. We would still have, however, $\langle v_1(t)v_1\rangle \to 0$ if the system is mixing.

When the system is not mixing, the limit $T \to \infty$ in the above integrals need not exist. We see, however, from equation (2.25) that for any finite system in a box

$$\lim_{T\to\infty} \int_0^T \langle f(t)f\rangle\, dt = 0, \text{ if it exists.} \tag{2.31}$$

This is so since

$$\langle F(T)F\rangle \le \langle F(T)F(T)\rangle^{\frac{1}{2}} \langle F^2\rangle \ , \tag{2.32}$$

so that when F is square integrable, $\langle F(T)F\rangle$ can either oscillate or approach zero.

The significance of these results for nonequilibrium statistical mechanics comes from the fact that the Kubo formulae [5] for transport coefficients all involve integrals over time, from $t = 0$ to $t = \infty$, of correlation functions of "fluxes" which can be written as Poisson brackets. These integrals will therefore, if they exist at all, be equal to zero in any finite system. (The important point of our theorem is that they do exist for mixing systems, which possibly include all real systems.) These formulae can therefore possibly yield transport coefficients only in the thermodynamic limit. In the thermodynamic limit, functions like q_1 need no longer be square integrable. Hence,even when $\langle f(t)f\rangle \to \langle f\rangle^2$ in such systems for square integrable functions f like v_1, $\langle v_1(t)v_1\rangle \to 0$ we can still define $\int_0^T \langle v_1(t)v_1\rangle\, dt = \langle q_1(T)v_1\rangle$, there is no reason to expect that $\langle q_1(t)v_1\rangle$ will also vanish (or even exist) as $t \to \infty$. (This will become clear in the examples given in the next section.)

One of the most important problems in statistical mechanics (if you care about rigorous results), at the present time, is therefore to investigate and hopefully to establish the existence in the thermodynamic limit of the time integrals used in the Kubo formulae. Unfortunately it seems impossible even to tackle this problem before one proves the existence of a time evolution in the thermodynamic limit. This has been established so far, for a general class of systems, only in one dimension, Lanford [8]. I shall come back to this later.

I shall conclude this section by defining still another class of systems; those with homogeneous Lebesgue spectrum. This will enable us to clarify somewhat the relation between the mixing properties of the flow and the properties of the Liouville operator L.

Definition C. Homogeneous Lebesgue Spectrum (Koopman). A system has a homogeneous Lebesgue spectrum if, when one diagonalizes L, every real number λ lies in the spectrum and has the same multiplicity and the spectral weight is just $d\lambda$. (This definition is copied from Wightman [1].) The space on which L acts here is the Hilbert space of complex-valued square integrable functions $f(x)$, $x \in S_E$.

Conditions A, B, and C are members of a hierarchy of increasingly stronger conditions on the flow (or the Hamiltonian H which generates it). It has been shown (see [3] for references) that

$$\text{Homogeneous Lebesgue spectrum} \longrightarrow \text{Mixing} \longrightarrow \text{Ergodicity}\,, \tag{2.33}$$

but not the converse. Mixing does, however, imply that L has no discrete eigenvalues other than zero. Such a property of L implies in turn that the system is at least weakly mixing, which implies ergodicity. A weakly mixing system is one in which [3]

$$\lim_{T\to\infty} \frac{1}{T} \int_0^T |\mu_o(A_t \cap B) - \mu_o(A)\mu_o(B)|\, dt = 0. \tag{2.34}$$

An eigenfunction corresponding to the eigenvalue zero is of course any constant of S_E. It follows already from ergodicity that the eigenvalue zero is simple; i.e., constants are the only eigenfunctions of L with eigenvalue zero. The converse is also true; i.e., if zero is a simple eigenvalue of L, then the flow is ergodic. This is indeed another way of stating the Birkhoff theorem, alluded to earlier, about the equivalence between ergodicity and the nondecomposability of S_E.

III. MODEL SYSTEMS

1. Two Particles in a One-Dimensional Box

It is not easy to find nontrivial yet simple examples of an ergodic Hamiltonian system. To give you something to do with your idle time here, consider a one-dimensional system consisting of two hard particles of masses m_1 and m_2 in a box of length $\mathcal{L}$ having rigid walls. The energy surface S_E is three-dimensional $0 \le q_1 \le q_2 \le \mathcal{L}$; $p_1^2/2m_1 + p_2^2/2m_2 = E$. I came here believing that this system is ergodic if $\theta = \cos^{-1}\left[1 - 2\,\frac{(m_1 - m_2)^2}{(m_1 + m_2)^2}\right]$ is not a rationale multiple of π. After talking with some people, I am now very doubtful. I also wonder what happens for more than two particles with different masses.

2. One-Dimensional Harmonic Oscillator

A trivial example of an ergodic system is a one-dimensional harmonic oscillator

$$H = \frac{1}{2}\, m^{-1} p^2 + \frac{1}{2}\, m\, \omega^2 q^2 \;. \tag{3.1}$$

The energy surface is a one-dimensional ellipse and certainly nondecomposable. The eigenfunctions of L are $(p \pm i\, m\, \omega q)^{\ell}$, where ℓ is an integer, with eigenvalues $\lambda_\ell = \ell\omega$. In terms of the action variables, (J, θ), the Hamiltonian is simply $H = \omega J$ and the flow is described by

$$\theta_t = \theta + \omega t \pmod{2\pi}; \quad J_t = J \ . \tag{3.2}$$

3. General Harmonic System

For a general harmonic system with N degrees of freedom (e.g., a harmonic crystal in one, two or three dimensions), we can always write the Hamiltonian in action-angle variables (J_i, θ_i), $i = 1, \ldots, N$,

$$H = \sum_{i=1}^{N} \omega_i J_i \ , \quad iL = \sum_{i=1}^{N} \omega_i \frac{\partial}{\partial \theta_i} \ , \tag{3.3}$$

where

$$J_i(t) = J_i, \quad \theta_i(t) = \theta_i + \omega_i t \quad \pmod{2\pi}. \tag{3.4}$$

All the J_i are constants of the motion, and therefore the system will not be ergodic on the $(2N-1)$-dimensional energy surface S_E for $N > 1$. This system may, however, still be ergodic on some M-dimensional subspaces of S_E, $M < N$. Such a subspace is a torus $\Pi(\{\theta\}_M)$ whose coordinates are the angle variables $(\theta_{i_1}, \ldots, \theta_{i_M}) = \{\theta\}$ for which the corresponding frequencies ω_{i_ℓ} are not resonant [3]

$$\sum_{\ell=1}^{N} n_{i_\ell} \omega_{i_\ell} \neq 0, \text{ for } n_{i_\ell} \text{ integers not all zero.} \tag{3.5}$$

What we mean here by ergodicity is that the flow induced on $\Pi(\{\theta\}_M)$ by the projection onto $\{\theta\}_m$ on the flow on S_E (really on $\{\theta\}_N$ since the J_i stay constant) is ergodic. (The measure $d\mu_o$ on $\Pi(\{\theta\}_M)$ is the usual Cartesian M-dimensional volume properly normalized.) In other words, any function f(x) which depends only on the variables $\{\theta\}_M$ is ergodic with respect to the microcanonical measure $d\mu_o$ on S_E,

$$\lim_{T\to\infty} \frac{1}{T} \int_o^T f(\{\theta_t\}_M) dt = \int_{S_E} f(\{\theta_M\})\, d\mu_o = \langle f \rangle \ . \tag{3.6}$$

The fact that <u>some</u> functions f(x) are ergodic even though the system is not ergodic may be true also in other than harmonic systems. A similar situation may happen also with regard to mixing and suggests the following definition.

4. Definition

The set of ergodic and mixing functions $\mathcal{E}$ and $\mathcal{M}$, respectively, is defined as follows:

$$f \in \mathcal{E} \text{ iff } \int |f|^2 d\mu_o < \infty \text{ and}$$

$$\bar{f}(x) \equiv \lim_{T\to\infty} \frac{1}{T} \int_o^T f(x_t) dt = \int f d\mu_o = \langle f \rangle \ ; \tag{3.7}$$

$$f \in \mathcal{M} \text{ iff } \int |f|^2 \, d\mu_0 < \infty \text{ and } \lim_{t \to \infty} \langle f(t) f \rangle = \langle f \rangle^2 . \tag{3.8}$$

The set $\mathcal{E}$ is clearly a linear subspace of L_2. By Khinchin's ergodic theorem, $\mathcal{M} \subset \mathcal{E}$.

5. Harmonic Crystals

We may summarize now a few of the known results about harmonic crystals in one, two, and three dimensions.

Beginning with Schrodinger, various authors [9] have studied, among other things, the velocity autocorrelation function of a specified particle in a harmonic crystal of N particles (using the canonical ensemble as the invariant measure). Since the velocity of the jth particle v_i depends, when expressed as a function of the action-angle variables $(\{J_i\}, \{\theta_i\})$, $i = 1, \ldots, Nd$ (d = dimensionality of space), not only on the θ_i but also on the J_i, it is not an ergodic function: $v_i \notin \mathcal{E}$ for fixed N. It is, however, possible to show that in the thermodynamic limit v_i is a mixing function. Mazur and Montroll [9] found that in this limit the velocity autocorrelation function decays as $t^{-d/2}$,

$$\langle v_i(t) v_i \rangle = \lim_{N \to \infty} \langle v_i(t) v_i \rangle_N \sim (\sin t)/t^{d/2} . \tag{3.9}$$

It is seen from equation (3.9) that $|\langle v_1(t) v_1 \rangle|$ is integrable in three dimensions but not in two or one dimension. This appears related (cf. discussion after equation [2.24]) to the Peierls result [10] that the mean square displacement $\langle q_i^2 \rangle$ of a particle at a fixed lattice site behaves (when the size of the crystal increases so that the boundaries of the crystals, where the particles are tied down to fixed positions, recede in all directions) like

$$\langle q_i^2 \rangle \sim \begin{cases} D_N , & \text{one dimension,} \\ \ln D_N , & \text{two dimensions,} \\ \text{constant,} & \text{three dimensions.} \end{cases} \tag{3.10}$$

Here D_N is the distance of the boundary from the jth site and approaches infinity. [Since in three dimensions each particle remains localized about its equilibrium position when $N \to \infty$, I expect that $\int_0^T \langle v_1(t) v_1 \rangle \, dt$ will go to zero as $t \to \infty$. This is what happens when particle one has a mass much larger than that of the other particles.]

While Mazur and Montroll established their results explicitly only for cubic lattices with nearest neighbor interactions, the essential feature of the system responsible for the decay of the velocity autocorrelations in the thermodynamic limit appears to be the absence of localized modes as $N \to \infty$. This "corresponds" to the Liouville operator L not having any discrete eigenvalues in this limit. (If L had this property for a finite system, then, as already mentioned at the end of Section II, the system would be at least <u>weakly mixing</u> which implies ergodicity.) We may therefore conjecture that whenever the spectrum of the **infinite** crystal is entirely continuous, all "local" functions are mixing and <u>cross-mixing</u>. We define a local function F as one which depends on the coordinates q_i and momenta p_i of a fixed finite number of particles $i = 1, \ldots, \ell$ and is square integrable,

$$\int |F(x_1, \ldots, x_\ell)|^2 \, d\mu_o \le K, \quad x_i = (q_i, p_i), \tag{3.11}$$

where $K < \infty$, as a constant independent of the size of the crystal characterized by N. If our conjecture is right, then, in the absence of localized modes for the infinite crystal,

$$\lim_{t \to \infty} \left\{ \lim_{N \to \infty} \langle F_\ell(t) G_j \rangle_N \right\} = \lim_{N \to \infty} \left[\langle F_\ell \rangle_N \langle G_j \rangle_N \right], \tag{3.12}$$

where F_ℓ and G_j are localized functions depending on ℓ and j coordinates and momenta, respectively.

The importance of isolated modes in determining the ergodic properties of local functions was exhibited very clearly by Cukier and Mazur [11], who investigated the ergodicity of the kinetic energy of a single "impurity," i.e., of a particle of mass M placed at position j in a harmonic chain of particles of mass m. They found that the function $p_j^2(t)$ will be ergodic as $N \to \infty$ if $M > m$ and is not ergodic if $M < m$. This difference is brought about by the existence of a localized mode when $M < m$ and its absence when $M > m$.

6. XY Model Spin System and General Quantum Systems

The inhibiting effect of an isolated eigenvalue (of the time evolution operator) on the relaxation of a local disturbance can also be seen explicitly in the work of Abraham et. al. [12] on the time evolution of a quantum spin system. They consider a one-dimensional system of spins with nearest-neighbor XY interactions without or with an external magnetic field h in the z-direction acting on the jth spin. The Hamiltonian of the system in these two cases is then

$$H_o = \tfrac{1}{4} \sum_{i=-N}^{N} \left[(1+\gamma)\sigma_i^x \sigma_{i+1}^x + (1-\gamma)\sigma_i^y \sigma_{i+1}^y \right], \tag{3.13}$$

$$H' = H_o + h\sigma_j^z \,. \tag{3.14}$$

Their results may then be extrapolated and interpreted [13] as showing that all "local" functions F and G which depend only on a finite set of spins $F(\sigma_{i_1}, \ldots, \sigma_{i_\ell})$, $G(\sigma_{i_1}, \ldots, \sigma_{i_k})$ will be mixing when $N \to \infty$:

$$\left.\begin{aligned} &\langle F(t) G \rangle - \langle F \rangle \langle G \rangle \to 0, \\ &\langle GF(t) G \rangle - \langle F \rangle \langle G^2 \rangle \to 0, \end{aligned}\right\} \quad \text{as } t \to \infty \tag{3.15}$$

if $H = H_o$ but not (generally) if $H = H'$. The approach to zero for a single spin correlation in the first case is as t^{-1}. The different behavior under the action of H_o and H' is due [13] to the presence of an isolated eigenvalue in the spectrum of H'.

This fits in with the general C* algebra formulation of the time evolution of infinite quantum systems [13]. It appears indeed that because of the discrete nature of the spectrum for all finite quantum systems confined to a bounded domain Λ there will be no mixing (decay of correlations) in such a system. The remarkable thing about Sinai's result is that it shows that finite classical systems can and do have purely continuous spectra. (Note that when Planck's constant $h \to 0$ the number of energy levels between some fixed E and E + ΔE becomes infinite.)

7. One-dimensional System of Hard Rods

An interesting model system which exhibits some "real fluid" -like properties is a one-dimensional system of hard rods with "diameter", a, and equal masses m = 1. The dynamics of this system consists entirely of an interchange of the velocities of pairs of neighboring particles upon collision and free motion between collisions. It is clear that when the diameter is zero, a = 0, then all functions f(x) which are symmetric in the coordinates and velocities (q_i, v_i) of all the particles, e.g.,

$$f(x) = \sum_{i=1}^{N} \varphi_1(q_i, v_i)$$

will have the same time evolution as the corresponding function in an ideal gas where the particles simply pass through each other. (For an ideal gas the q_i and v_i behave, except possibly for boundary conditions on the q_i, in exactly the same way as do the action-angle variables J_i and θ_i in a harmonic system [14].)

Hence the only "interesting" functions in a system of rods of zero diameter are those which depend on the coordinates and momenta of a specified set of particles, which we may consider to be "labeled," e.g., $\varphi_1(q_i, v_i)$, $\varphi_2(q_{i_1}, v_{i_1}, v_{i_2})$, etc. Different aspects of the time evolution of such labeled one-particle functions, say $\varphi_1(q_1, v_1)$, were investigated in detail by Jespen [15], Lebowitz and Percus [16], and Spitzer [17], while the time evolution of some symmetric one-particle functions (such as van Hove's coherent neutron scattering function) in a system of hard rods with finite diameters was investigated by Lebowitz, Percus, and Sykes [18]. I shall only mention here a few of those results (and some conjectures of mine) which are related to the problems we have been discussing here.

Let the particles be labeled with index i, i = -n, . . . , n, 2n + 1 = N, and confined to a box (or circle) of length $\mathcal{L}$. Then in the thermodynamic limit $N \to \infty$, $\mathcal{L} \to \infty$, $N/\mathcal{L} \to \rho$, the properly behaved functions [square integrable functions $\varphi_1(q_i, v_i)$, $\psi_1(q_\ell, v_\ell)$, $\langle \varphi_1{}^2 \rangle \le K$, $\langle \psi_1{}^2 \rangle < K$, independent of N, ?] are mixing,

$$\langle \varphi_1(t)\, \psi_1 \rangle \equiv \lim_{N \to \infty} \langle \varphi_1(t)\, \psi_1 \rangle_N \to \langle \varphi_1 \rangle \langle \psi_1 \rangle, \; t \to \infty . \tag{3.16}$$

In particular,

$$\langle v_1(t) v_1 \rangle \to 0, \qquad t \to \infty, \tag{3.17}$$

$$\int_0^\infty \langle v_i(t) v_i \rangle \, dt = (2\rho)^{-1} \langle |v_i| \rangle = D , \tag{3.18}$$

with D the self-diffusion constant. The mixing property in expression (3.17) is true only when $N \to \infty$, for it is possible to show explicitly in some cases that in a finite system the velocity autocorrelation function $\langle v_i(t) v_i \rangle_N$ does not approach zero [19]. When the rods have a finite diameter a, ρ in expression (3.18) is replaced by $\rho/(1 - \rho a)$. The same is true for other correlations of the form (3.16).

The form of the asymptotic approach to zero of $\langle v_1(t)v_1\rangle$ depends on the invariant measure $d\mu_o$ used. Since, as already mentioned, the entire dynamics of this system consists of an interchange of velocities during a collision, any ensemble density of the form

$$\rho_o(q_{-n}, v_{-n}, \ldots, q_n, v_n) = \mathcal{L}^{-N} \prod_{i=-n}^{n} h_o(v_i) - \mathcal{L}/2 \le q_i \le \mathcal{L}/2 , \qquad (3.19)$$

$$h_o(v) \ge 0,\ h_o(v_i) = h_o(-v_i),\ \int_{-\infty}^{\infty} h_o(v)\, dv = 1 , \qquad (3.20)$$

is invariant in time. It is furthermore required that the diffusion constant be finite, $\int |v|\, h_o(v)\, dv < \infty$, for the thermodynamic limit of the "process" to exist [17].

Several forms of $h_o(v)$ have been studied explicitly with the results that when $h_o(v)$ is the Maxwellian distribution [15]

$$h_o(v) = (\beta/2\pi)^{\frac{1}{2}} \exp[- v^2/2] , \qquad (3.21)$$

or [16]

$$h_o(v) = \tfrac{1}{2} c^2 (c^2 + v^2)^{-3/2} , \qquad (3.22)$$

then

$$\langle v(t)v_1\rangle \sim -t^{-3} \qquad t \to \infty . \qquad (3.23)$$

where $h_o(v)$ is discrete,

$$h_o(v) = \tfrac{1}{2} [\delta(v - c) + \delta(v + c)] , \quad c < 0 , \qquad (3.24)$$

then [16]

$$\langle v_1(t)v_1\rangle = c^2 \exp[- 2\rho c t] . \qquad (3.25)$$

Note that for the velocity distribution (3.22) v_1 is <u>not</u> square integrable. The time integral of $\langle v_1(t)v_1\rangle$ in equation (3.18) still exists, however, since $\langle v_1(t)v_1\rangle \sim |\ln t|$ as $t \to 0$.

Another result of interest is that the spatial density of a labeled particle which was specified initially (at $t = 0$) to be at a fixed position r' (i.e., the van Hove self-function of neutron scattering theory),

$$\rho G_s(r,t/r') = \lim_{N \to \infty} \langle \delta[q_s(t) - r]\, \delta(q_1 - r')\rangle_N = \rho G_s(r - r', t) , \qquad (3.26)$$

has the asymptotic form appropriate to a Brownian diffusion process

$$G_s(r, t) \sim (4\pi D t)^{-\frac{1}{2}} \exp[- r^2/4Dt] \qquad (3.27)$$

as $|r| \to \infty$, $t \to \infty$, $r/t^{\frac{1}{2}}$ finite. More precisely, if $q(t) = q_i(t) - q_i$ is the displacement of the jth particle, then in the thermodynamic limit, $A^{-\frac{1}{2}} q(At) \to z(t)$ as $A \to \infty$ with z(t) a standard Brownian diffusion process with D given in equation (3.18) [17].

Let me conjecture that in this system as in the harmonic crystal all square integrable local functions are mixing in the thermodynamic limit. (This seems to follow indeed from the work O. de Pazzes reported at this conference; Abstract, II.25. Similar results were also proven recently by Sinai [private communication].)

IV. THE LANFORD THEOREM

As you have heard from Griffith in his lecture, the rigorous study of the thermodynamic limit of equilibrium statistical mechanics has already achieved notable results. The comparable investigation of the infinite volume limit of nonequilibrium systems is much more difficult and has begun only recently.

Let $\{\underset{\sim}{q}, \underset{\sim}{p}\}$ represent the positions and momenta of a set (not necessarily finite) of particles of unit mass each. Then Newton's equations of motion have the explicit form

$$\dot{\underset{\sim}{q}}_i(t) = \underset{\sim}{p}_i(t), \quad \dot{\underset{\sim}{p}}_i(t) = \sum_{j \neq i} \underset{\sim}{F}\,[\underset{\sim}{q}_i(t) - \underset{\sim}{q}_j(t)] \ , \tag{4.1}$$

where F is the interparticle force. If we have a finite number of particles, then there is clearly a unique solution to this set of differential equations for all sets of initial conditions $\{\underset{\sim}{q}_i(0), \underset{\sim}{p}_i(0)\}$ (when F is suitably bounded). The existence of a meaningful solution to Newton's equations, i.e., the existence of a time evolution of the system, becomes, however, far from trivial when we consider a system consisting (from the beginning or in some limit) of an infinite number of particles. In such a system it is quite possible to begin with a perfectly reasonable set of initial values $\{\underset{\sim}{q}_i(0), \underset{\sim}{p}_i(0)\}$ and find after some finite time t that there are an infinite number of particles in a finite region of space and that the right side of equation (4.1) is infinite. We illustrate this with a simple example given by Lanford. Consider a system in which there are no forces, i.e., an ideal gas, and assume that at time zero $\underset{\sim}{p}_i = -\underset{\sim}{q}_i$ for each i; then at time $t = 1$ <u>all</u> the particles will be situated at the origin.

When the particles have hard cores, then it is possible (Ginibre) in two or three dimensions to construct "reasonable" initial configurations in which, through a cascade of collisions, some particle will have an infinite velocity after a finite time.

Thus, we need to find a class of initial conditions for which such catastrophes do not happen. In fact, since we are interested also in the equilibrium state of our systems, we would like to show that those classes of initial conditions which have nonzero probability of occurring in equilibrium do not give rise to such catastrophes. An even stronger desired result is to show that the time evolution of a part of the system contained in a fixed region of space Λ will, at any time t, be determined entirely by the state of the system at time $t = 0$ in the neighborhood of Λ (how large this neighborhood is will of course depend on t). This was indeed proven by Lanford [8] for one-dimensional systems. Assuming that $\underset{\sim}{F}$ has a finite range, $\underset{\sim}{F}(q) = 0$ for $q > D$, and that $\underset{\sim}{F}$ is bounded, $|\underset{\sim}{F}(q)| < K$, D and K some positive constants, Lanford proved the existence for all times of a "regular" solution of Newton's equations of motion for a "regular" initial configuration. A regular configuration is, roughly speaking, one in which the number of particles in a unit interval and the magnitude of the momentum of any particle in that interval have a bound of the form $\delta \log R$, where R denotes the distance of the interval from the origin. It was further proven by Lanford that, when the interparticle potential is positive or the activity is small, the set of nonregular configurations has probability zero in the equilibrium grand canonical ensemble.

Lanford's results have been extended recently to one-dimensional systems with hard cores [20]. (This is a "marriage" of Lanford's systems and the hard rod system discussed earlier.)

A question left open by these results is whether a state which at time $t = 0$ is described by a set of correlation functions can still be described by a set of correlation functions when $t \neq 0$. This was investigated by Gallavotti, Lanford, and Lebowitz [21], who proved that, for certain classes of initial states, the time-evolving state is described by correlation functions and that these correlation functions satisfy the BBGKY hierarchy in the sense of distributions.

The initial states we consider can be described as follows: Suppose that the system is in equilibrium at temperature β^{-1} and activity z under the influence of a pair potential and an external potential h which is localized in a finite region I_h. At time $t = 0$ we switch off the external field, and the system begins to evolve. We prove that if the correlation functions exist for the equilibrium state at $t = 0$ and if the Lanford theorem holds, then the system can always be described by a set of correlation functions which vary in time according to the BBGKY hierarchy. We are, however, unable to prove the convergence of the frugacity (or density) expansion for these functions at any $t > 0$ [22]. We are also unable to prove even that the time averaged correlation functions evolve toward the correlation functions which correspond to the equilibrium state at temperature β^{-1} and activity z (in the absence of an external field) as would be expected. We are, however, able to prove that the time-averaged correlation functions converge to a limit satisfying the stationary BBGKY hierarchy.

Note that when there are no interactions between the particles, i.e., for an ideal gas, the correlation functions we have been discussing here have (as indicated in section III) all the desirable mixing properties. This is due to the disappearance, through the unimpeded motion of the particles, of any local disturbance. The "approach to equilibrium" thus exhibited by an ideal gas is, however, different in an essential way from the irreversible behavior of real systems. The approach to equilibrium in real systems can be described by hydrodynamics and/or kinetic equations whereas the ideal gas does not obey Fourier's law of heat conduction or Fick's law of diffusion. (The model systems discussed in the last section are similar to the ideal gas in this respect.

The origin of these differences presumably lies in the existence of a mechanism (collisions) in real systems which acts locally to bring the system to a state of local equilibrium and the absence of such a mechanism in the ideal gas and in the model systems we have discussed. This difference is also exhibited when we consider systems in which the departure from the equilibrium state is "global." Consider, for example, a system whose state at $t = 0$ is specified by correlation functions whose spatial part is that obtained from a grand canonical (or canonical) ensemble at temperature T and frugacity z but whose velocity part is not Maxwellian (the mean kinetic energy is, however, specified by T). I would expect that in a real system, but clearly not in an ideal gas, these correlations would approach their true equilibrium values; e.g., the velocity distribution would become Maxwellian.

V. OPEN SYSTEMS

We shall now extend our analysis to systems whose time evolution is not given by a Hamiltonian flow in phase space. These are systems which are in contact with outside reservoirs. We shall be particularly interested in the "steady state" energy flux in a system in contact with heat reservoirs at different temperatures T_α. Following the general principles of statistical mechanics [4], we identify the observable properties of such a system with averages over a "suitable" phase-space ensemble. To obtain such a Gibbs ensemble we use a formalism developed in earlier work [23], [24], and look for the stationary solution of a generalized Liouville equation having the form

$$\frac{\partial \rho(x,t)}{\partial t} + (\rho, H) = \sum_{\alpha} \int [K_\alpha(x,x')\, \rho(x',t) - K_\alpha(x',x)\, \rho(x,t)]\, dx' . \qquad (5.1)$$

Here $K_\alpha(x,x')\, dx\, dt$ is the conditional probability that when the system is at the point x' in its phase space it will, due to its interaction with the αth reservoir, make a transition to the volume element dx, about x, in the time interval dt. It is assumed here that the reservoirs are "stationary" so that the K_α's are independent of time.

Equation (1.1) describes a stationary Markov process, and we may define the stochastic time evolution operator W^t, for $t > 0$, by

$$\rho(x,t) = W^t \rho(x,0) = \int W(x,t|x')\rho(x',0)dx, \quad t \geq 0, \qquad (5.2)$$

where $\rho(x,t)$ is the solution of equation (5.1) when the ensemble density at time zero is $\rho(x,0)$. The operators W^t form a semigroup

$$W^{t_1 + t_2} = W^{t_1} W^{t_2} , \quad t_1, t_2 \geq 0 , \qquad (5.3)$$

but are not unitary operators since the flow is not measure preserving. It is possible to show, under certain conditions on H and the K_α's [23], [24], generally satisfied by our systems, that as $t \to \infty$, $\rho(x,t)$ will approach (in some suitable sense) a stationary ensemble density $\rho_s(x)$ which is independent of the initial ensemble density $\rho(x,0)$,

$$\lim_{t \to \infty} W^t \rho(x,0) = \rho_s(x) , \qquad W^t \rho_s(x) = \rho_s(x) . \qquad (5.4)$$

When all the reservoirs have the same temperature this will be an equilibrium canonical ensemble, whereas for reservoirs at different temperatures this ensemble will represent a system in a steady nonequilibrium state through which heat is flowing. Define

$$W_t f = \int f(y)\, W(y,t|x)dy, \quad \langle f \rangle = \int f(y)\rho_s(y)dy,$$
$$\langle f(t)g \rangle = \langle (W_t f)g \rangle = \int [\int dy\ f(y)W(y,t|x)]\, g(x)\rho_s(x)\, dx . \qquad (5.5)$$

When equation (5.4) holds and $W(y,t|x) \to \rho_s(y)$ as $t \to \infty$, we have

$$W_t f \to \langle f \rangle , \quad \langle f(t)g \rangle \to \langle f \rangle \langle g \rangle \text{ as } t \to \infty . \qquad (5.6)$$

To obtain the energy flow into the system from each reservoir we multiply equation (5.1) by H and integrate over x to obtain

$$\frac{\partial\langle H\rangle}{\partial t} = \sum_{\alpha}\int\left\{\int K_{\alpha}(x,x')\ [H(x) - H(x')]dx\right\}\rho(x',t)dx' \equiv \sum_{\alpha} J_{\alpha'} \qquad (5.7)$$

where J_α is the average energy flux from the αth reservoir. In the steady state we have, of course, $\Sigma\ J_\alpha = 0$. Thus if the geometry is set up in such a way that the system is in contact with only two reservoirs--one "on the left" at a temperature T_L and "one on the right" at a temperature T_R with $T_L > T_R$--and if the system has a uniform "cross-section" S and "length" $\mathcal{L}$, then we expect that in the stationary state the heat flux $J = J_L = -J_R$ should, for macroscopic size systems, be related via Fourier's law to the average temperature gradient $(T_L - T_R)/\mathcal{L}$. More precisely, J should have the property that the quantity $\kappa(\mathcal{L}) \equiv (J/S)/[T_L - T_R)/\mathcal{L}]$ should approach a well defined limit κ when $\mathcal{L} \to \infty$. This κ, if it exists, we would identify with the heat conductivity of the system at temperature T when $T_L \to T_R \to T$.

This formalism has been applied [24] to a harmonic crystal with some particular forms of interaction with the heat reservoirs. The stationary nonequilibrium ensemble density for such a harmonic system was found to be a generalized Gaussian. The covariance matrix of this Gaussian was obtained there explicitly for a one-dimensional chain of equal masses with nearest neighbor interactions whose end atoms are in contact with heat reservoirs at temperature T_L and T_R. Identifying the number of particles in the chain with its length $\mathcal{L}$, it was found there that in the stationary nonequilibrium state $\kappa(\mathcal{L}) \sim \mathcal{L}$; i.e., the heat flux achieves a constant value, for fixed $T_L - T_{R'}$ independent of the length of the chain $\mathcal{L}$. A similar result obtains for any perfectly periodic harmonic crystal corresponding to an "infinite" heat conductivity, if one can speak of a heat conductivity in this case [25], [26].

Searching for a model system in which Fourier's law could be shown to hold, Casher and Lebowitz [27] investigated what happens in the same situation to a crystal whose atoms are not all of the same mass, with the different masses distributed at "random" [28]. We were unable to obtain a definite result for the asymptotic behavior of $k(\mathcal{L})$ but could show rigorously only that the heat flux J will not vanish as $\mathcal{L} \to \infty$ if the spectral measure of the infinite chain has an absolutely continuous part. Indeed, this is the reason why the heat flux in a periodic chain becomes independent of $\mathcal{L}$ as $\mathcal{L} \to \infty$.

We also showed, by using a theorem of Matsuda and Ishii [29], that for a random chain $J \to 0$ as $\mathcal{L} \to \infty$ with probability one with $\langle J\rangle > 0\ (\mathcal{L}^{-3/2})$, where $\langle J\rangle$ is the heat flow averaged over the random mass distribution. This may suggest that the eigenfrequencies of a disordered infinite chain are all isolated; but this is not so, as we show that the spectrum of an infinite chain in which the masses can have only two different values contains a nondenumerable infinity of points and is thus, in particular, not exhausted by a set of discrete eigenvalues having a denumerable number of accumulation points. This result is based on a proof that the cumulative frequency distribution of such a chain is continuous.

These results raise the possibility that the spectrum of a disordered chain may be of the singular continuous type; i.e., its continuous spectrum may have its support in a kind of Cantor set. They also raise the question of whether in other systems, too--e.g., hard spheres--the existence of transport coefficients in the infinite system may not require the absence of an absolutely continuous spectrum, i.e., the kind of spectrum Sinai proved exists for a finite system. On the other hand, as we have already seen, the irreversible decay of local disturbances requires the absence of localized bound states (corresponding to a point spectrum).

(Visscher has suggested that J may depend on the boundary conditions placed on the end atoms of the chain. This appears to agree with the finding of Rubin and Greer in a paper which will appear in J. Math. Phys.)

VI. METASTABLE STATES IN THE VAN DER WAALS-MAXWELL THEORY

Until now I have been talking about (a) time-dependent phenomena in isolated dynamical systems and (b) stationary nonequilibrium phenomena which are maintained in open systems by externally imposed gradients. I shall now speak briefly about a third kind of nonequilibrium situation: the metastable state of an isolated system [30].

We may characterize metastable thermodynamic states by the following properties:

(a) Only one thermodynamic phase is present.

(b) A system that starts in this state is likely to take a long time to get out. (6.1)

(c) Once the system gets out, it is unlikely to return.

In order to discuss the static and dynamic properties of metastable states in a rigorous fashion it is necessary to make precise the notion, inherent in all physical theories of metastability [31], of imposing a restriction on the system which keeps its density roughly uniform. In general, such a restriction may be represented by confining the configuration of the system to a suitable region R in configuration space. In order for this region to correspond to a metastable state, the restrictions defining it should correspond to the imposition of a roughly uniform density, in accordance with the criterion (6.1a), and it should also have properties corresponding to the conditions (6.1b) and (6.1c) mentioned earlier: if the dynamical state is initially in R, it is unlikely to escape quickly; and once it has escaped it is unlikely to return.

With this in mind we investigated the existence of such a region R for a system whose liquid vapor phase transition can be proven to be of the van der Waals type and is clearly understood [32]. The main feature of this system is that its interparticle potential v(r) can be cleanly divided into "short-range repulsive" and "long-range attractive" parts:

$$v(r) = q(r) + \gamma^{\nu} \varphi(\gamma r) \qquad (0 \le r < \infty), \tag{6.2}$$

where ν is the number of space dimensions and γ^{-1} is the range of the "Kac potential" $\gamma^{\nu}\varphi(\gamma r)$. We take here $\varphi(r) \le 0$ and set

$$\gamma^{\nu} \int \varphi(\gamma r)\, d^{\nu}r = \int \varphi(y)\, d^{\nu}y \equiv \alpha . \tag{6.3}$$

Under suitable conditions on q and φ, the thermodynamic limit of the Helmholtz free energy density (f.e.d.) at a given particle density ρ (the dependence on temperature is not displayed) is given in the limit $\gamma \to 0$ by [32]

$$f(\rho, 0^+) = \lim_{\gamma \to 0} f(\rho, \gamma) = CE[f_o(\rho) + \tfrac{1}{2}\alpha\rho^2] \ . \tag{6.4}$$

Here $f_o(\rho)$ denotes the free energy of the reference system--that is, the system whose interaction potential function is q instead of v--and the symbol CE indicates the convex envelope of the expression following it, i.e., the value of the maximal convex function whose value nowhere exceeds $f_o(\rho) + \frac{1}{2}\alpha\rho^2$. Since $\alpha \leq 0$, the function $f_o(\rho) + \frac{1}{2}\alpha\rho^2$ need not be convex even though $f_o(\rho)$ must be (see Fig. 1).

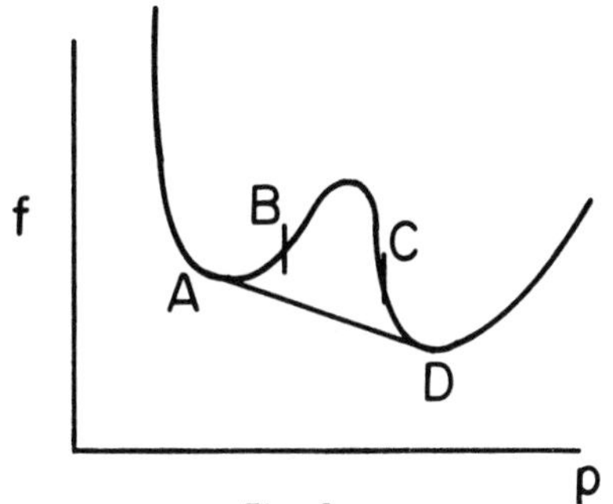

Fig. 1

The smooth curve in the figure is the graph of $f_o(\rho) + \frac{1}{2}\alpha\rho^2$, whose convex envelope is obtained by replacing the arc ABCD by the corresponding double tangent AD.

The parts of the curve which Maxwell associated with metastable states are the arcs AB, CD; these may be specified by the conditions

$$f_o(\rho) + \tfrac{1}{2}\alpha\rho^2 > f(\rho, 0^+) \tag{6.5a}$$

and

$$f_o''(\rho) + \alpha > 0 \ , \tag{6.5b}$$

where f_o'' denotes the second derivative of f_o.

To define our phase-space region R we start with a finite system in a domain Λ, of volume $|\Lambda|$, and divide Λ into cubical cells $\omega_1, \omega_2, \ldots, \omega_M$, each of volume $|\omega|$, $M|\omega| = |\Lambda|$. We define the dynamical variable n_i $(i = 1, \ldots, M)$ to be the number of particles in ω_i, and we choose two numbers ρ^- and ρ^+ such that $\rho^- < \rho^+$ and the condition

$$f_o''(x) + 2\alpha > \text{const.} > 0 \tag{6.6}$$

holds for all $x \in [\rho^-, \rho^+]$. Since α is negative, this is more restrictive than condition (6.5b), but the precise values of ρ^- and ρ^+ chosen are unimportant. In particular, they may be arbitrarily close. In accordance with the condition (6.1a) that the metastable state should correspond to a single thermodynamic phase, we now define R to be the set of all configurations compatible with M constraints:

$$\rho^- |\omega| < n_i < \rho^+ |\omega| \quad (i = 1, 2, \ldots, M) \ . \tag{6.7}$$

We now assume that at time $t = 0$ our system is in a state represented by a restricted grand canonical ensemble that is one constructed by selecting from a grand canonical ensemble those systems whose configuration is in R. The chemical potential of this grand ensemble is $f_o(\rho) + \alpha\rho$

with ρ satisfying condition (6.5a) and $\rho^- < \rho < \rho^+$. These conditions assure that the equilibrium state of this system has an average density which is not in (ρ^-, ρ^+) and hence the equilibrium probability of finding the system in R will be small (zero in the thermodynamic limit).

The choice of the grand canonical ensemble (rather than the canonical or microcanonical ensembles which are believed to be essentially equivalent to it) is a matter of convenience. It is used only for setting up the restricted ensemble at time 0; it does not imply that the system is open for times greater than 0. The dynamics of the system are assumed to be determined entirely by its own Hamiltonian (no interaction with the outside). The flow generated by this Hamiltonian will carry some systems initially in R outside R. If we let $\mu(R, t)$ be the measure of R at time t as in equation (2.6), i.e., the fraction of systems left in R at that time, then $\mu(R, 0) = 1$. We have that $p(R,t) = 1 - \mu(R,t)$ is the conditional probability for a system initially in R (i.e., in the metastable state) not to be in R at time t. It is then shown rigorously that for periodic boundary conditions, this conditional probability is at most λt, where λ is a quantity that goes to 0 in the limit

$$|\Lambda| >> \gamma^{-\nu} >> |\omega| >> r_o \ln |\Lambda| \,. \tag{6.8}$$

Here r_o is a length characterizing the potential q, and $x >> y$ means $x/y \to +\infty$. For rigid walls the same result is proved under somewhat more restrictive conditions. It is argued that a system started in the metastable state will behave (over times $<< \lambda^{-1}$) like a uniform thermodynamic phase with f.e.d. $f_o(\rho) + \frac{1}{2}\alpha\rho^2$, but that having once left this metastable state the system is unlikely to return.

The form of our upper bound on λ, is roughly $|\Lambda| \exp(-\Delta/kT)$ with Δ a positive "activation energy" proportional to $|\omega|$. Similar formulae for escape rates occur in other theories [32]. The main difference between these formulae and ours is that they are intended to be approximations to the true escape rate, whereas ours is a rigorous upper bound, but not necessarily a good approximation.

This upper bound was made possible by the fact that, in the limit we are considering, the range γ^{-1} of the Kac potential becomes very large compared with the other physical lengths, r_o and $\rho^{-1/\nu}$. This permits a clean separation of the effects of the Kac potential from those of the short-range potential q. This separation is accomplished by introducing an artificial new length $|\omega|^{1/\nu}$ satisfying the two conditions (both coming from inequality [6.8])

$$\begin{aligned} |\omega|^{1/\nu} &>> \rho^{-1/\nu} \ln |\Omega| \\ |\omega|^{1/\nu} &<< \gamma^{-1} \end{aligned} \tag{6.9}$$

and defining R through constraints (eq. [6.7]) on density variations over the length scale $|\omega|^{1/\nu}$.

In the canonical ensemble for a system of $N = \rho|\Lambda|$ particles the first condition in (6.9) ensures that there are enough particles in each cell to make a fluctuation from the average occupation number assumed in the metastable state which violates the "constraints," an unlikely event in the restricted equilibrium ensemble, and hence enables us to prove that the escape rate is small. The second condition ensures that any phase transition due to the Kac potential is completely suppressed by the constraints, and hence makes the state defined by these constraints a very unlikely one in the full equilibrium ensemble, if this ensemble predicts such a phase transition. This also ensures that this state is spatially uniform.

Since both parts of (6.9) are crucial to our treatment of metastability, the result that λ can be made as small as we please does not apply to real physical systems, for which there is no $\gamma \to 0$ limit. For a realistic potential, it may well be impossible to find regions R which have both arbitrarily small escape rates and arbitrarily small equilibrium probabilities--particularly in view of the apparent impossibility of analytically continuing the equilibrium free energies and correlation functions for such potentials into the domain of metastability [32], [33].

I would like to thank J. Percus, B. Simon, J. Sykes, and particularly R. Resibois for many valuable and clarifying discussions during the preparation of this talk.

REFERENCES

1. A. S. Wightman in Statistical Mechanics at the Turn of the Decade, ed. E. D. Cohen (Dekker, 1971).

2. Since I was unable to attend that symposium and Professor Uhlenbeck is "officially" retiring from teaching this year, I would like to dedicate this talk to him.

3. For references see [1] and the very nice book by V. I. Arnold and A. Avez, Ergodic Problems of Classical Mechanics (New York: Benjamin, 1968).

4. I would like to mention here the book by O. Penrose, Foundations of Statistical Mechanics (London: Pergamon, 1970), where a framework for describing the time-dependent behavior of macroscopic systems is developed and its relation to dynamics clarified (if not solved). For an account of the classical ergodic theory see I. Farquhar, Ergodic Theory in Statistical Mechanics (New York: Wiley, 1964).

5. H. Mori, Phys. Rev. **112**, 1829 (1958); L. Kadanoff and P. Martin, Ann. Phys. **29**, 419 (1963).

6. R. B. Griffith, J. Math. Phys. **6**, 1447 (1965).

7. Ja. Sinai, Statistical Mechanics Foundations and Applications; Proceedings of the I.U.P.A.P. Meeting, Copenhagen, 1966, ed. T. A. Bak (New York: Benjamin, 1967), p. 559; Russian Math. Sur. **25**, 137 (1970). The full proof of the theorem has not been published yet.

8. O. Lanford, Commun. Math. Phys. **9**, 176 (1968); **11**, 257 (1969).

9. G. Klein and I. Prigogine, Physica **19**, 74, 89, 1053 (1953); P. G. Hemmer, thesis, Trondheim, Norway, 1959; P. Mazur and E. Montroll, J. Math. Phys. **1**, 70 (1960); R. J. Rubin, J. Math. Phys. **1**, 309 (1960); **2**, 373 (1961), and references cited there.

10. R. E. Peierls, Quantum Theory of Solids (Oxford, 1955). E. W. Montroll, Proc. Third Berkeley Symposium on Math. Stat. and Prob. **3**, 209 (1957).

11. R. I. Cukier and P. Mazur, "The Ergodic Properties of an Impurity in a Harmonic Oscillator Chain" (preprint).

12. D. B. Abraham, E. Barouch, G. Gallavotti, and A. Martin-Lof, Phys. Rev. Letters **25**, 1449 (1970) and preprint; cf. also E. Barouch, B. McCoy, and M. Dresden, Phys. Rev. **A2**, 1075 (1970), and C. Radin, J. Math. Phys. **11**, 2945 (1970). See also discussion remark by G. G. Emch.

13. This interpretation of the results of Abraham et. al., was clarified for me by B. Simon. See also H. Narnhoffer, preprint I.H.E.S.

14. H. L. Frisch, Phys. Rev. **104**, 1 (1956).

15. D. W. Jepsen, J. Math. Phys. **6**, 405 (1965).

16. J. L. Lebowitz and J. K. Percus, Phys. Rev. **155**, 122 (1967).

17. F. Spitzer, J. Math. Mech. **18**, 973 (1969); "Random Processes Defined Through the Interaction of an Infinite Particle System," Lecture Notes in Mathematics, **89**, 201 (Berlin: Springer, 1969).

18. J. L. Lebowitz, J. K. Percus, and J. Sykes, Phys. Rev. **171**, 224 (1968).

19. J. L. Lebowitz and J. Sykes, to appear in J. Stat. Phys.

20. This was done recently by Sinai (private communication).

21. G. Gallavotti, O. E. Lanford III, and J. L. Lebowitz, J. Math. Phys. **11**, 2898 (1970).

22. For a Lorentz gas in which noninteracting particles--"wind" particles--move among and scatter from fixed hard spheres--"tree" particles--Gallavotti was able to prove (J. Math. Phys. **185**, 308 [1969]) that the time evolution of the wind particles' correlation functions can be described at any finite time by a convergent power series in the density of the tree particles when this density is sufficiently small.

23. P. G. Bergmann and J. L. Lebowitz, Phys. Rev. **99**, 578 (1955); J. L. Lebowitz and P. G. Bergmann, Ann. Phys. **1**, 1 (1957); J. L. Lebowitz, Phys. Rev. **114**, 1192 (1959); J. L. Lebowitz and A. Shimony, Phys. Rev. **128**, 1945 (1962).

24. Z. Rieder, J. L. Lebowitz, and E. Lieb, J. Math. Phys. **8**, 1073 (1967).

25. The explicit exact solution for two-dimensional perfect crystals with nearest as well as next nearest coupling was recently obtained by R. Helleman. The general form of the solution was treated earlier by H. Nakazawa.

26. Similar results for the perfect harmonic crystal were also obtained earlier by other investigators; cf. [9] and references in [24], and also R. Brout and I. Prigogine, Physica, **22**, 621 (1956); H. Nakazawa, Prog. Theor. Phys. **39**, 236 (1968); M. Bolsterli, M. Rich, and W. M. Visscher, Phys. Rev. **A1**, 1086 (1970).

27. A. Casher and J. L. Lebowitz, "Heat Flow in Regular and Disordered Harmonic Chains," to appear in J. Math. Phys.

28. (a) R. J. Rubin, J. Math. Phys. **9**, 2252 (1968); **11**, 1857 (1970), finds that the attenuation of a wave in a random chain of length $\mathcal{L}$ is proportional to $\exp[-\gamma\mathcal{L}]$. (b) For results of machine computations on such systems see D. N. Payton and W. M. Visscher, Phys. Rev. **156**, 1032 (1967); E. A. Jackson, J. P. Pasta, and J. F. Waters, J. Com. Phys. **2**,

207 (1968). Visscher and Payton are carrying out more computation now which should be available soon.

29. H. Matsuda and K. Ishii, Suppl. Prog. Theor. Phys. 45 (1970); cf. also H. Matsuda, T. Miyata, and K. Ishii, Suppl. To J. Phys. Soc. Japan, 26, 40 (1968).

30. O. Penrose and J. L. Lebowitz, "Rigorous Treatment of Metastable States in the van der Waals-Maxwell Theory," Stat. Phys. 3, 211 (1971).

31. J. Frenkel, Kinetic Theory of Liquids (Dover, 1955); N. G. van Kampen, Phys. Rev. 135A, 362 (1964); Physica, 48, 313 (1970); J. Langer, Ann. Phys. N. Y. 41, 108 (1967); M. E. Fisher, Physics, 3, 255 (1967); J. B. Jalickee, F. W. Wiegel, and R. J. Wezetti, J. Math. Phys. 11, 3168 (1970).

32. J. L. Lebowitz and O. Penrose, J. Math. Phys. 7, 98 (1966). See also D. J. Gates and O. Penrose, Comm. Math. Phys. 15, 255 (1970); 16, 231 (1970) for example of Kac potentials where (6.3) does not hold. (These papers contain references to the important pioneering work of van Kampen and Kac, Uhlenbeck and Hemmer in this field.)

33. O. Lanford and D. Ruelle, Comm. Math. Phys. 13, 194 (1969); also Langer and Fisher in reference [31] .

DISCUSSION

A. Siegert: Is there any indication in Sinai's proof that the limit is, for practical purposes, reached in a reasonable time, and not, for instance, in a time comparable with a Poincaré cycle?

B. Simon: The time it takes to reach the limit will depend on the particular observable that one is interested in. The autocorrelation function for a given observable differs from its final value in general by the Fourier transform of an L_1 function and therefore goes to zero. The approach may, however, be arbitrarily slow. Sinai's theorem involves showing that certain quantities approach zero exponentially fast, so it may be possible to prove that autocorrelation functions of interesting observables approach their asymptotic value exponentially; but obviously such an attempt must await the publication of Sinai's proof.

A. Scotti: Since it is reasonable to expect that Sinai's theorem for hard spheres will be generalized to physically more realistic potentials, and the considerations you have made on the limiting values of $\langle v(t)v(o)\rangle$ and $\int_o^T \langle v(t)v(o)\rangle\, dt$ depend only on the system being mixed, would you care to comment a little bit more on your philosophy regarding results that are valid only for a finite number of particles?

J. Lebowitz: I expect that for any fixed t, $\langle v(t)v(o)\rangle_N \xrightarrow[N \to \infty]{} \langle v(t)v(o)\rangle_\infty$, where the subscript indicates the size of the system.

B. Robertson: Dr. Lebowitz pointed out, equivalently, that the Fourier, Laplace transform of the Green-Kubo velocity autocorrelation function vanishes if the $\omega \to 0$ limit is carried out before the $k \to 0$ limit. This singular behavior does not occur in autocorrelation functions that include a suitable projection operator as given, for example, in J. Math. Phys. 11, 2482 (1970).

G. Emch: The kind of classical ergodic theory Professor Lebowitz just talked about has been (partly) extended to quantum situations, by Charles Radin and myself. As a particular application of that general theory, one can show exactly that certain infinite systems provide a thermal bath for all of their finite parts; specifically, local thermal deviations do decay to equilibrium. This result thus provides a proof of a specialized form of the second principle of thermodynamics. The XY model belongs to the class for which the above assertions (as well as some other such statements concerning local perturbations) hold.

[At that point a question was asked by M. Green on necessity versus sufficiency. The answer to that question was: What we have are certain conditions on the time evolution, which are sufficient to ensure the above behavior. What we say is that these conditions are in particular satisfied by the Hamiltonian of the XY model.]

H. Matsuda: Concerning the thermal conductivity of the isotopically disordered harmonic chain, Allen and Ford discussed it applying the Kubo formula. Although their formal expression for the thermal conductivity κ is right, they obtained a finite κ in the limit $N \to \infty$ which is found not correct. By making a correction to their results, Ishii and I obtained a κ which is proportional to $\sqrt{N}$. Apart from the question of the validity of the Kubo formula in this model, I believe that the above N-dependence of the Kubo thermal conductivity is the right answer.

W. Visscher: The thermal conductivity of a disordered harmonic chain does, in fact, depend strongly on the boundary conditions one assumes. For example, if the (0)th and (N + 1)th atoms are clamped, with the first and (N)th atoms weakly coupled to heat reservoirs at different temperatures, then, as Professor Matsuda says, the effective thermal conductivity goes as $N^{-\frac{1}{2}}$. On the other hand, if the ends of the chain are free, the effective thermal conductivity for large N goes as $N^{\frac{1}{2}}$. The reason for this is that in a disordered chain only the very low frequency normal modes contribute to the heat current because all the other modes are localized, and the coupling of the low frequency, long wavelength modes to the reservoirs is suppressed if the ends of the chain are clamped. If weak anharmonicities are introduced, exchange of energy between normal modes becomes possible, and the behavior of the thermal conductivity becomes more reasonable, namely, independent of N for long chains.

H. Wergeland: The peculiar nature of the energy transport in harmonic lattices to which Dr. Lebowitz alludes has of course given rise to a great many studies. Perhaps it is most strikingly brought out in the example chosen by Hemmer: When two contiguous parts of an infinite linear chain start with unequal temperatures, the temperature gradient at the junction will indeed gradually decrease to zero but the flow of heat across this point will increase to a constant value.

In this sense, therefore, one may say that the heat conductivity is infinite. Now, one can make more complicated harmonic models by distributing impurities (randomly or otherwise) in the lattice. And one can in this way obtain an averaged leveling of energy gradients which is different from simple elastic signals. Yet the transport of heat does not become freely diffusive. One cannot derive a purely parabolic Fourier equation along such lines because the dispersion function--however complicated one makes it--will always remain a causal one.

J. Harrison: The behavior of the linear chain of harmonic oscillators can be understood in terms of the electrical network analogy described in my contributed paper at this conference. By the described theorem of Haus and Adler, valid for linear circuits and hence for harmonic oscillators, the temperature "looking" toward one end of the chain will be invariant along the chain and will be the temperature of the far end. Thus the temperature source at one end looks into a sink characterized by the temperature at the other end. The heat transfer will therefore be proportional to the temperature difference, with the proportionality constant solely dependent upon the impedance "match" between the impedance of the temperature source and that of the sink as transformed along the chain. For long uniform chains this impedance factor will have a characteristic value averaged over a small frequency interval, which is independent of the length of the chain. For the chain with random masses, the average value of impedance of the sink as transformed over the length of the chain will tend to grow proportionally to the square root of length, modifying the heat transfer accordingly. A uniform thermal gradient varying inversely as the length can be recovered for both the uniform chain and the statistically homogeneous random chain by the introduction of a small amount of dissipative attenuation in the chain in a manner equivalent to the use of self-consistent heat reservoirs by Bolsterli, Rich, and Visscher.

G. Wannier: There is a development in experimental physics running parallel to the difficulties in the theory discussed here. Measurement of a coefficient of heat conduction is becoming increasingly difficult as crystals become more perfect and data more precise. The range in which the coefficient is determined by crystal imperfections is increasing. If the concept can be saved for perfect crystals, it will probably need more careful definition. Such a definition must take into account that part of the heat always travels with the velocity of sound.

II. Developments in Biology

5

NONLINEAR RATE PROCESSES, ESPECIALLY THOSE INVOLVING COMPETITIVE PROCESSES

Elliott W. Montroll

I. SOME IMPORTANT NONLINEAR RATE PROCESSES

Many important rate processes which govern the behavior of materials and organisms are described by nonlinear rate equations with and without random driving forces. The author and some of his colleagues have been trying to develop an intuition about these processes by investigating some models of them [1-5]. In this report we show the similarity of the basic rate equations of several superficially different processes and show how a few classes of solvable models can be constructed. Certain new results on models of competitive processes will be presented.

A rather general class of nonlinear processes is that which characterizes chemical kinetics of species which enjoy binary collisions. Consider the chemical rate equations

$$\frac{dc_n}{dt} = \sum_{\ell} k_n^{\ell} c_{\ell} + \sum_{\ell m} k_n^{\ell m} c_{\ell} c_m + F_n(t) , \qquad (1.1)$$

where the set of numbers $\{c_n\}$ represent concentrations or populations of various species identified by index n. The rate constants k_n^{ℓ} represent the rate of spontaneous transition from species ℓ into species n except for the special case $\ell = n$. The number k_n^n would usually (but not always) be negative and might represent the spontaneous disappearance rate of species n. Finally, the rate constants $k_n^{\ell m}$ would be associated with the formation of species n through collision of species ℓ and m. The function $F_n(t)$ is the rate at which species n is added or removed from the system. It might be a causal or random function of the time. It also might connect population growth or decline with fluid motions, in which case equation (1.1) would be coupled to hydrodynamic and thermodynamic equations. Coupled hydrodynamic and chemical processes can be expressed in terms of equations somewhat like (1.1) by interpreting the c_{ℓ}'s somewhat more broadly.

This research was partially supported by the Advanced Research Projects Agency of the Department of Defense and was monitored by ONR under contract N00014-67-A-0398-0005.

The numbers c_n might be interpreted as vectors whose components are identified with type of species, position, temperature, etc. The chemical processes in biological cells, as well as in the atmosphere, can be characterized by equations similar to (I.1). As is shown in Appendix I, one can make an appropriate transformation in the concentration variables which transforms equation (I.1) into (m, n, and s being indices and not powers)

$$\frac{dy^m}{dt} = \lambda_m y^m + \sum_{ks} \Gamma_{ns}^{m} y^n y^s \tag{I.2}$$

when there is no external driving force.

The basic equation for the investigation of the behavior of a viscous incompressible fluid is the Navier-Stokes equation

$$\underset{\sim}{u}_t + (\underset{\sim}{u} \cdot \nabla)\underset{\sim}{u} = -\rho^{-1}\nabla p + \nu\nabla^2 \underset{\sim}{u} + \underset{\sim}{F} \tag{I.3a}$$

with

$$\nabla \cdot \underset{\sim}{u} = 0 , \tag{I.3b}$$

where $\underset{\sim}{u}(r,t)$ is the velocity vector of a fluid element at point r at time t, ρ the density of the fluid, p(r,t) the pressure at (r,t), and ν the kinematic viscosity, and $\underset{\sim}{F}$ is a force driving the fluid. Note that by taking the divergence of equation (I.3a) and remembering equation (I.3b),

$$-\rho^{-1}\nabla^2 p = \nabla \cdot (\underset{\sim}{u} \cdot \nabla)\underset{\sim}{u} , \tag{I.3c}$$

which is analogous to Poisson's equation of electrostatics. One can determine p as a function of u. When this is inserted into equation (I.3a), the resulting equation depends only on u and not on p. If we make a Fourier analysis of the pressure field and the velocity field by letting

$$p(\underset{\sim}{r},t) = (2\pi)^{-3} \int p(\underset{\sim}{k},t)e^{i\underset{\sim}{k}\cdot\underset{\sim}{r}} d^3k , \tag{I.4a}$$

$$\underset{\sim}{u}(\underset{\sim}{r},t) = (2\pi)^{-3} \int \underset{\sim}{a}(\underset{\sim}{k},t)e^{i\underset{\sim}{k}\cdot\underset{\sim}{r}} d^3k , \tag{I.4b}$$

then it is easy to show from equations (I.3a)-(I.3c) that

$$\rho^{-1}k^2 p(\underset{\sim}{k},t) = -(2\pi)^{-3} \int [\underset{\sim}{k}' \cdot \underset{\sim}{a}(\underset{\sim}{k}-\underset{\sim}{k}',t)]\,[\underset{\sim}{k}\cdot\underset{\sim}{a}(\underset{\sim}{k}',t)]\, d^3k' , \tag{I.5a}$$

$$\underset{\sim}{k} \cdot \underset{\sim}{a}(\underset{\sim}{k},t) = 0 \tag{I.5b}$$

and finally that [5-7]

$$d\underset{\sim}{a}(\underset{\sim}{k},t)/dt = -\nu k^2 \underset{\sim}{a}(\underset{\sim}{k},t) + \underset{\sim}{F}(\underset{\sim}{k},t)$$
$$- i(2\pi)^{-3} \int d^3k'[\underset{\sim}{k}\cdot\underset{\sim}{a}(\underset{\sim}{k}',t)][\underset{\sim}{a}(\underset{\sim}{k}-\underset{\sim}{k}',t) - \underset{\sim}{k}\underset{\sim}{k}'\cdot\underset{\sim}{a}(\underset{\sim}{k}-\underset{\sim}{k}',t)/k^2] . \tag{I.6}$$

This equation is a special case of (I.1) (or, perhaps, rather [I.2]). In that notation,

$$k_n^{\ell} = -n^2 \nu \delta_{n,\ell} , \tag{I.7}$$

$\delta_{n,\ell}$ is the Kronecker delta. The $\underset{\sim}{k}$-space representation of the Navier-Stokes equation is somewhat simpler than the general rate equation (I.2) in that the binary collision rate constant depends only on two indices rather than three. The theory of turbulence would seem to have a mathematical structure similar to that of chemical kinetics, although the theory of neither of these processes has been developed very far.

While (I.6) is a vector equation, the analogous two-dimensional Navier-Stokes equation depends only on the scalar potential ϕ which is related to the flow field by

$$u_1 = -\partial\phi/\partial y \ , \quad u_2 = \partial\phi/\partial x \ . \tag{I.8}$$

If Φ is the Fourier transform of ϕ ,

$$\Phi(\underset{\sim}{k},t) = \int \phi(\underset{\sim}{r},t)e^{i\underset{\sim}{k}\cdot\underset{\sim}{r}}\, d^2r, \tag{I.9}$$

then it is easy to show [1] that

$$\frac{d\Phi}{dt} + \nu k^2\Phi = -ik^2(\underset{\sim}{k}\times\underset{\sim}{F})_z + (2\pi)^{-2}\int_{-\infty}^{\infty}\int (k'^2/k^2)[\underset{\sim}{k}\times\underset{\sim}{k}']_z\Phi(\underset{\sim}{k}',t)\Phi(\underset{\sim}{k}-\underset{\sim}{k}',t)\, d^2k' \tag{I.10}$$

where $[\underset{\sim}{a}\times\underset{\sim}{b}]_z \equiv a_xb_y - a_yb_x$,and F is the Fourier transform of the driving force.

A still simpler set of nonlinear rate equations is that of certain models of competition between biological species. The interaction between two species, one of which preys upon the other, has been modeled independently by Lotka [8,9] and Volterra [10,11]. Two species, 1 and 2, are to be considered such that in the absence of species 2, species 1 would increase exponentially in number; in the absence of species 1, species 2 would die out exponentially. Furthermore, it is postulated that through binary collisions, species 2 finds its food supply at the expense of species 1 which is eaten. The rate equations which describe this process are

$$\dot{N}_1 = k_1N_1 - \lambda_1N_1N_2 \ , \tag{I.11a}$$

$$\dot{N}_2 = -k_2N_2 + \lambda_2N_1N_2 \ . \tag{I.11b}$$

The constant λ_1 tells how rapidly species 1 would die out through encounters with species 2, and λ_2 is the rate at which species 2 increases due to encounters with species 1.

Volterra generalized the equations (I.11) to a set which describes the interaction of n species:

$$dN_i/dt = k_iN_i + \beta_i^{-1}\sum_{j=1}^{n} a_{ij}N_iN_j \ . \tag{I.12}$$

The first term describes the behavior of the ith species in the absence of others; when $k_i > 0$, the ith species is postulated to grow in an exponential Malthusian manner, with k_i as the "rate constant." When $k_i < 0$ and all other $N_j = 0$, the population of the ith species would die out exponentially. The quadratic terms in equation (I.12) describe the interaction of the ith species with all the other species. The ith term in the quadratic sum is proportional to the number of possible binary encounters N_iN_j between members of the ith species and members of the jth species. The constants a_{ij} might be either positive, negative, or zero. A positive a_{ij} tells us how rapidly encounters between i^{th} and j^{th} species will lead to an increase in N_i; a negative a_{ij} tells how rapidly these encounters will lead to a decrease in N_i, and a zero a_{ij} simply denotes the fact that i^{th} and j^{th} species do not interact. If during a collision between i^{th} and j^{th} species, j^{th} species increases, then i^{th} species decreases. Hence a_{ij} and a_{ji} have opposite signs. The positive quantities $\{\beta_i^{-1}\}$ have been named "equivalence" numbers by Volterra. During binary collisions of species i and j, the ratio of i's lost (or gained) per unit time to j's gained (or lost) is $\beta_i^{-1}/\beta_j^{-1}$. With this definition,

$$a_{ij} = - a_{ji} \ . \tag{I.13}$$

A detailed discussion of these equations can be found in references [4, 10, 11].

Another type of process which leads to somewhat similar nonlinear equations (though they are wave equations rather than rate equations) is the propagation of waves in anharmonic systems. A one-dimensional model which was studied by Fermi, Pasta, and Ulam [12] is

$$m\ddot{x}_i = \gamma(x_{i+1} - 2x_i + x_{i-1})[1 + \alpha(x_{i+1} - x_{i-1})], \quad i = 1, \ldots, N, \tag{I.14}$$

which, as $\alpha \to 0$, becomes the usual equation for a chain of springs and masses. The right-hand side has linear and quadratic terms just as our chemical rate equations (I.1). Machine solutions of this kind of equation have been used to investigate the manner in which an initial distribution of energy between various normal modes of the linear lattice ($\alpha = 0$) relaxes with time. The famous Fermi-Pasta-Ulam paradox concerning the nonachievement of equipartition was first discussed in terms of equation (I.14).

II. SOME STRATEGIES AVAILABLE FOR THE CONSTRUCTION OF SOLVABLE MODELS [1, 13]

Since the explicit solution of nonlinear coupled differential equations generally requires detailed numerical integrations, it is useful to find "solvable" models if one's aim is to develop an intuition rather than to give a precise analysis of some special phenomenon. This section is concerned with several strategies which are available for the construction of special solvable models.

Faltung-type integral equations of the form

$$dU(k,t)/dt = \int_{-\infty}^{\infty} S(k-k',t)U(k',t)\,dk' + \int\int_{-\infty}^{\infty} Q(k-k'-k'',t)\,U(k')\,U(k'')\,dk'\,dk'' \tag{II.1}$$

can be solved through the introduction of the Fourier transforms

$$u(\theta,t) = \int_{-\infty}^{\infty} U(k,t)e^{ik\theta}\,dk, \tag{II.2a}$$

$$s(\theta,t) = \int_{-\infty}^{\infty} S(k,t)e^{ik\theta}\,dk, \tag{II.2b}$$

$$q(\theta,t) = \int_{-\infty}^{\infty} Q(k,t)e^{ik\theta}\,dk. \tag{II.2c}$$

The functions S and Q are assumed to be known, while U is the unknown function. One finds

$$du(\theta,t)/dt = s(\theta,t)\,u(\theta,t) + q(\theta,t)[u(\theta,t)]^2 \ . \tag{II.3a}$$

This is easily reduced to quadratures by letting $u = 1/v$. Then

$$- dv/dt = sv + q \ , \tag{II.3b}$$

whose solution is

$$v(\theta,t) = v(\theta,a)\exp[-\int_a^t s(\theta,\tau)d\tau] - \int_a^t q(\theta,t_2)\, dt_2 \exp[-\int_{t_2}^t s(\theta,t)\, dt_1] , \tag{II.4}$$

so that

$$u(\theta,t) = u(\theta,a)/[I(\theta,t) - u(\theta,a)\, J(\theta,t)] , \tag{II.5}$$

where

$$I(\theta,t) = \exp[-\int_a^t s(\theta,\tau)\, d\tau] , \tag{II.6a}$$

$$J(\theta,t) = \int_a^t q(\theta,t')\exp[-\int_{t'}^t s(\theta,\tau)\, d\tau]\, dt' . \tag{II.6b}$$

By taking the Fourier inverse of equation (II.5), we find that

$$U(k,t) = \frac{1}{2\pi} \int_{-\infty}^{\infty} \frac{u(\theta,a)\exp(-ik\theta)\, d\theta}{I(\theta,t)-u(\theta,a)J(\theta,t)} , \tag{II.7}$$

so that the solution of equation (II.1) is reduced to quadratures. Two- and three-dimensional cases can be discussed in a similar way.

The Smoluchowski theory of colloid coagulation corresponds to the situation in which [14]

$$Q(K,t) = A\,\delta(K) , \tag{II.8}$$

where A is a constant and initially (at $t = a = 0$),

$$U(k,0) = \beta\delta(k-k_o) \quad \text{with } k_o > 0. \tag{II.9}$$

Another class of nonlinear equations whose members are solvable is that whose members are linearized by making a transformation in the dependent variables. One of the most elementary of these to be dealt with is

$$\partial u/\partial t = D\nabla^2 u + \lambda(u_x{}^2 + u_y{}^2 + u_z{}^2) , \tag{II.10}$$

which can be transformed into the diffusion equation (with $f_t \equiv \partial f/\partial t$)

$$f_t = D\nabla^2 f \tag{II.11}$$

in the variable f through the transformation

$$u(\underset{\sim}{r},t) = (D/\lambda)\log(\lambda f/D). \tag{II.12}$$

By defining the Fourier transform $U(\underset{\sim}{k},t)$ of $u(\underset{\sim}{r},t)$, it is easy to show that the k-space representation of equation (II.10) is

$$\partial U(\underset{\sim}{k},t)/dt = Dk^2 U(\underset{\sim}{k},t) + \lambda \int\int\int_{-\infty}^{\infty} (\underset{\sim}{k}-\underset{\sim}{k}')\cdot\underset{\sim}{k}' U(\underset{\sim}{k}',t)U(\underset{\sim}{k}-\underset{\sim}{k}',t)\, d\underset{\sim}{k}' . \tag{II.13}$$

The equation

$$u_t = Du_{xx} + \lambda u u_x \tag{II.14}$$

or

$$u_t = D[u_x + \tfrac{1}{2}(\lambda/D)u^2]_x \tag{II.15}$$

is transformed into the diffusion equation by the introduction of the new dependent variable f:

$$u = (2D/\lambda)f_x/f . \tag{II.16}$$

In terms of f,

$$u_x = (2D/\lambda)(ff_{xx} - f_x{}^2)/f^2$$

$$u_t = (2D/\lambda)(ff_{xt} - f_x f_t)/f^2 \equiv (2D/\lambda)(f_t/f)_x ,$$

so that equation (II.14) is equivalent to

$$([f_t - Df_{xx}]/f)_x = 0$$

or

$$f_t - Df_{xx} = A(t)f, \tag{II.17}$$

where A(t) is an arbitrary function of the time. The transformation

$$f = g \ \exp \int^t A(t) \ dt \tag{II.18}$$

yields $g_t = Dg_{xx}$. However, independently of the choice of A(t),

$$u = (2D/\lambda)(g_x/g) . \tag{II.19}$$

Hence with no loss of generality we can choose $A(t) \equiv 0$. The function f itself satisfies the diffusion equation, which is linear and solvable by well-known methods. The reader can easily construct other examples which can be solved in this manner. Equation (II.14) appears in Burger's model of turbulence [15, 16], and in the theory of thermal diffusion [17] and of separation cascades [18].

By generalizing the Volterra equations of the competition between species, one can construct a "solvable" set of equations. Consider the (γ, α) model

$$dN_i/dt = k_i N_i^{\gamma_i} + \beta_i^{-1} \sum_{j=1}^{n} a_{ij} N_i^{\gamma_i} (N_j^{\alpha_j} - 1)/\alpha_j , \quad i = 1, 2, \ldots, n, \tag{II.20}$$

which, in the special case $\gamma_i = \alpha_i = 1$ for all i is equivalent to the Volterra set if one introduces a new k_i' defined by

$$k_i' \equiv k_i - \beta_i^{-1} \sum_{j=1}^{n} a_{ij} \alpha_j^{-1} . \tag{II.21}$$

We again assume that $a_{ij} = - a_{ji}$. The set (II.20) has the alternative form

$$\beta_i (1 - \gamma_i)^{-1} \ dN_i^{1-\gamma_i}/dt = k_i \beta_i + \sum_{j=1}^{n} a_{ij}(N_j^{\alpha_j} - 1)/\alpha_j \tag{II.22a}$$

$$= \sum_{j=1}^{n} a_{ij}(N_j^{\alpha_j} - q_j^{\alpha_j})/\alpha_j , \tag{II.22b}$$

where the set $\{q_i\}$ is the equilibrium solution of equation (II.22a) obtained by setting $dN_i/dt = 0$ for all i:

$$k_i \beta_i + \sum_j a_{ij}(q_j^{\alpha_j} - 1)/\alpha_j = 0 . \tag{II.23}$$

We construct our required set of solvable models [5] by choosing $\alpha_j \equiv 1 - \gamma_j$ and defining

$$u_j = (N_j^{\alpha_j} - q_j^{\alpha_j})/\alpha_j \quad \text{or} \quad N_j = q_j(1 + \alpha_j u_j q_j^{-\alpha_j})^{1/\alpha_j} . \tag{II.24}$$

If $\alpha_i = 0$, then $u_i = \log(N_i/q_i)$ and $(N_i/q_i) = \exp(u_i)$. Generally, the equations for the new variables $\{u_i\}$ are linear with

$$\beta_i \, du_i/dt = \sum_{j=1} a_{ij} u_j \quad . \tag{II.25}$$

Of course, all our solvable equations whose solutions are obtained by the transformation (II.24) are of the more general form

$$\beta_i d\, N_i/dt = G_i(N_i)\,[\,\beta_i k_i + \sum_j a_{ij} F_j(N_j)] \ , \tag{II.26}$$

where the functions F and G are related by

$$[G_i(N_i)]^{-1} \ dN_i = dF_i(N_i) \ , \tag{II.27a}$$

i.e.,

$$[G_i(N_i)]^{-1} = dF_i(N_i)/dN_i \ . \tag{II.27b}$$

The case which was considered above was

$$F(N_i) = (N_i^{\alpha_i} - 1)/\alpha_i \ , \tag{II.28a}$$

$$G(N_i) = N_i^{1-\alpha_i} \ . \tag{II.28b}$$

Another interesting pair is

$$F(N_i) = [1 - \exp(\alpha_i N_i)]/\alpha_i \ , \tag{II.29a}$$

$$G(N_i) = \exp(\alpha_i N_i). \tag{II.29b}$$

The α_i's may be positive or negative.

III. ON THE GENERAL THEORY OF COMPETING SPECIES

Let us consider a more general model of competing species than that of Volterra, but subject to the restriction that time lags are not significant. With this limitation we suppose that

$$\frac{dN_i}{dt} = N_i G_i(N_1, N_2, \ldots, N_n) \ , \quad i = 1, 2, \ldots, n \ , \tag{III.1}$$

where we also postulate that the functions $\{G_i\}$ do not depend explicitly on the time. We further assume that an equilibrium exists in which none of the species vanish. That is, we assume that a set of positive equilibrium populations $(Q_1, \ldots, Q_n)$ exists such that

$$G_i(Q_1, Q_2, \ldots, Q_n) = 0 \ , \quad i = 1, 2, \ldots, n \ . \tag{III.2}$$

The functions G_i have a Taylor expansion about the equilibrium populations so that

$$G_i(N_i, \ldots, N_n) = \sum_j (N_j - Q_j)(\partial G_i/\partial N_j)_{eq}$$
$$+ \sum_{jk} (N_j - Q)(N_k - Q)(\partial^2 G_i/\partial N_j \partial N_k)_{eq} + \ldots , \tag{III.3}$$

where the subscript eq implies that the appropriate functions are to be evaluated for equilibrium populations. If we let

$$A_{ij} \equiv (\partial G_i/\partial N_j)_{eq} \quad \text{and} \quad A_{ijk} \equiv (\partial^2 G_i/\partial N_j \partial N_k)_{eq} \quad , \tag{III.4}$$

then

$$d \log N_i/dt = \sum_j A_{ij}(N_j - Q_j) + \sum_{jk} A_{ijk}(N_j - Q)(N_k - Q_k) + \ldots . \tag{III.5}$$

Note that if one neglects terms second order in the deviations from equilibrium, the resulting equations are similar to those of Volterra except that the a_{ij}'s are not necessarily antisymmetrical. General models such as equation (III.1) have been mentioned by a number of authors including McArthur [19] and Levine [20]. The F_i's may also be functions of other variables $z_1, \ldots, z_h$ and some special relations might exist between the N's and z's and each other.

The character of the second-order terms in equation (III.5) depends on the nature of the competitive process. When nonvanishing terms exist such that j and k are distinct and neither is equal to i, some type of "cooperation" exists between species. If $a_{ijk} < 0$ this would correspond to a coordinated attack of species j and k on i. If species i shows a gourmet instinct and prefers to eat species j and k together, than a_{ijk} would be positive. In the absence of such effects, the only second-order terms in equation (III.5) would correspond to $j = k$ or to either k or j being equal to i, or $i = k = j$.

One of our solvable nonlinear models can be made to resemble a more general model to within first-order terms in the deviations from equilibrium by proceeding in the following manner. We consider the "solvable" set of equations

$$d \log N_i/dt = N_i^{\alpha_i} \, [k_i + \sum_{j=1}^{n} a_{ij}(N_j^{\alpha_j} - 1)/\alpha_j] \tag{III.6}$$

or

$$d \log N_i/dt = N_i^{\alpha_i} \sum_{j=1}^{n} a_{ij}(N_j^{\alpha_j} - q_j^{\alpha_j})/\alpha_j . \tag{III.7}$$

If we expand the populations $\{N_i\}$ about the equilibrium points $\{q_i\}$, we find that

$$d \log N_i/dt = \sum_j a_{ij} q_i^{-\alpha_i} q_j^{\alpha_j - 1} (N_j - q_j) + O(N_j - q_j)(N_k - q_k) . \tag{III.8}$$

The first-order term on the right can be identified with that in equation (III.5) by making the following connections. First choose each $q_i \equiv Q_i$ so that systems have the same equilibrium values. Then choose the set $\{a_{ij}\}$ so that

$$a_{ij} = q_i^{\alpha_i} q_j^{1-\alpha_j} A_{ij} \quad , \tag{III.9}$$

where the α_i's are yet to be specified. From the relations

$$k_i = \sum a_{ij}(q_j^{\alpha_j} - 1)/\alpha_j \tag{III.10}$$

the $\{\alpha_i\}$ are yet to be defined. The choice

$$\alpha_1 = \alpha_2 = \ldots = \alpha_n = 0$$

automatically gives the correct constant term on the right-hand side of equation (III.5) even without expanding around the equilibrium points. Another way of choosing the α_i's is to, in some sense, make the second-order terms on the right-hand side of equation (III.8) represent the right-hand second-order terms in equation (III.5). In the absence of cooperative and gourmet effects, this can be done well either in a least square sense or in a term-by-term fit when the graph of the interaction is fairly loose.

IV. INFLUENCE OF RANDOM DRIVING FORCES ON INTERACTING POPULATION VARIABLES

In this section we generalize equation (III.1) to include the effect of noise on population growth rates. We consider the process characterized by the rate equations

$$dN_i/dt = N_i[G_i(N_1, N_2, \ldots, N_n) + F_i(t)] , \qquad \text{(IV.1)}$$

where $F_i(t)$ is a random function which we shall postulate to be generated by a Gaussian random process such that

$$\langle F_i(t)\rangle = 0 \text{ and } \langle F_i(t_1)F_j(t_2)\rangle = \sigma_{ij}^2\, \delta(t_1 - t_2). \qquad \text{(IV.2)}$$

Equation (IV.1) can be considered as a model of other complex situations besides competition between living species. The growth of many quantities such as production levels and the number of competing businesses of various types, as well as population figures, may be described by equation (IV.1).

We introduce the variables

$$v_i = \log N_i \qquad \text{(IV.3)}$$

so that

$$dv_i/dt = G_i(v_1, v_2, \ldots, v_n) + F_i(t) . \qquad \text{(IV.4)}$$

Then, in view of the postulated properties, (IV.2), of $G_i(t)$, the joint probability distribution function $P(v,t)$ can be shown by standard methods to satisfy the Fokker-Planck equation

$$\frac{\partial P}{\partial t} = \sum_i \frac{\partial(PG_i)}{\partial v_i} + \tfrac{1}{2} \sum_{ij} \sigma_{ij}^2 \frac{\partial^2 P}{\partial v_i \partial v_j} . \qquad \text{(IV.5)}$$

Since experience with the "many-body" Fokker-Planck equation has not been as extensive as that with the many-body Schrodinger equation, it would be useful to attempt to put it in that form. Feynman diagrams, cluster integral expansions, second quantization, etc., would then be applicable. As we shall see below, this is generally not possible, but models can be constructed for which it is.

Let us introduce a new dependent variable $\psi(v,t)$ through

$$P(v,t) = \psi(v,t)\exp[J(v)] , \qquad \text{(IV.6)}$$

where the function $J(v)$ is an arbitrary function of $\{v_i\}$ and will be defined later. Then equation

(IV.5) becomes

$$\frac{\partial \psi}{\partial t} = \frac{1}{2} \sum_{ij} \sigma_{ij}^2 \frac{\partial^2 \psi}{\partial v_i \partial v_j} - \psi \sum_i \left(\frac{\partial J}{\partial v_i} + \frac{\partial}{\partial v_i} \right) (G_i - \frac{1}{2} \sum_j \sigma_{ij}^2 \frac{\partial J}{\partial v_j})$$

$$- \sum_i \frac{\partial \psi}{\partial v_i} [G_i - \frac{1}{2} \sum_j (\sigma_{ij}^2 + \sigma_{ji}^2) \frac{\partial J}{\partial v_j}] . \tag{IV.7}$$

All terms containing first derivatives in this equation vanish if one defines J as the solution of the set of equations

$$G_i = \frac{1}{2} \sum_j (\sigma_{ij}^2 + \sigma_{ji}^2) \partial J/\partial v_j , \quad i = 1, 2, \ldots, n . \tag{IV.8}$$

When J is related to the G_i through equations (IV.8), then (IV.7) becomes

$$2 \frac{\partial \psi}{\partial t} = \sum_{ij} \sigma_{ij}^2 \frac{\partial^2}{\partial v_i \partial v_j} - \psi \sum_{ij} \sigma_{ji}^2 \left(\frac{\partial^2 J}{\partial v_i \partial v_j} + \frac{\partial J}{\partial v_i} \frac{\partial J}{\partial v_j} \right) . \tag{IV.9}$$

This equation has the same mathematical form as the many-body Schrodinger equation for particles with tensor masses if one interprets the quantity

$$\Phi(v_1, \ldots, v_n) \equiv \sum_{ij} \sigma_{ji}^2 \left(\frac{\partial^2 J}{\partial v_i \partial v_j} + \frac{\partial J}{\partial v_i} \frac{\partial J}{\partial v_j} \right) \tag{IV.10}$$

as a many-body potential. Since t, rather than it, appears on the left-hand side of equation (IV.9), the equation is more reminiscent of the Bloch equation of statistical mechanics.

Since the set of defining equations (IV.8) for J are linear in $\partial J/\partial v_i$, one can solve them for $\{\partial J/\partial v_i\}$ to find that

$$\partial J/\partial v_i \equiv X_i(v) = H_i(G_1, G_2, \ldots G_n) . \tag{IV.11}$$

Note that

$$dJ = \frac{\partial J}{\partial v_i} dv_1 + \ldots + \frac{\partial J}{\partial v_n} dv_n \equiv X_1 dv_1 + \ldots + X_n dv_n . \tag{IV.12}$$

The solution of this "total differential equation" yields the required function J.

In preparation for the examination of the general case, let us first examine the special situation with

$$\sigma_{ij}^2 \equiv \sigma_i^2 \delta_{ij} . \tag{IV.13}$$

Then

$$G_i = \sigma_i^2 \partial J/\partial v_i , \quad i = 1, 2, \ldots, n . \tag{IV.14}$$

Clearly, if this set of equations is to have a consistent solution, the fact that

$$\partial^2 J/\partial v_i \partial v_j = \partial^2 J/\partial v_j \partial v_i \tag{IV.15}$$

implies that we require

$$\sigma_i^{-2} \partial G_i/\partial v_j = \sigma_j^{-2} \partial G_j/\partial v_i , \tag{IV.16}$$

which would not always be the case. When it is true, then equation (IV.9) reduces to

$$2\frac{\partial\psi}{\partial t} = \sum_i \sigma_i^2 \frac{\partial^2\psi}{\partial v_i^2} - \psi\sum_i\left(\frac{\partial G_i}{\partial v_i} + G_i^2/\sigma_i^2\right) . \qquad \text{(IV.17)}$$

In general, the necessary and sufficient condition that the equation in (n + 1) variables

$$X_1 d\psi_1 + X_2 d\psi_2 + \ldots + X_{n+1} d\psi_{n+1} = 0 \qquad \text{(IV.18)}$$

have a solution of the form

$$\phi(x_1, x_2, \ldots, x_{n+1}) = \text{constant} \qquad \text{(IV.19)}$$

is that the x_i's satisfy the equation

$$X_\nu\left\{\frac{\partial X_\mu}{\partial x_\lambda} - \frac{\partial X_\lambda}{\partial x_\mu}\right\} + X_\mu\left\{\frac{\partial X_\lambda}{\partial x_\nu} - \frac{\partial X_\nu}{\partial x_\lambda}\right\} + X_\lambda\left\{\frac{\partial X_\nu}{\partial x_\mu} - \frac{\partial X_\mu}{\partial x_\nu}\right\} = 0$$

$$(\lambda, \mu, \nu = 1, 2, \ldots, n+1) . \qquad \text{(IV.20)}$$

We interpret $x_i \equiv v_i$ if $i = 1, 2, \ldots, n$ and $J \equiv x_{n+1}$ while $X_{n+1} \equiv -1$. The total number of such equations is n(n - 1)(n - 2) of which (n - 1)(n - 2) are independent.

The discussion of equation (IV.18) when equation (IV.20) is not satisfied is known as Pfaff's problem [27]. In general, a total differential equation in 2n or 2n - 1 variables is equivalent to a system of not more than n algebraic equations. Such a set of auxiliary equations does not seem to be appropriate for our problem. When n = 1, there is of course no difficulty. In that case equation (IV.14) is the ordinary equation

$$G = \sigma^2\, dJ/dv \quad \text{so that} \quad J(v) = \int_a^v G(e^v)\, dv$$

and equation (IV.17) becomes [4]

$$2\frac{\partial\psi}{\partial t} = \sigma^2\frac{\partial^2\psi}{\partial v^2} - \psi\left\{\frac{\partial G}{\partial v} + \sigma^{-2}G^2\right\} . \qquad \text{(IV.13)}$$

The cases

$$G(N) = k[1 - (N/\theta)] \quad \text{and} \quad -k\log(N/\theta)$$

arise in the Verhulst and Gompertz models of population growth. These are discussed in reference [4] where it is shown that in the Gompertz case the resulting equation is the Bloch equation for the harmonic oscillator potential if t is interpreted to be $\beta = 1/kT$. The Verhulst case corresponds to the Morse potential.

The Fokker-Planck equation (IV.5) for one of the solvable models ((II.22) with $\alpha_i = 1 - \gamma_i$) in the presence of a random forcing term becomes mathematically equivalent to the Fokker-Planck equation of a system of coupled oscillation. This is discussed in reference [5].

V. ON A CONNECTION BETWEEN THE VOLTERRA MANY-SPECIES EQUATIONS AND THE LOTKA-VOLTERRA TWO-SPECIES EQUATIONS

A classic result of the Lotka-Volterra equations for two competing species (I.11) is the existence of periodic solutions. When the initial populations are close to their equilibrium levels, these periodic solutions are essentially sinuosoidal. They become more and more spiked as initial populations deviate more from the equilibrium values. Generally they do not approach equilibrium in either case, but the populations merely oscillate around the equilibrium values.

A result referred to in reference [4] concerning the many-species Volterra model is that the solution is almost periodic and that the normalized autocorrelation function of the population of a given species spends most of its time in a narrow band of width $n^{-\frac{1}{2}}$ around zero; n being the total number of species. That is, oscillations tend to become damped out.

We now consider certain graphs which represent the interaction between many species and show how a small number of them exhibit periodic population variations while the others enjoy more random and smaller amplitude variations. We will also show that mathematically this effect is analogous to localized modes of vibration around certain types of defects in crystals. We restrict our discussion to one of our solvable models of population competition.

The graph of interactions which we select is one which characterizes (2n + 2) species which influence each other in the manner specified in figure 1. Species j eats (j - 1) and is eaten by

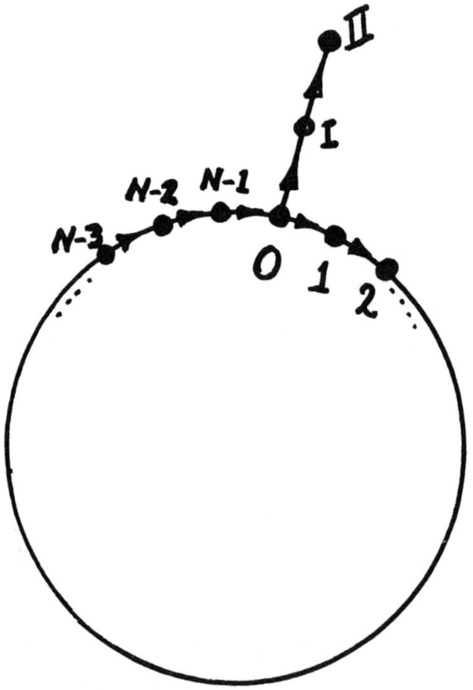

Fig. 1. A circle of prey and predators. Species j preys upon j - 1 and is preyed upon by j + 1. The side chain which contains 1 and 2 is analogous to the Volterra-Lotka two-species situation. The chain of N species plays the role of the infinite nourishment bath for J in the Volterra-Lotka two-species model. The arrows indicate the direction of material flow.

(j + 1) so that the mass flow proceeds along a ring in the direction indicated by the arrows. We also include the two anomalous species 1 and 2 such that species 1 requires 0 for its survival while species 2 eats only 1. We define $Y_j(t)$ to be the population of the jth ring species at time t while those of the loose-end species 1 and 2 are respectively Z_1 and Z_2. We model the prey-predator relationship between the various species by one of our solvable sets of rate equations (II.22b) with all $\alpha_i \to 0$.

The appropriate rate equations are then

$$Y_i^{-1}\, dY_i/dt = -\lambda_i + \sum_k a_{ik} \log Y_k \quad i = 1, 2, \ldots, N-1, \tag{V.1a}$$

$$Y_o^{-1}\, dY_o/dt = -\lambda_o + \sum a_{ok} \log Y_k - b_1 \log Z_1, \tag{V.1b}$$

$$Z_1^{-1}\, dZ_1/dt = b_1 \log Y_o - b_2 \log Z_2, \tag{V.1c}$$

$$Z_2^{-1}\, dZ_2/dt = -\mu + b_2 \log Z_1. \tag{V.1d}$$

Now let us assume that the steady-state solutions of these equations are $\{Q_k\}$, P_1, and P_2 which satisfy

$$\lambda_i = \sum_k a_{ik} \log Q_k, \tag{V.2a}$$

$$\lambda_o = \sum a_{ok} \log Q_k - b_1 \log P_1, \tag{V.2b}$$

$$b_1 \log Q_o = b_2 \log P_2, \tag{V.2c}$$

$$\mu = b_2 \log P_1. \tag{V.2d}$$

Then, if we let

$$y_i = \log(Y_i/Q_i), \qquad z_i = \log(Z/P_i), \qquad i = 1, 2 \tag{V.3}$$

our basic equations (V.2a) - (V.2d) then have the structure

$$dy_i/dt = \sum a_{ik} y_k \qquad \text{if } i \neq 0, \tag{V.4a}$$

$$dy_o/dt = \sum a_{ok} y_k - b_1 z_1, \tag{V.4b}$$

$$dz_1/dt = b_1 y_o - b_2 z_2, \tag{V.4c}$$

$$dz_2/dt = b_2 z_1. \tag{V.4d}$$

Furthermore, if we let

$$y_i = f_i e^{i\omega t} \qquad \text{and} \qquad z = v e^{i\omega t}, \tag{V.5}$$

then

$$i\omega f_i - \sum_k a_{ik} f_k = 0, \tag{V.6a}$$

$$i\omega f_o - \sum a_{ok} f_k = -b_1 v_1, \tag{V.6b}$$

$$i\omega v_1 = b_1 f_o - b_2 v_2 \,, \tag{V.6c}$$

$$i\omega v_2 = b_2 v_1 \,. \tag{V.6d}$$

The variable v_1 is related to f_o by

$$v_1 = -\, i\omega b_1 f_o/(\omega^2 - b_2^2) \tag{V.7}$$

so that equations (V.6a) and (V.6b) can be combined in the single equation

$$i\omega f_j - \sum_k a_{jk} f_k = i\omega b_1^2 f_o \delta_{jo}/(\omega^2 - b_2^2) \,. \tag{V.8}$$

This equation is reminiscent of the type of equation one must deal with in investigation of the effect of defects on lattice vibrations; indeed, we treat it in the same manner that one treats the lattice vibration equations. We define G_j as the Green's function of the network without the loose-end fragment. It is the solution of the equation

$$LG_j(\omega) \equiv i\omega G_j(\omega) - \sum_k a_{jk} G_k(\omega) = \delta_{j,0} \,. \tag{V.9}$$

The solution of equation (V.8) is easily verified to be

$$f_j = i\, b_1^2 f_o G_j/(\omega^2 - b_2^2) \,. \tag{V.10}$$

The frequency ω which is as yet unspecified is obtained by setting $j = 0$. Then ω is the solution of the algebraic equation

$$\omega^2 - b_2^2 = i\omega b_1^2 G_j(\omega) \,. \tag{V.11}$$

In order to obtain an understanding of the above formulae, we examine a special case of our ring of interactions in which

$$a_{jk} = a(\delta_{k,j+1} - \delta_{k,j-1}) \quad \text{and} \quad f_j \equiv f_{2N+j} \,. \tag{V.12}$$

Then equation (V.8) becomes

$$i\omega f_j - a(f_{j+1} - f_{j-1}) = i\omega b_1^2 f_o \delta_{j,0}/(\omega^2 - b_2^2) \,. \tag{V.13}$$

The Green's function solution of

$$i\omega G_j - a(G_{j+1} - G_{j-1}) = \delta_{j,0} \quad \text{with } G_j \equiv G_{2N+j} \tag{V.14a}$$

is

$$G_j = \frac{1}{2Ni} \sum_{k=0}^{2N-1} \frac{\exp(2\pi ijk/2N)}{\omega - 2a \sin(2\pi k/2N)} \tag{V.14b}$$

As $N \to \infty$

$$G_j = \frac{1}{2\pi i} \int_0^{2\pi} \frac{\exp(i\theta j)\, d\theta}{\omega - 2a \sin\theta} = (U^{|j|}/\omega)[1 - (2a/\omega)^2]^{-\frac{1}{2}} \exp(j-1)\pi i \,, \tag{V.14c}$$

where

$$U = (\omega/2a)\{1 - [1 - (2a/\omega)^2]^{\frac{1}{2}}\} \tag{V.14d}$$

since

$|U| < 1, \quad G_j \to 0 \quad \text{as } |j| \to \infty$

In our special case we combine equation (V.10) with equation (V.14c) to find the equation which defines the frequency, ω, of what we shall find to be localized oscillations; that is, periodic changes in populations of species "near" species zero in our species graph. Then

$$(\omega^2 - b_2^2)f_j = i\omega b_1^2 f_o[U^{|j|}/\omega]\,[1 - (2a/\omega)^2]^{-\frac{1}{2}} \exp[\tfrac{i\pi}{2}(j - 1)] \qquad (V.15)$$

where j is now measured relative to 0 so that it has values $j = 0, \pm 1, \pm 2, \ldots$, in the limit as $N \to \infty$. When $j = 0$, equation (V.15) becomes

$$[1 - (2a/\omega^2)]^{\frac{1}{2}} = b_1^2(\omega^2 - b_2^2)\ . \qquad (V.16)$$

There is a positive root ω_1^2 of this equation when $b_2^2 > 2a$, as can be seen from figure 2 where

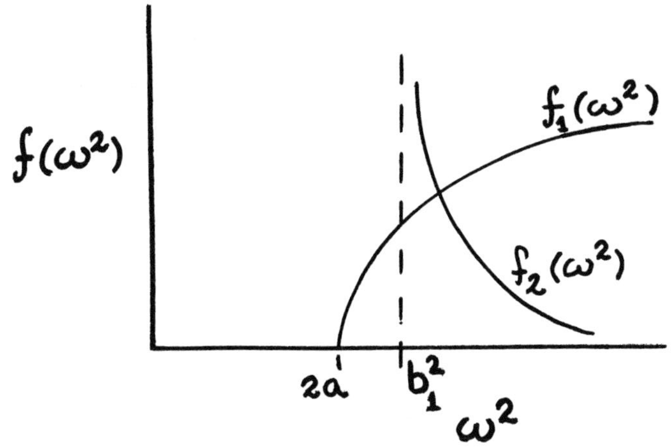

Fig. 2. The frequency of localized oscillations near species 1 and 2 in the food web is obtained.

we have plotted on the same figure the functions f_1 and f_2 defined by

$$f_1(\omega^2) = [1 - (2a/\omega^2)]^{\frac{1}{2}}, \qquad (V.17a)$$

$$f_2(\omega^2) = b_1^2/(\omega^2 - b_2^2)\ . \qquad (V.17b)$$

The value of ω^2 at which the two curves intersect yields the frequency of oscillation of the populations of species 1 and 2. The amplitudes V_1 and V_2 are related and given by equations (V.6d) and (V.7). Equation (V.6d) implies that the populations of 1 and 2 are out of phase. Finally, when ω^2 is determined, it can be seen from equations (V.14) and (V.15) that the amplitudes f_j of the oscillations of frequency ω of the populations of the species converted to form the ring of interactions decays exponentially as one moves from species zero in either direction along the ring. The component of frequency ω is thus "localized" on the interaction graph so that it is strong only for those species which lie near 1 and 2 on the graph. Those species which are far away from 1 and 2 oscillate in the random manner discussed in reference

[4] while 1 and 2 can have a periodicity component in their population variation curve as is characteristic of the Volterra-Lotka two-species model. This behavior is similar to that of localized lattice vibration modes which exist near some types of defects in crystal.

VI. SOLITONS

It is well known that every twice-differentiable function of the form

$$f(x,t) \equiv f(x - ct) \tag{VI.1}$$

is a solution of the linear wave equation

$$f_{tt} = c^2 f_{xx} \; . \tag{VI.2}$$

Such a solution preserves its form as it propagates in the positive x-direction with velocity c. While nonlinear wave equations do not have this property, they sometimes have special solutions which exhibit the propagating wave structure. Such solutions are sometimes called solitary waves or solitons. They play an important role [21-23] in the Fermi-Pasta-Ulam problem in which coupled anharmonic springs and masses do not achieve an equipartition of energy into the normal modes of the corresponding linearized problem. The aim of this section is to present several examples of solvable nonlinear processes in which solitons can be exhibited.

Let us return to our equation (II.20) with all $\alpha_i \equiv 0$ and with all $\gamma_i = 1$:

$$\beta_i \, dN_i/dt = \beta_i k_i N_i + N_i \sum_j a_{ij} \log N_j \; ; \tag{VI.3}$$

so that

$$\beta_i \, d\log(N_i/q_i)/dt = \sum_j a_{ij} \log(N_j/q_j) \; , \tag{VI.4}$$

where $\{q_i\}$ is the set of equilibrium populations which satisfy equation (II.23) with all $\alpha_i \equiv 0$. Again let the graph of our system be a circle with nearest-neighbor interactions with the $\{a_{ij}\}$ given by equation (V.12). Then, if we let all β_i be equal to β and set $\tau = t/\beta$, our equations have the form (we form our ring with 2N variables)

$$dv_j/d\tau = a(v_{j+1} - v_{j-1}) \; , \qquad j = 1, 2, \ldots, 2N \; , \tag{VI.5}$$

$$v_j = \log(N_j/q_j), \quad \text{and} \quad v_j \equiv v_{j+2N} \; . \tag{VI.6}$$

Our set of equations has the soliton solution of the form

$$v_j = v \sin(j\theta_k \pm \omega_k \tau) \; , \tag{VI.7}$$

where

$$\theta_k = 2\pi k/2N \quad \text{and} \quad \omega_k = 2a \sin \theta_k \; . \tag{VI.8}$$

The solution (VI.7) with the plus sign of course corresponds to a sine wave which propagates in the negative direction with velocity $v = \omega_k/\theta_k$ while that with the minus sign represents a similar wave propagated in the positive direction with the same velocity.

The sum of the functions which represent waves going in opposite directions is also a solution of the linear equations (VI.5). Such a solution is

$$v_j = k[\sin(j\theta_k + \omega_k\tau) + \sin(j\theta_k - \omega_k\tau)] \tag{VI.9}$$

where k represents the amplitude of the waves. When this is transformed to the population variables, one finds

$$N_j(\tau) = q_j \exp\{k[\sin(j\theta_k + \omega k\tau)]\}\exp\{k[\sin(j\theta_k - \omega k\tau)]\} . \tag{VI.10}$$

In the regime of large k, the sine wave solution of the linear equation is sharpened for those values of $(j\theta_k \pm \omega_k\tau) > 0$ and flattened when $(j\theta \pm \omega_k\tau) < 0$. The exponentials are plotted in figure 3 with $k = 10$. When the two exponentials are out of phase by 180°, the product of the

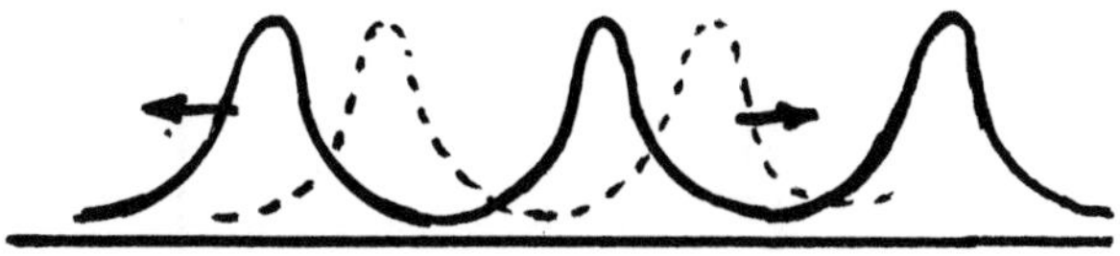

Fig. 3. Form of soliton wave solution of our nonlinear rate equations.

peak of one and the trough of the other is 1, whereas in a regime in which two troughs coincide the product is e^{-20}; it has the value e^{20} when both factors are in phase. We consider long-wavelength solitons in figure 4 and note the behavior of two peaks "a" and "b" which are moving toward each other. When they collide, the variable v_j goes wild, changing its value by a factor of e^{20}. After the collision, both solitons continue on their way as though the collision had never occurred. The solitons preserve their character in a manner analogous to that of stable particles in nature. We now consider another type of soliton problem.

The problem of combined population growth and diffusion has been discussed by a number of authors, including those listed in references 1 through 5. The process is sometimes characterized by the differential equation [24,25]

$$u_t = D\nabla^2 u + kuF(u) \ , \tag{VI.11}$$

where $u \equiv n/\theta$ with $n(r,t)$ being the population per unit area at point r at time t and θ the saturation population per unit area. D is the population diffusion constant, k a population growth rate constant, and ∇^2 the two-dimensional Laplacian. The function F(u) determines the manner in which the population reaches its saturation level. A typical form for F(u) might be

$$F(u) = (1 - u^\alpha)/\alpha \ , \tag{VI.12}$$

where α equals 1 for a Verhulst saturation term and $\alpha = 0$ for the Gompertz saturation scheme. The one-dimensional form of equation (VI.11) with $F(u) = (1 - u)$ was first discussed by R. A. Fisher [24]. We limit ourselves here to some brief remarks about the one-dimensional case.

Let us consider

$$u_t = Du_{xx} + kuF(u) \ , \tag{VI.13}$$

where F(u) is a decreasing function of u for $0 < u < 1$ which has the property $F(1) = 0$. It can easily be verified that the choice

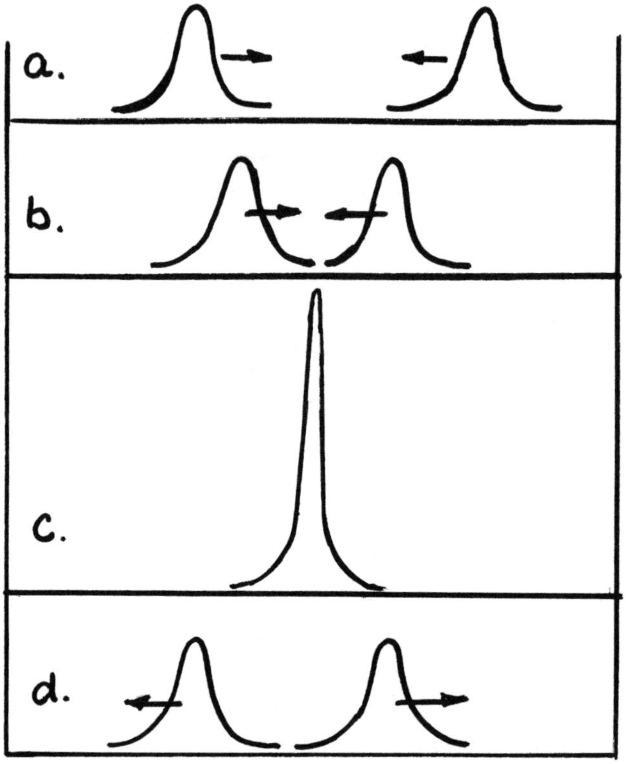

Fig. 4. When our two solitons approach each other as in (a), they eventually overlap as in (c) with tremendous distortion. Then they emerge from the entanglement in their original form as in (d).

$$F(u) = (1 - u)[1 + \epsilon(1 - 2u)] \qquad \text{(VI.14)}$$

has a solition solution

$$u(w) = \tfrac{1}{2}\{1 - \tanh(\alpha w)\} = (1 + e^{2\alpha w})^{-1} \qquad \text{(VI.15)}$$

with

$$w = (x - vt). \qquad \text{(VI.16)}$$

The values of α and v are related to k, D, and ϵ with

$$v = (kD/\epsilon)^{\frac{1}{2}}, \quad \alpha = (k\epsilon/4D)^{\frac{1}{2}}. \qquad \text{(VI.17)}$$

Note that in the solution (VI.15)

$$u(x - vt) = \tfrac{1}{2}\{1 - \tanh\ \alpha(x - vt)\}$$
$$= \begin{cases} 1 & \text{if } x \ll vt \\ 0 & \text{if } x \gg vt, \end{cases} \qquad \text{(VI.18)}$$

so that the saturated population front moves to the right with velocity v. The quantity $u(x - vt) = 1/2$ at the front $x = vt$. The limit case $\epsilon = 0$ would yield $F(u) = (1 - u)$ as found in Fisher's equation. However, the fact that $v \to \infty$ and $\alpha \to 0$ indicates that equation (VI.15) is not a soliton solution of the Fisher equation.

The soliton solution of Fisher's equation, which must satisfy

$$\frac{Dd^2u}{dw^2} + v\,\frac{du}{dw} + ku(1-u) = 0 \ , \tag{VI.19}$$

has not been exhibited in closed form, but its general character has been investigated numerically by Fisher and others. The way in which an initial population distribution relaxes has not been discussed analytically for any reasonable choice of F(u), although machine solutions have been obtained for Fisher's equation.

The author [3] has found a closed form solution of

$$u_t = D\{u_{xx} + [\frac{1-G'(u)}{G(u)}]u_x{}^2\} + kG(u)\ . \tag{VI.20}$$

When one chooses $G(u) = u(1-u)$, the soliton solution with the property (VI.18) is

$$u(x,t) = \left[1 + \exp\{\alpha[x - (\alpha D + k\alpha^{-1})t]\}\right]^{-1} \tag{VI.21}$$

with the wave front velocity

$$v = (\alpha D - k\alpha^{-1}). \tag{VI.22}$$

The above choice of G(u) also leads to an interesting initial value problem in which the population is initially peaked at the origin with a Gaussian-like distribution

$$u(x,0) = \eta/[\eta - (1-\eta)\exp(x^2/2a^2)] \tag{VI.23}$$

with $0 < \eta < 1$ as the highest population level (as a fraction of the way to the saturation value $u = 1$). It is shown that for $t > 0$

$$u(x,t) = \eta a\{\eta a + (1-\eta)(a^2+2Dt)^{\frac{1}{2}}\exp[-kt + x^2/2(a^2+2Dt)]\}^{-1} \tag{VI.24}$$

With increasing t, the population gets saturated in the neighborhood of $x = 0$ and the saturation wave propagates to both right and left with velocity

$$dx/dt \sim \pm\, 2(kD)^{\frac{1}{2}} \qquad \text{as} \qquad t \to \infty\ . \tag{VI.25}$$

Hence two soliton waves have been created. They propagate in opposite directions.

J. Canosa [26] has recently given an interesting general discussion of soliton solutions of nonlinear rate equations.

APPENDIX: A TRANSFORMATION OF THE GENERAL EQUATION OF BINARY COLLISION KINETICS

Note that if we multiply the rate equation (I.1) by the elements of a matrix C which diagonalizes the matrix $A = (a_i{}^j)$, i.e., such that

$$\sum_{ij} c^{mi} a_i{}^j c_{nj} = \delta_{mn}\lambda_n$$

with

$$\sum_m c_{mj} c^{mi} = \delta_{ij}\ , \qquad \sum_m c_{jm} c^{km} = \delta_{jk}\ ,$$

then equation (I.1) becomes

$$\frac{d}{dt}(x_i c^{mi}) = \sum_{j} c^{mi} a_i^{\ j} x_j + \sum_{k\ell} c^{mi} a_i^{\ k} x_k x_\ell + \sum_{i} c^{mi} F_i .$$

Clearly $\{c^{mi}\}$ is the set m of elements of the inverse of $c = (c_{mi})$. Now let

$$x_i = \sum_{m} c_{mi} y^m .$$

Then

$$\frac{dy^m}{dt} = \sum_{ijn} c^{mi} a_i^{\ j} c_{nj} y^n + \sum_{k\ell sn} c^{mi} a_i^{\ k\ell} c_n y^n + G_m$$

or

$$\frac{dy^m}{dt} = \lambda m y^m + \sum_{k\ell} \Gamma_{ns}^{\ m} y^n y^s + G_m$$

where

$$\Gamma_{hs}^{\ m} = \sum_{k\ell} c^{mi} a_i^{\ k\ell} c_{nk} c_{s\ell} , \quad G^m = \sum_{i} c^{mi} F_i .$$

REFERENCES

1. E. W. Montroll, Boulder Lectures in Theoretical Physics, 10A, 531 (1967).
2. E. W. Montroll, Contemporary Physics (Trieste), 2, 273 (1969).
3. E. W. Montroll, J. Appl. Prob. 4, 281 (1967).
4. N. S. Goel, S. C. Maitra, and E. W. Montroll, Rev. Mod. Phys. 43, 241 (1971).
5. E. W. Montroll, in Some Mathematical Problems in Biology, vol. 4, ed. J. Cowan (1971), p. 101.
6. L. Onsager, Nuovo Cim. Suppl. 6, 279 (1949).
7. C. C. Lin and W. H. Reid, Hdb. der Phys. 8, 2, 438 (1963).
8. A. J. Lotka, Proc. Nat. Acad. Sci. 6, 410 (1920).
9. A. J. Lotka, Elements of Mathematic Biology 1924 (New York: Dover, 1956).
10. V. Volterra, Léçon sur la théorie Mathematiques de la lutte pour la vie (Paris, 1931).
11. U. D'Ancona, The Struggle for Existence (Leiden: Brill, 1954).
12. E. Fermi, J. Pasta, and S. Ulam, Collected Works of Enrico Fermi, 2 (Chicago, 1965), 978.
13. E. W. Montroll, Energetics in Metallurgical Phenomenon, 3 (Gordon and Breach, 1967), 123.
14. M. Smoluchowski, Ann. Phys. 48, 1103 (1915); J. Phys. Chem. 92, 129 (1917).
15. J. M. Burgers, Proc. Acad. Sci. Amsterdam, 43, 2 (1940); 53, 247 (1950).
16. J. D. Cole, Quart. J. Appl. Math. 9, 225 (1951).
17. C. D. Majumdar, Phys. Rev. 81, 844 (1951).

18. E. W. Montroll and G. F. Newell, J. App. Phys. **23**, 184 (1952).

19. R. H. MacArthur and M. L. Rosenzweig, American Naturalist **97**, 209 (1963).

20. S. A. Levine, American Naturalist **104**, 413 (1970).

21. N. Zabusky and M. D. Kruskal, Phys. Letters, **15**, 240 (1965).

22. M. Toda, J. Phys. Soc. Japan, **22**, 431 (1967); **23**, 501 (1967).

23. E. W. Montroll, Proc. of Texas Spring School on Stat. Mech., 1970 (in press).

24. R. A. Fisher, Ann. Eugen. **7**, 355 (1937).

25. A. Kolmogorov, I. Petrovsky, and N. Piscounov, Bull. de l'Univ. d'Etat a Moscou (Ser. Internat.) **AL**, 1 (1937).

26. J. Canosa, preprint.

27. E. L. Ince, Ordinary Differential Equations 1926 (New York: Dover, 1956).

DISCUSSION

N. Van Kampen: If I understand you correctly, your equations describe only the rate of change of the average numbers of individuals. They do not contain the fluctuations about these averages due to the finiteness of the populations.

E. Montroll: That is correct.

G. Nicolis: 1. There have been several attempts to text experimentally the validity of the Volterra-Lotka model in artificial ecosystems. The surprising result is that in almost all experimental conditions the final state corresponded to a complete extinction of one of the species. Could this imply that in real ecosystems with coexistence of many species, the interaction processes are very different and much more complicated than in Volterra-Lotka model? For instance, one could suggest, following an old idea by Kolmogoroff, that the observed ecological stability is best interpreted in terms of models predicting limit cycle type of oscillation rather than oscillations arising in a conservative system.

2. The statistical mechanical analogy of the Volterra-Lotka system implies that the distribution of fluctuations of any size around the steady state is a time-independent canonical form. We have carried out a master equation study of the problem of two populations imbedded in a statistical bath of other populations. The results of this analysis (G. Nicholis, Adv. Chem. Phys. **19**, 209 [1971]) show that for small fluctuations one recovers the limiting form of Kamer's canonical distribution. For finite fluctuations, however, the system cannot attain a steady state but evolves slowly in phase space in a quasi-periodic trajectory. Would you also agree that the canonical distribution form is not justified for large fluctuations?

E. Montroll: There are a number of important factors omitted from the Volterra-Lotka equations when systems are far from equilibrium. When some of these are included, systems tend to have limit cycles. Others lead to an eventual approach to equilibrium. Which of these is appropriate for a given situation depends on the various parameters of the problem. I believe that the canonical distribution is still valid for large fluctuations in systems which involve a large number of species such that each species is influenced by a finite number of other species as the total number of species becomes infinite.

I. Prigogine: It is obvious that the chemical equations represent only average equations and that supplementary assumptions have to be introduced to study the fluctuations around the macroscopic time-independent or time-dependent states predicted by the classical equations. Generally the assumption is made (see, e.g., D. Mac Quappie, Supplementary Review Series in Applied Probabilities [Methuen, London, 1967]) that we may formulate a master equation in concentration space from which then the fluctuations may be derived (this is equivalent to introducing an appropriate Langevin friction term). This procedure has also been applied by Nicolis and Babloyantz (J. Chem. Phys. **51**, 6, 2632, 1969). It leads to satisfactory results when linear processes such as

$$A \rightleftarrows X \rightleftarrows Y \rightleftarrows \ldots \rightleftarrows F \qquad (1)$$

are considered. However when nonlinear processes such as

$$A + X \rightleftarrows X, \quad X + Y \rightleftarrows Y, \quad Y \rightleftarrows B , \qquad (2)$$

are involved, it seems that one has to be much more careful. Indeed, in most situations (certainly in the case of classical reactions and also in many ecological problems) one has _two_ time scales. One corresponds to elastic collisions; the second (the longer one), to inelastic collisions. In other words there is a mechanism of regulation inside each species itself. This is _not_ taken into account by the usual techniques and may lead to quite different results for the fluctuations (see a forthcoming paper in Proc. Nat. Acad. Sci., USA, 1971 by Nicolis and Prigogine).

PHASE TRANSITIONS AS CATASTROPHES

Rene Thom

I. INTRODUCTION

A. Metabolic Field

1. Definition

Let $E \longrightarrow B$ be a differentiable fiber space (E, B differentiable manifolds), with a compact fiber M^n. Suppose we are given in E a vector field X which is vertical, that is, tangent to the fibers $p^{-1}(x)$ [$\dot{p}(X) = 0$ for all $x \in E$]. At each fiber M_x, $x \in B$, X defines a dynamical system (M_x, X_x); hence the dynamical system (M, X) can be regarded as a field of local dynamical systems (M_x, X_x) (as, in the original field theory, a field was regarded as a continuum of local oscillators). This whole situation will be called a metabolic field above B.

2. Asymptotic Regime of a Dynamical System: Attractors

A subset $F \subset M$ of the dynamical system (M, X) is said to be an attractor if (a) F is a closed invariant set; (b) every trajectory which leaves F is in F; (c) there exists a fundamental system of neighborhoods U_α of F in M such that $f_t(U_\alpha) \subset U_\alpha$ for all $t > 0$; and (d) almost any trajectory of X in F is dense in F. Any attractor admits a local Liapunov function--that is a function $G: U \longrightarrow R$ which is strictly increasing along the trajectories of X in a neighborhood of F.

3. Examples of Attractors

(i) The one-point attractor, where $X = (X_i) = (\Sigma\, a_i^{\,i}\, x_i)$ where the square matrix $(a_i^{\,i})$ has all its eigenvalues with real part negative.

(ii) The "generic" closed trajectory.

Definition. The set $U \subset M$ of points m with $\omega(m) \subset F$ is called the basin of the attractor F.

4. Structurally Stable Attractors

An attractor F of (M, X) is said to be structurally stable if for any X' sufficiently near X, the flow X' admits an attractor F' homeomorphic to F through a ϵ-homeomorphism h_2: F $\longrightarrow$ F'.

Note.--We do not require this ϵ-homeomorphism to map a trajectory of X onto a trajectory of X'.

5. Conjecture

For almost all vector fields X in $\mathcal{D}(M)$, the space of all flows on M, the flow X admits only a finite number of attractors, and almost all trajectories of X tend to one of these attractors.

Remark.--The present status of this conjecture--which may be looked on (in S. Smale's terminology) as a generalized spectral theorem, is still very uncertain; this does not affect the following definitions, as the previous examples of attractors (and many others) are known to be structurally stable. The above conjecture only provides a more general setting for our model.[1]

B. Morphological Processes

Let E $\longrightarrow$ B be a metabolic field on B. Given any point b $\subset$ B, let us suppose that the corresponding local dynamical field admits a structurally stable attractor. Then all nearby fibers have a homologous attractor. A state of the metabolic field is theoretically defined by a section σ of the fiber map p: E $\longrightarrow$ B. More precisely, we may suppose that two fiber-asymptotically equivalent sections σ, σ' define the same "state." We denote by $\bar{\sigma}$ this asymptotic state (attractor) of the section.

Any point b where $\bar{\sigma}$ is continuous and $\bar{\sigma}(b)$ is a structurally stable attractor is called a regular point of the process. Where $\bar{\sigma}$ is not continuous, we have a catastrophe point; the set of catastrophe points is closed. Suppose that we are given a global metabolic field E $\longrightarrow$ B. A point b $\in$ B, is called a bifurcation point, if at this point $\bar{\sigma}(b)$ is not structurally stable. Hence the flow X_b belongs to a subset (Σ) of $\mathcal{D}(N)$, the so-called bifurcation subset, consisting of flows which have at least one structurally unstable attractor.

According to the generalized spectral conjecture above, (Σ) is nowhere dense. Moreover, the set (Σ) has at least partially a stratified structure; it is locally defined by "semianalytic" equations, and inherits a partition in submanifolds (strata) having smoothness properties of their tangent planes along their boundary (Whitney stratifications). Hence the notion of general position of a map of a manifold with respect to Σ may be defined.

Definition. A metabolic field is said to be generic if the corresponding map: F: B $\longrightarrow$ $\mathcal{D}(M)$ is transversal to the bifurcation set Σ.

C. Bifurcation and Catastrophe

Suppose that at a point x $\in$ B, the attractor $\bar{\sigma}(x)$ becomes structurally unstable. By a slight variation of the field (M_x, X_x), several attractors $(A_1 \ldots A_K)$, the basins of which are

[1]This conjecture is now known to be false. Sheldon Newhouse (Proceedings of the Bahia Symposium on Dynamic Systems, Aug. 1971) did provide a counter example.

adherent to the vanishing basin of $\sigma(x)$, may succeed it. In general, there will be for $t > t(x)$ a partition of a neighborhood of x between all the new attractors. Each attractor A_i will be dominant over a domain W_i of U [i.e., $\bar{\sigma}(x') = A_i$ if $x' \in W_i$]. The set of points where $\bar{\sigma}$ jumps from an attractor A_i to another A_j will be called a set of conflict points.

If x is a bifurcation point, then in general there are "conflict" points near x. We may assume that at a conflict point $y \in B$ the section $\bar{\sigma}(y)$ is in general undefined and discontinuous. Roughly speaking, two domains D_i, D_j are separated by a hypersurface of catastrophe (conflict) points, which plays the role of a shock wave; it is a locus of discontinuity in the phenomenological properties of the medium; hence it defines a "morphology."

In any morphological theory the fundamental problem is to give rules (or laws) describing the catastrophe set. The fact that (frequently) catastrophes are generated by bifurcations gives only a qualitative insight into the topological structure of the conflict set. If we want a quantitative model describing this set, then the problem becomes more difficult, as it encompasses (1) the evolution of shock waves and (2) the qualitative and quantitative description of interactions between shock waves.

II. RULES OF COMPETITION

A. Maxwell's Convention

1. The Rule

Suppose that the "inner stability" of all local attractors (of a field X) can be described by some global potential $V: M \longrightarrow R$ (this would be the case if $X = -\text{grad}_M V$ for a fixed Riemann metric on M). Then Maxwell's rule states:

For a field X on M, the dominant regime is the minimum $m_i \in M$ with absolute minimal value $V(m_i)$.

Despite its obvious shortcomings, to be discussed later, Maxwell's rule has the advantage of being simple and allowing a complete qualitative description of the catastrophe set.

2. Description of the Catastrophe Set According to Maxwell's Rule

If we admit Maxwell's rule and the transversality assumption on the bifurcation subset in the space of potentials $\{V\}$, then the catastrophe set has a polyhedral (stratified) structure. It contains two kinds of strata:

1. Bifurcation strata. These correspond to points x where the corresponding minimum $\bar{\sigma}(x)$ is no longer quadratic. Simplest example: $V_x = u^4$ in one variable.

2. Conflict strata. These are defined by $V(m_1) = V(m_2)$, where m_1 and m_2 are two distinct minima of lowest value.

The topology of conflict strata is quite easy to describe: it is given on n-space by a set of at least one equality $X_i = X_j$, where X_i, X_j are the two smallest of a set of at most $(n+1)$ linear independent forms X_i. Hence it has the topology of the first derived barycentric subdivision of the simplex. For instance, in two variables, the usual zero-dimensional conflict stratum is the triple point.

Hence it satisfies Gibbs's phase rule.

3. Bifurcation Strata

At a bifurcation point x, we introduce a mathematical notion: the idea of a "universal unfolding" of a singularity. Given a singular point 0 of f [$f_{x_i}(0) = 0$ for all i], if this point is algebraically isolated (this means that the ideal generated by the f_{x_i} contains a power of the maximal ideal m_o, i.e., any monomial x_ω of sufficiently high degree may be written as $x_\omega = \Sigma c_i f_{x_i}$, where the c_i are C^∞), then all nearby deformations of f are equivalent, up to a change of coordinates, to a function extracted from a family

$$F_u(x) = f(x) + \Sigma u_i g_i(x)$$

called the universal unfolding of the singularity f.

Using this result, one gets on the space of parameters u_i a semialgebraic set Σ of bifurcation set of points where the differentiable type of F varies.

For any deformation $G = G(x, \lambda)$ of f(x), we get an equivalent family by a conveniently chosen map: $(\lambda) \longrightarrow (u_i)$, and the catastrophe set of G is the counterimage of Σ by φ. If φ is transversal onto Σ, we get for $\varphi^{-1}(\Sigma)$ a stratified set with semi-algebraic local models. The dimension ν of this algebraic unfolding space is equal to the maximal number of nondegenerate critical points U_k [rank of $R((x_i))/\text{Ideal } f_{x_i}$] of a conveniently chosen generic deformation of f (Milnor's number of the singularity).

Hence, if the dimension of R^n, the base space, is less than this number, we have less than n minima near any bifurcation point. Hence Gibbs's phase rule is also valid near any bifurcation point. In fact, among the ν critical points there are at most $\frac{1}{2}\nu + 1$ minima, as, according to Morse's theory, one needs $\mu - 1$ points of index one to put μ minima into a connected contractible graph of separatrices. Hence there are at most $\frac{1}{2}\nu + 1$ regimes in competition near any bifurcation stratum. Call cut-locus K the closed set in the function space $\mathcal{D}$ of all local potentials which is the complementary set of the set A of functions which admit a simple absolute minimum. Then the conflict strata are counterimages of K and contain in their adherence some bifurcation strata.

Example: The Riemann-Hugoniot catastrophe. $V = \frac{1}{4}x^4$. The universal unfolding is $V = \frac{1}{4}x^4 + \frac{1}{2}ux^2 + vx$, $V_x = x^3 + ux + v$. (See fig. 1.) The bifurcation stratum is the origin $(u,v = 0)$, the conflict stratum the half-axis $(u < 0, v = 0)$. This is the simplest example of the classical critical point as it appears in the van der Waals equation; the half-axis $(u < 0, v = 0)$ corresponds to the separatrix between phases in the (p,T) = plane, V being the chemical potential $c = \int pdv$.

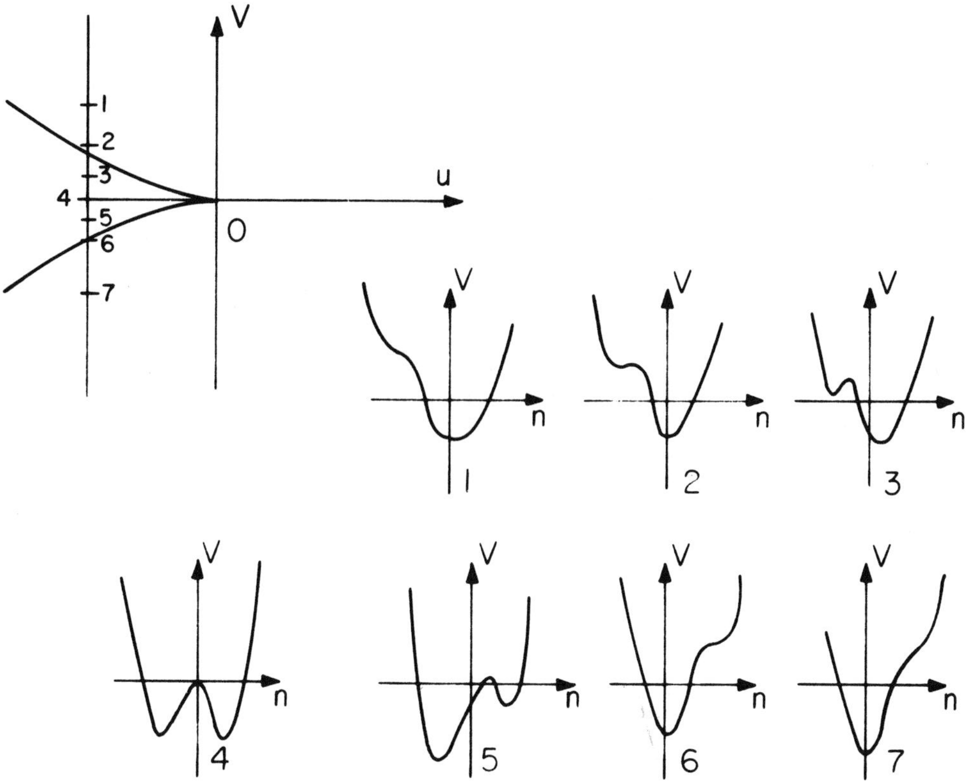

Fig. 1

B. Defects of Maxwell's Rule

Because of its static character, Maxwell's rule is in general not valid.

1. A perhaps better approximation would be to take as a measure of the stability of a minimum the (relative) height of the lowest threshold on the frontier of its basin. If we admit that the true dynamical system is not a gradient system, but is a material point subjected to the potential defined by V and to stochastic perturbations affecting its kinetic energy, then such an assumption would be motivated. But here again, as pointed out to me by J. Hubbard, a threshold on the basin's frontier is of no importance if on the other side there is a shallow basin where the representative point stays for only a very short time before returning to its original basin. A precise mathematical formulation would require an evaluation of the average time of stay T(v) in a basin (or its other neighborhoods) after crossing a threshold with velocity v.

2. Moreover, Maxwell's rule does not take into account the phenomenon of delay. If we arbitrarily move the point $x \in B$ across a catastrophe hypersurface, there is little chance that the representative point $y \in \bar{\sigma}(x)$ will immediately "know" that there is a lower minimum than itself. Hence a regime may still dominate as a "metastable" equilibrium, a local minimum whose height

is above the absolute minimum. In some sense the "fold" bifurcation strata give the maximum possible domain for a given regime (situation of perfect delay). Quite frequently, two conflicting regimes may give rise by an auxiliary coupling to two velocity fields in B. In this way we may get a hysteresis cycle (see fig. 2 in case of two conflicting regimes associated to a "cusp" [Riemann-Hugoniot catastrophe]).

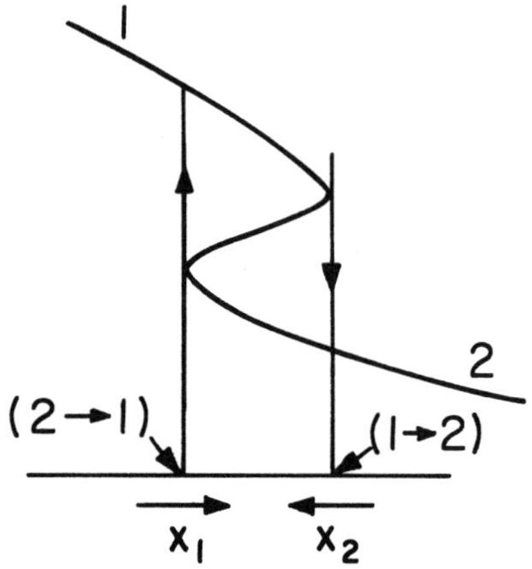

Fig. 2

III. ANALYSIS OF THE CRITICAL REGION OF A FLUID: A MODEL BASED ON A COMPETITION BETWEEN PHASES

Experiment shows that, in the neighborhood of the critical point of a fluid, the van der Waals equation leads to wrong results. It is well known (see M. Fisher [1]) that the van der Waals equation predicts for the coexistence curve a parabolic curve in $(T - T_C)^{\frac{1}{2}}$, and experiment reveals that $\rho_L - \rho_G$ (difference on an isobar) is of the form $C(T - T_C)^{1/3}$. Moreover, the two branches of the coexistence curve are related by the law of rectilinear diameter:

$$\rho_L + \rho_G = \rho_C[1 + a(T_C - T)/T_C] .$$

The critical isotherm, instead of being of the form $p - p_C = A(\rho - \rho_C)^3$ is more faithfully represented by $p - p_C = C(\rho - \rho_C)^{4,2}$. Here we give a model which explains these features by a basically simple idea, namely, that in the critical region the "temperature" is ill defined and is subject to some kind of "disease," of singularity; this singularity is itself bound to a critical point (in the usual sense) of a classical Hamiltonian describing globally the thermodynamic behavior of the fluid.

Let us consider a classical dynamical system, given by a phase space M^{2n} and a Hamiltonian $H: M^{2n} \longrightarrow R$. Given a value E of the energy, the energy hypersurface $H^{-1}(E)$ has a Liouville measure m(E). We define the entropy of the system as log m, its temperature by $T^{-1} = (d \log m)/dE$. In most thermodynamic systems (like the perfect gas), the specific heat, γ, is constant. Hence m is of the form $m = E^{\alpha}$, $\alpha > 0$. Thus the curve $S = \log m(E)$ is convex, which insures that there exists only one thermodynamic state corresponding to a value of T. If one takes an arbitrary

Hamiltonian system, defined by H, then there is no reason why its "formal temperature" should be identified with the classical thermodynamic temperature. In particular, the curve log m(E) = S(E) may not be convex (there are even worse systems, with negative temperatures). But if we assume that our system may be put in ergodic coupling with itself (another exemplar of itself), thus giving a compound system with the same thermodynamic function S(E), then the convexity requirement is essential. In particular, if the function S(E) has a double tangent, then between the abscissa e_1, e_2 we have an abrupt change, leading from one "phase" (below e_1) to another (above e_2). The difference $e_2 - e_1$ is, for instance, the latent heat of evaporation (see fig. 3).

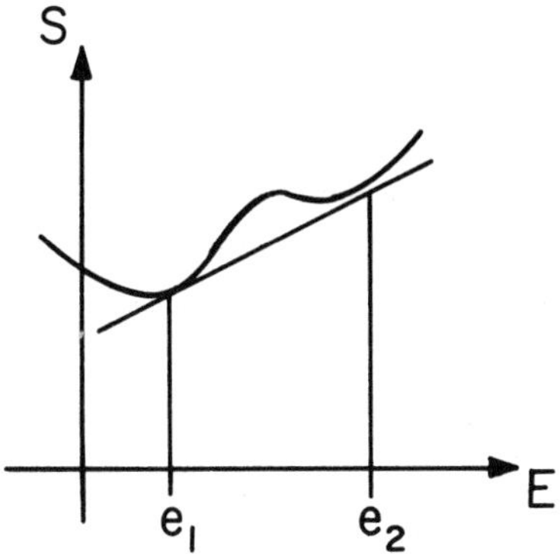

Fig. 3

This double-tangent phenomenon may occur if the Hamiltonian H goes through an ordinary critical value corresponding to a quadratic critical point. For instance, for an oscillator of one dimension, a saddlepoint in (p, q) space for H gives rise to the situation of figure 3, where we have equality of the Liouville measures (the lengths) of the level curves $xy = \pm \epsilon$. Hence the curve m(E) should have a singularity like that in figure 4 with negative temperature above the critical value c. If we want the two planes of figure 5 to become identical, we have let e_1, e_2

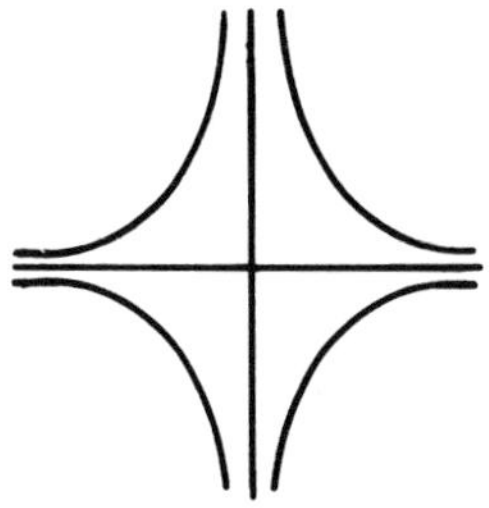

Fig. 4

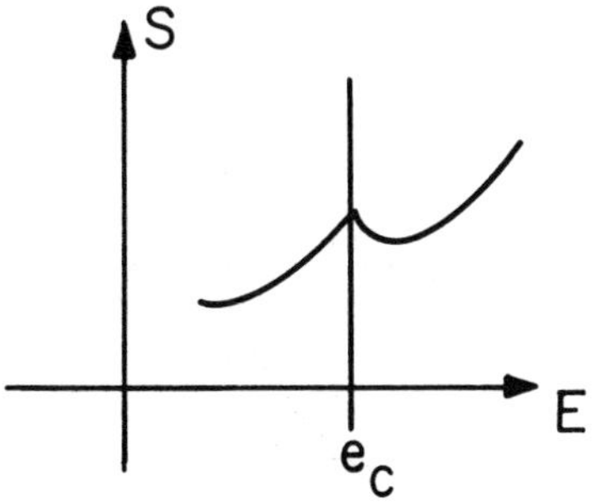

Fig. 5

tend toward a specific e_c; then the graph S(E) has to get a fourth-order contact of the type (see fig. 6)

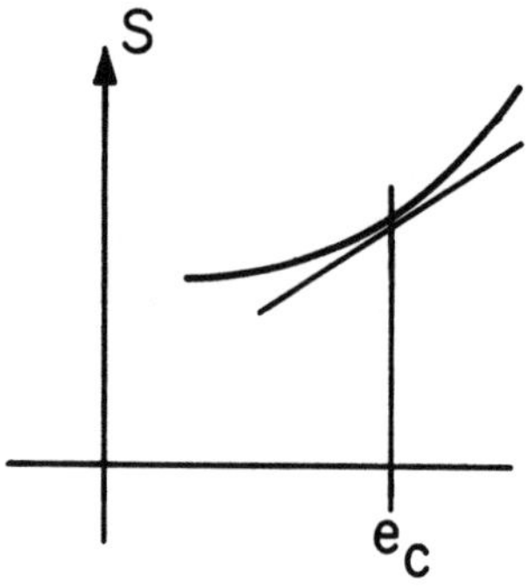

Fig. 6

$$S - S_c = (\frac{1}{T_C})\ (E - E_c) + K(E - E_c)^4 . \qquad (A)$$

This double-tangent condition is well known, and is used, for instance, with Maxwell's convention, to determine the chemical potential as a function of E. But the idea of using the same construction for the temperature itself (for a mass of fluid at constant pressure) does not seem to have been considered.

Let us consider a deformation of the curve

$$f = \tfrac{1}{4}x^4 .$$

We may suppose that the derivative $f' = x^3$ is deformed into

$$f' = x[x - \alpha][x - \beta], \quad \alpha < 0, \quad \beta > 0$$

as, by translation, one may assume that one of the roots of f' is zero (see fig. 7). Hence, by integration,

$$f = \tfrac{1}{4}x^4 - 1/3\ \Delta x^3 + px, \text{ where } \Delta = \alpha + \beta,\ p = \alpha\beta.$$

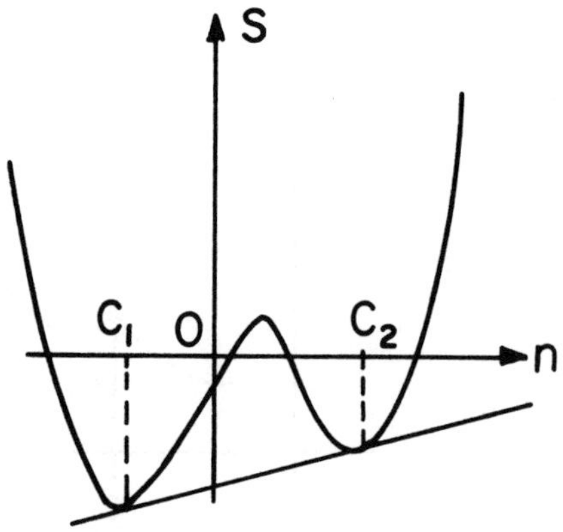

Fig. 7

To determine the slope of the double tangent, we cut f by $y = \lambda x + \mu$; hence

$$\tfrac{1}{4}x^4 - 1/3\, sx^3 + \tfrac{1}{2}px^2 = (\lambda x + \mu)\ . \qquad (B)$$

Let c_1, c_2 be the two double roots. We get as relations between roots and coefficients of equation (B)

$$s = 2(c_1 + c_2)\ ,$$

$$p = c^2 + c_2^2 + 4c_1 c_2\ ,$$

$$\lambda = 2(c_1 c_2^2 + c_2 c_1^2) = 2s\ c_1 c_2 (c_1 + c_2)\ ;$$

hence

$$c_1 + c_2 = \tfrac{1}{2}s,\ \ p = \tfrac{1}{4}s^2 + 2\, c_1 c_2,\ \ c_1 c_2 = \tfrac{1}{2}(p - \tfrac{1}{4}s^2)\ ,$$

$$\lambda = \tfrac{1}{2}s\,(p - \tfrac{1}{4}s^2)\ .$$

Provided that the deformation of f is such that α, β are linear functions of the deformation parameter t, then $(c_1 - c_2)^2 = (c_1 + c_2)^2 - 4c_1 c_2 = \tfrac{1}{4}s^2 - 2(\tfrac{1}{4}ps^2) = 3/4\, s^2 - 2p$ is a quadratic function in t, hence $|c_1 - c_2|$ is also linear in t. But then λ, which is the slope of the double tangent, is a cubic-order term in t. But through Clapeyron's formula,

$$c_2 - c_1 = q = \left(\frac{\partial S}{\partial V}\right)_p (V_2 - V_1)\ ,$$

we may identify the E-axis with the V-axis. This explains why $\rho_L - \rho_G$ is of the form $A(T - T_C)^{1/3}$. But unfortunately this simple reasoning does not explain the law of the rectilinear diameter. To explain this law we have to use a more complete analysis. The parabola γ^c $(dp/dv)_T = 0$ of the van der Waals equation is projected on the (p, T) plane as a semicubic parabola γ having a cusp at the critical point C. This cubic parabola should represent--theoretically--the extreme limit curve reachable by metastable states (like superheated liquid, or supersaturated gas) (see figs. 8 and 9). But this equilibrium is in general broken much earlier.

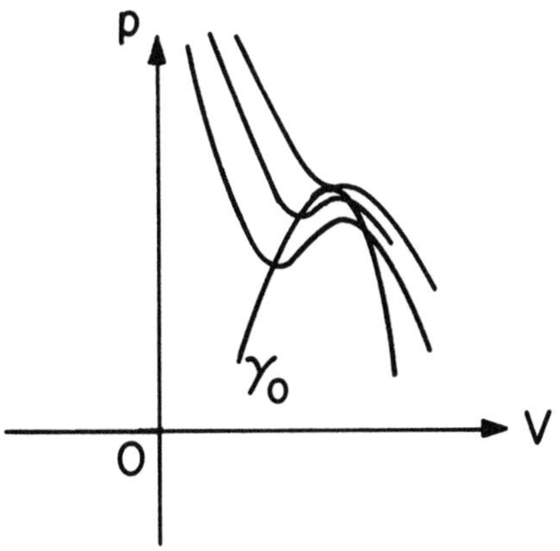

Fig. 8

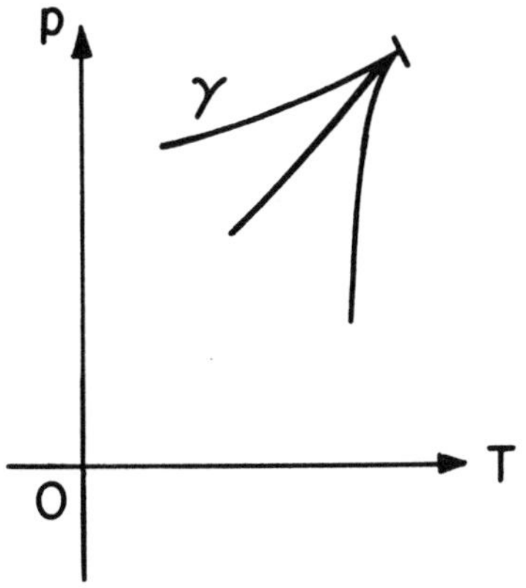

Fig. 9

If we consider that on the (p, T) plane we have a metabolic field defined by grad c, where c is the chemical potential $c = \int p\,dv$, then at a point of the "fold curve γ," the curve c(v) has a flex point with horizontal tangent (fig. 10). Inside the curve γ, we have a small basin associated to the "vanishing" phase (fig. 11). The basic idea is that a "phase" is described by a material

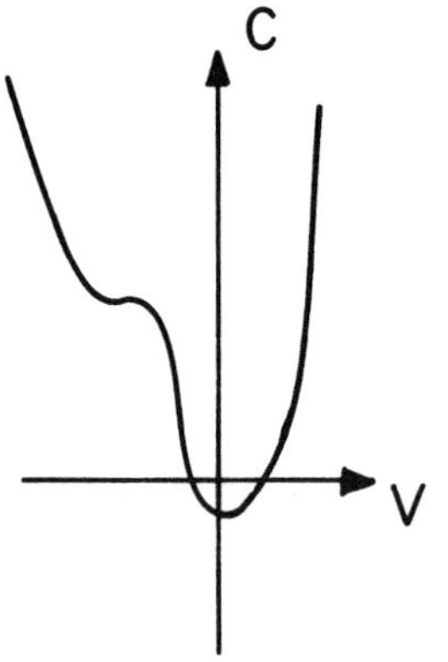

Fig. 10

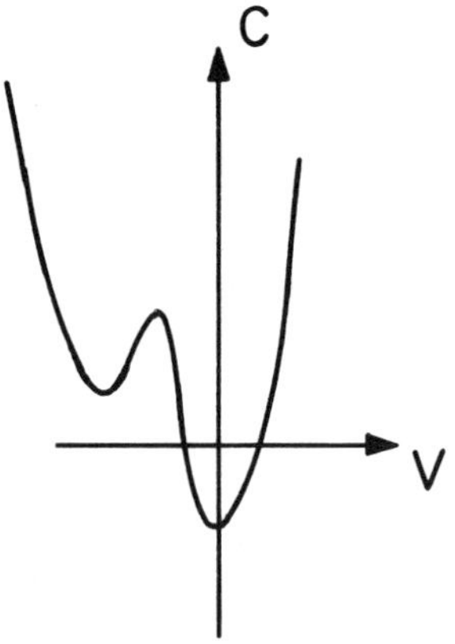

Fig. 11

point oscillating in this potential well. At low temperatures, the kinetic energy of the representative point (which symbolizes fluctuations due to thermal agitation of molecules) is very small. Hence the dynamic behavior is practically the one given by the gradient field grad c, and the whole situation is described (relatively) accurately by the van der Waals equation. At higher temperatures, however, if the kinetic energy of the point reaches an average value $\bar{h}$, then if d is the relative height of the threshold with respect to the minimum, total instability will develop when $z = d - \bar{h} = 0$.

The fundamental idea of the model is that the phase's vanishing stability may be conveniently described by an (ideal) oscillator, whose Liouville measure μ measures the "stability" of the vanishing phase. For such a μ, we shall assume that μ is a function of z alone, as it vanishes for $z = 0$. For reasons which will appear later, we shall assume that the Liouville measure starts with a quadratic term, with a Taylor expansion of the form

$$\mu_i = A_i z^2 + B_i x^3 + \ldots,$$

each phase ($i = 1, 2$) having its own coefficients A_i, B_i. The quantity μ may then be considered a normal coordinate to the branch of γ. With this kind of model, the total curve γ limiting the existence of metastable states is not γ, but a curve γ_1 approximately parallel to γ inside γ. If completely drawn, such a curve has the appearance of a Descartes leaf--it describes a loop around the critical point (cf. fig. 12). Such a figure is quite easy to obtain around a cuspidal caustic in geometrical optics (see Plate I in [2]). Theoretically one might say that the inside of the loop describes a kind of "hole" in the (p, T) plane, a hole of "forbidden states." But it might be argued that the boundary of the loop does not mean anything from a physical point of view.

In such a figure, the true critical point is no longer the cusp of γ, but the double point 0 of γ_1. From 0 we get a half-curve describing the equilibrium between the two phases, the

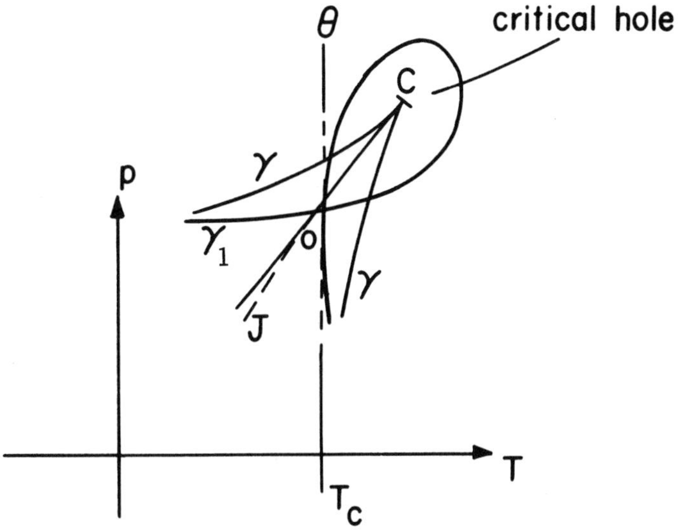

Fig. 12

coexistence curve J; this curve is determined by the condition of "thermal" equilibrium between the two "vanishing stability" oscillators of the two phases. Now, the critical point 0 is such that the corresponding van der Waals c(v) curve has two equal minima; hence near the threshold, on both sides, we have a local equation of type

$$Z = ax^2 + bx^4 + \ldots ,$$

where x can be interpreted as $\rho - \rho_c$.

The slope of the separatrix J in 0 is determined by the condition that the projecting coefficients $\cos\theta_1$, $\cos\theta_2$ associated with the normals to the branches of γ_1 are such that in the functions

$$(\cos\ \theta_1)^{-1}\ [A_1Z^2 + B_1Z^3],\quad (\cos\ \theta_2)^{-1}\ [A_2Z^2 + B_2Z^3]$$

we have

$$\frac{A_1}{\cos\theta_1} = \frac{A_2}{\cos\theta_1} = K$$

(equality of lower-order terms). Now, let x_G, x_L be the values $\rho_G - \rho_C$, $\rho_L - \rho_C$ along J. We get for the two Liouville measures (with new notations for the constants)

$$\mu_G = K(ax_G{}^2 + bx_G{}^4)^2 + K_1(ax_G{}^2 + bx_G{}^4)^3 ,$$

$$\mu_L = K(ax_L{}^2 + bx_L{}^4)^2 + K_2(ax_L{}^2 + bx_L{}^4)^3 .$$

Hence

$$\mu_G = K\frac{x_G{}^4}{4} + K_{1'}\frac{x_G{}^6}{6} + \ldots ,\quad \mu_L = K\frac{x_L{}^4}{4} + K_{2'}\frac{x_L{}^6}{L} + \ldots .$$

Considering the x_i as local coordinates for the energy e, we have the first integral condition $x_G - x_L$ = constant for p = constant, expressing $e_G + e_L$ = constant. Hence, by derivation, and considering that $x_L = v_L - v_c$ is negative, and if T stands for $1 - \frac{T}{T_c}$;

$$T = K x_G^3 + K_1' x_G^5 + \ldots, \quad -T = K x_L^3 + K_2' x_L^5 + \ldots.$$

Then, discarding terms of order higher than 5,

$$T = K x_G^3 \left(1 + \frac{K_1}{K} x_G^2\right), \quad T = -K x_L^3 \left(1 + \frac{K_2'}{K} x_G^2\right),$$

so we get at first approximation $x_G = (T/K)^{1/3}$ and

$$x_G^3 = \frac{T}{K}\left(1 + \frac{K_1'}{K} x_G^2\right)^{-1} .$$

Extracting the cube root,

$$x_G = \frac{T}{K}^{1/3} \left(1 - \frac{K_1'}{3K}\left(\frac{T}{K}\right)^{2/3}\right) = \left(\frac{T}{K}\right)^{1/3} - \frac{K_1'}{3K^2}\left(\frac{T}{K}\right),$$

$$x_L = -\frac{T}{K}^{1/3} \left(1 - \frac{K_2'}{3K}\left(\frac{T}{K}\right)^{2/3}\right) = -\left(\frac{T}{K}\right)^{1/3} + \frac{K_2'}{3K^2}\left(\frac{T}{K}\right).$$

Hence we get the law of the rectilinear diameter. The justification for this construction lies in the fact that the "vanishing stability" oscillator may be identified for a mass of liquid with the quadratic term of the Taylor expansion near the critical point:

$$S(E) - S(C) = \frac{1}{T_C}(E - E_C) + A(E - E_C)^4 + A_1(E - E_C)^6 + \ldots, \qquad (A)$$

where as usual the energy E has to be locally identified with specific volume $x_L = V_L - V_C$. This also gives a result with respect to the critical isotherm. The transversal coordinate to the gas branch of γ_1, the Liouville measure μ_G of the "vanishing stability" oscillator of the gaseous phase, may be identified up to a scalar factor to $p - p_C$. Hence, along the critical isotherm for $p < p_C$,

$$p - p_C = [K_1(x_G - x_C)^4 + K_2(x_G - x_C)^6 + \ldots],$$

an expansion which does not look incompatible with the exponent 4, 2 $\pm$ 0, 1 found by experimentalists. This is true only for the $(p - p_C < 0)$ = branch; as the $(p - p_C > 0)$ = branch runs near the critical hole of "forbidden states," it should be significantly flatter than the gaseous branch.

This model has to be considered as the easiest mathematical way to get to the phenomenological results describing the coexistence curve. The problem of justifying this construction from an orthodox statistical mechanical viewpoint (involving reduction to a molecular mechanism) is naturally entirely open. (Personally, I suspect this would not be easy and perhaps finally not so important.)

IV. APPLICATION TO BIOLOGY

The notion of "phase," when sufficiently generalized, may be of great interest in biology. It was observed, long ago, that it is very difficult to specify the phase of living matter. Is it liquid? Or solid? Or both? Or neither? The answer is not easy. Quite likely, the notion of phase will have to be generalized to contain the notion of cellular differentiation. As was pointed out by Delbrück in 1949 [3], we may regard cellular differentiation as a local stable regime of the metabolism. If we consider the local states of metabolic activity as a metabolic field in the sense of section I, then any histological differentiation may be considered as defined by an attractor of the local metabolism. The bifurcation problem involved by the change of these attractors is not well known mathematically. It is quite obvious that we may get, by a proper choice of attractors, situations quite different from those described by Maxwell's rule for gradient fields (for instance, generalized catastrophes as defined in [2]). The basic idea of "universal unfolding" still keeps some validity. But we do not know how to construct them. Moreover the degeneracies of an attractor to a lower-dimensional one (phenomenon of resonance) may give rise to very complicated topologies for the unfolding space: it might have, for instance, the topology of a homogeneous space. That way there would be a possibility of describing the complicated morphologies observed in embryological development, like neurulation or heart formation. The totality of all possible cellular differentiations may be put in a large "phase diagram." Embryological development can be considered as the evolution of a two sphere S^2 (or D^3) across this phase diagram. Each differentiation is described by an attractor: the topology of this attractor (its inner metabolic structure) is intrinsically related to its biological meaning, the physiological function of the corresponding tissue. This may give a clue to understanding the finalistic and regulative processes in biology, as well as the basic procedures through which the "genetic code" ensures the stability of the spatio-temporal development of a living being.

REFERENCES

1. M. E. Fisher, The Nature of Critical Points (Lectures in theoretical physics, Vol. 2), (Boulder: Univ. of Colorado Press, 1964).

2. R. Thom, "Topological models in biology," Topology, 8, 313-36 (July 1969).

3. M. Delbrück, Unités biologiques douées de continuité génétique (Colloque C. N. R. S., Paris 1949, p. 33).

DISCUSSION

I. Prigogine: The interesting considerations of R. Thom have to be supplemented by other physical considerations to lead to a better understanding of the concept of structure, especially for biological systems.

The most detailed analysis of order made in physics refers to equilibrium situations in the thermodynamic sense. The basic principle is then the Boltzmann order principle which finds its quantitative formulation in the Boltzmann factor $\exp(-E/kT)$. The Boltzmann factor expresses a competition between the energy E of a state and the temperature. This competition on the molecular level is then amplified to the macroscopic level when phenomena such as phase transitions, ferromagnetism, or partial miscibility are involved.

One of the most interesting problems of statistical physics and thermodynamics is to extend the concept of order to nonequilibrium situations. The remarkable feature is that in the presence of well-defined nonlinearities (such as may occur in hydrodynamics or in specific classes of catalytic chemical reactions), a new type of order may appear at sufficient distance from thermodynamic equilibrium. This new type of order appears through amplification of certain types of fluctuations and may be briefly called "order through fluctuation" (for more details and references see the monograph by P. Glansdorff and I. Prigogine, "Thermodynamic Theory of Structure, Stability and Fluctuations," to appear Wiley Interscience 1971, French edition to appear Masson 1971). To make this a little more specific, consider a system of chemical reactions describing the conversion of a set of initial reactants $\{A\}$ to the final products $\{F\}$ and let $\{X, Y, \ldots\}$ represent the values of the concentrations of the intermediates. It is understood that the system $\{X, Y, \ldots\}$ is open to $\{A\}$, $\{F\}$ and subject to time-independent conditions. It is convenient to measure the deviation from equilibrium by the value of some parameter R which depends on the ratios $\{A\}/\{F\}$ as well as on other parameters such as temperature. Due to the nonlinearity of the equations of evolution of the intermediates, there is in general more than one steady state. At thermodynamic equilibrium $R = R_{equ}$, the correct solution being the one minimizing the free energy.

Imagine now that R deviates more and more from R_{equ}. For $|R - R_{equ}$ "small," the equilibrium solution first changes smoothly and we obtain a branch referred to as the "thermodynamic branch." We have now obtained explicit thermodynamic criteria which express conditions which have to be satisfied in order that the thermodynamic branch becomes unstable at a given distance from equilibrium. (The so-called excess entropy production has to vanish.) The system then leaves the thermodynamic branch and goes into a new state which now becomes stable.

Briefly speaking, the breakdown of the thermodynamic branch corresponds to the violation of Boltzmann's order principle. Beyond this instability new types of structures appear, we have so-called dissipative structures. These are really new sites of matter maintained through continuous exchanges of energy and matter with the outside world. Such dissipative structures are characterized by a coherent time-space behavior, in contrast with the usual thermodynamic behavior near equilibrium.

Detailed studies show that essential mechanics of living systems work "beyond" the thermodynamic instability (see I. Prigogine and G. Nicolis, Quarterly Rev. of Biophysics, to appear 1971, where references may be found). Using these concepts, M. Eigen has even been able to give an interpretation of the evolution of "competing" biological molecules toward some type of genetic code (M. Eigen, to appear Naturwissenschaften, 1971).

7

STOCHASTIC MODELS OF NEUROELECTRIC ACTIVITY

Jack D. Cowan

1. INTRODUCTION

Theories of how the brain works take many different forms. Some are essentially combinatorial, and are posed in terms of computer science. The hypothesis is made that certain neural nets function as pattern detectors, or as simple memory stores, for example. Model nerve nets are then devised to carry out such tasks in an efficient manner. This approach has been used recently in a very interesting way to provide a series of models for cerebral and cerebellar cortex, and for the hippocampus [1-3]. The end product is a series of predictions about the functional characteristics of nerve cells and synapses, and an interpretation of some of the anatomical details of neural nets. Models of this kind serve a very useful purpose, but they are difficult to test physiologically, since they do not generate predictions about neural dynamics, i.e., about the spatio-temporally organized patterns of activity recorded in brains, which evidently underlie signal transmission. To serve this purpose, theories are required which emphasize the temporal aspects rather than the purely combinatorial aspects of the electrical activity of cells and populations. In this paper I wish to give a short account of some recent investigations, the intention of which is to provide a theory of this activity.

Histological studies of the brain show it to be a densely packed net of some 10 billion nerve cells, and countless satellite cells, the whole occupying a container of some 2 thousand cubic centimeters. Studies carried out some twenty years ago [4] led to the conclusion that any theory of neural nets must depend upon statistical considerations in which the connections of individual cells are not nearly so important as the global pattern of connectivity in populations. More recent investigations have led to a modification of this view [5-6]. Nerve cell fibers do not make connections haphazardly, but appear to be directed to specific groups of cells. If the statistical view is to remain tenable, it must be that of a structured and heterogeneous population, perhaps

---

Supported in part by the Alfred P. Sloan Foundation and the Otho S. A. Sprague Memorial Institute.

subdivided into reasonably homogeneous populations, each of which is connected in a particular fashion to other homogeneous populations. Somewhat related conclusions about the statistical nature of neuroelectric activity can be derived from a study of the electrical responses of cells and populations. Only averages appear to be reliable indicators of neural activities. At the cellular level, characteristic discharge patterns are enormously variable, especially in unanesthetized animals [7]. Similarly, population records show considerable variability in both the "spontaneous" and "evoked" cases [8]. There is thus an appearance of randomness in most if not all neural nets, both in the fine details of interconnection and in the individual records of activity. Our problem is to give some expression to this randomness in developing a semistatistical theory of neural activity. In this paper only the temporal consequences of such a theory will be considered; very important considerations concerning neural activity which is both spatially and temporally organized will not be covered here.

2. STATISTICS OF CELLULAR ACTIVITY

Microelectrode records of the discharge patterns of single cells show that a majority of the cells fire continually at a low rate, with occasional <u>bursts</u> of activity, and that stimulation changes such patterns in a variety of ways. Histograms of the intervals between successive spikes show a wide variety of shapes, characteristic of various random point processes. If plotted on a long time scale (see fig. 1), such histograms can be fitted by a random burst process [9]. Other cells fire in regular bursts, the frequency and amplitude of which are controlled by stiumli [10]. To

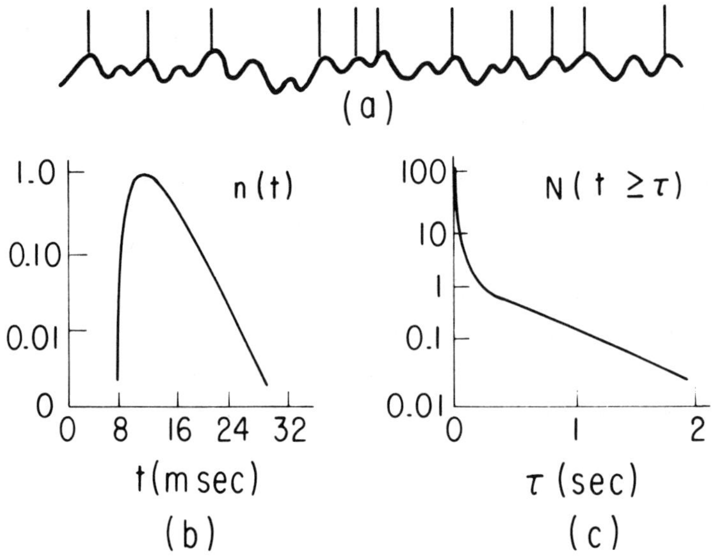

Fig. 1

construct a model which will do justice to such data we start with a simplified electrical model of the cell membrane which dates back almost to the beginning of the century [11]. This consists of a linear RC circuit responding to superimposed current pulses. The resulting voltage drives an amplitude sensitive pulse generator, so that if the voltage exceeds a certain threshold, a pulse is generated which propagates to other cells, and also resets the membrane voltage (see fig. 2).

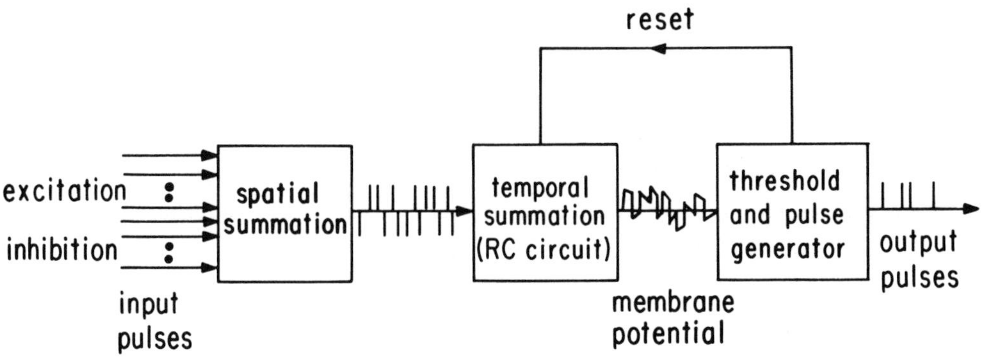

Fig. 2

Various elaborations of this are possible--for example, the intensity of incoming current pulses may be made a function of the membrane voltage, reflecting voltage dependent conductance changes produced in the membrane by synaptic excitation. On the assumption that the incoming pulse patterns are essentially "shot noise," the variation in membrane potential may be represented as a Markov process [12]. In the simplest case, when incoming pulses produce either unit positive jumps (excitation) or unit negative jumps (inhibition), a "birth-and-death" process results [13]. Let $p(v,t|v_o)$ be the transition probability density for membrane potential, τ = RC the time constant of the membrane, and $p(v'|v,t)$ the cumulative distribution of jump sizes. Then

$$\frac{\partial P}{\partial t} = \frac{\partial}{\partial v}\left[\frac{vP}{\tau}\right] - \lambda_e[P(v,t|v_o) - P(v-1,t|v_o)] + \lambda_i[P(v+1,t|v_o) - P(v,t|v_o)], \quad (2.1)$$

where λ_e and λ_i are the rate parameters of the incoming excitation and inhibition. At the threshold voltage $\theta(t)$, the process terminates and a spike is generated. The appropriate boundary condition is therefore

$$P(v'|\theta(t),t) = \begin{cases} 0 & (v' \neq 0) \\ 1 & (v' = 0)\,. \end{cases} \quad (2.2)$$

It will be seen that the problem is a conventional one, that or a random walk to an absorbing barrier. The walk, however, consists of randomly occurring jumps superimposed on a continuous drift away from the threshold, which acts as an absorbing barrier. As long as the time constant τ remains finite, solutions cannot be obtained in any simple fashion, even in the case where the threshold $\theta(t)$ is a constant. Various approximations have been introduced which have the effect of converting the continuous drift into a discontinuous one, thereby making the process a discrete random walk, the moments of which can be obtained in a reasonably compact form [14,15]. A somewhat more tractable model is obtained if the diffusion approximation is introduced [16, 17]. It is assumed that hundreds of synapses are continually activated in such a way that the total activity seen by an individual cell may be represented as the difference of two shot-noise processes, an excitatory one with rate parameter λ_e and an inhibitory one with rate parameter λ_i. If $(\lambda_e - \lambda_i)$ is sufficiently small, and λ_e and λ_i sufficiently large, and if the jump sizes are sufficiently small compared to $\theta(t)$, then the variation in membrane potential may be represented by the Ornstein-Uhlenbeck equation [18]

$$\frac{\partial P}{\partial T} = \frac{\partial}{\partial x}[xP] + \frac{\partial^2 P}{\partial x^2} \tag{2.3}$$

where

$$x = (v - \mu\tau)/(\tfrac{1}{2}kt)^{\frac{1}{2}}, \quad T = t/\tau,$$

μ is the drift parameter, and k the variance parameter of the incoming Gaussian delta-correlated noise, (sic.) white-noise. The boundary conditions now become

$$P(-\infty, T|x_o) = P[\epsilon(T), T|x_o] = 0$$

and (2.4)

$$P(x, 0|x_o) = \delta(x - x_o).$$

Once again explicit solutions for this problem cannot be obtained very easily, although when the boundary $\epsilon(T) = [\theta(T) - \mu\tau]/(\frac{1}{2}kt)^{\frac{1}{2}}$ is independent of time, the Laplace transform of the first passage density can be obtained in closed form [19, 20]. In certain special cases time-dependent perturbation theory can be used to provide approximate analytical solutions of the problem [21].

Let $w = x - \epsilon(T)$, then equation (2.3) becomes

$$\frac{\partial P}{\partial T} = \frac{\partial}{\partial w}[wP] + \frac{\partial^2 P}{\partial w^2} + f(T)\frac{\partial P}{\partial w} \tag{2.5}$$

subject to the conditions

$$P(-\infty, T|w_o) = P(0, T|w_o) = 0, \quad P(w, 0|w_o) = \delta(w - w_o), \tag{2.6}$$

where

$$f(T) = \dot{\epsilon}(T) + \epsilon(T). \tag{2.7}$$

Equation (2.5) represents a time-dependent perturbation of the Ornstein-Uhlenbeck equation, "switched-on" at T = 0, with boundary conditions on the half-line $(-\infty, 0)$. This will be recognized as a problem concerning quantum-mechanical harmonic oscillators. As is well known, one may write an integral equation equivalent to (2.5) [22]. Let $P_o(w, T|w_o)$ be the Green's function solution of equations (2.5) and (2.6), when f(T) = 0. Then

$$P(w, T|w_o) = P_o(w, T|w_o) + \int_0^T \int_{-\infty}^0 d\tau\, dw'\, P_o(w, T-\tau|w')\, f(\tau)\frac{\partial}{\partial w'} P(w', \tau|w_o), \tag{2.8}$$

which may be solved by iteration. The function of interest for the neural problem is the first-passage density

$$\mathcal{P}(0, T|w_o) = -\frac{\partial}{\partial T}\int_{-\infty}^0 dw'\, P(w', T|w_o). \tag{2.9}$$

This is identical with the probability density of inter-spike interval provided the membrane potential resets of w_o each time the threshold is reached. It can be shown [21] that the solution of equations (2.8) and (2.9) is

$$\mathcal{P}(0,T|w_o) = \frac{|-w_o|}{(2\pi)^{\frac{1}{2}}} e^{-T}(1-e^{-2T})^{-3/2} \exp\left[-\frac{w_o^2 e^{-2T}}{2(1-e^{-2T})}\right]$$

$$+\left(-\frac{\partial}{\partial T}\right) e^{-T}(1-e^{-2T})^{-1/2} \exp\left[-\frac{w_o^2 e^{-2T}}{2(1-e^{-2T})}\right]$$

$$\times (2/\pi)^{\frac{1}{2}} \int_0^T d\tau e^{\tau} f(\tau) \operatorname{erf}\left[-\frac{\beta}{(2\alpha)^{\frac{1}{2}}}\right] + \ldots, \qquad (2.10)$$

where

$$\alpha = (e^{2(T-\tau)}-1)^{-1} + (1-e^{-2\tau})^{-1}, \quad \beta = w_o e^{-\tau}(1-e^{-2\tau})^{-1}.$$

It follows that if f(T) is sufficiently small, an approximate solution is obtained of the solution of the Ornstein-Uhlenbeck process on the curved barrier $\epsilon(T)$. Now

$$f(T) = \dot{\epsilon}(T) + \epsilon(T) = [\dot{\theta}(T) + \theta(T) - \mu\tau]/(\tfrac{1}{2}kt)^{\frac{1}{2}}.$$

Thus $f(T) = 0$ whenever $\theta(T) = \mu\tau$, or when $\theta(T) = \mu\tau + (\theta_o - \mu\tau)e^{-T}$. In both cases $\mathcal{P}(0,T|w_o) = \mathcal{P}_o(0,T|w_o)$ [23, 24, 25]. A more general case is $\theta(T) = \theta_\infty + (\theta_o - \theta_\infty)e^{-\delta T}$, equivalent to $f(T) = [(\theta_\infty - \mu\tau) + (\theta_o - \theta_\infty)(1-\delta)e^{-\delta T}]/(\tfrac{1}{2}kt)^{\frac{1}{2}}$, where $0 \le \delta \le 1$. The threshold time-constant is about 40 msec compared to a membrane time-constant of about 5 msec, corresponding to a δ of 0.25. Evidently $\mathcal{P}(0,T|w_o) \sim \mathcal{P}_o(0,T|w_o)$ whenever $(\tfrac{1}{2}kt)^{\frac{1}{2}}$ is sufficiently large. For a cell receiving excitation from several hundred fibers under physiological conditions, this voltage is of the order of 2-3 mV, compared to an equilibrium threshold θ_∞ of about 10 mV. Thus when $\mu\tau \sim 7$ mV or more, f(T) is at most 1. Computed results are comparable with experimental data (see fig. 3). Of course, such a model neglects many of the details of cellular morphology and physiology, but it does provide some insight into the nature of neural variability. The conclusions to be drawn from this study are that any unsynchronized excitation, or else intrinsic fluctuations in cell thresholds, can serve to produce the variability observed in cellular discharges.

Given an approximation to the inter-spike interval density, one can use the Laplace transform to obtain moments of the density, and also to obtain such quantities as the event density, the conditional probability density that the membrane voltage reaches threshold at the instant t. The conditional event density is the sum of convolutions of the first passage density:

$$N(0,T|w_o) = \mathcal{P}(0,T|w_o) + \mathcal{P}(0,T|w_o) \otimes \mathcal{P}(0,T|w_o) + \ldots, \qquad (2.11)$$

so that

$$N_L(0,S|w_o) = \mathcal{P}_L(0,S|w_o)[1 - \mathcal{P}_L(0,S|w_o)]^{-1}. \qquad (2.12)$$

This expression is well known from renewal theory [26]. The instantaneous frequency is simply the reciprocal of the first moment of $\mathcal{P}(0,T|w_o)$, the mean inter-spike interval. Numerical

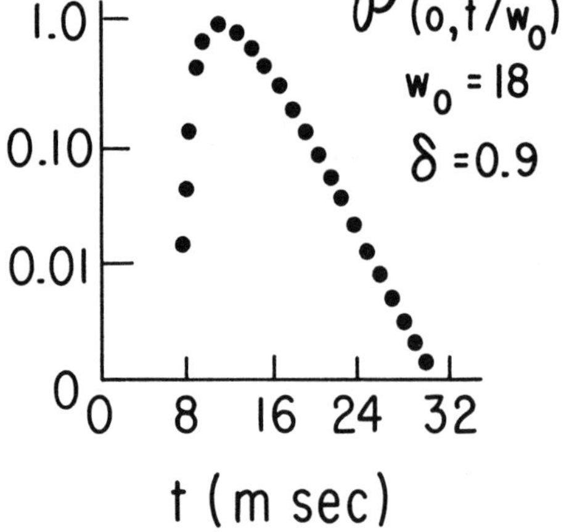

Fig. 3

calculations of this quantity indicate that there is a monotonic relationship between the intensity of incoming excitation, and the instantaneous frequency. Let $N = \tau n(w_0)$ be the average number of impulses within a time constant. Let $M = (\lambda_e - \lambda_i)\tau/\theta_\infty$ be the equilibrium voltage built up in the membrane, i.e., $\mu\tau$ in units of threshold. Let S be the rms noise voltage in threshold units, i.e., $(kt/2\theta_\infty)^{\frac{1}{2}}$. Then by curve fitting [27] (see fig. 4),

$$N = c \exp [\alpha(M - N) + \beta S^2] . \tag{2.13}$$

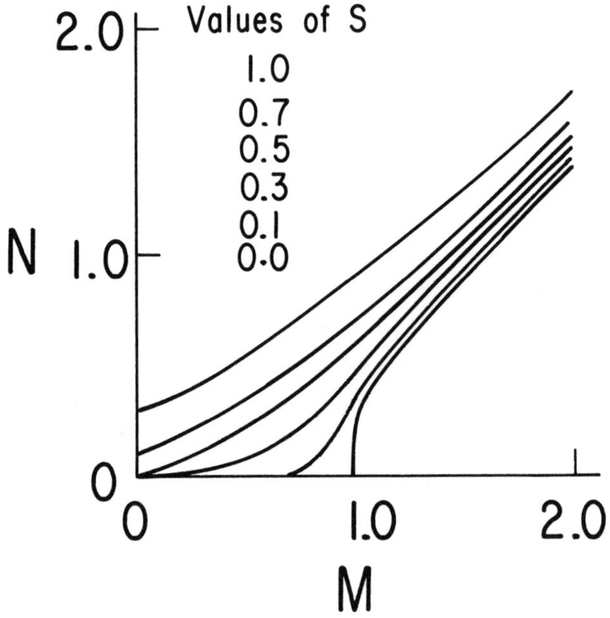

Fig. 4

3. STATISTICS OF NEURAL NET ACTIVITY

Given some insight into single cell variability, it is natural to turn to a consideration of the activity in nets. The appropriate analysis is in terms of two cell interactions. The simplest equation turns out to be a mixed Markov equation containing both drift and diffusion terms, and also jump terms containing time-varying parameters. Such an equation is not tractable, and it is therefore necessary to develop a more simplified model of the neural response. Equation (2.13) provides the necessary starting point. It can be rewritten as

$$N_i = \exp\left[\sum_j^n \alpha_{ij} N_j - \delta_i\right] \tag{3.1}$$

for the i(th) (secondary) cell, receiving excitation from n (primary) cells, where the α_{ij} are defined in terms of synaptic conductances, membrane time constants, and neural thresholds [27]. If the cells also have dead times or absolute refractory periods of duration r before they become sensitive to excitation after the emission of an impulse, one obtains the equation

$$N_i = (1 - rN_i)\exp\left[\sum_j^n \alpha_{ij} N_j - \delta\right]. \tag{3.2}$$

Thus

$$N_i = L\left[\sum_j \alpha_{ij} N_j - \delta_i\right], \tag{3.3}$$

where L[x] is the "logistic" function of population growth, $e^x(1 + re^x)^{-1}$. Alternately one has [28]

$$\ln\left[\frac{N_i}{1 - rN_i}\right] = \sum \alpha_{ij} N_j - \delta_i \;. \tag{3.4}$$

To obtain an equation of motion from this equation it suffices to recall that there are conduction and synaptic delays of various kinds, all of which can be subsumed under a single time-constant of the order of a few milliseconds at most. The result is an equation of the form

$$\left(\frac{d}{dT} + 1\right) \ln\left[\frac{N_i}{1 - rN_i}\right] = \sum_j \alpha_{ij} N_j - \delta_i \;, \tag{3.5}$$

where T is measured in units of the time-constant. If one adds an external input to each cell, and incorporates various constants into the terms, equation (3.5) may be rewritten as

$$\frac{dx_i}{dT} + x_i(1 - x_i) \ln \frac{x_i}{1 - x_i} = \left(\epsilon_i + \frac{1}{\beta_i} \sum_j \alpha_{ij} x_j\right) x_i(1 - x_i) \;, \tag{3.6}$$

where $x_i = rN_i$, and where ϵ_i contains the external input as a parametric excitation. In a certain sense this equation represents a generalization of the McCulloch-Pitts nerve-cell model for individual pulse emission with an all-or-nothing threshold, to a model for pulse-train emission with a continuous nonlinear characteristic.

It can be shown that for sufficiently large ϵ_i, the term $x_i(1 - x_i)\ln[x_i/(1 - x_i)]$ is small compared to $\epsilon_i x_i(1 - x_i)$, so that it can be neglected [29]. Equation (3.6) therefore reduces to

$$\frac{dx_i}{dT} = (\epsilon_i + \frac{1}{\beta_i} \sum_j \alpha_{ij} x_j) x_i(1 - x_i) , \tag{3.7}$$

where the coupling coefficient matrix $|\alpha|$ is as yet unspecified. To do this, one must consider what would be an appropriate anatomy to incorporate. It has been argued elsewhere that equation (3.2) might serve as a model of the interactions between neural nets in the cerebral neocortex and in the thalamus, and similar interactions in the cerebellum [28]. Large stellate cells in the thalamus are supposed to excite pyramidal cells in the neocortex. In turn these cells are supposed to inhibit the stellate cells. Both populations are supposedly embedded in a net of interneurons whose function is to maintain the appropriate synaptic coupling coefficients α_{ij} by some regulatory process (see fig. 5). Finally there is an inhibitory input to the cortical pyramids and a corresponding excitatory input to the thalamic counterparts. Such inputs may be external, or they may be taken to be internal, in which case the thalamic cells may be said to act as pacemakers, driving the passive cells of the neocortex. The coupling coefficients are supposedly maintained so that

$$\alpha_{ij} + \alpha_{ji} = 0, \quad i \neq j ; \tag{3.8}$$

i.e., all things being equal, the i(th) and the j(th) cells are in equilibrium. It also follows from equation (3.7) that

$$\alpha_{ij} < 0 . \tag{3.9}$$

It should be obvious that what has been set up is an analogue of the well-known Lotka-Volterra equations for predator-prey interactions between species, in which thalamic cells "feed" predatory cortical cells. This topic has been extensively investigated [30-33], and many of the recent results apply directly to the neural equations above, with trivial modifications.

Let q_i be the stationary states of the net, such that

$$\epsilon_i \beta_i + \sum_j \alpha_{ij} q_j = 0 . \tag{3.10}$$

The transformation

$$v_i = \ell n \frac{x_i/q_i}{(1 - x_i)} \tag{3.11}$$

may be used to reduce equations (3.7) to the equivalent form

$$\frac{dv_i}{dT} = \sum_j^n \gamma_{ij} \frac{\partial G}{\partial v_j} , \tag{3.12}$$

where $\gamma_{ij} = \alpha_{ij}/\beta_i \beta_j$, and where

$$G = \sum_j \beta_j \, [\ell n(1 + q_j e^{v_j}) - q_j v_j] . \tag{3.13}$$

The function G is a Liapunov function for the system [34], for

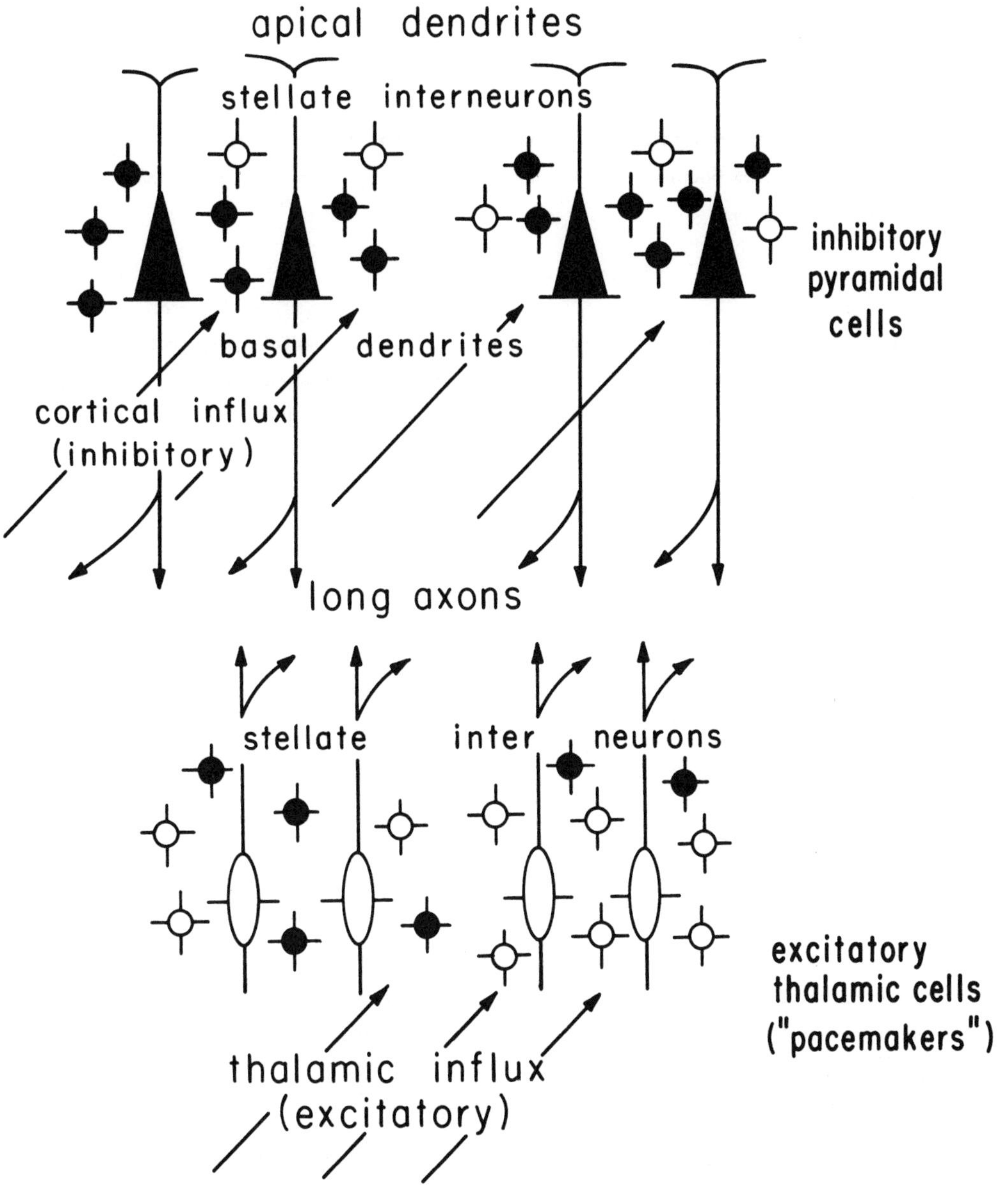

Fig. 5. Thalamo-cortical circuits.

$$\frac{dG}{dT} = \sum_i \frac{dv_i}{dT}\frac{\partial G}{\partial v_i} + \frac{\partial G}{\partial T} = \sum_i \sum_j \alpha_{ij} q_i q_j \left[\frac{\exp(v_i)}{1 + q_i \exp(v_i)} - 1\right]\left[\frac{\exp(v_j)}{1 + q_j \exp(v_j)} - 1\right] < 0$$

and G is evidently a positive definite function of v_i. Thus the net activity is asymptotically stable, and any oscillations about the stationary states gradually damp out. It is now assumed that there are intrinsic fluctuations associated with the activities of cells which can be represented by Gaussian delta-correlated noise, and that the variance of such diffusion terms is proportional to $-\alpha_{ii}$; i.e.,

$$k_i + \alpha_{ii} = 0. \tag{3.14}$$

This is not unreasonable; the greater the feedback or dissipation intrinsic to neural activities, the greater the consequent fluctuations. One may therefore write a Langevin equation for the neural activities of the form

$$dv_i = \sum_j \gamma_{ij} \frac{\partial G}{\partial v_j} dT + dB(T), \tag{3.15}$$

where dB(T) is a Langevin term with variance parameter k_i. Such systems have been analyzed before in the Lotka-Volterra context [32,33]. For the particular assumptions made here it is easily shown that the associated Fokker-Planck equation for the probability density $f(v_1, v_2, \ldots, v_{2n}, T)$ is

$$\frac{\partial f}{\partial T} = \sum_i \left[\frac{\partial}{\partial v_i} \left[\sum_j \gamma_{ij} \frac{\partial G}{\partial v_j} f \right] + \frac{k_i}{2} \frac{\partial^2 f}{\partial v_i^2} \right], \tag{3.16}$$

the equilibrium solution of which is

$$f(v_1, v_2, \ldots, v_{2n}) = \exp[-\beta G]. \tag{3.17}$$

This is an example of the fluctuation-dissipation theorem, for although the system is nonlinear, the dissipation is linear in the variable x_i. Moreover, since G is a sum function, for each individual variable

$$f(v_i)\, dv_i \sim \exp[-\beta G_i]\, dv_i. \tag{3.18}$$

This result is extremely interesting because it provides some justification for the use of the Gibbs ensemble in the theory of neural nets. It is also possible to obtain the above equilibrium density in the following simple fashion. Suppose a single cell exists in a random environment the effects of which are represented by Gaussian delta-correlated noise. Suppose that the cell is differentially sensitive to such fluctuations, in that it weights them by a factor proportional to its own activity. Let this factor be $x_i(1 - x_i)$. Then the cell's activity may be represented by the nonlinear Langevin equation:

$$dx_i = -\beta x_i\, dT + [x_i(1 - x_i)k]^{\frac{1}{2}}\, d\xi_o(T), \tag{3.19}$$

where $\xi_o(T)$ is standardized Gaussian delta-correlated noise, with zero mean and unit variance. Once again it is easily shown that the equilibrium density for this process is given by expression (3.18) [35]. As a matter of fact the transition density for this process is easily obtained and provides an approximate if rather crude means for estimating the first few moments of the transient response patterns of nerve cells.

Such results suggest a Gibbs ensemble treatment of neural nets. The main results have been presented elsewhere, in considerable detail [28], with a slightly different interpretation of the variable x. The equilibrium density takes the form

$$p(x_i)\,dx_i = \frac{x_i^{p-1}(1 - x_i)^{q-1}dx_i}{B(p,q)} \quad , \tag{3.20}$$

where B(p,q) is Euler's β-function and $p = \beta_i q_i/\theta$, $q = \beta_i(1 - q_i)/\theta$, where θ is the analogue of temperature in the net. It is a measure of the amplitude of fluctuation of activity and is given by

$$\theta = \beta_i \frac{\langle (x_i - q_i)^2 \rangle_{av} / x_i(1 - x_i)}{1 - \langle (x_i - q_i)^2 \rangle_{av} / x_i(1 - x_i)} \quad . \tag{3.21}$$

The smaller θ is, the more regular the activity; the larger θ is, the more "burst-like" it is. The equilibrium density can be used to provide a variety of statistics relating to net activity. Among these are the "interval" density (as seen through some kind of averaging), the amplitude density of membrane potential (seen through the same averager), the mean rate at which fluctuations in activity occur, and certain correlation functions for the joint activities of coupled cells. Thus the interspike interval density is given by

$$p(\tau_i)\,d\tau_i = \frac{r^{-1}(r/\tau_i)^{p+1}(1 - r/\tau_i)^{q-1}}{B(p,q)} \quad . \tag{3.22}$$

The parameters p and q are themselves functions of the parameters q_i and θ, which reflect the cell's environment and its interaction with other cells. Some examples of this are shown in figure 6. It will be seen that the distribution is approximately the Wiebull distribution characteristic

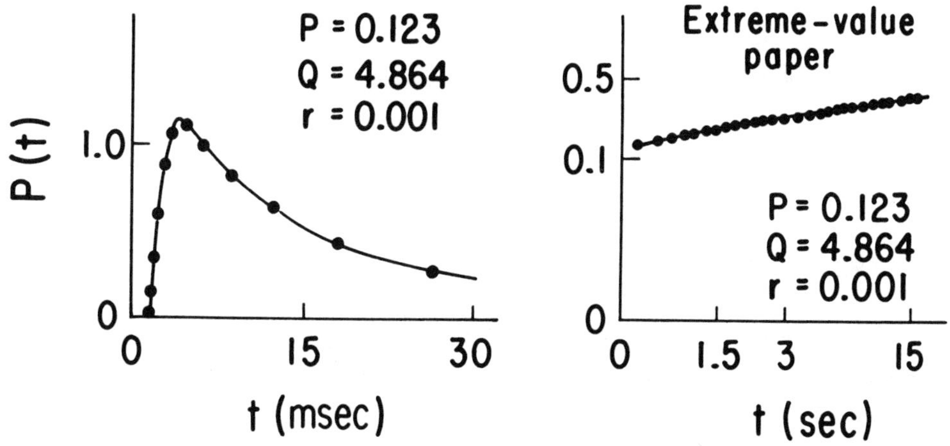

Fig. 6

of extreme-value statistics. The amplitude density of membrane potential can be obtained by way of the equilibrium equation

$$\ln \frac{x_i}{1 - x_i} = \frac{\alpha}{\tau} v_i + \frac{\beta}{\tau} \quad , \tag{3.23}$$

from which it follows that the variable

$$u_i = \exp[\frac{\alpha}{T} v_i + \frac{\beta}{T}]$$

has the density

$$p(u_i)\, du_i = \frac{u_i^{p-1}}{(1+u_i)^{p+q}} \tag{3.24}$$

This will be recognized as a β-density of the second kind, the tails of which are of Pareto type [36], so that linear combinations of such random variables have amplitude densities, the tails of which are also of Pareto type. This is important, for it permits the prediction of relationships between single unit activities and evoked membrane potentials and the electroencephalograms recorded by way of macroelectrodes (EEG), on the obvious assumption that to first order such potentials are the linear combination of cell-membrane potentials.

Another statistical variable of some interest is the mean crossing rate $\langle |\dot{x}_i|\, \delta(x_i - x)\rangle_{av}$, the ensemble average of the rate of crossing of the amplitude x by the i(th) variable x_i. This expression can be derived exactly from the Gibbs theory. It is

$$\langle |\dot{x}_i|\, \delta(x_i - x)\rangle_{av} = 2|\gamma_{ij}|\, x(1-x)p(x)\, q_i(1-q_i)p(q_i)\, \theta \tag{3.25}$$

in the case of two cells. More complicated expressions can be derived for larger nets. This expression provides a means for the estimation of γ_{ij}, the effective interaction coefficients of neural nets. Alternately,

$$\gamma_{ij} = \theta^{-1} \langle v_i \dot{v}_j \rangle_{av} \tag{3.26}$$

so that γ_{ij} is proportional to the cross-correlation of the current driving the j(th) cell and the voltage built up in the i(th) cell.

As the final topic in this section the effects of random parametric excitation on nets will be considered. It is assumed that the input coefficient ϵ_i has superimposed on it a Gaussian delta-correlated fluctuation, the effect of which is to produce random variations in the stationary states q_i. The equations of motion are as before, (3.12). However,

$$G = G_o + \sum_i \frac{\partial G}{\partial q_i} \delta q_i = G_o + G_1, \tag{3.27}$$

where G_o is the unperturbed Liapunov function of (3.13). By using known results for nonlinear Langevin equations [37], the following Fokker-Planck equation can be obtained for the transition density $f[v_1(T), v_2(T), \ldots, v_{2n}(T) | v_1(0), \ldots, v_{2n}(0)] = f(\bar{v}|\bar{v}_o)$. Let

$$\Omega_o = - \sum_i \frac{\partial}{\partial v_i} [\sum_j \gamma_{ij} \frac{\partial G_o}{\partial v_j}] , \tag{3.28}$$

$$\Omega_1 = - \sum_i \frac{\partial}{\partial v_i} [\sum_j (\frac{\gamma_{ij}}{\delta q_j}) \frac{\partial G_1}{\partial v_j}] . \tag{3.29}$$

Let $\bar{v}^o$ (T" - T') be the value of $\bar{v}$ at the instant T' - T" before 0, due to the unperturbed motion;

then

$$\frac{\partial f(\bar{v}|\bar{v}_o)}{\partial T} = \left\{ \Omega_o(\bar{v}) + \int_{-\infty}^{T} \langle \Omega_1[\bar{v}(T)]\, \Omega_1[\bar{v}^{\circ}(T'' - T)] \rangle_c \, dT \right\} f(\bar{v}|\bar{v}_o) \,, \tag{3.30}$$

where $\langle\ \rangle_c$ is the cumulant. It follows that

$$\frac{\partial f}{\partial T} = - \sum_i \left\{ \frac{\partial}{\partial v_i} \left[\sum_j \gamma_{ij} \frac{\partial G_o}{\partial v_j} f \right] + \sum_k \frac{\partial^2}{\partial v_i \partial v_k} \left[\sum_j \frac{\gamma_{ij}\gamma_{kj}}{\delta q_j^2} \left(\frac{\partial G_1}{\partial v_j} \right)^2 f \right] \right\} \tag{3.31}$$

and, using equations (3.13) and (3.27), that [38]:

$$\frac{\partial f}{\partial T} = - \sum_i \left\{ \frac{\partial}{\partial v_i} \left[\sum_j \gamma_{ij} \beta_j \left(\frac{q_j^{o} e^{v_j}}{1 + q_j^{o} e^{v_j}} - q_j^{o} \right) f \right] \right.$$

$$\left. + \sum_k \frac{\partial^2}{\partial v_i \partial v_k} \left[\sum_j \gamma_{ij}\gamma_{kj} \beta_j^2 \left(\frac{e^{v_j}}{(1 + q_j^{o} e^{v_j})^2} - 1 \right)^2 f \right] \right\} . \tag{3.32}$$

Given the equilibrium solution of this equation, the final phase-space density is

$$\Phi(\bar{v}) = \int d\bar{v}_o \, \exp[-\beta G(\bar{v}_o)] \, f(\bar{v}|\bar{v}_o) \,. \tag{3.33}$$

However, the equilibrium solution to equation (3.32) has not yet been found. A rough insight into what is likely to occur can be obtained from the one-dimensional nonlinear Langevin equation:

$$dx = x(1 - x)\, d\xi_o(T) \,, \tag{3.34}$$

the equilibrium transition density of which is

$$f(x|x_o)dx_o = \left(\frac{x}{1-x} \right)^p \left(\frac{x_o}{1 - x_o} \right)^{-p-1} dx_o \,. \tag{3.35}$$

This distribution is singular in x, tending to a δ-function as $x \to 1$. The equilibrium density $\Phi(x)$ behaves in similar fashion. The conclusion to be drawn from this is that many cells coupled together will spend most of their time either completely off, or else firing at their maximum rates. This suggests that the effect of parametric excitation on the net is (paradoxically) to decrease somewhat the overall variability of the activity, and to produce more staccato, burst-like activity in the net.

4. NEURAL FIELD THEORY

The results of section 3 indicate that a very particular set of assumptions is required to obtain what is essentially Hamiltonian dynamics from neural nets. In fact, only the coupling relationship $\alpha_{ij} + \alpha_{ji} = 0,\ i \neq j;\ \alpha_{ii} = 0$, will serve. Any other law will produce either damped behavior--for example, whenever $\alpha_{ii} < 0$, given the antisymmetry of α_{ij}--or else undamped behavior whenever $\alpha_{ii} > 0$. The general conditions can be expressed in various ways, for example, by means of Liapunov theory [34]. The (locally) unstable cases are of great interest, for since the x's are bounded between 0 and 1, limit cycles are found in such cases. An example is seen in the system [39] :

$$\frac{dx_1}{dt} = (5.2 + 14x_1 - 18x_2)\ x_1(1 - x_1)\ ,$$

$$\frac{dx_2}{dt} = (-4.8 + 18x_1 - 4x_2)\ x_2(1 - x_2) \qquad (4.1)$$

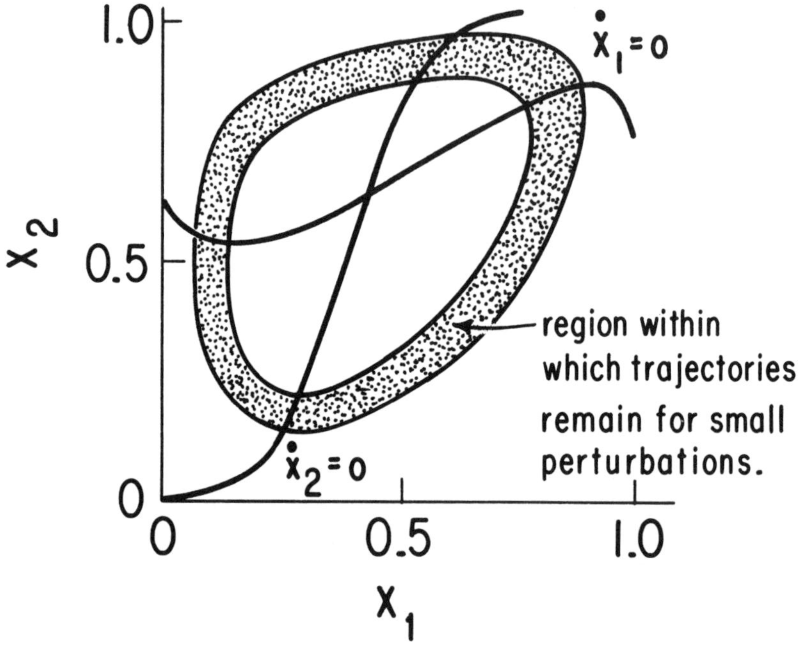

Fig. 7

(see fig. 7). Such results suggest that it would be advantageous to set up a simple formalism which would permit the treatment of all these different cases, in a sufficiently comprehensive fashion. The methods of nonequilibrium statistical mechanics could probably be used to some advantage, except that the sheer complexity of the details would probably vitiate the utility of such an approach.

From a theoretical point of view it seems reasonable to move from a detailed local description of cellular interactions to a more global one in which only the gross statistical features of connectivity are incorporated. Such statistical field theories have been investigated in the past [40-42]. In what follows, a recent attempt will be sketched, the main intention of which is to incorporate both excitatory and inhibitory cell populations in a satisfactory manner [43]. Only the simplest problem will be considered in this paper, that of a homogeneous net of cells, each of which has a fixed threshold θ_i and a membrane response function h(t). The cells are assumed to be randomly distributed with a volume or surface density ρ. Only one-dimensional problems will be considered here. Let $\xi(x)$ therefore be the mean number of fibers from cells in an infinite plane of unit thickness which synapse on a given cell at a distance x from the plane. The function $\xi(x)$ thus represents the global pattern of connectivity of the net. At any instant a proportion of the cells ρS are sensitive and respond to incoming excitation, and the remaining ρR cells have just fired, and are used or refractory. Let F(x,t) represent the proportion of cells becoming active per unit time at the point s, at the instant t; i.e., F(x,t) is the rate at which cells become active. F(x,t) will be called the "activity" of the net, and in the field-theoretic description, such activity forms the "sources" of the field. The field "excitation" is given by

$$\psi(x,t) = \int_{-\infty}^{\infty} F\left(X, t - \frac{|x - X|}{v}\right) \xi(x - X)\, dX \ , \tag{4.2}$$

where v is the velocity of conduction of pulses in the net. The function $\psi(x,t)$ is evidently the rate of arrival of impulses to cells in the plane x, from all other cells in the net. Such excitation in turn generates new sources, so that in general

$$F(x,t) = \psi[F(x,t-\tau),\ \psi(x,t-\tau)] \ , \tag{4.3}$$

where $\psi[\ ]$ is some nonlinear functional of the sources and excitation, and where τ is a delay which takes account of various retardations.

In what follows only temporal activity will be considered, so that the net will be assumed to be isotropic, and the excitation uniform over the net. Then

$$\psi(x,t) = F(t) \int_{-\infty}^{\infty} \xi(x - X)\, dX = kF(t) \ . \tag{4.4}$$

It is also assumed that the excitation varies slowly compared to the membrane time-constant, so that the mean integrated excitation built up in the cells is given by

$$\int_0^t \psi(x,T)\ h(t-T)\, dT = k\ F(t) \int_0^t h(t-T)\, dT.$$

The function h(t) is taken to be a rectangular step of duration s and unit amplitude, whence

$$\int_0^t \psi(x,T)\ h(t-T)\, dT \sim s\ k\ F(t) \ . \tag{4.5}$$

The duration s corresponds to the so-called period of latent addition of the cell. If s is sufficiently small, the proportion of cells which receive at least threshold excitation is substantially independent of whether or not the cells concerned have previously exceeded the threshold. Thus the activity at the instant t is proportional to the rate at which sensitive cells reach threshold at the instant $t - \tau$, and this rate equals the product of two terms, the proportion of cells which are sensitive at $t - \tau$, multiplied by the proportion of cells which receive at least threshold excitation at $t - \tau$. The first term is given by

$$1 - \int_{t-\tau-r}^{t-\tau} F(t')\ dt' \ ,$$

where r is the refractory period of the cells. The second term is essentially the event density (see section 2). It follows from equation (3.3) that the second term is

$$L[\,skF\,] \quad ,$$

so that equation (4.3) takes the form

$$F(t) = \left[1 - \int_{t-\tau-r}^{t-\tau} F(\lambda)\, d\lambda\right] L\,[\,skF(t-\tau)\,] \;. \tag{4.6}$$

If the refractory period r is also of the order of s, then

$$F(t) = [1 - rF(t-\tau)]\; L\;[\,skF(t-\tau)\,] \;;$$

or expanding about zero, assuming the retardation to be small,

$$\frac{\partial F}{\partial T} = -F + (1 - rF)\; L\;[\,skF\,] \;, \tag{4.7}$$

where $T = t/\tau$.

If the net is assumed to consist of two distinct types of cells, excitatory and inhibitory [44], then equation (4.7) gives rise to a pair of coupled equations, of the form

$$\begin{aligned} \frac{\partial E}{\partial T} &= -E + (1 - rE)\; L_e\,[\,s(k_1E - k_2I + P_1)\,]\;, \\ \frac{\partial I}{\partial T} &= -I + (1 - rI)\; L_i\,[\,s(k_3E - k_4I + P_2)]\;, \end{aligned} \tag{4.8}$$

where P_1 and P_2 represent external sources of excitation. These equations are evidently closely related to equations (3.5) which were introduced to deal with the local aspects of net activities. Indeed, starting from equation (3.3) it would be equally appropriate to take as the basic equation

$$\left(\frac{d}{dT} + 1\right) N_i = L\,[\,\sum_j \alpha_{ij} N_j - \delta_i\,] \;, \tag{4.9}$$

in which case the step from a net to a field representation of neural activity becomes obvious.

The dynamics of equations (4.8) are of considerable interest, and are closely related to those of equations (4.9) and (3.5), especially when the refractory period r is small. The stationary states are easily obtained from solutions of the equations.

$$\begin{aligned} \ell n\,\frac{E}{1-(r+1)E} &= sk_1E - sk_2I + sP_1 \;, \\ \ell n\,\frac{I}{1-(r+1)I} &= sk_3E - sk_4I + sP_2 \;. \end{aligned} \tag{4.10}$$

These give rise to various types of singularities in the (E, I) plane, ranging from isolated to multiple stable foci, or else to isolated or multiple unstable foci, separated by limit cycles. Some examples are shown in figure 8. The transient behavior of the function E - I is of particular interest, since it is a measure of the net excitatory activity in the field. This activity ranges from rapidly damped oscillations to maintained limit-cycle oscillations, the particular form being set by the parameters k_i and P_i. Such a function is of particular interest because it provides a model for the generation of spontaneous and evoked potentials. Another important property of these equations is that the frequencies found in limit-cycle oscillations are monotonically related

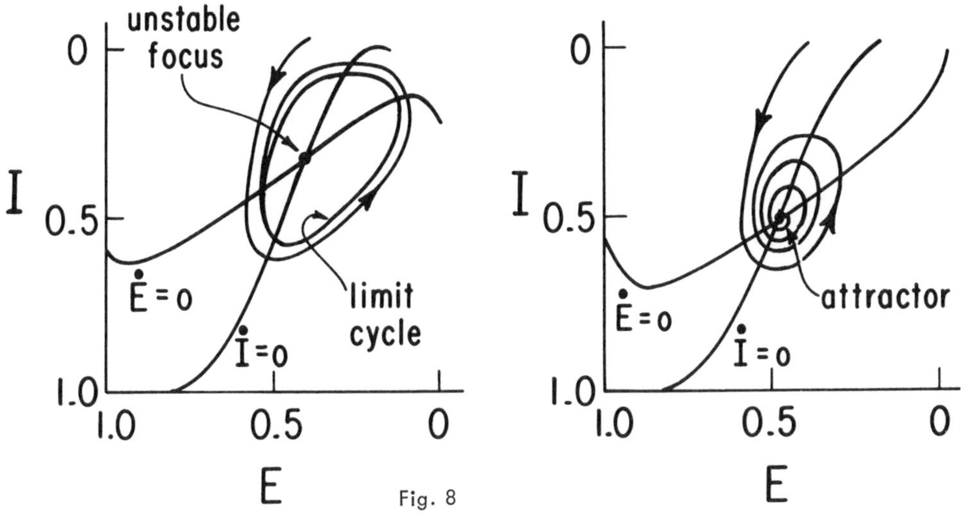

Fig. 8

to the strengths of incoming excitation P_i, in much the same way as is the rate of pulse emission in individual nerve fibers [45]. There is thus coding of stimulus intensity into maintained frequencies of burst activity within the underlying net. The difference between this kind of coding and the burst-coding obtained in section 3 from the Hamiltonian systems is, of course, that the limit-cycle activity is stable, and small perturbations of the parameters will not change the dynamics, except at certain boundaries. Equations (4.8) are therefore "structurally stable" [46], and the stability results from the existence of dissipative rather than conservative dynamics. Structurally stable systems are surely more relevant to neural net theory than conservative, structurally unstable, systems, although it does not follow that the use of the canonical density for averaging is precluded, since as has been indicated, such densities can be obtained from slightly dissipative noisy systems.

As a final point, the connection between the nonlinear field theory which has been outlined and such consequences of neural activity as perception, learning, and memory will be considered. As has been noted, the structural properties of the field dynamics are all highly dependent on parameters, and although the field is structurally stable with respect to small perturbations of the parameters, there are evidently boundaries in the parameter-space which separate the different dynamical possibilities. Any perturbation which takes the system across such boundaries will switch the system from one dynamical regime to another. Such changes are now called "catastrophes." It follows that if the parameters of the field are modifiable, then in principle long-lasting changes can give rise to catastrophes, and so one might expect to observe the sudden emergence or quenching of maintained oscillatory activity in the field, as a result of long-lasting modifications of synaptic connectivity. Now the outcome of the combinatorial approach to neural modeling is a set of predictions concerning which cells are excitatory or inhibitory, and which synapses are modifiable, as well as some account of the overall level of activity to be expected in certain nets, but not a detailed set of predictions about the specific kinds of changes to be found in the overall patterns of activity. Nonlinear field theory, which incorporates modifiable parameters and some of the known details of neuroanatomy, appears to be rich enough to provide a method by which the global effects of local changes in the nervous system can be understood, at least in simple terms.

REFERENCES

1. D. Marr, J. Physiol. 202, 437 (1969).
2. D. Marr, Proc. Roy. Soc. London, B, 176, 161 (1970).
3. D. Marr, Phil. Trans. Roy. Soc. London, B, 262, 23 (1971).
4. D. A. Sholl, The Organization of Cerebral Cortex (London: Methuen, 1953).
5. R. M. Gaze, The Formation of Nerve Connections (New York: Academic, 1970).
6. M. Jacobson, Developmental Neurobiology (New York: Holt, Rinehart, and Winston, 1970).
7. G. L. Gerstein and N. Y.-S. Kiang, Biophysical J. 1, 1, 15 (1960).
8. M. A. Brazier, Acta Physiol. Neerlandica, 6, 692 (1957).
9. D. R. Smith and G. K. Smith, Biophysical J. 5, 1 (1965).
10. G. F. Poggio and L. J. Viernstein, J. Neurophysiol. 6, 517 (1964).
11. L. Lapique, L'Excitabilité en Fonction du Temps (Paris: Hermann, 1926).
12. C. E. Molnar, thesis, Massachusetts Institute of Technology (1966).
13. R. B. Stein, Biophysical J. 5, 173 (1965).
14. M. ten Hoopen, Biophysical J. 6, 4, 435 (1960).
15. N. S. Goel, N. Richter-Dyn, and J. R. Clay (in press).
16. S. E. Fienberg, Kybernetik, 6, 227 (1970).
17. R. M. Capocelli and L. M. Ricciardi, Kybernetik, 8, 214 (1971).
18. G. E. Uhlenbeck and L. S. Ornstein, Phys. Rev. 36, 823 (1930).
19. A. J. F. Siegert, Phys. Rev. 81, 617 (1951).
20. B. K. Roy and D. R. Smith, Bull. Math. Biophys. 31, 2, 341 (1969).
21. J. D. Cowan, in preparation.
22. R. P. Feynman and A. R. Hibbs, Quantum Mechanics and Path Integrals (New York: McGraw-Hill, 1965).
23. M. C. Wang and G. E. Uhlenbeck, Rev. Mod. Phys. 17, 323 (1945).
24. W. M. Siebert, Quart. Progr. Rept., M.I.T., R.L.E., 94, 281 (1969).
25. H. Sugiyama, G. P. Moore, and D. H. Perkel, Mathematical Biosciences 8, 323 (1970).
26. D. R. Cox, Renewal Theory (London: Methuen, 1962).
27. P. I. M. Johannesma, thesis, University of Nijmegen (1969).
28. J. D. Cowan, Lectures on Mathematics in the Life Sciences 2, ed. M. Gerstenhaber (Providence, R.I.: A. M. S., 1970).

29. J. D. Cowan and E. H. Kerner, unpublished.

30. V. Volterra, Lecons sur la theorie mathematique de la lutte pour la vie (Paris: Gauthier-Villars, 1931).

31. E. H. Kerner, Bull. Math. Biophys. 19, 121 (1957); 21, 217 (1959); 23, 141 (1961).

32. E. Leigh, Lectures on Mathematics in the Life Sciences 1, ed. M. Gerstenhaber (Providence, R. I.: A.M.S., 1968).

33. N. S. Goel, S. C. Maitra, and E. W. Montroll, Rev. Mod. Phys. 43, 231 (1971).

34. J. P. LaSalle and S. Lefshetz, Stability by Liapunov's Direct Method (New York: Academic, 1965).

35. J. D. Cowan, unpublished.

36. W. Feller, Introduction to Probability Theory and its Applications, II (New York: Wiley, 1966).

37. R. Kubo, Lecture Notes on Statistical Mechanics, University of Chicago, (1963).

38. J. D. Cowan and T. Sinnema, unpublished.

39. J. D. Aplevich, thesis, University of London (1967).

40. R. L. Beurle, Phil. Trans. Roy. Soc., B, 240, 669 (1956).

41. J. S. Griffith, Bull. Math. Biophys. 25, 111 (1963); 27, 187 (1965).

42. M. ten Hoopen, Cybernetics of Neural Processes, ed. E. R. Caianiello (Rome, Italy: CNR, 1965).

43. H. Wilson and J. D. Cowan, Biophysical J. 1, 1 (1972).

44. J. C. Eccles, The Physiology of Nerve Cells (Baltimore: Johns Hopkins Press, 1957).

45. R. B. Stein, Proc. Roy. Soc. London, B, 167, 64 (1967).

46. R. Thom, Topology, 8, 313 (1969).

DISCUSSION

M. Green: Is it possible to deal with identical neurons as well as networks?

J. Cowan: Yes, and I do so; but one must realize that a single neuron is very complicated--perhaps as complicated as a 704 computer. There is certainly a great amount of work going on at the individual nerve cell level, but the object of the present studies is to obtain an understanding of large-scale neural activity.

G. Nicolis: It seems quite reasonable indeed that, by their very stability, limit cycles should be important in the theory of nervous activity (for a few comments in the same direction, see, e.g., R. Lefever and G. Nicolis, J. Theoret. Biology, 31 [1971]). I would like to ask Prof. Cowan if he can think of experimental tests which would prove that rhythmic activity is best interpreted in terms of limit cycle rather than Volterra-Lotka-type oscillations. I imagine that autocorrelation functions could be very sensitive to the type of periodic behavior and therefore also good candidates for such experimental tests.

J. Cowan: There are various phenomena connected with so-called alpha bursting which hold promise for a determination of the dynamical nature of nervous activity. There is also the system which controls walking, the properties of which appear to fall into the limit-cycle class, and which appears to be eminently suitable for an experimental investigation.

O. Gurel: 1. What is the stability from the outside of the limit cycle? 2. Did you investigate the region of an asymptotic stability corresponding to the Liapunov function obtained by you? If so, did you compare it with the region bounded by the limit cycle?

J. Cowan: We have not investigated these questions, since we are interested only in qualitative, not quantitative, answers, as yet.

J. Levett: You have given some indications of thalamocortical interactions. Have you applied your approach to the visual or other systems? For example, in the visual system, ganglion cells output spike potentials which drive higher centers. Likewise the retinal ganglion cells are driven by more peripheral cells than photoreceptors and bipolars (as well as others) which seem to have no action potentials associated with them. The retina also receives centrifugal fibers on feedback from higher centers. Would you comment on this system in the light of your talk?

J. Cowan: We are presently investigating the properties of what is essentially a model of some of the neural populations you mentioned. Preliminary studies by Dr. Wilson and myself indicate that the nonlinear field model furnishes an economical description of important features of the visual system.

E. M. Simon: For electrophysiological experiments the electrical discharge pattern has a lot of fine structure. This detail in spectral pattern should contain microscopic information regarding the nature of nerve transmission. Therefore, physicists should investigate the correlation function, etc., and do more careful spectral analysis to arrive at more detailed models.

J. Cowan: To my knowledge little substantive work has been done in this direction.

CELL MIGRATION AND THE CONTROL OF DEVELOPMENT

Morrel H. Cohen and Anthony Robertson

I. INTRODUCTION: DEVELOPMENT AND ITS CONTROL

In the development of multicellular animals, in particular the Metazoa, a fertilized egg transforms into an adult. Its mass can increase by a factor of 10^{10}. One cell can become 10^{12} cells, grouped into 10^{2} cell types and arranged into complex structures. At the cellular level, the transformation is effected by at least six basic processes: nuclear or cell division, cell growth, cell movement, intercellular contact formation (with or without cytoplasmic fusion), cellular secretion, and cell differentiation including cell death. These are the basic building blocks of development when it is viewed from above the cellular level. From that viewpoint, they are unitary and intrinsically simple events. It is their control which leads to the highly standardized adult form characteristic of a particular genome. To understand Metazoan development, we must therefore understand developmental control processes.

The basic processes occur according to varying schedules at various parts of the embryo. Their integration into reliable development implies the existence of a system controlling both the sequences and positions of events. There must be spatial and temporal control of the processes of development at the multicellular level. This must proceed through cellular interactions involving the cytoplasm. However, it is difficult to make inferences directly from experience as to the nature of developmental control systems. The facts of development are so complex and so diverse as to require a unifying framework before they can be selected for ordering. Models of developmental control systems sufficiently subtle and general are a necessary first step, and these are beginning to emerge [1-4].

We call a collection of coupled or interacting cells, controlled by the same system, a field. In a quite simple yet general model [3], we have supposed there to be a subset of cell state variables S, relevant to the control of development. As the cells are coupled, one possibility

---

Supported in part by the Alfred P. Sloan Foundation and the Otho S. A. Sprague Memorial Institute.

is that S is constant and time independent throughout the field, $S = S_o$. The field could not then be guided toward development of ordered heterogeneity by S. Instead, the steady and uniform state of the field S_o is unstable with respect to a state $S = S(\underset{\sim}{r}, t)$ which depends on cell position $\underset{\sim}{r}$ within the field and on time t because of the interplay between the internal dynamics of each cell and the coupling between neighboring cells. The position dependence of S provides positional information [4] or a map [2] for the cells in the field, and the time dependence of S provides a clock [2].

Models for the control of development can be classified according to the nature of the position and time dependence of S. Variation with $\underset{\sim}{r}$ can, broadly speaking, be graded or periodic and, with t, quasi-static or periodic, giving four categories in all. For example, in the simplest quasi-static gradient model, positional information (a map) is provided by a gradient in concentration of a morphogen set up by diffusion from source to sink cells [5]. Values within the gradient are read by cells whose fate they determine. There is no clock in this model. Gradients are undoubtedly important in the control of development, although some difficulties are encountered in setting up gradients that regulate in order to produce size-independent patterns or that produce spatially periodic structures. These related matters have been discussed by Turing [6], by Prigogine, Nicolis, and coworkers [7], by Wolpert [4], by Crick [5], and by Cohen [2,3].

Alternatively, Goodwin and Cohen [1] have constructed a periodic gradient model, the phase-shift or thunder-and-lightning model. In this model, there is a periodic propagation of a brief signal throughout a developing tissue from a pacemaker, or highest-frequency, region. Position is read as phase difference between two events, either two signals propagating at different velocities [1] or one propagating event and one local event [2]. The signals are provided by oscillators within cells which are autonomous unless the cells are coupled by, for example, low-resistance junctions. When the cells are coupled, a signal from a high-frequency oscillator entrains those of lower frequency, stimulating them to signal earlier than they would have if autonomous. In this way a propagating wave can be set up. Its propagation depends on the existence of both a frequency gradient and a period of refractoriness after stimulation. Details may be found in Goodwin and Cohen [1] (see also section V.B). This kind of model naturally provides a clock; indeed, it is based on one.

The first task faced in analyzing the system controlling a particular set of developmental events is deciding which category of model to use. Categorization of the spatial dependence of the control system is normally easy: a graded pattern presumably developed under the control of a gradient, etc. One cannot, however, decide on the time dependence of controlling agents, quasi-static or periodic, on a post hoc basis. Continual monitoring is necessary, with the interval between successive observations short compared to the period, to distinguish quasi-static from periodic phenomena. Before experiments capable of distinguishing the two categories can be designed, one must know at least the order of magnitude of the expected period T.

Goodwin and Cohen [1] have argued that T should be of order minutes. For the control system to operate effectively as a clock, T must be small on the developmental time scale T_D. For it to provide positional information, it must be possible to resolve many subintervals of time within T. Since time resolution is limited to the molecular time scale T_μ, we have

$$T_\mu < T < T_D , \qquad (I.1)$$

which is satisfied by putting

$$T \simeq (T_\mu T_D)^{\frac{1}{2}} , \qquad (I.2)$$

As T_D is of order hours and T_μ of order seconds, equation (I.2) implies that T is of order minutes.

Division and differentiation are too slow to provide information on a time scale of minutes. Contact formation involves brief events, isolated in time, and is therefore unsuitable. Movement, certain kinds of specialized growth, and secretion, however, are all continuous processes. They can be monitored by time-lapse microphotography and quantitatively analyzed. Specialized growth, as of nerve axons, and secretion are often movement related. The main focus of this paper is therefore on morphogenetic movement and its special role in revealing the dynamical character of developmental control.

However, it would be gratuitous to suggest time-lapse studies of cell migration as the means of classifying developmental control systems as quasi-static or periodic were there not already evidence for the existence of periodic morphogenetic movements and other periodic developmental events. The known examples have already been collected [3] and are reproduced in Table 1 with a few additions [8-19].

Having made the point in this section that observations of cell migration or movement can reveal fundamental aspects of the dynamics of developmental control systems, we go on, in section II, to describe kinds and examples of morphogenetic movement. A qualitative discussion of the movement of isolated cells is given in section III. A new quantitative theory which explores the effects of inhibition on the movement of single cells and on chemotaxis is reported briefly in section IV. The development of the cellular slime molds illustrates well the points made in sections I-IV and is of considerable interest in itself. The life cycle of these slime molds is described in section V, emphasis being put on morphogenetic movement and its control. Finally, in section VI, parallels between slime-mold development and that of more complex organisms are pointed out and speculations are made on the existence of a homology between functional and developmental systems.

II. KINDS AND EXAMPLES OF MORPHOGENETIC MOVEMENT

We have grouped the morphogenetic movements known to us into five categories:

1. Local rearrangements.--Such movements occur, for example, during early cleavage and the subsequent formation of the blastula. Each cell remains in its original neighborhood; small, local movements during and immediately after each cleavage or division result ultimately in the shell structure of the blastula. This is well illustrated for the sea cucumber by figure 87 of Balinsky [20].

Table 1

Periodic Events During Development

Organism	Event	Period	Observer
Cellular Slime Molds:			
D. discoideum	Aggregation	5 m	Shaffer [8]
D. mucoroides	Aggregation	5 m	Arndt [9]
D. rosarium	Aggregation	5 m	Konijn, unpublished
D. purpureum	Aggregation	Minutes	A. Robertson, unpublished
P. violaceum	Aggregation	1.5 m	Cohen and Robertson [10]
P. pallidum	Aggregation	1.5 m	A. Robertson, unpublished
D. discoideum	Migration	5 m	Robertson [11]
D. discoideum	Culmination	5 m	Robertson [11]
Amphibia:			
Triturus alpestris	Closure of neural fold	5 m	Selman [12] and private communication
Ambystoma mexicanum	Closure of neural fold	30 m	Selman [12] and private communication
Bufo vulgaris	Movement of tug cells	10 m	Sirakami [13]
Xenopus	Circus movement, disaggregated gastrulae cells	90 s	Holtfreter [14] A. S. G. Curtis private communication
Chick	Migration of heart mesoderm	10 m	R. L. De Haan private communication
	Movement of area pellucida-opaca boundary	2 m	R. L. De Haan private communication
	Membrane ruffling in fibroblasts	3 m 30-60 s (37° C)	Ingram [15] A.S.G. Curtis private communication
	Gross muscular activity from day 8	Minutes	Hamburger et al. [16]
	Growth of neurons	Minutes	Weiss [17]
Honey bee	Gross muscular activity in larvae	Minutes	Du Praw [18]
Locust	Chitin deposition	15-30 m	Neville [19]

2. Free movements of individual cells.--Cells initially in close contact with their neighbors can become detached and move onto the surface of a tissue or into the fluid of an interior cavity. Subsequent movements are those of a free cell either over a surface or through fluid. Aggregative movements of the cellular slime molds over a substrate are prime examples; these will be discussed in later sections. Another example is formation of hair-follicle rudiments by a similar process of aggregation over the surface of the epithelium, as illustrated for the mouse in figure 215 of

Balinsky [20]. Chick-heart mesoderm cells migrate over the surface of the epiblast to form the heart rudiments (Table 1). Examples of cells moving freely through a cavity are the primary mesenchyme cells in the sea urchin during gastrulation (see below), and the mesodermal and endodermal cells migrating from the primitive streak (fig. 150, Balinsky [20]) of the chick embryo.

3. Coordinated movements of cell groups.--These can be one-dimensional--i.e., movements of strings or streams of cells--or two-dimensional--i.e., movements of sheets of cells. Those cellular slime molds with periodic aggregative momements, of Table 1 and section V, exhibit stream formation and movement. Far more common in development is the movement of sheets of cells. A process called epiboly (Balinsky [20], p. 218) occurs in fish. In it, a sheet of cells sweeps down over the yolk of the egg. In amphibia and other organisms the next stage of development beyond the blastula is the gastrula, as shown for a frog in figure 137 of Balinsky. The process of gastrulation is initiated by an inward dimpling of the blastula at the blastopore, followed by an invagination which is accompanied by large-scale migrations and foldings of sheets of cells forming specific portions of the outer surface of the blastula into specific interior positions (fig. 140, Balinsky [20]). Neurulation follows gastrulation. During it a region of negative curvature is first established on the surface of the gastrula; crests rise at its boundaries and then close to form the neural tube in movements which have been observed to be periodic in at least two cases (Table 1) (cf. figs. 141 and 143 of Balinsky [20]; fig. 12.1 of Trinkhaus [21]). The movements, grossly speaking, are those of sheets. Further details of the movements underlying both gastrulation and neurulation are discussed below.

4. Cell processes, with cell body stationary.--These are most beautifully illustrated during gastrulation of the sea urchin. After the initial invagination of the blastula and detachment of the primary mesenchyme cells into the blastocoel (the cavity within the blastula), secondary mesenchyme cells differentiate on the inner surface of the bulge and extend long, pseudopodal processes while themselves remaining stationary (Gustafson and Wolpert [22], Gustafson [23]). These extend across the blastocoel, divide into finer, filopodal processes, and anchor by making specific contacts at cell junctions on the opposite wall (Gustafson [23]). The original processes contract, pulling the bulge into the blastocoel and forming the primitive gut and coelom. The process occurs in amphibia, and has been observed to be periodic in <u>Bufo vulgaris</u> (Table 1, ref. 10). The growth of axons proceeds in a remarkably similar fashion, resulting, however, in a permanent fiber, and not an organelle used once for contraction. Periodicity is observed in axonal growth as well (Table 1, ref. [17]).

5. Changes in shape leading to morphogenetic movements.--The change in sign of the curvature of the blastula surface in the initiation of gastrulation and of the gastrula surface in the initiation of neurulation are caused by changes in cell shape. In neurulation the cells become progressively more bottle shaped, fatter at the inner surface than at the outer. As close contact is maintained between lateral neighbors, this leads immediately to the negative curvature (Trinkhaus [21], fig. 12.2). This "bottling," in turn, is caused by formation and contraction of

fibers running around the cell under its outer surface.

All of the movements 1-5 listed above strongly resemble the movements of free amoeboidal cells, with varying emphasis on different aspects of the entire complex series of subevents comprising a single motile event. Indeed, in case 5 only one aspect, a particular kind of contraction, is involved. Accordingly, a natural next step in understanding morphogenetic movement would be to understand the amoeboidal movement of free cells. Because the cellular slime molds have both a free-moving amoeboidal phase and a multicellular development of considerable interest, we have focused on amoeboidal movement in these organisms, particularly in Dictyostelium discoideum and D. minutum. We report some of our observations in the next section and go on to quantitative analysis in the subsequent section.

III. AMOEBOIDAL MOVEMENT IN THE CELLULAR SLIME MOLDS

The amoebae of Dictyostelium discoideum, a species of cellular slime mold, have a vegetative or feeding phase during which they move over their substrate, find bacteria by chemotaxis [24], and feed on them. Their shape is normally quite irregular because of the mechanism of their movement described below. Nevertheless, they are, on average, isotropic, with the diameter of their periphery in contact with the substrate being about 10 μ. They are classified as amoebae [25], and their motion is typically amoeboid.

Their movement appears to us to be a consequence of a random sequence of separately identifiable motile events. Each motile event is quite complex. We have observed via time-lapse microphotography the following set of subevents to occur within each motile event. We frequently see movement start with the extension of a slender, tapered filopod 0.5 μ in average diameter and 5-10 μ long. We conjecture that filopod extension is normally the first step, although we have not yet demonstrated this conclusively. Next, the filopod waves about. Nothing further occurs unless the tip of the filopod anchors itself to the substrate. This appears to initiate pseudopod formation, in which cytoplasmic flow causes an abrupt and rapid ballooning out of the cell membrane around the filopod, which disappears from view as the pseudopod is formed. At this point the significant movement of the center of gravity of the cell takes place, and it becomes very difficult to separate the process into its components without having made quantitative measurements of the relative motion of different parts of the cell. There appear to be simultaneously a contraction of the filopod and a cytoplasmic flow into it. This process appears to be followed by a transverse contraction of the cell at the pole opposite to the site of filopod initiation, which impels the cytoplasm into the laterally expanded and axially contracted pseudopodal region.

As movement proceeds, filopods frequently reappear at the rear end of the amoeba, still anchored to the substrate. They may either pull away from the cell, be dragged along behind the cell, or be retracted or resorbed; all cases have been observed. This continuity of filopod anchoring during movement and the observation of as many as five or six pseudopods simultaneously active suggests that membrane dynamics are quite complex during movement. Either a crystalline ordering of the membrane subunits with a low resistance to shear deformation or a liquid crystal ordering which permits flow is indicated. Experiments with marked membranes would be profitable.

The direction of a movement step is basically the direction of contraction and flow of the cell body. That in turn is the direction of the axis of the pseudopod, which in turn is the direction in which the filopod was originally extended. We have observed that the filopod emerges normal to the membrane with an rms deviation of 6° [26]. Thus a motile event initiated at a certain point on the membrane with the emergence of a filopod in sum consists of a movement step in the direction of the normal to the membrane periphery at that point.

Our observations also suggest that at any moment there are specific sites distributed around the membrane periphery at which filopod formation, and hence an entire movement step, can be initiated. The number of these is probably no more than the ratio of the membrane perimeter to the filopod base diameter, about 30 to 60. Using this idea and the observations reported above, one can construct a simple and appealing model of chemotaxis for both the vegetative and the aggregative cells [10].

We shall consider aggregative cells first. These, as discussed in section V, have passed through a period of differentiation called interphase and have become polar. Movement is initiated only over about one-third of the membrane periphery at one end of the cell. Moreover, in fully differentiated cells, movement occurs only in response to a chemotactic agent, 3'5' cyclic AMP (cAMP). The cAMP is released periodically by a neighboring cell, probably in a brief burst, and cannot remain in solution for more than about l second before it is degraded by an extracellular enzyme [27].

On the other hand, the movement step initiated by the cAMP endures about 100 seconds. We conclude that the motile event is initiated by a superthreshold concentration of cAMP [10]. At the point on the membrane periphery which first encounters the threshold concentration, the membrane periphery osculates the threshold concentration contour. The membrane normal coincides with the direction of the signaling neighbor, and movement automatically proceeds toward the source of the chemotactic signal.

The chemotactic movement of amoebae is more complex. First, in the absence of a chemotactic agent, there is a certain spontaneous probability per unit time that a movement step is initiated at each site. This is the same for all sites. The number of pseudopods observed simultaneously is never more than of order 10 per cent of the estimated number of sites, and these tend to be well spaced. Therefore, we suppose here, and elaborate in the next section, that once pseudopod formation is initiated, filopod formation is inhibited in neighboring sites. A chemotactic agent can be expected to increase (decrease) the probability per unit time that a given site initiates filopod formation. A nonuniform distribution of chemotactic agent leads to movement up (down) the concentration gradient as in the model of Keller and Segel [28] or of Patlak [29]. However, the influence of inhibition, ignored by earlier authors, leads to a closer connection with experiment, as discussed in the next section.

IV. AN EXCITATION-INHIBITION MODEL OF THE MOVEMENT OF ISOLATED CELLS

The discussion of amoeboidal movement in section III should convey at the very least that it is a complex, multistage process. The various categories of morphogenetic movement described in section II appear to have some or all of the elements of free amoeboidal movement elaborated in that section. Moreover, the constraints imposed on cells moving in coordinated groups by coupling to their neighbors make the movement of single cells far easier to analyze. It seems sensible, therefore, to essay a quantitative theory of free amoeboidal movement (two-dimensional movement over a surface) as a preliminary to the more difficult quantitative study of morphogenetic movement.

Morphogenetic movements of categories 2, 3, and 4 are those which relate more closely to free amoeboidal movement. These are all directed movements: movements from specific origins toward specific goals. They may or may not proceed by chemotaxis, but chemotaxis provides a suitable example of the directed movement of free amoebae and we study it here. By chemotaxis is meant movement toward (positive chemotaxis) or away from (negative chemotaxis) a source of a chemical attractant or repellant [24]. For definiteness we consider the case of an attractant only.

The classical theory of amoeboidal movement, as illustrated in the work of Keller and Segel [28] and of Patlak [29], may be expressed within our present frame of reference as follows. Let the number of amoebae per unit area of substrate be $N(\underset{\sim}{r},t)$ at point $\underset{\sim}{r}$ and time t, presuming that the distance scale over which N varies is large relative to the amoeba separation and, therefore, the amoeba size. Suppose that within each amoeba there is a well-defined marker for its position, e.g., its center of gravity or the center of its nucleus. The movement process described in section III results in displacements $\underset{\sim}{\rho}$ of the marker from its initial position $\underset{\sim}{r}$. These occur with a probability $W(\underset{\sim}{r},t;\underset{\sim}{\rho})d^2\rho$ per unit time, of a step $\underset{\sim}{\rho}$ into $d^2\underset{\sim}{\rho}$. W would be isotropic in $\underset{\sim}{\rho}$, and time and position independent in the absence of the chemotactic agent, the concentration of which is $C(\underset{\sim}{r},t)$ in molecules per unit area.

The amoebae thus undergo a random walk characterized by a drift velocity

$$\underset{\sim}{v}(\underset{\sim}{r},t) = \int d^2\rho\ \underset{\sim}{\rho}\ W(\underset{\sim}{r},t;\underset{\sim}{\rho}) \tag{IV.1}$$

and a diffusion coefficient

$$\underset{\sim}{D}(\underset{\sim}{r},t) = 1/2 \int d^2\rho\ \underset{\sim}{\rho}\underset{\sim}{\rho}\ W(\underset{\sim}{r},t;\underset{\sim}{\rho})\ . \tag{IV.2}$$

Consequently, the amoeba density obeys the equation

$$\frac{\partial N}{\partial t}(\underset{\sim}{r},t) = \underset{\sim}{\nabla}\underset{\sim}{\nabla}:\ [\underset{\sim}{D}(\underset{\sim}{r},t)\,N(\underset{\sim}{r},t)] - \underset{\sim}{\nabla}\cdot[\underset{\sim}{v}(\underset{\sim}{r},t)N(\underset{\sim}{r},t)]\ . \tag{IV.3}$$

To proceed further, one must know the dependence of W on $C(\underset{\sim}{r},t)$. We suppose that there are specific sites, closely spaced around the periphery of the amoeba membrane near the substrate, from which filopods emerge and initiate the movement step. Let the frequency of initiation of movement steps from a site at point $\underset{\sim}{r}'$ per unit length of membrane periphery, g, depend only on the concentration of attractant $C(\underset{\sim}{r}', t)$ at that point:

$$g(\underset{\sim}{r}',t) = g[C(\underset{\sim}{r}',t)]. \qquad (IV.4)$$

Let the probability distribution of step lengths $f(\underset{\sim}{r}', \underset{\sim}{\rho})$ from a site at $\underset{\sim}{r}'$ be unaffected by the presence of the attractant. Let ds be an element of the periphery and $\underset{\sim}{a}(s)$ be the displacement of the point $\underset{\sim}{r}'$ on the periphery from the position of the marker, $\underset{\sim}{r}$. It follows that

$$W(\underset{\sim}{r},t;\underset{\sim}{\rho}) = \int ds\, g\{C[\underset{\sim}{r} + \underset{\sim}{a}(s), t]\}\ f[\underset{\sim}{r} + \underset{\sim}{a}(s), \underset{\sim}{\rho}] . \qquad (IV.5)$$

We suppose that a typical amoeba radius a is small compared with the distances over which C varies appreciably with $\underset{\sim}{r}$, expand, and keep only lowest orders in a in evaluating W, $\underset{\sim}{v}$, and $\underset{\sim}{D}$:

$$C[\underset{\sim}{r} + \underset{\sim}{a}(s), t] = C(\underset{\sim}{r},t) + \underset{\sim}{a}(s) \cdot \underset{\sim}{\nabla} C(\underset{\sim}{r},t) . \qquad (IV.6)$$

We now suppose that the step-length distribution is independent of angle and that the direction distribution is sharply peaked about the direction of the local normal, $\hat{n}(s)$:

$$f[\underset{\sim}{r} + \underset{\sim}{a}(s), \underset{\sim}{\rho}] = f_1(\rho)\ f_2[\psi(s)] , \qquad (IV.7)$$

where $\psi = \cos^{-1}\ \hat{n}(s)\cdot\hat{\rho}$ and f_2 is sharply peaked about $\psi = 0$. This gives

$$W(\underset{\sim}{r},t;\underset{\sim}{\rho}) = W_o(\underset{\sim}{r},t;\underset{\sim}{\rho}) + W_1(\underset{\sim}{r},t;\underset{\sim}{\rho}), \qquad (IV.8)$$

where

$$W_o(\underset{\sim}{r},t;\underset{\sim}{\rho}) = g[C(\underset{\sim}{r},t)]\ f_1(\rho) \int ds\, f_2\,[\psi(s)] \qquad (IV.9)$$

and

$$W_1(\underset{\sim}{r},t;\underset{\sim}{\rho}) = \frac{dg[C(\underset{\sim}{r},t)]}{dC}\ f_1(\rho)\ \underset{\sim}{\nabla}\ C(\underset{\sim}{r},t) \cdot \int ds\ \underset{\sim}{a}(s)\ f_2[\psi(s)] . \qquad (IV.10)$$

The membrane configuration is random; equations (IV.9) and (IV.10) must therefore be averaged over membrane configurations as well. We replace the average over configurations by an average, typical, or most probable configuration, which we take as circular, with radius a. The integrals (IV.9) and (IV.10) can immediately be carried out with the results:

$$W_o(\underset{\sim}{r},t;\underset{\sim}{\rho}) = \nu[C(\underset{\sim}{r},t)]\ \frac{f_1(\rho)}{2\pi} , \qquad (IV.11)$$

$$W_1(\underset{\sim}{r},t;\underset{\sim}{\rho}) = \frac{d\nu\,[C(\underset{\sim}{r},t)]}{dC}\ a\underset{\sim}{\nabla}\ C(\underset{\sim}{r},t) \cdot \frac{\hat{\rho} f_1(\rho)}{2\pi} \qquad (IV.12)$$

where

$$\nu[C(\underset{\sim}{r},t)] = 2\pi a g\,[C(\underset{\sim}{r},t)] \qquad (IV.13)$$

is the total step frequency for an amoeba at $\underset{\sim}{r}$,t.

We can now evaluate equations (IV.1) and (IV.2) with the result that, to lowest order in a, $\underset{\sim}{v}$ is given by W_1 and $\underset{\sim}{D}$ by W_o:

$$\underset{\sim}{v} = A\,\underset{\sim}{\nabla} C ; \qquad (IV.14)$$

$$\underset{\sim}{D} = D\,\underset{\sim}{I} ; \qquad (IV.15)$$

$$A = 2(\nu'/\nu)\,D', \qquad \nu' = d\nu/dC ; \qquad (IV.16)$$

$$D = \tfrac{1}{4}\,\nu\,\langle\rho^2\rangle, \qquad \langle\rho^2\rangle = \int d\rho\ \rho^3\ f_1(\rho); \qquad (IV.17)$$

$$D' = \tfrac{1}{4}\,\nu a\,\langle\rho\rangle, \qquad \langle\rho\rangle = \int d\rho\ \rho^2\ f_1(\rho) . \qquad (IV.18)$$

We suppose that the effect of the chemotactic agent is to stimulate the production of filopods, i.e., to increase ν monotonically from some spontaneous value ν_o in its absence to a saturating ν_m, as in figure 1. The quantity ν' is therefore positive, and the amoebae drift up the concentration

k

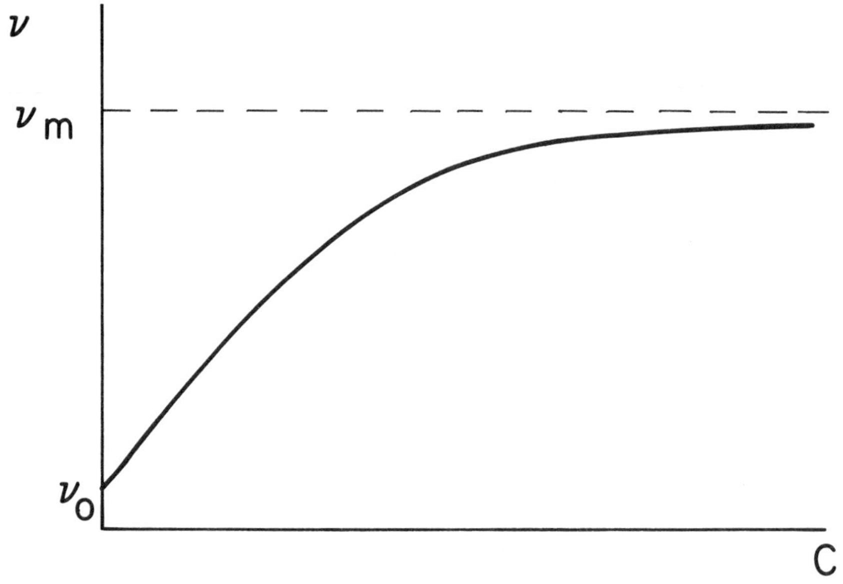

Fig. 1. Hypothetical dependence of movement step frequency ν on concentration C of chemotactic agent. ν_o is the frequency of spontaneous activity, and ν_m is the maximum frequency at which movement steps can be stimulated by the attractant.

gradient, according to equations (IV.14) and (IV.16). Moreover, when C becomes large enough that $\nu \to \nu_m$ and ν'/ν becomes small, chemotaxis ceases. Both results are in accord with experiment.

Nevertheless, the theory sketched above has two serious deficiencies. First, we note that D, a direct measure of the randomness of the movement, increases monotonically with C. In the aggregation of D. minutum, however, Gerisch's observations show that the random component of the movement of an individual amoeba decreases dramatically as the aggregate, the source of attractant, is approached [30]. We have repeated Gerisch's observations and have found the same effect. Moreover, in those morphogenetic movements we have observed to be well directed, the random component is significantly smaller than would be consistent with the above theory. Movements during gastrulation and epiboly and migration of precardiac mesoderm in the chick (cf. time-lapse film of R. L. De Haan) are good examples. Second, there are relatively few pseudopods occuring simultaneously (1-6 typically), and these are well spread out over the membrane periphery. This suggests that there is an interaction between sites, which is missing from the above theory, and that it is inhibitory in nature. After a movement step is initiated at one site, movement steps are inhibited for a certain period T_I at neighboring sites.

The problem of many interacting sites is too difficult to be the subject of this initial study of the role of inhibition in amoeboidal movement. We treat the problem of a two-site amoeba moving in one dimension (fig. 2). Suppose the step length to be sharply distributed

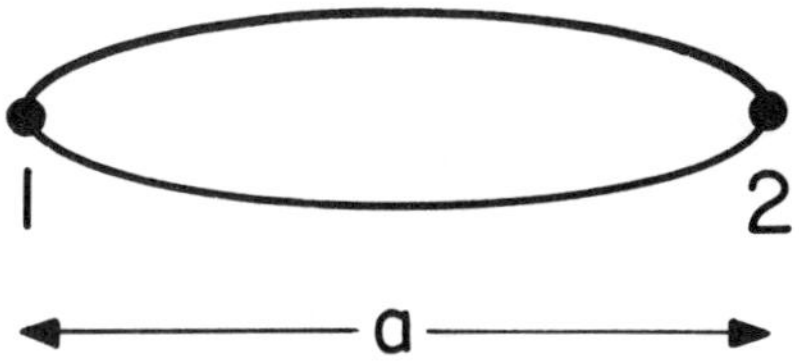

Fig. 2. An amoeba of length a with two sites 1 and 2 at which movement steps are initiated. Movement is one-dimensional parallel to the axis joining the sites in steps - u for 1 and u for 2.

about u. Then, the expressions for D and C corresponding to equations (IV.14) - (IV.18) are

$$D = \tfrac{1}{2}(\nu_1 + \nu_2)\, u^2 \qquad \text{(IV.19)}$$

and

$$v = (\nu_1 - \nu_2)u = \frac{d\nu}{dC}\frac{dC}{dx}\, au \,, \qquad A = \nu' au \qquad \text{(IV.20)}$$

where ν_1 and ν_2 are the excitation frequencies at the two sites and a is the separation between them.

Before adding inhibition, it is necessary to go over to a more detailed statistical characterization of the random excitation processes at each site than is provided by specifying the mean rate $\bar{\nu}$. Let τ be the interval between the initiation of two successive movement steps at each site. Let $P(\tau)$ be the interval distribution. It follows that

$$\bar{\nu} = \frac{1}{\bar{\tau}} \,, \quad \bar{\tau} = \int_0^\infty P(\tau)\, \tau \, d\tau \,. \qquad \text{(IV.21)}$$

We suppose there to be a refractory period T_r. No step can follow another in less time than T_r, so that

$$P(\tau) = 0 \,, \quad \tau < T_r \,. \qquad \text{(IV.22a)}$$

We suppose the excitation frequency to be ν so that

$$\frac{dP}{d\tau} = -\nu P \,, \quad t > T_r \,, \qquad \text{(IV.22b)}$$

or

$$P = \nu \exp[-\nu(t - T_r)] \,, \quad t > T_r \,. \qquad \text{(IV.22c)}$$

From equation (IV.22) it follows that

$$\bar{\nu} = \frac{\nu}{1 + \nu T_r} \,, \qquad \text{(IV.23)}$$

so that $\bar{\nu}$ increases from ν to the limiting value of $1/T_r$ as the excitation frequency passes from $\ll 1/T_r$ to $\gg 1/T_r$ with increasing attractant concentration.

Now let us suppose that inhibition is present. One site cannot initiate a movement step if the other site has done so, until a time T_I has passed, when the inhibited site becomes excitable again. Equations (IV.22) must be modified to read

$$P(\tau) = 0, \quad \tau < T_r , \tag{IV.24a}$$

$$\frac{dP(\tau)}{d\tau} = -\nu P(\tau) \; \mathbb{P}(t) , \quad \tau > T_r , \tag{IV.24b}$$

where $\mathbb{P}(\tau)$ is the probability that the site is sensitive, i.e., that the other site has not been active within a time T_I. The coupled pair of equations (IV.24) for sites 1 and 2 can be solved exactly for the two interval distributions. The process is tedious and the expressions for the solutions complicated. The analysis will be published elsewhere, and only the results will be quoted here. First, no essential change in the motion results when $T_I \le T_r$. Once again, an increase in concentration of chemotactic agent leads to an increase in the random component of the movement. However if T_I exceeds T_r, and if ν_1 exceeds ν_2, then site 1 can, with some appreciable probability, initiate a second movement step while site 2 is still inhibited. Thus site 1 can continue to initiate steps at a mean frequency approaching $1/T_r$ and therefore quite periodically, while ν_2 remains inactive for long periods. Steady motion up the gradient results. The detailed conditions required are

$$\nu_{1,2} T_r \gg 1, \quad \nu_{1,2}(T_I - T_r) > 1 ,$$

and (IV.25)

$$(\nu_1 - \nu_2)(T_I - T_r) \gtrsim 1 ,$$

from the detailed analysis, and examination of the interval distributions confirms the above picture.

In conclusion, it appears likely that, through inhibition, nonpolar cells can have directed movement without a strong random component even when excited by a high concentration of attractant.

V. THE CELLULAR SLIME MOLDS

The cellular slime molds exhibit the six cellular processes of section I in their development in particularly simple ways, often well separated in space and in time. The existence of developmental control systems has been demonstrated for them and some of the features elucidated, including the presence of both categories of time dependence, smooth and periodic. Their life cycle is simple and suitable for investigation by time-lapse microphotography, and events during it are readily quantifiable. Accordingly, we, following many other developmental biologists, have selected them for experimental studies of the control of development. We here review briefly the available relevant information. For further information see Bonner's book [25] and the review articles of Shaffer [31] and of Gerisch [30]. We intercalate our own view of the theoretical interpretation.

A. Life Cycle

More is known about Dictyostelium discoideum than any other species; we therefore describe its life cycle (see fig. 3). Given certain environmental conditions of humidity and chemical composition, the elliptical, polysaccharide-coated spores germinate [32]. The process

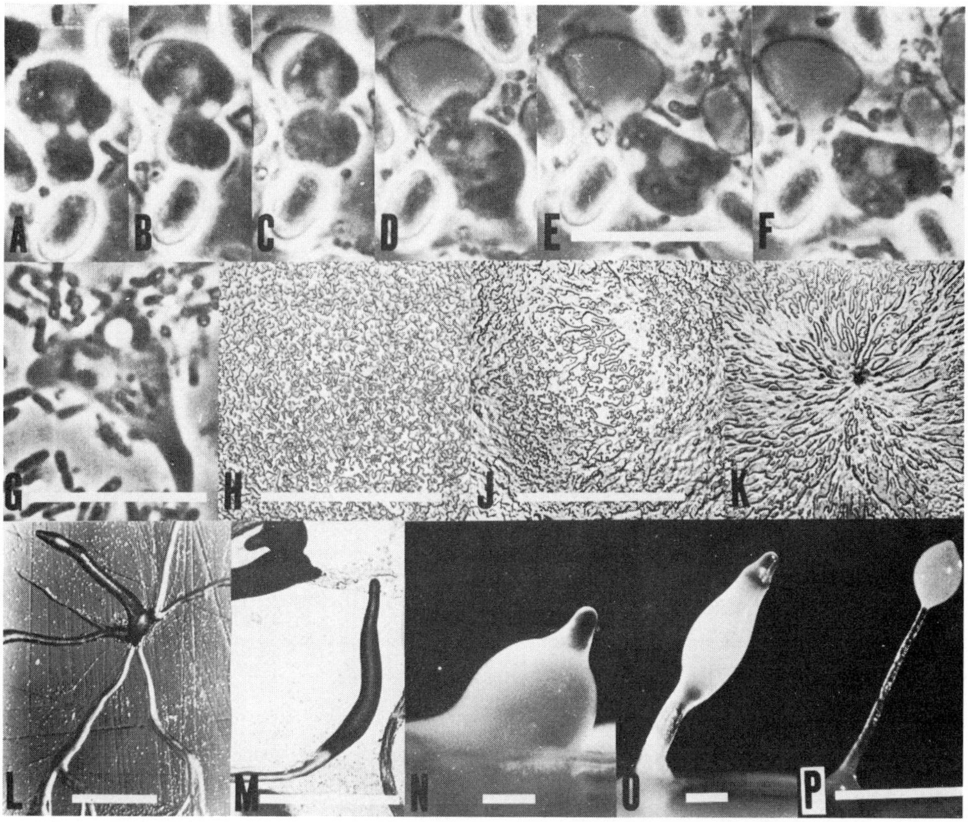

Fig. 3. Life cycle of D. discoideum (photographed and prepared by David Drage).

A-F, spore germination; G, vegetative amoeba containing partially digested bacteria and showing nucleus and contractile vacuole; H, field of amoebae in early interphase 1 hour after centrifugation; J, 6 hours later, showing beginning of aggregation; K, 2 hours after J, showing aggregate with tip and streams; L, late aggregate with slug leaving, 6 hours later than K; M, migrating slug with slime sheath collapsing behind; N, early culmination about 1 hour from end of migration; O, fruiting body erecting about 3 hours after N; P, 8 hours after O, fruiting body with stalk, spherical spore mass, and tip differentiating into spores.

Length marks: A-F 10 μ; G, 10 μ; H, 1 mm; J and K, 1 mm; L, ½ mm; M, ½ mm; N, 1/10 mm; O, 1/10 mm; P, ½ mm.

has been described in some detail [33,34]. Amoebae, 5-10 μ in diameter, emerge and move in the aqueous film on the surface of their substrate. If suitable bacteria are present, these are ingested and digested, and the amoebae can grow and divide. So long as food is available, growth and division continue [35]. The food bacteria are found by chemotaxis, and we think that absence of food, which triggers the next stage in the developmental cycle, is signaled by the absence of the chemotactic agent released by the bacteria. This agent can therefore be considered to have a trophic influence, inhibiting further development. Shaffer has shown that amoebae, separated by an agar block from bacteria, will nonetheless be inhibited from passing through interphase and therefore from aggregating [36]. When food is exhausted, the amoebae continue to move for about half an hour [37]. As in the feeding stage, pseudopod production can apparently occur at any point on the surface membrane--the amoebae do not show any signs of polarity. The amoebae then become stationary, passing through what has been called "interphase" for six to eight hours [38]. We interpret this stage as a genuine phase of differentiation during which there is considerable internal reorganization. At the end of interphase the amoebae are polar [27,39]: the polarity is most obviously indicated during movement, when it is plain that there is a front, pseudopod-forming end, and a rear end from which pseudopods are never projected. There are other ultrastructural signs of polarity, but few details are properly known [25].

In a post-interphase "field" of amoebae some individuals, apparently distributed at random [40], become centers toward which neighboring amoebae move. After some initial perturbations, the aggregation pattern stabilizes and the amoebae go on to form roughly hemispherical aggregates, containing a few hundred to perhaps 100,000 individuals [41]. Some cells at the top of each aggregate begin to secrete a muco-polysaccharide slime which forms a cone inverted over the aggregate. A group of cells forms a noticeable tip, which persists through the rest of the life cycle. Amoebae continue to move in toward the aggregation center. Vertical movement is restrained by the slime sheath, which hardens and becomes attached to the substrate at the periphery of the aggregate, except under the tip where it is formed. There, the inward movement of the cells can convert to vertical movement of the aggregate, forming an elongating vertical column, which eventually becomes unstable and falls over, crawling off as the slug, toward light, heat, or humidity, still surrounded by the slime sheath. After a period of migration, the slug stops. Its tip, which is of smaller diameter than its body and is normally held clear of the substrate, rotates to the top of the cell mass which forms a slightly squashed sphere. At the end of migration the front third of the slug contains cells (apart from those in the tip) which are destined to form a stalk; and the rear two-thirds, cells which (apart from a small, posterior group of cells which form the base plate) are destined to become spores [41,42]. This spatial arrangement is preserved when the slug stops, but with the axis rotated from horizontal to vertical. A rough sphere is formed in which the slug tip sits on top of a mass or prestalk cells, which is on top of the pre-spores, which in turn cover the base plate. Stalk cells are then extruded downward from just below the tip, simultaneously undergoing their final differentiation into a vacuolated, cuboidal form. The stalk meets and attaches to the base plate; extrusion continues so that the spore mass is lifted up by the stalk as it forms. The tip remains well defined until the formation of the stalk is complete, and then it in turn differentiates into spores [43] (see also our observation of Bonner's film [44]) and is absorbed into the spore mass. This marks the end of the developmental part of the life cycle, which began with the onset of interphase. The spore mass dries, and the spores can eventually be blown away, to germinate when they reach a suitable environment. Or, alternatively, in a moist environment, the stalk collapses and the spores germinate locally, the duration of development presumably allowing time for the food bacteria to regenerate [45]. As far as is known, there is no stage of true sexual reproduction; the existence of genetic recombination, and its possible mechanism, remain uncertain [25].

B. Organizing Waves

Before proceeding with our analysis of aggregation, it is convenient to introduce the concept of an organizing wave. As mentioned in section I, in the Goodwin-Cohen [1] model of a developmental control system, cells within a tissue are considered to be oscillators that can be autonomous in isolation, or entrained when coupled to cells of higher autonomous frequency. In the actual tissue, a gradient of autonomous frequency is supposed to exist away from a highest-frequency, or pacemaker, region. The oscillations can be complex biochemical events of duration comparable to their period. One aspect, the signaling aspect, is supposed brief. During it a signal is produced which can be transmitted to neighboring cells. On being signaled, a sensitive cell emits its own signal after a fixed time delay. A cell is refractory, i.e., insensitive, for some fraction of its autonomous period after signaling. This guarantees the outward propagation of a wave of signaling from the pacemaker, which can entrain all cells within the tissue with total periods longer than the pacemaker period and refractory periods shorter.

Such waves, entailing pacemakers, autonomous frequencies, refractoriness, and domains of entrainment were termed organizing waves by Goodwin and Cohen. A closely related theory was first introduced by Wiener and Rosenblueth [46] in connection with the functional control of the heart and has since been extensively developed for wave propagation in excitable media [47]. Such waves occur in the gut as well [48], and have long been familiar in nervous tissue [49, 50]. Physical analogues have been produced by Zaikin and Zhabotinski [51, 52] in the form of waves of chemical reaction propagating through an aqueous solution. One function of such waves is the synchronization of all local clocks within a tissue to the global time provided by the pacemaker. In the Goodwin-Cohen model [1] and in the single-event model [2], they provide positional information as well. In D. discoideum and the other slime molds listed in table 1, they control morphogenetic movement and possibly other processes of development.

C. Analysis of Aggregation

At the end of interphase aggregation centers form. These are probably initially those single cells which first begin to secrete an attractant, cyclic AMP [53, 54], but very rapidly the centers become small clusters of cells. In some species there are special "founder cells" [55]--that is, those cells which secrete attractant are noticeably different in morphology and become so after feeding ceases--but there is no evidence that this is so for D. discoideum. It is a remarkable fact that the density of aggregates, and hence the number of successful centers in a given area, is independent of population density over a wide range [56]. The number of centers increases to a maximum and subsequently decreases; this is because some centers are "taken over" by their neighbors, a phenomenon which we consider in more detail later.

In any case, our interpretation of the phenomena observed in films of aggregating amoebae of D. discoideum [10, 27] is as follows: some cells, distributed at random, begin to emit periodic pulses of an attractant which has been identified as cyclic AMP. The attractant molecules diffuse across the substrate surface in the aqueous film and reach neighboring amoebae. If these are presented with a suprathreshold extracellular concentration of cyclic AMP, they respond in quite a complex way. After an approximately 15 second delay, a signaled amoeba emits its own pulse of cyclic AMP, then initiates an all-or-nothing step of movement toward the signal source. The step lasts for about 100 seconds, and during this period the amoeba is refractory to further stimulation. Refractoriness guarantees the outward propagation of the signal from the signaling center. It also defines the upper limit of the signaling frequency. Gerisch has measured this and finds that the shortest period between pulses is about 2 minutes [30], approximately the refractory period. A centrifugal wave of inward movement is propagated and is repeated every time the center pulses. The pulse period ranges up to 5 minutes.

There is a critical density of amoebae below which aggregation cannot occur [10] because the signal amplitude is reduced both by diffusion and by the action of an extracellular phosphodiesterase [57, 58]. The critical density reported in the literature corresponds to an interamoeba spacing of about 70 μ [59]. It would take 2 seconds for a molecule the size of cyclic AMP to diffuse this far, while the rate of propagation of the signal, even at higher densities, is so slow [30] that we have been forced to postulate the intracellular delay of about 15 seconds as the rate-limiting step in signal propagation. With hindsight we can see that the delay is probably inevitable if temporal relations are to be preserved even when the amoebae are in contact, as in the later stages of development. The success of signal propagation as a pulse also depends on the removal of cyclic AMP from the environment, for otherwise the postulated receptors which detect a suprathreshold cyclic AMP concentration would saturate, and the system as a whole would lose directionality (cf. the discussion of section IV). We have also shown that the AMP supply must be recycled if sufficient cyclic AMP is to be made during aggregation [10]. It is therefore likely that the amoebae can take up 5' AMP from their environment, as it is produced by the action of phosphodiesterase on cyclic AMP. The fairly constant propagation velocity, due to the intracellular delay, is possibly also important in determining territory size, for territories are delineated by the overlapping of refractory phases [1]. This forms refractory boundaries through which the aggregative signal cannot propagate. In qualitative terms, this property, together with the possibility of entrainment of one center by its close neighbor, is the most important determinant of territory size. In time-lapse films (Bonner, 1960; Drage, unpublished; Gerisch, 1963) one can see an initial period in which neighboring centers entrain and lose entrainment, together with the formation of refractory boundaries between centers that are sufficiently far apart. After this initial period of internecine strife the number of centers remains quite constant. Similar results have been found by V. Nanjundiah (unpublished) in a one-dimensional simulation.

During the early stages of aggregation, streams may form. These occur where population density is locally relatively high, so that amoebae first move toward their neighbors and then toward the center, proving that each amoeba relays the signal [31]. When they come into contact, centripetal streams form. Because the intracellular delay is rate limiting, propagation velocity does not increase in the streams, although the amoebae may well signal each other directly via intercellular junctions of which there is some evidence in recent electron micrographs (Drage, Robertson, and Cohen, unpublished). The entry of an amoeba into a stream demonstrates polarity well [60]: the front (receptor and motile) end moves in and makes contact with the rear (signaling) end of the nearest anterior neighbor, while the rear (signaling) end is contacted by the front ends of the nearest posterior neighbors. Polarity is therefore preserved [61]. This preservation is, we suggest, important to both signal propagation and movement in response to the signal. Of course, amoebae within a population will vary considerably with respect to properties like metabolic rate, autonomous pulse frequency, rate of movement, and so forth. It seems reasonable that values of the three properties mentioned should be positively correlated, and indeed that those amoebae which pass through interphase most rapidly and therefore first begin to signal are also those with, on the average, the highest autonomous frequencies. They would therefore be able to entrain most of their neighbors, as those will have lower autonomous frequencies. It is also possible that a "driven" cell is inhibited from pulsing autonomously [1]. A third factor that would assist stable entrainment is the Doppler shift in received signal that would be experienced by those amoebae that took larger steps than the average in moving toward the center [27]. This shift would allow entrainment of amoebae that were autonomously faster than the center, because they would "see" an artificially high center frequency.

D. Analysis of Morphogenetic Movement

Dictyostelium discoideum amoebae move in several situations: as free individuals; as individuals in contact within streams and late aggregates; as part of an organism, the slug; and in erecting the fruiting body. We suggest that all these types of movement involve the same fundamental processes, and, in particular, that the movement of the slug is an inevitable consequence of restricting the chemotactically responsive amoebae within a slime sheath and by antero-posterior intercellular connections.

We have observed time-lapse films of D. discoideum moving freely on a moist agar surface (films taken by David Drage). As discussed in more detail in section II, the formation of a pseudopod is prefaced by the extension of filopods, $\frac{1}{2}\mu$ or less in diameter. These anchor at their tips and a "primary pseudopod" forms as a bulge at the filopod base, extending along the filopod and then enlarging into a secondary pseudopod which appears to contract, pulling the amoeba forward. When the polar aggregating amoebae form streams, pseudopod formation on most of the anterior surface is inhibited by intercellular contacts with the posterior surfaces of the amoeba in front. Micrographs show that primary pseudopods of reduced size can form in the interstices between neighboring amoebae, and it is possible that these help to anchor the stream to the substrate. In any case, movement within streams is visibly periodic, controlled by the chemotactic signals from the aggregation center, or pacemaker. We postulate that the normal chemotactic response of early aggregation continues, with the restrictions that secondary pseudopods are inhibited by a mechanism like contact inhibition, and that the amoebae are actually joined by antero-posterior junctions. The chemotactic response then consists of the production of primary pseudopods, followed by contraction of the cell body and then relaxation and withdrawal of the pseudopods. This sequence spreads outward as a wave along the stream, and can be seen as a wave of increased optical density at the site of contraction. The contraction, with the aid of anchorage by primary pseudopods, pulls the peripheral portion of the stream centrally. All our visual observations support the postulate that the mechanism of stream movement is fundamentally the same as that for movement by separate amoebae. In particular, the velocity of the wave again implies an intracellular delay time of about 15 seconds.

As amoebae move toward the aggregation center, they pile up, and a column is erected, apparently because the movement of the amoebae is constrained by slime secreted onto the aggregate surface. The slime forms a continuous restraining sheath on the surface of the column, except for the tip, where it is either liquid or simply not present. The column continues to increase in height until it becomes unstable and falls over, forming a prone "pseudoplasmodium," or slug [41]. The amoebae within the slug are still enclosed within the slime sheath, except at the tip [61,62]. They move as a body in the direction of the tip; and, as they move, the slime sheath collapses behind. The length of the slug does not change significantly during migration--the amoebae within definitely move en masse, not as free individuals. Nothing is known about the mechanism of slug movement, but our observations on time-lapse films of migrating slugs have led us to the following qualitative picture.

The tip cells, which have special histochemical properties [63], and which are probably those cells that formed a tip in the later stages of aggregation, act as pacemakers. They signal autonomously, releasing periodic pulses of cyclic AMP, exactly as they had done during aggregation. Within the slug there are probably intercellular contacts [64], possibly tight junctions, and it is likely that the cyclic AMP pulse can pass through these, as seems probable in the stream. (Some cyclic AMP is known to be released extracellularly, however, although less than from an equivalent number of aggregating amoebae [65].) The signal propagates backward along the slug with the same parameters, and in particular the same intracellular delay, as before. We have observed a backward-propagating wave of activity in moving slugs; its velocity and time course are consonant with an intracellular delay of 15 seconds and a refractory period of about 100 seconds. We therefore think that the same chemotactic response is produced by the signal; each amoeba, when signaled, produces its own signal after a delay and then extends primary pseudopods, which may assist in anchoring the slug to the slime sheath, and therefore fixing its position with respect to the substrate on which the slime sheath lies. Intercellular contacts between signaling and receptor poles inhibit secondary (and most primary) pseudopod formation, but do not inhibit the cycle of contraction and relaxation, which, as the active layer of amoebae is anchored within the slime sheath, pushes the anterior part of the slug forward, and pulls the posterior part toward the active region. We have also observed (R. Lymn, unpublished micrographs) that interlocking pseudopods may hold the amoebae together in the slug. These might afford a means of lateral binding of the strings of cells in the slug. The region of greatest contraction can be seen as a bulge in the slug, and it is with respect to this bulge that the anterior and posterior parts of the slug are moving at any moment. We are at present working on the quantitative implications of this model of movement; the qualitative ones fit our observations and those data already available in the literature [2].

The movements of fruiting body formation are also hardly described, let alone understood. However, we have again observed (from our own films and one taken by Bonner [44]) that they have a periodic component, with a frequency similar to that of the aggregation movements and the signaling within the slug. The fruiting body is erected by the downward extrusion of cells from just below the tip of the cell mass, which is organized as outlined in section V.A. Extrusion often appears to be periodic, with a period of about 5 minutes. As the cells are extruded downward, they differentiate into stalk cells. It is interesting that Bonner has shown that a high concentration of cyclic AMP will also induce differentiation to the stalk-cell type, and it is therefore interesting to speculate that the downward-extruding stalk passes through a region of high concentration of cyclic AMP. We should, however, mention that it is possible to induce stalk differentiation by other agents which have effects on the outer surface of the plasma membrane (G. Gerisch, unpublished observation).

E. Sorting Out and Differentiation

The final products of differentiation are two cell types: stalks and spores. Possibly one should count the pacemaker and base plate cells as types as well, although the former certainly become spores and the latter are morphologically similar to stalk cells. (We are not considering the change from feeding to aggregative amoebae at the moment, because all amoebae that enter an aggregate have been through this change. In our view, however, it is a genuine differentiation [38,64].) The final product of our account should be a description of how these cell types arise, and how their spatial distribution is controlled. Unfortunately, little is known about either process. The available evidence suggests that in D. discoideum differentiation does not begin until the later stages of aggregate formation: the tip cells then become histochemically distinct [65,43]. There is also some suggestion (D. Drage) that they are the first cells to show obvious prespore vacuoles [66]. This is significant because the tip cells ultimately become spores, and because they might well be the most "advanced" cells in the aggregate; that is, they are probably among those cells that finished interphase first, and that have a relatively high metabolic rate.

As the slug migrates, the prespore cells, those in the rear two-thirds, show an increasing number of prespore vacuoles [66,64]. The vacuoles may therefore be used as a criterion for recognizing prespore cells in the slug. However, the slug regulates; even if the rear half of a slug is isolated, it will culminate to produce a fruiting body with roughly normal proportions [41]. Presumably the pacemaker tip must redifferentiate and the prestalk cells must be produced from cells that were destined to become spores. One can therefore argue that neither the prespore nor prestalk cells are yet properly differentiated, as they can be interconverted. Indeed, the same could be said of the spores themselves, which can become feeding amoebae, and then presumably either stalk or spore under appropriate conditions. However, there are no data allowing us to say whether cell division is necessary before a spore can become either of the cell types.

"Differentiation" can be followed in several ways: one is vital staining [42]. A variety of vital dyes shows differential staining in the slug. The boundaries between differently staining regions are where they would be expected in the light of the final distribution of stalk and spore cells. Other histochemical techniques, for example measurement of the uptake of radioactive tracers from the food bacteria [43], indicate a similar discontinuous distribution of cell properties in the slug. How does this distribution arise? Do all the cell types preexist, and then sort out, or do they differentiate according to their environmental circumstances, and, in particular, according to their position in the slug?

The idea of sorting out prestalk and prespore cells from an initially random distribution is quite old [25]. Bonner suggested that, in the slug, relatively fast cells moved to the front, while relatively slow cells fell back. Stalk cells would therefore be derived from the relatively fast population, spores from a slower population. However, the evidence that there is significant sorting out is apparently equivocal [67] and perhaps should not be accepted until it is possible to follow the courses of individual cells. Furthermore, amoebae from the front of a disaggregated slug move at the same rate as those from the rear [61], and it would have to be possible to regulate relative velocity if velocity were in fact the determinant of cell state.

Takeuchi [43] has convincingly shown that values of several cell properties, including specific gravity, are randomly distributed among a population of preaggregation amoebae, but that they are already sorted out into a gradient in the slug. Again, the available evidence indicates that sorting out has occurred by the time the slug is formed, and has probably occurred by the end of aggregation. It is germane to this point that the migration stage is not essential, but may indeed be lost altogether under certain environmental conditions. We therefore feel that sorting out occurs before slug formation, but that differentiation, the interpretation of position, occurs within the slug, or within the conus if a slug stage is missing. If the slug stage is missing, time is still needed for differentiation within the conus before culmination can proceed.

F. Discussion

It is possible to make many models that would satisfy the requirements for regulation within the slug, and that would supply positional information for interpretation by amoebae within the slug or conus. We do not propose to examine these in detail: reference may be made to Cohen's paper [2]. However, one such model, the single-event model, at least illustrates the possibilities. We assume that the periodic signal which controls morphogenetic movement throughout the developmental cycle of D. discoideum is also used to supply positional information. There is good evidence that amoebae initially distributed at random sort into a gradient, that differentiation is a result of interpretation of this gradient, and that regulation of the gradient is possible. One way of supplying positional information would be to propagate a periodic signal, which acts as a time base, or global clock, and compare its phase with that of local clocks. In the case of D. discoideum a "local clock" would be the intrinsic oscillator of an amoeba, having a characteristic autonomous frequency, and having sorted into appropriate position in a gradient of autonomous frequencies. Thus, if a signal is propagated from a pacemaker, or highest-frequency, region, and can stimulate amoebae down the gradient to produce their own signals, its phase can be compared with that of each local oscillator. As the frequency gradient is descended, local phase will be increasingly retarded with respect to the phase of the pacemaker signal. Local phase difference can therefore be used as a measure of position within a frequency gradient. This is because each cell in the frequency gradient is stimulated to signal earlier than it would have done had it been allowed to complete its autonomous cycle which began the last time it was signaled. Regulation implies regulation of the frequency gradient, as this is the supplier of positional information. This could be brought about in several ways; one possibility is an active transport mechanism that moves the signaling compound, possibly cyclic AMP, or its degraded form 5' AMP, up the frequency gradient. (This is probably necessary in any case as a considerable amount of signaling substance must be lost from the tip cells.) However, it is idle to speculate about detailed mechanisms at this stage; suffice it to say that the periodic process used for the control of morphogenetic movement could also be a prototypic developmental control system. A pertinent observation is that in two cases in which a slug was trisected we observed a gradient of regulation rate. The front third reorganized and fruited first, the rear third last (Bonner's 1960 film [44]). This is consistent with the presence of a frequency gradient; indeed, there was a gradient from front to rear in the frequency of the periodic component of the movements of the fruiting bodies produced after regulation, suggesting incomplete regulation of the original frequency gradient. Whitfield also describes a similar gradient of regulation rates when cells reorganized from a disaggregated slug [68]. The details of the control process which seem to be important are: its periodicity; its use of an extracellular transmitter; the 15-second intracellular delay which leads to a roughly constant signal velocity throughout development; the polarity of the aggregation-competent amoebae, which must be preserved for signaling to be successful; and the continuation of the process throughout the developmental cycle.

Additional evidence that the tip is an organizing center for development is provided by tip-transplant experiments. Slug tips grafted onto another slug take over that position of the slug posterior to the graft if polarity is preserved [25]. We have transplanted the tips of fruiting bodies (D. Wonio, unpublished) onto fields of postinterphase amoebae, where they become centers of aggregates which go on to fruit without a slug phase.

Insofar as they have been studied, all of the slime molds listed in table 1 which show periodic aggregative phenomena resemble *D. discoideum* far more than those which aggregate slowly. There are interesting correlations of morphology with category of control system. All periodic aggregants show streams; all smooth aggregants such as *D. minutum* do not. One can show that streams must occur if each amoeba acts as a secondary signal source. The amoebae of smooth aggregants therefore do not relay the signal before they reach the aggregate. *Dictyostelium discoideum* can be converted to a smooth aggregant from a periodic aggregant by mutation [30], implying that periodic and quasi-static control systems can be of comparable complexity. Moreover, preaggregation patterns observed in late interphase in *D. discoideum* (so-called clouds) [31] resemble aggregates of *D. minutum* in that they are round and without streams (cf. fig. 3). These and other observations indicate that different elements of the developmental control system and of the competence to perform the developmental processes emerge at different times during interphase.

The arguments in the preceding paragraph show that it is possible to infer features of a developmental control system by direct observation of morphology. We therefore feel that our gradually emerging understanding of a developmental control process will provide an extremely powerful tool for analyzing the development of more complex embryos.

REFERENCES

1. B. C. Goodwin and M. H. Cohen, J. Theoret. Biol. **25**, 49 (1969).

2. M. H. Cohen in Some Mathematical Questions in Biology, ed. J. D. Cowan (Providence, R.I.: American Mathematical Society, 1971), **3**, 3.

3. M. H. Cohen, Symp. Soc. Exp. Biol. no. 25, 455 (University of Cambridge Press, 1971).

4. L. Wolpert, J. Theoret. Biol. **25**, 1 (1969).

5. F. H. C. Crick, Nature, **225**, 420 (1970).

6. A. M. Turing, Phil. Trans. Roy. Soc. B, **237**, 37 (1952).

7. See, e.g., the review by I. Prigogine and G. Nicolis entitled Biological Order, Structure, and Instabilities, Quart. Rev. Biophy. **4**, 107 (1971).

8. B. M. Shaffer, Am. Naturalist, **91**, 19-35 (1957).

9. A. Arndt, Archiv fur Entwicklungs-merchanik, **136**, 681-744 (1937).

10. M. H. Cohen and A. Robertson, J. Theoret. Biol. **31**, 119 (1971).

11. A. Robertson, in Some Mathematical Questions in Biology, ed. J. D. Cowan (Providence, R.I.: American Mathematical Society, 1971), 4, 48.

12. G. Selman, J. Embryol. Exp. Morph. 6, 448-465 (1958).

13. K. I. Sirakami, Memoirs of the Faculty of Liberal Arts and Education, No. 10 (Yamanashi University, Japan, 1959).

14. J. Holtfreter, J. Morph. 79, 27 (1946); Ann. N.Y. Acad. Sci. 49, article 5 (1948).

15. V. M. Ingram, Nature, 223, 641 (1969).

16. V. Hamburger, M. Balaban, R. Oppenheim, and E. Wenger, J. Exp. Zool. 1959, 1-14 (1964).

17. P. Weiss, Dynamics of Development (New York: Academic Press, 1968).

18. E. J. DuPraw, Analysis of Embryonic Development in the Honey-bee, I, a film (Library of Congress) (1963).

19. A. C. Neville, Biol. Rev. 42, 421-441 (1967).

20. B. I. Balinsky, An Introduction to Embryology (3rd ed.; Philadelphia: W. B. Saunders Co., 1970).

21. J. P. Trinkhaus, Cells into Organs (Englewood Cliffs, N.J.: Prentice-Hall, 1969).

22. T. Gustafson and L. Wolpert, Exp't'l Cell Res. 24, 64 (1961).

23. T. Gustafson, Exp't'l Cell Res. 32, 570 (1963).

24. J. Adler, Science, 166, 1588 (1969).

25. J. T. Bonner, The Cellular Slime Molds (Princeton, 1967).

26. D. Wonio, unpublished observations.

27. M. H. Cohen and A. Robertson, J. Theoret. Biol. 31, 101 (1971).

28. E. F. Keller and L. A. Segel, J. Theoret. Biol. 30 (1970).

29. C. S. Patlak, Bull. Math. Biophys. 15, 311, 431 (1953).

30. G. Gerisch, Current Topics in Developmental Biology, 3, 157 (1968).

31. B. M. Shaffer, Adv. in Morphogenesis, 2, 109 (1962).

32. D. A. Cotter and K. B. Raper, Proc. Nat. Acad. Sci. U.S.A., 56, 880 (1966).

33. D. A. Cotter and K. B. Raper, J. Bacteriol. 96, 1680 (1968).

34. D. A. Cotter, L. Y. Mirua-Santo, and H. R. Hohl, J. Bacteriol. 100, 1020 (1969).

35. G. Potts, Flora, 91, 281 (1902).

36. B. M. Shaffer, J. Cell. Sci. 1, 391 (1966)

37. E. W. Samuel, Develop. Biol. 3, 317 (1961).

38. J. T. Bonner, Symp. Soc. Exp. Biol. 17, 341 (1963).

39. J. T. Bonner, Biol. Bull. 99, 143 (1950).

40. B. M. Shaffer, Quart. J. Microscope Sci. 98, 393 (1957).

41. K. B. Raper, J. Elisha Mitchell Scient. Soc. 56, 241 (1940).

42. J. T. Bonner, Am. Naturalist, 76, 79 (1952).

43. I. Takeuchi in Nucleic Acid Metabolism, Cell Differentiation and Cancer Growth, ed. E. V. Cowdrey and S. Seno (London: Pergamon Press, 1969).

44. J. T. Bonner, a film "Fruiting in the cellular slime molds" (1960).

45. D. J. Drage, unpublished observations.

46. N. Wiener and A. Rosenblueth, Arch. Inst. Cardiologia de Mexico, 16, 105 (1940).

47. V. I. Krinskii in Systems Theory Research, vol. 20, ed. A. A. Lyapunov (New York: Plenum, 1971).

48. V. I. Krinskii in Medical Physiology, ed. V. B. Mountcastle (St. Louis: Mosby, 1968).

49. K. S. Cole, Membrane, Ions and Impulses (Berkeley: University of California Press, 1968).

50. R. L. Beurle, Phil. Trans. Roy. Soc. B, 240, 609 (1957).

51. A. N. Zaikin and A. M. Zhabotinsky, Nature, 225, 535 (1970).

52. A. Winfree, time-lapse film (1970).

53. J. T. Bonner, J. Exp. Zool. 106, 1 (1947).

54. J. T. Bonner, D. S. Barkley, E. M. Hall, T. M. Konijn, J. W. Mason, A. O'Keefe, and P. B. Wolfe, Develop. Biol. 20, 72 (1969).

55. B. M. Shaffer, J. Exp. Biol. 38, 833 (1961).

56. J. T. Bonner and M. R. Dodd, Biol. Bull. 122, 13 (1962).

57. B. M. Shaffer, Nature, 171, 975 (1953).

58. Y. Y. Chang, Science, 160, 57 (1958).

59. T. M. Konijn and K. B. Raper, Develop. Biol. 3, 725 (1961).

60. B. M. Shaffer in Primitive Motile Systems in Cell Biology, ed. R. D. Allen and N. Kamiya (New York: Academic Press, 1964).

61. D. F. Francis, Ph.D. thesis, University of Wisconsin (1965).

62. D. R. Garrod, J. Cell. Sci. 4, 981 (1969).

63. H. Mine and I. Takeuchi, Annual Report of Biological Works of the Faculty of Science of Osaka University, 15, 97 (1967).

64. Y. Maeda and I. Takeuchi, Development, Growth and Differentiation, 3, 232 (1969).

65. J. T. Bonner, J. Exp. Zool. 110, 259 (1949).

66. H. R. Hohl and S. T. Hamamoto, J. Ultrastructure Res. 26, 442 (1969).

67. P. Farnsworth and L. Wolpert, in press (1971).

68. F. E. Whitfield, Exp. Cell. Res. 36, 62 (1964).

DISCUSSION

J. L. Jackson: You spoke of a signal. Do you know what the mechanism for the transmission of the signal is and could you say something about it?

M. H. Cohen: We are familiar in detail with the signal in only one case, the aggregation of Dictyostelium discoideum just shown in the slides and film. There the signal consists of a molecule, cyclic AMP, released periodically in brief bursts into the aqueous film over the surface of the substrate by randomly distributed cells which become centers of aggregation. When a super-threshold concentration reaches a neighboring amoeba by diffusion, the latter responds by a movement step of about 20 μ lasting 100 seconds. In addition, 15 seconds after signal reception, the receiving amoeba releases its own pulse of cyclic AMP. The original cell is refractory and does not respond to the sign. The net result is an outward-propagating wave of inward movement.

R. Thom: 1. How can you be sure that a periodic phenomenon has a true control function in the developmental process and is more than an irrelevant temporal symmetry-breaking process?

2. How can a phase-shift model explain regulation of development in Dictyostelium, where you said that global regulation occurs up to a factor of 10^5 in the number of cells?

M. H. Cohen: 1. One is sure that a periodic phenomenon has a true control function only after the phenomenon has been observed actually controlling a developmental process. In D. discoideum, morphogenetic movement, a developmental process, is demonstrably controlled by the periodic propagation of pulses of an attractant during early aggregation. Evidence for the continuity of the same control system throughout later development exists but has not yet been analyzed in detail. In the other organisms showing the periodic phenomena listed in table 1, no direct evidence for the existence of a periodic control system is known.

2. That range of regulation does not present a difficulty in principle to the phase-shift

model. More to the point is the absence of any evidence for a second periodic event. For that reason we favor the single-event model, for which that range of regulation also presents no difficulty in principle.

E. M. Simon: Many of your interesting pictures appear to behave like many phenomena encountered in hydrodynamics, especially turbulent phenomena. Perhaps use of the Navier-Stokes equation can serve as a model to better explain such morphogenetic patterns. There have been, for example, computer experiments using the Lagrangian form of the N-S equation in which the surface boundary of aerodynamic turbulence formation generates patterns similar to those in your slides.

M. H. Cohen: Such an approach does not seem directly useful for this problem. However, the general concepts of instabilities and bifurcation of solutions of systems of equations include the hydrodynamic phenomena you speak of and have provided useful ways of thinking about morphogenesis and pattern formation of biology. Specific examples involving chemical kinetics and transport have been studied by Turing, Prigogine, Nicolis and others. I have endeavored to show (Proc. 25 S.E.B. Symposium, 1971) that the dynamics of coupled cells in an embryonic field leads to equations of similar form with similar consequences.

G. Nicolis: 1. It is interesting to point out that the Zhabotinski reaction may give rise to quite different phenomena, depending on the boundary conditions. For instance, when the reaction occurs in a long thin vessel (practically a one-dimensional system), one first observes a local inhomogeneity which then propagates in different directions until it stabilizes to a time-independent steady state in which the different chemicals are spatially separated. One has therefore here an example of a space dissipative structure (M. Herschkowitz-Kaufman). On the other hand, if the reaction occurs in a layer of unstirred solution one observes propagating wavefronts of the type shown by Professor Cohen. When two waves initiating from different centers meet, it may happen that instead of annihilation one observes an entrainment of the low-frequency wave to the higher frequency.

2. In the case of slime-mold aggregation in a static gradient, Keller and Segel have been able to relate aggregation to the acrasin metabolism in the cell and to show that aggregation is the result of an instability of the homogeneous cell and acrasin distribution. Is it reasonable to expect that the aggregation corresponding to a periodic acrasin evolution could be interpreted in similar terms, provided inhibition is taken into account in the acrasin metabolism?

3. We have studied, recently, model chemical systems, giving rise to chemical waves beyond a nonequilibrium instability of the uniform steady-state solution (M. Herschkowitz-Kaufman and G. Nicolis, submitted to J. Chem. Phys.). Under certain conditions, the velocity of propagation of the wave parts is between one and two orders of magnitude larger than the

propagation due to diffusion. One may expect therefore that chemical waves provide a very efficient mechanism for the transmission of information over large distances.

III. Generalized Hydrodynamics

9

MICROSCOPIC DESCRIPTION OF THE LINEARIZED HYDRODYNAMIC MODES

Pierre Resibois

1. THE ROLE OF HYDRODYNAMIC MODES IN STATISTICAL PHYSICS

Let us start our discussion by considering the Laplace-Fourier transform of the linearized hydrodynamic equations of macroscopic physics [1,2]:

$$-i\omega n_{\mathbf{q}}(\omega) + i\,\mathbf{q}\cdot n\,\mathbf{g}_{\mathbf{q}}(\omega) = n_{\mathbf{q}}(t=0) \qquad \text{(continuity equation)}\,, \tag{1}$$

$$-i\omega \mathbf{g}_{\mathbf{q}}(\omega) + i\mathbf{q}\,\frac{1}{n}\left(\frac{\partial p}{\partial n}\right)_T n_{\mathbf{q}}(\omega) + i\mathbf{q}\,\frac{1}{n}\left(\frac{\partial p}{\partial T}\right)_n T_{\mathbf{q}}(\omega)$$

$$+\frac{\eta}{n}\left\{\mathbf{q}\left[\mathbf{q}\cdot\mathbf{g}_{\mathbf{q}}(\omega) + q^2\mathbf{g}_{\mathbf{q}}(\omega)\right] + \frac{(\zeta - 2\eta/3)}{n}\left[\mathbf{q}\left(\mathbf{q}\cdot\mathbf{g}_{\mathbf{q}}(\omega)\right)\right]\right\} = \mathbf{g}_{\mathbf{q}}(t=0)$$

(Navier-Stokes equations) (2)

$$-i\omega T_{\mathbf{q}}(\omega) + i\,\mathbf{q}\cdot\frac{T}{nC_v}\left(\frac{\partial p}{\partial T}\right)_n \mathbf{g}_{\mathbf{q}}(\omega) + \frac{\kappa}{nC_v}\,q^2\,T_{\mathbf{q}}(\omega) = T_{\mathbf{q}}(t=0)$$

(temperature equation). (3)

In these equations, n, p, and T respectively denote the equilibrium density, temperature, and specific heat at constant volume; η and ζ are the shear and bulk viscosity, and κ is the thermal conductivity. Moreover $n_{\mathbf{q}}(\omega)$, $\mathbf{g}_{\mathbf{q}}(\omega)$, and $T_{\mathbf{q}}(\omega)$ are the Fourier-Laplace transforms of the local particle density (we consider a one-component fluid), local momentum (we take the mass $m = 1$), and local temperature. For instance, one has:

$$n_{\mathbf{q}}(\omega) = \int_0^\infty dt\; e^{i\omega t} \int d^3r\; e^{-i\mathbf{q}\cdot\mathbf{r}}\;\delta n(\mathbf{r},t)\,, \tag{4}$$

where $\delta n(\mathbf{r},t)$ is the density fluctuation at point $\mathbf{r}$ and time t; similar definitions hold for $\mathbf{g}_{\mathbf{q}}(\omega)$ and $T_{\mathbf{q}}(\omega)$.

Although the problem of macroscopic hydrodynamic modes can be formulated in various ways, perhaps the most direct one is to consider the (non-Hermitian) eigenvalue problem associated with the left-hand side of equations (1)-(3); in the small-q limit, one readily finds the following eigenvalues (assuming for simplicity that $\underline{q}$ is oriented along the x-axis):

$$\lambda_y = \lambda_z = -q^2/\eta \ , \tag{5a}$$

$$\lambda_{\pm} = \pm \ icq - q^2 \, \Gamma , \tag{5b}$$

$$\lambda_T = -q^2 \, \kappa/n \, C_p \ ; \tag{5c}$$

where

$$\Gamma = \tfrac{1}{2}\left[\frac{4\eta/3 + \zeta}{n} + \left(\frac{1}{C_v} - \frac{1}{C_p}\right) \frac{\kappa}{n} \right] \tag{6}$$

and c is the sound velocity

$$c^2 = \frac{C_p}{C_v} \left(\frac{\partial p}{\partial n} \right)_T \ . \tag{7}$$

Clearly λ_y and λ_z describe the viscous damping of shear viscosity modes, $\lambda_{\pm}$ describe damped sound-wave propagations, and λ_T corresponds to the diffusion of the thermal mode. The corresponding left and right eigenfunctions are easily computed, but we shall not need these here.

Since the early days of statistical mechanics, a central problem has been to obtain a microscopic derivation of these linearized hydrodynamic equations together with explicit formulae for the transport coefficients appearing in them. Moreover, the extension of these equations to larger values of q and ω--where no simple macroscopic description holds--is the object of the so-called generalized hydrodynamics, which has undergone great development in recent years, as will be discussed by other speakers at this conference.

In the present lecture we want to emphasize the converse viewpoint, namely, the importance that this concept of hydrodynamic modes has for our understanding of microscopic phenomena. Indeed, by an interesting feedback mechanism, hydrodynamic modes have been used in statistical physics in at least two important respects.

First, the concept has served in finding a microscopic definition of transport coefficients (the so-called autocorrelation function formulae) by identifying the equations of motion for the average value of the microscopic operators describing the five conserved quantities (in a state taken initially to be in local equilibrium) with the macroscopic equations of motion (1)-(3). This is, for instance, the point of view adopted by Mori [3, 5] and by Kadanoff and Martin [2] .

Second, the concept has also been used in the analysis of a series of problems involving long-wavelength divergences. Leaving aside the case of long range forces, let us cite:

(i) Problems connected with the nonanalytic density expansion of transport coefficients; we should mention the work of Pomeau [6] who introduced the long-wavelength propagation of hydrodynamic modes in the treatment of the diverging three-body collision operator in two dimensions. Similar considerations have been used recently in order to interpret the slow power-law decay of the velocity correlation function in dense gases [7-12] in two and three dimensions.

(ii) The critical behavior of transport coefficients; here again, the long-distance propagation of hydrodynamic modes plays a crucial role in the understanding of the singularities of transport coefficients close to the critical point. This point has been emphasized in the work of Fixman [13],

Kawasaki [14], and Kadanoff and Swift [15].

(iii) The long-wavelength, low-frequency, behavior of the Van Hove time-dependent correlation function [16]; this behavior is again determined from the macroscopic hydrodynamic equations [1-3] together with the well-known thermodynamics of fluctuations [12, 16, 17].

Although all these works share the feature of using hydrodynamic concepts in problems of statistical physics, the level of sophistication introduced in these theories is quite variable. For example, Van Hove [16] and Mountain [17] assume, with no more than heuristic justification, that macroscopic laws can be used to describe the time-dependent density-density correlation function for large distances and long times. In the works of Mori [3-5], Kawasaki [14], and Kadanoff and Swift [15], microscopic hydrodynamic modes are explicitly constructed in terms of the complete equilibrium N-particle distribution function.

Finally, in the work of Pomeau [6] and of Dorfman and Cohen [10], microscopic hydrodynamics modes are defined in terms of a one-particle distribution function, but for dilute (or moderately dense) gases only.

Having stressed the importance of hydrodynamical modes in statistical physics, we now want to show that, even for strongly coupled fluids, statistical hydrodynamic modes--and their corresponding eigenvalues--can be explicitly constructed in terms of one-particle distribution functions. The usefulness of such a tool becomes evident once it is realized that a full understanding of the aforementioned divergence problems requires a detailed many-body analysis; yet such an analysis can generally be carried through only if one can think, and calculate, in terms of one-body quantities. To illustrate this point, let us simply recall the importance of "one-body propagators" in any diagrammatic analysis of the N-body problem (see, for example, Abrikosov et al. [18] vs. Prigogine [19] and Resibois [20] for two very different formulations in terms of graphs).

The remainder of this paper is organized as follows: in section II, we first briefly recall the linearized generalized Boltzmann equation (LGBE) in a strongly interacting system, together with a few remarks on the traditional way transport coefficients are calculated from it. We then show that this transport problem can also be formulated from the eigenvalue problem associated with the LGBE; this eigenvalue problem is treated by standard perturbation calculus, to second order in the uniformity parameter q, starting from the basis formed by the eigenfunctions of the homogeneous LGBE. The usual transport coefficients then appear as the q^2 coefficients of the five eigenvalues which tend to zero when $q \to 0$. Although obtained from the LGBE only, these transport coefficients involve their correct potential contributions, to arbitrary order in the interaction, as can be shown by the formal identity between the expressions obtained here and the well-known correlation function formulae. The corresponding eigenfunctions give, when calculated to lowest order in q, the required statistical expression for the hydrodynamical modes. Of course, it is not the aim of the present paper to give a detailed technical report of the calculations; we shall thus be very sketchy and we refer the interested reader to the original papers [21, 22]. Finally, in section III we discuss some possible applications and developments of the method. However, it should be recognized that, in this field, the most promising problems remain to be solved and we shall thus be very brief, indicating a few directions of work which are either under active investigation (for example, the kinetic behavior of a gas with long-range forces) or which can possibly be opened by the present method (for example, an N-body analysis of the mode-mode coupling approach to critical transport phenomena).

2. STATISTICAL HYDRODYNAMIC MODES AND THE LINEARIZED GENERALIZED BOLTZMANN EQUATION

It is nowadays a well-known fact that, within the frame of a theory linearized around absolute equilibrium, the Fourier transform of the one-particle distribution function, denoted by $f_q(v,t)$, obeys the following exact non-Markovian kinetic equation [19-21] (from now on, we drop vector notation):

$$\partial_t f_q(v,t) + iqv_x f_q(v,t) = \int_0^t G_q(v,\tau) f_q(v,t-\tau) + \mathcal{D}_q(v,t) \tag{8}$$

Here the kernel $G_q(v,\tau)$ is a non-Markovian collision operator and $\mathcal{D}_q(v,t)$ is an inhomogeneous term describing the effect of the initial correlations on the time evolution of the one-particle distribution function. Equation (8) can be derived by a variety of methods, in particular the perturbation method of Prigogine and coworkers [19, 20], the Bogolubov streaming-operator method [23], and the Zwanzig [24] projection-operator technique.

In the present review, we shall not need the explicit form of $\mathcal{D}_q(v,t)$ and $G_q(v,t)$ except for a few general properties which we shall mention without proof; these proofs can be established starting from any of these formalisms.

In transport theory, one is interested in the long-time and small-q behavior of equation (8); in this case, one can show that equation (8) reduces to the following (Markovian) linearized generalized Boltzmann operator (LBGE) [21]:

$$\partial_t f_q(v,t) + iqv_x f_q(v,t) = C_q^{\ell} f_q(v,t) . \tag{9}$$

Here the linearized collision operator C_q^{ℓ} is defined by

$$C_q^{\ell} = i\psi_q(v;o) + i(\Omega_q - 1)\left[-qv_x + \psi_q(v;o)\right] \tag{10}$$

where $\psi_q(v;z)$ is the Laplace transform of the kernel $G_q(v,\tau)$:

$$G_q(v,\tau) = -\frac{1}{2\pi i}\oint dz\, e^{-izt}\, \psi_q(v;z) \tag{11}$$

and Ω_q is defined by the following implicit relation:

$$\Omega_q = 1 + \sum_{n=1}^{\infty} \frac{(-1)^n}{n!}\left(\frac{\partial^n \psi_q}{\partial z^n}\right)_{z=0} \left\{\Omega_q\left[-v_x q + \psi_q(v;o)\right]\right\}^{n-1} \Omega_q . \tag{12}$$

We refer the reader to the literature for the transformations leading from equation (8) to equation (9) [19-21, 23].

Before attacking the problem of statistical hydrodynamic modes, let us digress briefly and mention the traditional approach to the calculation of transport coefficients in a dense system [25]. In a first step, one solves equation (9) by a Chapman-Enskog type of procedure, for the double purpose of (1) evaluating the kinetic part of the transport coefficients and (2) determining the two-particle distribution function, which is expressed as a functional of f_1:

$$f_2(r_1, v_1; r_2, v_2; t) = f_2(r_1, v_1; r_2, v_2 | f_1) . \tag{13}$$

From f_2, the potential part of the transport coefficients is then computed. Thus, in this usual approach, the LGBE, equation (9), is insufficient to determine completely the transport coefficients, including their potential part: the functional relation (13) is also required. Let us point out that, once this latter relation is known, the collision operator that appears in the right-hand side of equation (8) can of course be readily calculated through the BBGKY (Bogoliubov, Born, Green, Kirkwood and Yvon) equation:

$$\partial_t f_1 + v_1 \frac{\partial f_1}{\partial r_1} = n \int dr_2 dv_2 \frac{\partial V}{\partial r_{12}} \frac{\partial}{\partial v_1} f_2(r_1, v_1; r_2, v_2; t) . \tag{14}$$

Yet, the converse is of course not true: if C_q^{ℓ} is given, we know only the particular combination of f_2 which appears in the right-hand side of equation (14), but f_2 itself cannot be determined. And we shall see later, in our formalism it is nevertheless sufficient to know C_q^{ℓ} to second order in q in order to calculate the exact transport coefficients, to arbitrary order in the interaction.

Let us now consider the eigenvalue problem associated with the LGBE equation (9). We write:

$$(C_q^{\ell} - iqv_x) \psi_n^q(v) = \lambda_n^q \psi_n^q(v) , \tag{15}$$

and we expand C_q^{ℓ} in powers of the uniformity parameter:

$$C_q^{\ell} = C_o^{\ell} + q C_o^{\ell'} + \frac{q^2}{2!} C_o^{\ell''} , \tag{16}$$

where $C_o^{\ell'} = (\partial C_q^{\ell} / \partial q)_{q=0}$, etc. Similarly, we expand the eigenvalues and the eigenfunctions:

$$\lambda_n^q = \lambda_n^o + q \lambda_n^{(1)} + q^2 \lambda_n^{(2)} + \ldots , \tag{17}$$

$$\psi_n^q(v) = \psi_n^o(v) + q \psi_n^{(1)}(v) + \ldots . \tag{18}$$

Treating q as a smallness parameter, we get thus a perturbation scheme for which the natural basis is of course the eigenfunction of the homogeneous collision operator $C_o^{\ell} = i(\Omega_o \psi_o)$. This immediately leads to two difficulties:

(1) The operator C_o^{ℓ} is not symmetric [26], and we will be obliged to work with a biorthonormal set of eigenfunctions. Using the obvious abstract vector notation

$$\langle v | f \rangle \equiv f(v) , \tag{19}$$

we have thus to consider the eigenvalue problem [27]

$$C_o^{\ell} | \varphi_n \rangle = \lambda_n^o | \varphi_n \rangle \tag{20}$$

together with the adjoint problem

$$\langle \bar{\varphi}_n | C_o^{\ell} = \lambda_n^{o} \langle \bar{\varphi}_n | \ . \tag{21}$$

As is well known, one can make the sets $|\varphi_n\rangle$ and $\langle\bar{\varphi}_n|$ biorthonormal:

$$\langle \bar{\varphi}_n | \varphi_{n'} \rangle = \delta_{nn'}^{K_r} \tag{22}$$

Here the following definition of the scalar product has been chosen:

$$\langle f | g \rangle = \int dv \ [\varphi^{eq}(v)]^{-1} \ f(v) g(v) \ , \tag{23}$$

where $\varphi^{eq}(v)$ is the Maxwellian.

Due to simple properties of the operator $i\psi_o$ (which is symmetric and admits of the five collision invariants 1, $\underline{v}d$, $v^2/2$ as eigenfunctions with zero eigenvalue) and of Ω_o, it is possible to construct explicitly the five right- and left-eigenfunctions of C_o^{ℓ} with zero eigenvalue. We shall not display them explicitly and we shall merely denote them by $|\varphi_\alpha\rangle$ and $\langle\bar{\varphi}_\alpha|$ where here, as in the following, $\alpha = 1, 2, \ldots, 5$. In particular, as was pointed out previously in a different context [26], $|\varphi_5\rangle$ describes the one-particle kinetic energy, while $\langle\bar{\varphi}_5|$ corresponds to the total one-particle energy.

Moreover, although we cannot calculate them explicitly, we shall assume that the eigenvalues λ_n^{o} with $n \neq (\alpha)$ have no accumulation point at the origin, in order to make sure that transport coefficients can be defined [22] .

(**2**) Because of the five-fold degeneracy of $\lambda_\alpha^{o}(=0)$ we cannot immediately start the perturbation calculus with equations (15)-(18): we first have to remove this degeneracy by solving exactly the first-order problems [29]:

$$\left[C_o^{\ell} + q(C_o^{\ell'} - iv_x)\right] | \psi_r^{o} \rangle = \lambda_r^{(1)} | \psi_r^{o} \rangle \quad [r \in (\alpha)] \ , \tag{24a}$$

$$\langle \bar{\psi}_r^{o} | \left[C_o^{\ell} + q(C_o^{\ell'} - iv_x)\right] = \lambda_r^{(1)} \langle \bar{\psi}_r^{o} | \quad [r \in (\alpha)] \ , \tag{24b}$$

in the subspaces respectively spanned by $|\varphi_\alpha\rangle$ and $\langle\bar{\varphi}_\alpha|$.

This again can be done exactly in terms of the equilibrium properties of the system, using some simple properties of the operator $C_o^{\ell'}$; for example, we find:

$$|\psi_1^{o}\rangle = 2^{-\frac{1}{2}} \left[\frac{(kT)^{\frac{1}{2}}}{c} |\varphi_1\rangle + |\varphi_{v_x}\rangle + (\tfrac{3}{2} kT)^{\frac{1}{2}} \frac{(\partial p/\partial T)_n}{ncC_v} |\varphi_5\rangle \right] \tag{25}$$

with the corresponding eigenvalue $\lambda_1^{(1)} = ic$. In equation (25), the coefficients have their usual thermodynamic meaning; in particular, c is the adiabatic sound velocity (see equation [7]).

These new eigenfunctions $|\psi_r^{o}\rangle$ (and $\langle\bar{\psi}_r^{o}|$), together with

$$|\psi_n^{o}\rangle \equiv |\varphi_n\rangle \quad (\text{and } \langle\bar{\psi}_n^{o}| = \langle\bar{\varphi}_n|) \ , \quad n \notin (\alpha) \ , \tag{26}$$

provide us with a biorthonormal basis which we assume to be complete. They form a zeroth approximation from which the perturbation scheme defined in equations (15)-(18) can proceed in the standard way.

In particular, it is easy to obtain

$$\lambda_\alpha^{(2)} = - \sum_{n \neq (\alpha)} \langle \bar{\psi}_\alpha^{\,o} | (C_o^{\ell'} - iv_x) | \psi_n^{\,o} \rangle \frac{1}{\lambda_n^{\,o}} \langle \bar{\psi}_n^{\,o} | (C_o^{\ell'} - iv_x) | \psi_\alpha^{\,o} \rangle$$

$$+ \frac{1}{2} \langle \bar{\psi}_\alpha^{\,o} | C_o^{\ell''} | \psi_\alpha^{\,o} \rangle . \tag{27}$$

It is our claim that these five $\lambda_\alpha^{(2)}$ are identical to the q^2-part of the eigenvalues (5a,b,c). For example, using the explicit form of $|\psi_1^{\,o}\rangle$ and $\langle \bar{\psi}_1^{\,o}|$, we have shown that:

$$\lambda_1^{(2)} = - \frac{1}{2n} [(4\eta/3 + \zeta) + (1/C_v - 1/C_p)\kappa] \tag{28}$$

with microscopic expressions for the transport coefficients η, ζ, and κ which agree with the well-known autocorrelation function formulae [2-5] to all orders in the coupling constant. Although the proof of this equivalence follows very closely previous work by the author [30] and by Nicolis-Severne [31], it involves a fairly careful analysis of the many-body aspects of the problem and will not be treated here.

This result suggests that the eigenfunctions $|\psi_\alpha^{\,o}\rangle$ (and $\langle \bar{\psi}_\alpha^{\,o}|$) are, to lowest order in the uniformity parameter, the microscopic one-body analogue of the macroscopic hydrodynamic modes; for example, equation (25) describes a propagating sound-wave mode and the four other $|\psi_\alpha^{\,o}\rangle$ (and $\langle \bar{\psi}_\alpha^{\,o}|$) describe, respectively, a second sound mode, two shear-viscosity modes, and one entropy-fluctuation mode.

An important by-product of our calculation is that exact transport coefficients, including their potential part, can be obtained by solving the one-particle LGBE to second order in the uniformity parameter. Nowhere in this calculation is there any appeal to the functional relation (13) which is needed in the traditional approach. More precisely, a careful investigation of the various quantities which appear in equation (27) shows that in order to compute exact transport coefficients it is sufficient to know the single one-particle operator $\psi_q(v; z)$, defined by equation (8), in the immediate vicinity of the point $z = q = 0$; we have proved indeed that

$$\psi_o(v; o), \quad \psi_o'(v; o) \left\{ \equiv \left[\partial \psi_q(v; o)/\partial q \right]_{q=0} \right\},$$

$$\psi_o''(v; o), \quad \left(\frac{\partial \psi_o}{\partial z}\right)_{z=0}, \quad \left(\frac{\partial \psi_o'}{\partial z}\right)_{z=0}, \quad \text{and} \quad \left(\frac{\partial^2 \psi_o}{\partial z^2}\right)_{z=0}$$

are the only quantities which explicitly enter into equation (27).

Before closing this section, let us point out the recent paper of P. Martin and D. Forster [32] who have developed, independently of the present work, a formalism which treats linear response functions in terms of the eigenmodes of the generalized collision operator. Although their calculation is presently limited to a weak-coupling description of the collision operator, their conclusions are identical to ours in the hydrodynamic limit. Yet the technique used by these authors is very different from the one presented here, and it might be more powerful for the study of the departure from hydrodynamic behavior, i.e., for larger wavenumbers and frequencies.

3. APPLICATIONS AND DEVELOPMENTS OF THE THEORY

The central result of the preceding section is that in formulating the LGBE as an eigenvalue problem, we have explicitly constructed five (right- and left-) eigenfunctions which, to all order in the interaction, describe hydrodynamic modes in terms of one-body distributions. The corresponding eigenvalues give microscopic expressions for transport coefficients which agree with the correlation function formulae. Although the idea of getting hydrodynamic modes, and the corresponding transport coefficients, by an expansion of a transport equation in the uniformity parameter is not new--it was used, for example, in the theory of sound propagation in dilute gases [34] and in the mode-mode coupling approach to critical transport phenomena [15]--no such approach was previously developed in terms of one-body distributions for a strongly interacting system.

Though equation (27) offers in principle new microscopic expressions for transport coefficients, it is clear that the present method has not been devised for an explicit calculation of these coefficients. In particular, a tentative density expansion of these formulas would lead to the same divergence difficulty as the more usual methods [35]. On the contrary, the usefulness of our method lies in the fact that, whenever an expression of the type $(i\omega - iqv_x + C_q^{\ell})^{-1}$, involving the inverse linearized generalized Boltzmann operator, is encountered in a microscopic calculation, it need not be computed but can be replaced by

$$\lim_{q\to 0} \left(\frac{1}{i\omega - iqv_x + C_q^{\ell}}\right) = \sum_{r\in(\alpha)} |\bar{\psi}_r^{\,o}\rangle \frac{1}{i\omega + q\lambda_r^{(1)} + q^2\lambda_r^{(2)}} \langle\psi_r^{\,o}| , \quad (29)$$

where the transport coefficients appearing in $\lambda_r^{(2)}$ are replaced by the approximation adequate to the problem at hand.

Let us now briefly review a few problems where such a situation is encountered:

a) Low-wavenumber, low-frequency limit of the Van Hove correlation function.--The Van Hove function is defined by

$$G(r - r';t) = \frac{1}{n} \sum_{i,j} \int dr^N dv^N \, \delta(r - r_i) \exp(-iLt) \, \delta(r' - r_j) \, \rho_N^{eq}(r^N, v^N) , \quad (30)$$

where ρ_N^{eq} denotes the canonical equilibrium distribution.

As already mentioned in the introduction, heuristic macroscopic arguments indicate that for $q, \omega \to 0$, the Laplace-Fourier transform of equation (30) is given by [2, 16, 67]:

$$G_{q,\omega} \underset{q,\,\omega\to 0}{=} nkT\,\chi_T \Big[\left(1 - \frac{C_v}{C_p}\right) \frac{1}{-i\omega + q^2\kappa/n\,C_p} + \frac{1}{2}\sum_{\pm} \left(\frac{C_v}{C_p}\right) \frac{1}{-i\omega \pm icq + q^2\Gamma} \Big] , \quad (31)$$

where χ_T is the isothermal susceptibility.

Though there is little doubt that equation (31) is correct, it is nevertheless gratifying that a fully microscopic justification of this result can be given within the present formalism [36]. It would take us too far astray to give here the detailed derivation, and we shall limit ourselves in indicating the general idea: one first remarks that equation (30) can be written:

$$G(r,t) = \frac{1}{n} \int d^3v \, \bar{f}_1(r,v;t) \, , \tag{32}$$

where the one-body distribution $\bar{f}_1(r,v;t)$ is defined by

$$\bar{f}_1(r_1,v_1;t) = N \int dr^{N-1} \, dv^{N-1} \exp(-iLt) \sum_i \delta(-r_i) \, \rho_N^{eq} \, . \tag{33}$$

One then shows that $\bar{f}_1(r,v;t)$ obeys a kinetic equation of the type (8). In the small-q, small - ω limit, the solution of this kinetic equation is expressible in terms of the operator $(i\omega - iqv_x + C_q^{\ell})^{-1}$ which is treated as in equation (29); this leads then quite directly to the required result, equation (31).

b) Kinetic behavior of the van der Waals fluid and mode-mode coupling.--Much work has been devoted to the equilibrium properties of the van der Waals fluid, where the intermolecular potential is

$$V(r) = V^{SR}(r) + w(r) \, . \tag{34}$$

Here $V^{SR}(r)$ denotes a short-range potential and $w(r)$ is chosen to be a Kac long-range potential

$$w(r) = \gamma^3 \, v(\gamma r) \, , \tag{35}$$

whose range γ^{-1} is very long compared with the range of V^{SR}.

For example, one can show that the following expansion holds for the pressure [37]:

$$p = p^{SR} + \frac{n^2}{2\,kT} \int d^3r \, v(r) + \sum_{n=3}^{\infty} \gamma^n \, p^{(n)} \, , \tag{36}$$

leaving aside possible nonanalyticity for finite γ when T_c is approached.

It is of course tempting to develop a similar model for nonequilibrium properties. Assuming that the transport properties due to V^{SR} are known, we ask for the corrections due to the long-range potential. This model has at least a double interest:

1) Far from the critical point, it can serve as an alternative to the Rice-Alnatt theory [38], which assumes that the potential $w(r)$ is weak but essentially of finite range; here, on the contrary, the long-range character of the weak attractive part of the molecular interactions is explicitly taken into account.

2) Close to the critical point, the van der Waals fluid behaves classically when $\gamma \to 0$ [39] and, in parallel, its dynamical behavior will possibly be simpler than realistic systems.

As mentioned above, the basic operator in kinetic theory is $\psi_q(v;z)$ (see equation [11]) in the vicinity of $z = q = 0$. With the potential (34)-(35), we have thus to look at the γ behavior of $\psi_q(v;z\,|\,\gamma)$ when γ becomes small.

To simplify the problem, preliminary investigations were made in the strict homogeneous limit $q = 0$ [30]. Because the potential $w(r)$ is "weak," it is tempting to treat it by a naive perturbation method around the reference hard-core fluid; yet, when this is done, one immediately finds that this perturbation expansion is divergent. It is out of place to give here the details of the analysis, but the physical reason for this divergence is easy to understand. Indeed, in a perturbation description, the motion of the particles between the interactions is treated as free; the corresponding propagator is well known [19, 20]:

$$G_q^{(o)}(z) = \frac{1}{(z - iqv_x)} \quad . \tag{37}$$

Yet, if the potential is long range, the important values of q in equation (37) are small [$q \sim O(\gamma)$], which means that we take the particle as freely propagating over long distances. However, we are in a dense system and particles are continuously interacting; in the $q \to 0$ limit, we should--instead of equation (37)--consider the propagation of each particle in the presence of all the others. This means replacing equation (37) by

$$G_q^{(o)}(v; z) \longrightarrow G_q(v; z) = \frac{1}{(z - iqv_x + C_q^{\ell})} \ , \tag{38}$$

which, in the small-q and small-z limit can be treated according to equation (29).

Once this kind of "hydrodynamic renormalization" is performed i.e.,

$$\psi_o(v; z \mid \gamma, G_q^{o}) \longrightarrow \tilde{\psi}_o(v; o \mid \gamma, G_q) \ , \tag{39}$$

one finds indeed that $\tilde{\psi}_o$ can be expanded in power of γ:

$$\tilde{\psi}_o = \psi_o(v; o \mid w(r) \equiv 0) + \sum_{n=1}^{\infty} \gamma^n \tilde{\psi}_o^{(n)}(v; o \mid G_q) \ . \tag{40}$$

Note that there are no γ^o contributions due to w(r), in contrast to equation (36).

The theory outlines here is still in a preliminary stage, and its success will ultimately depend on the possibility of performing explicit calculations with this model. Yet we already find an interesting feature in equation (39); namely, we have an example of a <u>microscopic mode-mode description of the collision process</u>. A similar description is already known in the simpler problem of the diverging collision operator (three-body in two-dimensions, four-body in three-dimensions) of moderately dense gases [6, 10]. The case of the van der Waals fluid is sufficiently complex to indicate the way toward a similar formulation in the critical behavior of realistic fluids [15]. Yet, in this latter case, we are still at the level of pure conjecture.

Finally, let us point out that the hydrodynamic theory presented here can be extended with no great pain to fluid mixtures; the only new feature is the appearance of a diffusion mode together with the well-known thermal-diffusion coupling [41]. Similarly, there is no difficulty of principle in extending the above perturbation calculus to higher order in q. Yet the method presented here makes sense only if there is a sharp separation between the transport modes (which tend to zero when $q \to 0$) and the relaxation modes. Thus, for large q-values there is not much that is useful in the above formulation, and, as already mentioned, the theory of P. Martin and D. Forster [32] might be more powerful for the study of the nonhydrodynamic regime.

It is gratefully acknowledged that, in the course of the work presented here, the author has benefited from fruitful discussions with Professor J. Lebowitz, Professor G. Nicolis, and Dr. J. Piasecki.

REFERENCES

1. L. Landau and E. Lifshitz, Fluid Mechanics (London: Pergamon Press, 1963).

2. L. Kadanoff and P. Martin, Ann. Phys. 24, 419 (1963).

3. H. Mori, Phys. Rev. 112, 1829 (1958).

4. H. Mori, Progr. Theoret. Phys. (Japan), 28, 763 (1962).

5. H. Mori, Progr. Theoret. Phys. (Japan), 33, 423 (1965).

6. Y. Pomeau, Phys. Letters, 27A, 601 (1968); and Ph.D. thesis (unpublished), Universite de Paris (1969).

7. B. Alder and T. Wainwright, Phys. Rev. A1, 18 (1970).

8. K. Kawasaki, Phys. Letters, 32A, 379 (1970).

9. M. Ernst, E. Hauge, and J. Van Leeuwen, Phys. Rev. Letters, 25, 1254 (1970).

10. R. Dorfman and E. Cohen, Phys. Rev. Letters, 25, 1257 (1970).

11. Y. Pomeau, preprint (1970).

12. R. Zwanzig and M. Bixon, Phys. Rev. A2, 2005 (1970).

13. M. Fixman, J. Chem. Phys. 36, 310 (1961).

14. K. Kawasaki, Phys. Rev. 150, 291 (1966).

15. L. Kadanoff and J. Swift, Phys. Rev. 166, 89 (1968).

16. L. Van Hove, Phys. Rev. 95, 249 (1954).

17. R. Mountain, Rev. Mod. Phys. 38, 205 (1966).

18. A. Abrikosov, L. Gorkov, and I. Dzyaloskinsky, Methods of Quantum Field Theory in Statistical Physics (Englewood Cliffs, New Jersey: Prentice-Hall, 1963).

19. I. Prigogine, Nonequilibrium Statistical Mechanics (New York: Interscience, 1962).

20. P. Resibois, in Many Particle Physics, ed. E. Meeron (New York: Gordon and Breach, 1966).

21. P. Resibois, J. Stat. Phys. 2, 21 (1970).

22. P. Resibois, Bull. Cl. Sci. Ac. Roy. Belg. 56, 160 (1970).

23. E. Cohen, in Fundamental Problems in Statistical Mechanics, Vol. 2, ed. E. Cohen (Amsterdam: North Holland Publ. Co., 1968).

24. R. Zwanzig, Phys. Rev. 124, 983 (1961).

25. See, for example, S. Choch and G. Uhlenbeck, The Kinetic Theory of Dense Gases (unpublished, University of Michigan, 1958).

26. L. Garcia-Colin, M. Green, and F. Chaos, Physica, 32, 450 (1966).

27. See, for example, P. Dennery and A. Krzywicki, Mathematics for Physicists (New York: Harper and Row, 1967).

28. The role of this assumption appears clearly in the recent work of L. Sirovich and K. Thurber, J. Math. Phys. 10, 239 (1969) in the study of sound propagation; see also L. Sirovich, Phys. Fluids, 6, 10 (1963) and E. Uhlenbeck and G. Ford, Lectures in Statistical Mechanics (Providence, R.I.: American Math. Soc. 1963). Let us also point out that we are presently unable to say anything about the nature of the expansions (16)-(18) in the uniformity parameter q.

29. See, for example, A. Messiah, Mecanique Quantique, Vol. 2 (Dunod, 1960).

30. P. Resibois, J. Chem. Phys. 41, 2979 (1964).

31. G. Nicolis and G. Severne, J. Chem. Phys. 44, 1477 (1966).

32. D. Forster and P. Martin, Phys. Rev. A2, 1575 (1970).

33. See also Cl. George, Bull. Cl. Sci. Acad. Roy. Belg. 53, 623 (1967).

34. See G. Uhlenbeck and G. Ford, loc. cit.

35. See, for example, M. Ernst, L. Haines, and J. Dorfman, Revs. Mod. Phys. 41, 296 (1969), and references quoted there.

36. P. Resibois, Physica, 49, 591 (1970).

37. See, for example, J. Lebowitz, G. Stell, and S. Baer, Phys. Fluids, 6, 1282 (1964).

38. See, for example, S. Rice and P. Gray, The Statistical Mechanics of Simple Liquids (New York: Interscience, 1965).

39. See, for example, E. Stanley, Introduction to Phase Transition and Critical Phenomena (Oxford, 1971).

40. J. Piasecki and P. Resibois (unpublished).

41. G. Vasu and P. Resibois (unpublished).

DISCUSSION

R. Balescu: I should like to clarify a confusion which arose in the discussion about the meaning of "Markovian" and "non-Markovian" equations. The general kinetic equation presented by Resibois in the form

$$\partial_t f_q(t) + iqv_x f_q(t) = C_q^{\ell} f_q(t) \tag{a}$$

can also be written explicitly in the following form

$$\partial_t f_q(t) = i\Omega_q [\psi_q(o) - qv_x] f_q(t) , \tag{b}$$

where $\psi_q(o)$ is the limit as $z \to 0^+$ of an "irreducible" collision operator, and Ω_q is a complicated operator defined in terms of ψ_q and its derivatives at $z = 0$. If $G_q(t)$ denotes the inverse Laplace transform of $\psi_q(z)$, the same collision operator can be written in the equivalent form

$$\partial_t f_q(t) + iqv_x f_q(t) = \int_0^{\infty} d\tau \; G_q(\tau) f_q(t - \tau) , \tag{c}$$

which expresses the collision process in "non-Markovian" form. If the time delay in $f_q(t - \tau)$ were neglected in equation (b), equation (c) would reduce to (b) with $\Omega_q = 1$. The presence of the operator Ω_q precisely accounts for the finite duration of the collision process. It is also Ω_q which makes Resibois's collision operator nonsymmetric.

I made this remark in order to stress the fact that the "non-Markovian" behavior relevant to the transport coefficients is in fact fully retained in the collision operator C_q^{ℓ}.

P. Gray: Does Professor Resibois intend to give the impression that there is no systematic treatment of the Van Hove function, beginning from microscopic variables, and leading to the known hydrodynamic limit?

P. Resibois: I think there is no treatment at the level of kinetic equations, except for model systems.

P. Gray: I think there is: the work of Achasu and Duderstadt, for instance, and my own contributed paper.

P. Resibois: Their work is based on a formulation in terms of a projection operator, and does not make a microscopic analysis of their properties.

P. Gray: I consider that I have made a microscopic derivation because I am able to express the result in terms of generalized transport coefficients, which are identified with correlation function expressions.

P. Resibois: As is well known, in projection-operator techniques, one does not really reduce the N-body problem (involving, for example, a reduced distribution function), but one hides--although often elegantly and in a useful manner--the difficulty of the problem in a "projected" Hamiltonian whose properties are, to the best of my knowledge, largely unknown. This difficulty is completely avoided in the approach presented here.

B. Robertson: Dr. Resibois suggested that his method could be used in calculating "microscopic" quantities. This suggestion should be considered with caution. For example, his method cannot be used to calculate even the long-time behavior of correlation functions that include a projection operator. The reason for this is that the projection operator subtracts out the hydrodynamic motion at each stage of the time development. For this to happen it is necessary only to define the projection operator in terms of a thermodynamically complete set of variables. Then the resulting correlation functions are the correct ones to use for obtaining the transport coefficients. Further details are given in J. Math. Phys. **11**, 2482 (1970).

P. Resibois: In answer to this remark, I just want to stress again that I have discussed here the role that hydrodynamical modes play in the microscopic calculation of various quantities, in particular the Green-Kubo type of correlation function formulae. Various generalizations of these formulae, which have recently been proposed, have not been considered.

H. Mori: In the quantum-mechanical case, don't you have any difficulty in defining the local temperature to derive the temperature equation?

P. Resibois: Although the quantum-mechanical version of this theory has not been explicitly worked out, the formal analogy between the classical and quantum-mechanical form of the transport equation in normal fluids leads me to the belief that no difficulty should occur in the former case; in particular, the definition of a local temperature should derive naturally from the construction of the entropy eigenmode.

K. Matsuno: In a strongly interacting system, do you think that it is necessary to take into account an effect of coherent scatterings such as "quasi-bound state" in the small-wave-number region in addition to noncoherent scatterings, that is, hydrodynamic excitations?

P. Resibois: As far as I can see, the extension of the present theory to more complicated systems will automatically include any kind of excitations provided that there exists a hydrodynamical time scale larger than any other characteristic time of the problem.

D. Forster: Our work which you have mentioned (Forster and Martin, Phys. Rev. A4, 1575 [1970]) is based on a description of linearized kinetic theory in terms of correlation functions. It thus describes only small fluctuations from the start, but it describes those in a manner which is applicable to all, not just long, times and wavelengths. We have explicitly looked at a weakly coupled fluid; for it, most of the fundamental assumptions of your theory (e.g., concerning the spectrum of the collision operator) can be proved. However, making similar assumptions for arbitrary coupling, we have also obtained linearized hydrodynamics for the general case, in a manner similar to yours. The advantage of the correlation-function formulation, as I see it, is this: it allows us to establish those conditions on the nonlocal collision operator which guarantee the correct behavior both at long times and small wave-vector hydrodynamics and at short times (sum rules). For approximate calculations in the intermediate regime, e.g., in terms of models and interpolation schemes, this should be useful.

P. Resibois: As I mentioned in my talk, I do indeed believe that your method might be more adequate for arbitrary frequency; on the other hand, my calculations give an expression of the hydrodynamic mode in terms of equilibrium properties--and identify the eigenvalues with the Green-Kubo expression for transport coefficients for arbitrary coupling; this turns out to be extremely useful in many applications.

NUMBER AND KINETIC ENERGY DENSITY FLUCTUATIONS IN CLASSICAL LIQUIDS

Aneesur Rahman

1. INTRODUCTION

The last few years have seen a keen revival of interest in the phenomena related with the variation of the density of a system in space and time. This is due to advances in statistical mechanics on the theoretical side, to the availability of reactors on the experimental side, and to the possibility of using digital computers for large-scale numerical "experiments."

In a schematic way, without any effort at being exhaustive, one can recognize the landmarks in the theoretical development as the work of Kadanoff and Martin [1], Mori [2], and of Zwanzig [3]. The following review of the present situation is entirely based on the generalized Langevin equation approach which is the approach of Mori (implicitly) and of Zwanzig (explicitly).

The time development of a dynamical variable, say D, is determined by the Liouville operator, L, by the equation

$$D(t) = \exp(iLt)\ D(0)\ . \tag{1}$$

In statistical mechanics one is interested in the time dependence of the autocorrelation A(t) given by

$$A(t) \equiv \langle D^*(\tau)D(t+\tau)\rangle_{av} = \langle D^*(0)D(t)\rangle_{av}\ , \tag{2}$$

where $\langle\ .\ .\ .\ \rangle_{av}$ indicates averaging over the initial conditions D(0) using an appropriate equilibrium ensemble of states.

A simple example is the case of a polar medium in which each molecule carries a permanent dipole moment, $\underset{\sim}{\mu}$. In such a case the autocorrelation $\langle\underset{\sim}{\mu}(0)\cdot\underset{\sim}{\mu}(t)\rangle_{av}$ is directly related to the dielectric relaxation in the medium. In the present review we will not be concerned with this autocorrelation; it is mentioned here just to emphasize that the method of analysis presented in the following paragraphs has wide applicability. Here we shall restrict ourselves

---

Based on work performed under the auspices of the U. S. Atomic Energy Commission.

to a detailed consideration of two very special dynamical variables, namely (i) $\sum_{j=1}^{N} \delta(\underset{\sim}{r} - \underset{\sim}{r}_{j,t})$, where $\underset{\sim}{r}_{j,t}$ is the position vector at time t of particle labeled j in an N-particle system and (ii) $\sum_{j=1}^{N} \underset{\sim}{v}_{j,t}^{\,2}\, \delta(\underset{\sim}{r} - \underset{\sim}{r}_{j,t})$, where $\underset{\sim}{v}_{j,t}$ is the velocity of particle j at time t. These variables are respectively the number density (N.D) and the kinetic energy density (K.E.D.) (in appropriate units) at point $\underset{\sim}{r}$ at time t. The work of Van Hove [4] has shown that the first quantity leads to an autocorrelation, denoted by $G(\underset{\sim}{r},t)$, which is directly connected with the experimental technique of neutron inelastic scattering. A variable analogous to the second occurs, for example, in the study of hydrodynamic phenomena, where one is concerned with the coupled space-time fluctuations of number and energy density in the system.

2. DEFINITION OF OSCILLATORY AND NONOSCILLATORY VARIABLES

We shall first mention, without giving detailed proofs, certain elementary properties of autocorrelations of the type A(t), for the case of classical mechanics; slightly different statements need to be made for quantum mechanical systems.

$A(t) \equiv \langle D^*(0)D(t)\rangle_{av}$ is a real, even function of time with a Taylor expansion which we shall write as

$$A(t) = A_o - A_2 t^2/2! + A_4 t^4/4! \; . \; . \; . \; . \qquad (3)$$

Obviously $A_o \geq 0$ (being equal to $\langle D^*(0)D(0)\rangle_{av}$). However since, e.g., [5], $\ddot{A}(0) = -\langle \dot{D}^*(0)\dot{D}(0)\rangle_{av}$, $A_2, A_4, \ldots \geq 0$.

We shall assume that as $t \to \infty$ the correlations are dissipated so that $A(t \to \infty) \to 0$. Further, the integrals $I_o = \int_0^\infty A(s)ds$ and $I_1 = \int_0^\infty sA(s)ds$ will be assumed to exist with the additional property $I_o \geq 0$.

We shall classify dynamical variables D(t) as (1) nonoscillatory variables, those for which $I_o > 0$; and (2) oscillatory variables, those for which $I_o = 0$, with the auxiliary requirement that $t \int_0^t A(s)ds$ also $\to 0$ as $t \to \infty$.

The distinction is obviously unambiguous. The nomenclature "oscillatory variable" for the case $I_o = 0$ is quite natural because for I_o to be zero the function A(t), starting with its positive initial value A_o, must take on negative values at some time or the other such that the total area under A(t) becomes exactly zero. When $I_o > 0$, the autocorrelation may or may not take on negative values, and even if it does, the negative values are not enough to balance the positive values.

At this stage it will be useful to give a concrete example of these two types of variables.

If $\dot{x}$ is the velocity of a particle in the x-direction, then according to the Einstein equation the constant of self-diffusion is given by $\int_0^\infty \langle \dot{x}(0)\dot{x}(t)\rangle_{av}\, dt$. Hence for all systems (except those which are confined within impenetrable boundaries) the velocity of a particle is a nonoscillatory variable. However, the force, or $\ddot{x}$, is an oscillatory variable in all cases. This follows from the equation $\langle \dot{x}(0)\dot{x}(t)\rangle_{av} = \langle \dot{x}^2\rangle_{av} - \int_0^t (t-s)\langle \ddot{x}(0)\ddot{x}(s)\rangle_{av}\, ds$, which, on going to $t = \infty$, implies $\int_0^\infty \langle \ddot{x}(0)\ddot{x}(t)\rangle_{av}\, dt = 0$, and $\langle \dot{x}^2\rangle_{av} + \int_0^\infty t\,\langle \ddot{x}(0)\ddot{x}(t)\rangle_{av}\, dt = 0$.

3. THE LANGEVIN EQUATION

In this section we shall briefly recall the well-known Langevin equation before proceeding to its extension in the next section.

According to the Langevin equation the dissipation of the autocorrelation A(t) takes place as

$$\dot{A}(t) = -\alpha A(t) \text{ or } A(t) = A_o \exp(-\alpha|t|), \quad \alpha > 0 . \tag{4}$$

An expansion of A(t) in even powers of t is not possible. However, one can still use equation (4) provided no conclusions are drawn from it about the short time behavior of the autocorrelation A(t). From equation (4) we get

$$I_o = \int_0^\infty A(t)dt = A_o/\alpha > 0 . \tag{5}$$

Thus, if A(t) obeys the Langevin equation, the corresponding D(t) is a nonoscillatory variable according to the criterion formulated in the previous section.

An equation analogous to the above but applicable to an oscillating variable is

$$\dot{A}(t) = -\omega_o^2 \int_0^t A(s)ds - \alpha A(t), \quad \alpha \text{ and } \omega_o^2 > 0 , \tag{6}$$

or

$$A(t) = A_o - \omega_o^2 \int_0^t (t-s)A(s)ds - \alpha \int_0^t A(s)ds . \tag{7}$$

For A(t) to vanish at $t = \infty$ we must have $I_o = 0$ such that $t \int_0^t A(s)ds \to 0$ with $t \to \infty$, thus making A(t) the autocorrelation of an oscillating variable D. Notice also that, A_o, ω_o^2 being positive and $A_o + \omega_o^2 I_1 = 0$, we must have $I_1 < 0$.

This equation represents a simple harmonic oscillator when $\alpha = 0$, but an increase in the value of α does not modify the oscillating nature of the variable in the sense prescribed in the previous section. We also note in passing that the equation for A(t) is simply that of a damped harmonic oscillator, namely, $\ddot{A} + \alpha\dot{A} + \omega_o^2 A = 0$ with the constraint $\dot{A}(t=0) = -\alpha A_o$.

4. THE GENERALIZED LANGEVIN EQUATION FORMALISM

Notice that in the equations written in the previous section the dissipation of the autocorrelation is produced by the term $-\alpha A(t)$ and is thus completely known at time t without any information about the behavior of A(t) at previous times.

The insight we have gained from the work of Zwanzig enables us to state that a generalization of these equations is to make the dissipative process non-Markovian, i.e., to include, using an appropriate weighting function, all the values of A(t) from A(0) to A(t) in determining the amount of dissipation at time t. A physically meaningful weighting function will be expected to vanish for large times.

For the nonoscillatory variable we get

$$\dot{A} = - \int_0^t K(s)A(t-s)ds\,, \qquad K_0 > 0\,. \tag{8}$$

[The previous simpler, but inexact, equation (4) is regained by taking K(t) a function going to zero so rapidly that the integral above is essentially $\propto A(t)$ except when $t \to 0$.]

If the Taylor expansion for K(t) is written as $K_0 - K_2t^2/2! + \ldots$, we get $K_0 = A_2/A_0$; and integrating both sides from 0 to ∞ gives

$$A_0 = I_0 \int_0^\infty K(s)ds\,. \tag{9}$$

Thus equation (8) is applicable only if A(t) is the autocorrelation of a nonoscillatory variable; otherwise $\int_0^\infty K(s)ds$ would diverge.

Notice that a knowledge of A_0, A_2, and I_0 determines K_0 and $\int_0^\infty K(t)dt$.

In an analogous manner we write for the oscillatory case, denoting the autocorrelation by B and the weighting function by L to avoid confusion with the previous case,

$$\dot{B} = -\omega_0^2 \int_0^t B(s)ds - \int_0^t L(s)B(t-s)ds\,, \qquad L_0 \text{ and } \omega_0^2 > 0\,, \tag{10}$$

or

$$B = B_0 - \omega_0^2 \int_0^t (t-s)B(s)ds - \int_0^t L(s)ds \int_0^{t-s} B(u)du\,, \tag{11}$$

so that now $L_0 = (B_2/B_0) - \omega_0^2$, and

$$B_0 + \omega_0^2 J_1 = 0\,, \tag{12}$$

where $J_1 = \int_0^\infty tB(t)dt$.

Notice that a knowledge of B_0, B_2, J_1 determines ω_0^2 and L_0, but the area under L(t) is not constrained by this information. Also the frequency, ω_0, given by equation (12), is determined in terms of the initial value B_0 and an integral, J_1, involving all the values of B(t).

The situation represented by this equation will tend toward harmonic oscillations in two circumstances. If L(t) is constant $= L_0$, then we get pure harmonic oscillations of frequency $= (B_2/B_0)^{\frac{1}{2}}$. However, if L(t) decays from initial value L_0 in a time very large compared to $(B_0/B_2)^{\frac{1}{2}}$, we get a damped oscillatory behavior with a spectrum showing a maximum at a frequency $<(B_2/B_0)^{\frac{1}{2}}$ and a width inversely proportional to the lifetime of L(t).

It will be necessary and convenient quite often to consider not the autocorrelation itself but its Laplace-Fourier transform:

$$\tilde{A}(\omega) = \int_0^\infty e^{-i\omega t} A(t)\,dt \quad . \tag{13}$$

For the case represented by equation (8),

$$\tilde{A}(\omega) = \frac{A_o}{i\omega + \tilde{K}(\omega)} \quad , \tag{14}$$

in which $\tilde{K}(\omega)$ is usually described as a frequency-dependent friction constant. Equation (4) would have given $\tilde{A}(\omega) = A_o/(i\omega + \alpha)$.

For the oscillatory case of equation (10), we get

$$\tilde{B}(\omega) = \frac{B_o}{i\omega + \omega_o^2/i\omega + \tilde{L}(\omega)} \quad . \tag{15}$$

The oscillatory nature shows itself up explicitly in the term $\omega_o^2/i\omega$ which makes $\tilde{B}(\omega = 0) = 0$, this being the condition for B(t) to be oscillatory.

The dissipative functions of the type K(t) and L(t) are usually referred to as memory functions for the respective autocorrelations. We shall use this nomenclature in the following sections.

5. GENERALIZED LANGEVIN EQUATION FOR A NONOSCILLATORY A(t) AND ITS SECOND DERIVATIVE

It is easily proved that if $A(t) = \langle D(0)D^*(t)\rangle_{av}$, then [5] $\ddot{A}(t) = -\langle \dot{D}^*(0)\dot{D}(t)\rangle_{av}$, i.e., $-\ddot{A}$ is the autocorrelation of the derivative of D(t). But, A(t) being an even function of t, $\dot{A}(0) = 0$. Hence

$$A(t) = A_o + \int_0^t (t-u)\ddot{A}(u)\,du \quad , \tag{16}$$

and going to $t = \infty$ we see that $\ddot{A}(t)$ is an oscillatory autocorrelation, irrespective of the nature of A(t).

Now, let A(t) be nonoscillatory obeying equation (8). Then $K_o = A_2/A_o$ and $\int_0^\infty K(s)ds = A_o/\int_0^\infty A(s)ds$; and since $\ddot{A}$ is oscillatory, let

$$\frac{d}{dt}\ddot{A} = -\omega_o^2 \int_0^t \ddot{A}(s)ds - \int_0^t L(s)\ddot{A}(t-s)ds \quad , \tag{17}$$

where $\omega_o^2 = A_2/A_o = K_o$, and $L_o = A_4/A_2 - A_2/A_o$. It is then easy to show that

$$\dot{K}(t) = -\int_0^t L(s)K(t-s)ds \quad , \tag{18}$$

i.e., <u>L is the memory function of the memory function of A</u>.

The above statements, in terms of Fourier-Laplace transforms, are equivalent to the following:

$$\tilde{A}(\omega) = \frac{A_o}{i\omega + \tilde{K}(\omega)} = \frac{A_o}{i\omega + \omega_o^2/[i\omega + \tilde{L}(\omega)]} \quad , \tag{19}$$

which leads to the continued fraction approach of Mori [2]. Writing $B = -\dot{A}$, we get

$$\tilde{B}(\omega) = \frac{B_o}{i\omega + \omega_o^2/i\omega + \tilde{L}(\omega)} \qquad [\equiv i\omega A_o + \omega^2\tilde{A}(\omega)] , \tag{20}$$

where $B_o = A_2$, $\omega_o^2 = A_2/A_o$, $L_o = B_2/B_o - \omega_o^2 = A_4/A_2 - A_2/A_o$, and $\tilde{L}(\omega = 0) = (\omega_o^2/A_o)\int_0^\infty A(t)\,dt$.

It is now clear that the strength of this formalism lies in pinning down the unknown function L(t) rigorously, in terms of the initial time behavior of A(t) (A_o, A_2, A_4) on the one hand and the area under A(t) from 0 to ∞ on the other. The success of the procedure will then depend on the extent to which L(t) is a "simple" function of time; if it turns out that L(t), together with the above constraints, cannot be guessed with any degree of certainty, then one can claim that the formalism outlined above is inadequate for understanding the nature of the autocorrelation A(t) and hence of the dynamical variable D(t).

6. FLUCTUATIONS OF DENSITY AND KINETIC ENERGY IN LIQUIDS

The introductory material presented above will now be applied to the fluctuations of density and kinetic energy in a liquid.

We define the number density (N.D.) and the kinetic energy density (K.E.D.) in the system as

$$N(\underset{\sim}{r},t) = N^{-\frac{1}{2}} \sum_{j=1}^{N} \delta(\underset{\sim}{r} - \underset{\sim}{r}_{j,t}) , \tag{21a}$$

$$M(\underset{\sim}{r},t) = N^{-\frac{1}{2}} \sum_{j=1}^{N} \underset{\sim}{v}_{j,t}^{\,2}\, \delta(\underset{\sim}{r} - \underset{\sim}{r}_{j,t}) , \tag{21b}$$

respectively. In these definitions $\underset{\sim}{r}_{j,t}$ and $\underset{\sim}{v}_{j,t}$ are the position and velocity of particle j at time t. [N is the total number of particles and is not to be confused with $N(\underset{\sim}{r},t)$ defined in equation (21a).]

The space Fourier transforms are given by

$$N(\underset{\sim}{\varkappa},t) = N^{-\frac{1}{2}} \sum_{j=1}^{N} \exp(i\underset{\sim}{\varkappa}\cdot\underset{\sim}{r}_{j,t}) , \tag{22a}$$

$$M(\underset{\sim}{\varkappa},t) = N^{-\frac{1}{2}} \sum_{j=1}^{N} \underset{\sim}{v}_{j,t}^{\,2} \exp(i\underset{\sim}{\varkappa}\cdot\underset{\sim}{r}_{j,t}) . \tag{22b}$$

Notice that $dN(\underset{\sim}{\varkappa},t)/dt = iJ_{\parallel}(\underset{\sim}{\varkappa},t)$, where $J_{\parallel}$ is the longitudinal component of the current density, namely, $N^{-\frac{1}{2}} \sum_{j=1}^{N} (\underset{\sim}{\varkappa}\cdot\underset{\sim}{v}_{j,t})\exp(i\underset{\sim}{\varkappa}\cdot\underset{\sim}{r}_{j,t})$.

The autocorrelations of $N(\underset{\sim}{\kappa},t)$, $J_{\parallel}(\underset{\sim}{\kappa},t)$, and $M(\underset{\sim}{\kappa},t)$ will be denoted by $F(\underset{\sim}{\kappa},t)$, $G(\underset{\sim}{\kappa},t)$, and $H(\underset{\sim}{\kappa},t)$, respectively. These are given by the following expressions:

$$F(\underset{\sim}{\kappa},t) = \langle N^{-1} \sum_j \sum_\ell \exp[i\underset{\sim}{\kappa} \cdot (\underset{\sim}{r}_{\ell,o} - \underset{\sim}{r}_{j,t})]\rangle_{av} , \tag{23a}$$

$$G(\underset{\sim}{\kappa},t) = \langle N^{-1} \sum_j \sum_\ell (\underset{\sim}{\kappa} \cdot \underset{\sim}{v}_{\ell,o})(\underset{\sim}{\kappa} \cdot \underset{\sim}{v}_{j,t}) \exp[i\underset{\sim}{\kappa} \cdot (\underset{\sim}{r}_{\ell,o} - \underset{\sim}{r}_{j,t})]\rangle_{av} , \tag{23b}$$

$$H(\underset{\sim}{\kappa},t) = \langle N^{-1} \sum_j \sum_\ell \underset{\sim}{v}_{\ell,o}^{\,2} \underset{\sim}{v}_{j,t}^{\,2} \exp[i\underset{\sim}{\kappa} \cdot (\underset{\sim}{r}_{\ell,o} - \underset{\sim}{r}_{j,t})]\rangle_{av} . \tag{23c}$$

Henceforth we shall write F, G, H, and various other quantities as functions of $\kappa = |\underset{\sim}{\kappa}|$, since we shall be concerned only with isotropic systems.

Following the convention introduced in the earlier part of this presentation we shall write, e.g., H_o, $-H_2$, H_4, etc., to denote derivatives of $H(\underset{\sim}{\kappa},t)$, at $t = 0$, of order 0, 2, etc., in the manner of equation (3).

The following is a summary of the properties of the various derivatives of interest, and includes the definitions of the frequencies ω_o, ω_ℓ, ω_T and Ω which will be used later.

(i) F_o, usually denoted by $S(\kappa)$, is the structure factor given in terms of the pair correlation function g(r) by the equation

$$S(\kappa) = 1 + \frac{N}{V} \int d\underset{\sim}{r} \; [g(r) - 1] \; (\sin \kappa r/\kappa r) . \tag{24}$$

$$\text{(ii)} \quad F_2 \equiv G_o = \kappa^2 \langle \underset{\sim}{v}^2 \rangle_{av}/3 = \kappa^2 k_B T/M = \omega_o^{\,2} F_o \quad (\text{Def. of } \omega_o^{\,2}) . \tag{25}$$

$$\begin{aligned} \text{(iii)} \quad F_4 &\equiv G_2 = F_2 [3F_2 + \Omega^2(0) - \Omega^2(\kappa)] \\ &= \omega_\ell^{\,2} F_2 \quad (\text{Def. of } \omega_\ell^{\,2}) , \end{aligned} \tag{26}$$

where

$$\Omega^2(\kappa) = \frac{N}{V} \int d\underset{\sim}{r} \; g(r) (\cos \kappa x) (1/M) \, \partial^2 \varphi/\partial x^2 , \tag{27}$$

where φ is the potential of interaction between the particles. Notice that

$$\Omega^2(0) = \langle a^2 \rangle_{av} / \langle v^2 \rangle_{av} , \tag{28}$$

where

$$\underset{\sim}{a} = d\underset{\sim}{v}/dt .$$

$$\text{(iv)} \quad H_o = \frac{15}{9} \langle \underset{\sim}{v}^2 \rangle_{av}^{\;2} + [S(\kappa) - 1] \langle \underset{\sim}{v}^2 \rangle_{av}^{\;2} , \tag{29}$$

where the first term is obtained out of the diagonal terms in the double summation which defines $H(\underset{\sim}{\kappa}, t = 0)$; the second term is obtained from the nondiagonal terms using the fact that positions and velocities in a classical system are statistically independent. Simplifying the above expression for H_o gives $H_o = [6 + 9S(\kappa)] (k_B T/M)^2$.

$$(v) \quad H_2 = [12\,\Omega^2(0) + 35\varkappa^2 k_BT/M]\,(k_BT/M)^2$$
$$= \omega_T^2\, H_o \text{ (Def. of } \omega_T^2), \tag{30}$$

which is most easily obtained by considering the autocorrelation of $dM(\varkappa,t)/dt$, namely $N^{-\frac{1}{2}} \sum_i (2\underset{\sim}{v}_i \cdot \underset{\sim}{a}_i + i\underset{\sim}{\varkappa} \cdot \underset{\sim}{v}_i\, \underset{\sim}{v}_i^2) \exp(i\underset{\sim}{\varkappa} \cdot r_i)$ and using the fact that only the diagonal terms are nonzero in the double sum arising out of the product of two single sums. The term $12\Omega^2(0)\,(k_BT/M)^2$ is then the value of $4\langle(\underset{\sim}{v}_i \cdot \underset{\sim}{a}_i)^2\rangle_{av}$, and $35\varkappa^2(k_BT/M)^3$ is that of $\langle(\underset{\sim}{\varkappa} \cdot \underset{\sim}{v}_i)^2 \underset{\sim}{v}_i^4\rangle_{av}$.

Finally, we shall write $\pi S(\varkappa,\omega)$, $\pi C(\varkappa,\omega)$, and $\pi T(\varkappa,\omega)$ for the Laplace-Fourier transforms, so that

$$S(\varkappa,\omega) = \frac{1}{\pi}\int_0^\infty e^{-i\omega t}\, F(\varkappa,t)\, dt \ , \tag{31a}$$

$$C(\varkappa,\omega) = \frac{1}{\pi}\int_0^\infty e^{-i\omega t}\, G(\varkappa,t)\, dt = [i\frac{\omega}{\pi}F_o + \omega^2 S(\varkappa,\omega)]\ , \tag{31b}$$

$$T(\varkappa,\omega) = \frac{1}{\pi}\int_0^\infty e^{-i\omega t}\, H(\varkappa,t)\, dt \ . \tag{31c}$$

It follows that F_o, F_2, F_4, H_o, H_2 are just the appropriate moments of S, C, and T, i.e., integrals of the type $\int_{-\infty}^{+\infty} \omega^{2n}\, f(\varkappa,\omega)\, d\omega$.

Some limiting properties of these moments are of interest. As $\varkappa \to 0$ we have $F_o \equiv S(\varkappa) \to (N/V)k_BT\chi_T = k_BT/MC_T^2$, where C_T is the isothermal sound velocity; $\omega_o^2 \to \varkappa^2 C_T^2$; $\omega_\ell^2 \to \varkappa^2 C_3^2$ [6]; $\omega_T^2 \to 2\,\Omega^2(0)/[1 + 1.5S(0)]$; and since S(0) for most liquids is of the order of 3 to 5 x 10^{-2}, the value of ω_T^2 within a few percent is simply $2\Omega^2(0)$ [and thus can be obtained from the properties of the single-particle velocity autocorrelation function, $\langle \underset{\sim}{v}(0) \cdot \underset{\sim}{v}(t)\rangle_{av}$].

For large $\varkappa$, on the other hand, one gets $F_o = 1$, $\omega_\ell^2 - 3\omega_o^2 = \Omega^2(0)$, $H_o = 15(k_BT/M)^2$, and $\omega_T^2 = (7/3)\omega_o^2$, while ω_o^2 increases indefinitely as $\varkappa^2 k_BT/M$. For an ideal gas ω_ℓ^2 becomes equal to $3\omega_o^2$.

The memory-function formalism presented in the previous section enables us to write

$$\pi S(\varkappa,\omega) = S(\varkappa)\left[i\omega + \frac{\omega_o^2}{i\omega + (\omega_\ell^2 - \omega_o^2)\tilde{K}_c(\omega)}\right]^{-1} , \tag{32a}$$

$$\pi C(\varkappa,\omega) = \frac{\omega_o^2 S(\varkappa)}{i\omega + (\omega_o^2/i\omega) + (\omega_\ell^2 - \omega_o^2)\tilde{K}_c(\omega)} = [i\omega S(\varkappa) + \omega^2 \pi S(\varkappa,\omega)], \tag{32b}$$

$$\pi T(\varkappa, \omega) = \frac{H_o}{i\omega + \omega_T^2 \tilde{K}_T(\omega)} \tag{32c}$$

where $K_c(t)$ denotes the memory function for the current autocorrelation $G(\varkappa, t)$:

$$\tilde{K}_c(\omega) = \int_0^\infty e^{-i\omega t} K_c(t)\, dt, \qquad K_c(t=0) = 1 \ , \tag{33}$$

and similarly for $K_T(t)$, the memory function for $H(\varkappa, t)$.

The purpose of the data presented in the following paragraphs is to investigate the nature of $K_c(t)$ and $K_T(t)$. Both these are of course functions of $\varkappa$ as well.

7. DENSITY AND KINETIC ENERGY FLUCTUATIONS IN A LIQUID-SODIUM-LIKE SYSTEM

Using an effective interionic pair potential for liquid sodium derived by Shyu et al. [7], the author has made detailed molecular-dynamics calculation for a system of 500 particles which is expected to simulate the motion of ions in liquid sodium at 397° K with a density of 0.927 g cm^{-3}. Some of the relevant details will be given elsewhere.

The trajectories generated in this calculation were used as primary data for the evaluation of the autocorrelations F, G, H (equations [23]) for $\varkappa = 0.23$ Å^{-1}. The following values were found for the quantities that determine F_0, F_2, . . . H_2 (equations [24]-[30]):

$$S(\varkappa) = 0.031, \quad \tau\omega_o = 4.17, \quad \tau\omega_\ell = 6.10, \quad \tau\Omega(0) = 13.7 \ , \tag{34}$$

where τ, the unit of time, is 0.845×10^{-12} sec.

The functions $F(\varkappa,t)$, $G(\varkappa,t)$, and $H(\varkappa,t)$ are shown in figure 1; and the real part of the Laplace-Fourier transforms defined by equations (31), in figure 2.

The accuracy of the data shown in figure 1 decreases as t increases. An estimate of the errors in all three curves is as follows.

At $t/\tau = 1$, the values shown are uncertain by above ± 0.05. The variation in $H(\varkappa,t)$ between $t/\tau = 0.1$ and $t/\tau = 0.2$ is quite genuine. At $t/\tau = 2.5$, however, the values of F and G shown in the figure are comparable to the uncertainty. This happens in H already at $t/\tau = 1.8$.

In figure 2 there are variations of intensity which arise out of the usual termination problems in numerical Fourier transformations. These are indicated by cutting across the spurious maxima and minima with a dashed line. A rough estimate of the overall accuracy of the transform is the amplitude of the spurious variations around the dashed line.

a. $F(\varkappa,t)$ and $G(\varkappa,t)$

The first remarkable feature is that the N.D. fluctuations clearly show a propagating wave with an identifiable frequency at about $5.3\,\tau^{-1}$. Since the value of $\varkappa$ is 0.23 Å^{-1}, this gives a velocity of sound = 2.7×10^5 cm sec^{-1}. (In liquid Na at 371° K the value is 2.5×10^5 cm sec^{-1}.)

Second, the spectrum contains a maximum at $\omega = 0$ as well, implying, as in the case of the hydrodynamic regime, the presence in the system of two distinct relaxation processes for

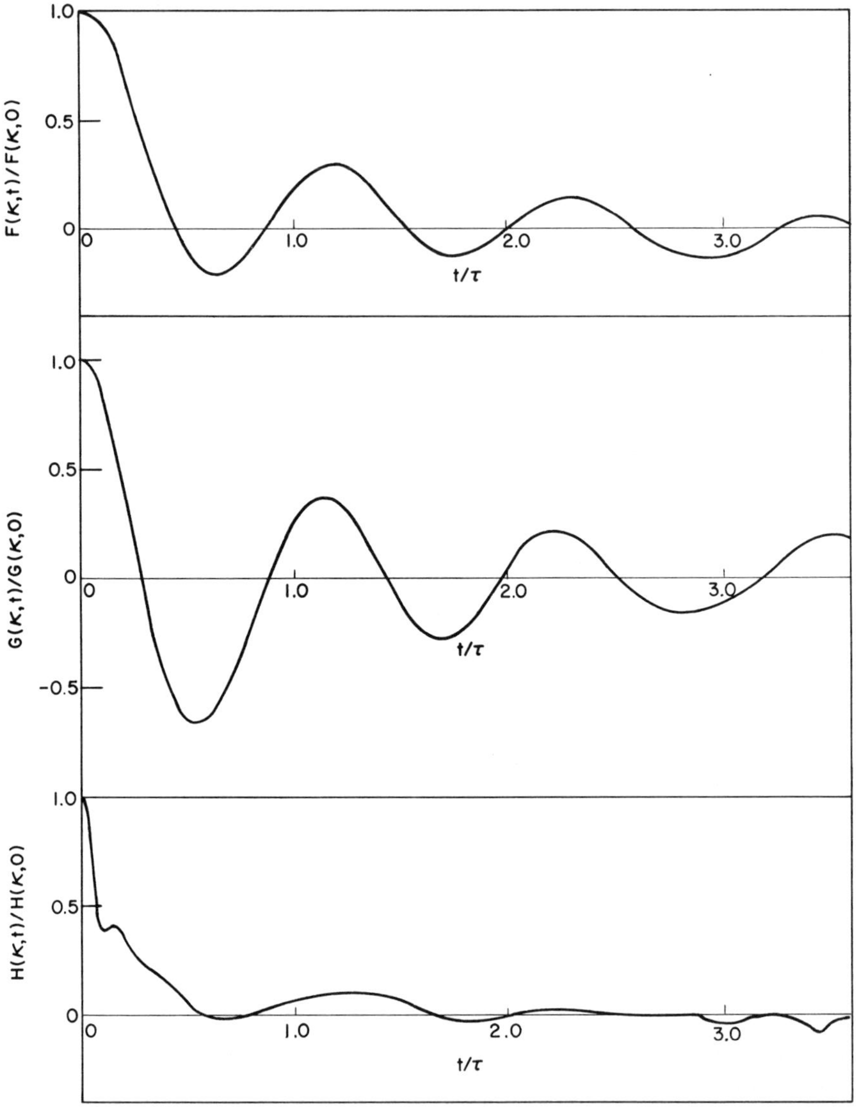

Fig. 1. Number, current, and kinetic-energy density fluctuations, F, G, H, respectively, at $\varkappa = 0.23\ \text{Å}^{-1}$. Statistical errors dominate beyond $t/\tau = 2.5$ for the top two curves and $t/\tau = 1.8$ for the bottom curve. The variation in the latter at $t/\tau = 0.1$ is <u>not</u> statistical in origin.

N.D. fluctuations of this wavelength. It should be noted that a calculation of this type was first performed by Kurkijarvi et al. [8], whose work has remained unpublished except for a brief mention of the results [9].

It is shown in the Appendix that in the hydrodynamic limit the memory function for $G(\varkappa,t)$, namely $K_c(t)$, consist of two parts: one part has a relatively fast decay and is due to

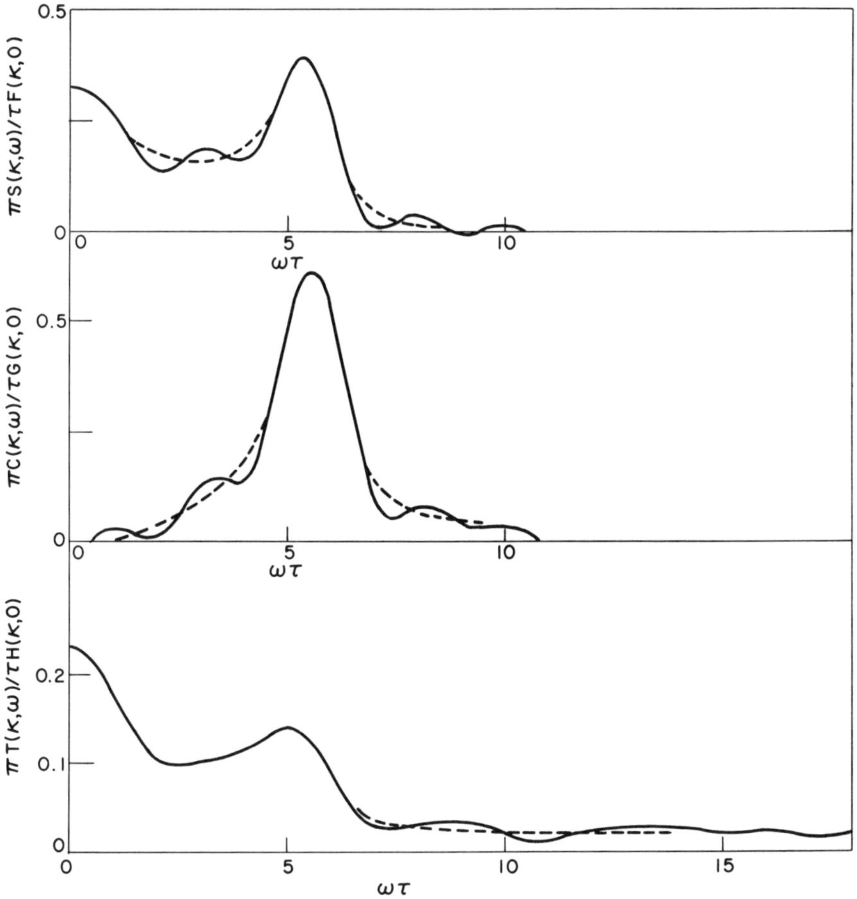

Fig. 2. Cosine transforms of data shown in fig. 1. The spurious intensity variations are pointed out by drawing a smooth curve through them (shown by ---). The peak in $S(\varkappa, \omega)$ at $\omega\tau = 5.3$ gives velocity of sound $= 2.7 \times 10^5$ cm sec^{-1}.

memory effects in dissipation from viscosity while the other arises out of dissipation due to heat diffusion. It is also shown in the Appendix (see equation [A5]) that the relative weights of these two parts of $K_c(t)$ in the hydrodynamic limit are proportional to $(\omega_\ell^2 - \gamma\omega_o^2)$ and $(\gamma - 1)\omega_o^2$, where γ is the ratio of specific heats C_p/C_v or C_s^2/C_T^2, C_s and C_T being the usual sound velocities. Using the value of C_s given by the peak in $S(\varkappa, \omega)$ (fig. 2) and of C_T given by equating ω_o^2 to $\varkappa^2 C_T^2$, we get $\gamma \simeq 1.5$. (In liquid Na, at 371° K, the value of γ is quoted as 1.1.)

Since the data at $\varkappa = 0.23$ Å^{-1} show two distinct peaks in the N.D. fluctuations, we shall assume a hydrodynamic form for $K_c(t)$, namely,

$$(\omega_\ell^2 - \omega_o^2)K_c(t) = (\omega_\ell^2 - \gamma\omega_o^2)\exp(-\mu^2 t^2) + (\gamma - 1)\omega_o^2 \exp(-\nu^2 t^2). \qquad (35)$$

It is then easy to show that the coefficient G_4 in the small time expansion of G(t) is given by

$$G_4/G_o = \omega_\ell^4 + 2(\omega_\ell^2 - \gamma\omega_o^2)\mu^2 + 2(\gamma-1)\omega_o^2\nu^2 . \qquad (36)$$

We also have

$$\pi S(\kappa, \omega=0) = \frac{S(\kappa)}{2}(\omega_\ell^2 - \omega_o^2)\tilde{K}_c(\omega=0)$$
$$= \frac{S(\kappa)}{\omega_o^2}\,\tfrac{1}{2}\pi^{\frac{1}{2}}\left(\frac{\omega_\ell^2-\gamma\omega_o^2}{\mu} + \frac{(\gamma-1)\omega_o^2}{\nu}\right) . \qquad (37)$$

The two unknowns μ, ν can now be found by determining G_4/G_o [from the initial values of $G(t)/G_o$ shown in fig. 1] and $\pi S(\kappa, \omega=0)$ shown in figure 2.

Using $G_4/G_o = 3000\,\tau^{-4}$, $\gamma = 1.5$, $\pi S(\kappa, \omega=0) = 0.32\,\tau$, one gets $\mu\tau = 8.34$ and $\nu\tau = 1.77$.

Figure 3 shows $F(\kappa,t)$ and $G(\kappa,t)$ obtained with the memory function given by equation (35). The data points from figure 1 are shown as circles and the agreement is fairly satisfactory.

The manner of calculating μ and ν from equations (36) and (37) is, however, not fully satisfactory. First, the value of γ cannot be determined unambiguously because of the spurious intensity variations in $S(\kappa, \omega)$ in figure 2. Second, and for the same reason, $S(\kappa, \omega=0)$ is not well determined to the extent of about 10%. Third, G_4/G_o is hard to determine from a numerical table of values of $G(\kappa,t)/G(\kappa,0)$, and the value $G_4/G_o = 3000\,\tau^{-4}$ is uncertain by about 20%.

Hence it is interesting to remark here that by using γ, μ, ν as least-squares fitting parameters for $F(\kappa,t)$ one gets $\gamma = 1.45$, $\mu\tau = 8.57$, $\nu\tau = 1.63$ as the best values, which are indeed quite close to the values directly calculated from the data by using equations (36) and (37).

To recapitulate, we have used, as input, the short time behavior of the autocorrelation $F(\kappa,t)$ and the area under it, to construct a memory function containing two distinct frictional processes of very different decay constants ($\mu/\nu \sim 5$); this memory function gives a satisfactory representation of the data.

b. $H(\kappa, t)$

Unlike the N.D. autocorrelation, the situation regarding the K.E.D. autocorrelation is quite difficult. Figure 1 shows the initial fast decay of this autocorrelation arising out of the high value of $\Omega^2(0)$. This is reversed quite suddenly at about $t/\tau = 0.1$. It is worth emphasizing that the minimum and maximum between $t/\tau = 0.1$ and $t/\tau = 0.2$ is not due to insufficient statistics.

The short time behavior of $H(\kappa,t)$ is shown again in figure 4 together with the memory function $K_T(t)$. The oscillation of $K_T(t)$ with a large negative value at about $t/\tau = 0.075$ is unmistakable; in fact, a function of the form $\exp(-\lambda^2 t^2)\cos(ft)$ with $\lambda\tau \cong 11$ and $f\tau \simeq 30$ fits the function $K_T(t)$ quite well up to about $t/\tau = 0.15$.

As we have seen in the introductory sections, the memory function is a more general description of dissipative processes than that contained in a simple Langevin equation. A large negative value of the memory function implies a tendency toward an increased correlation, or at

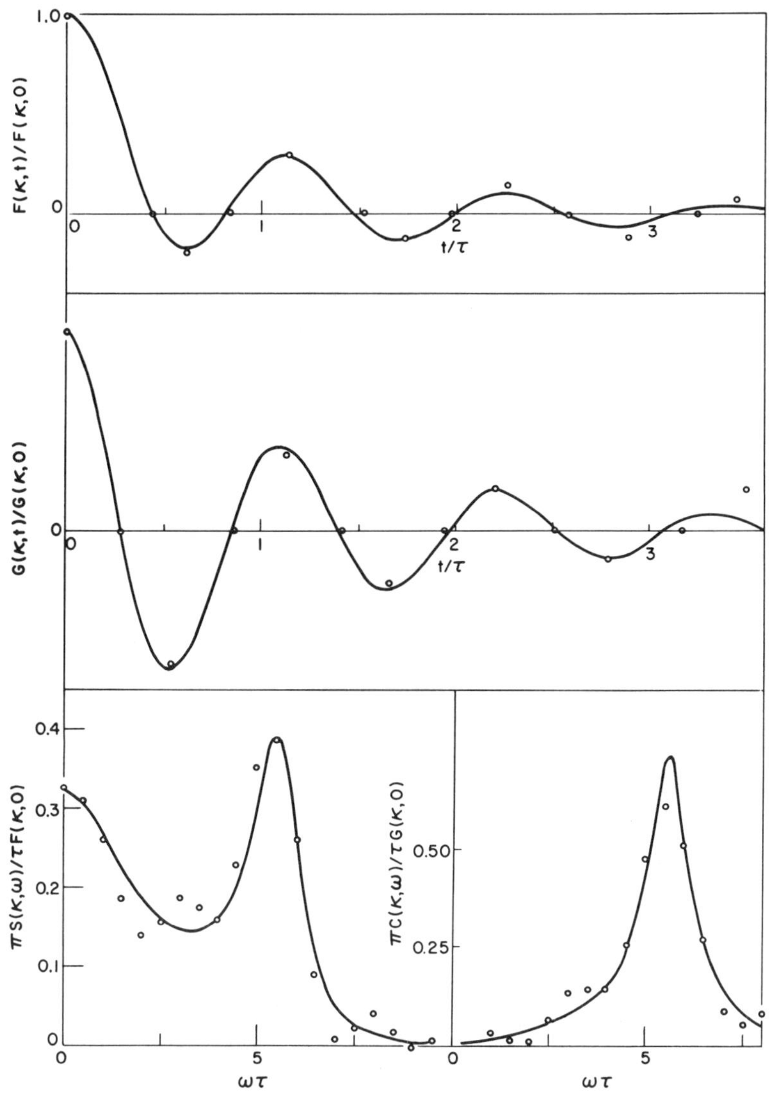

Fig. 3. F(κ,t) and G(κ,t) shown on top are obtained with the two-Gaussian memory function in equation (35). The cosine transforms are shown at the bottom. In all four curves o o o o indicate the data (shown in figs. 1 and 2).

least a momentary delay in the continued decay of the autocorrelation.

The time scale $t/\tau \simeq 0.1$, over which the autocorrelation shows a sudden change in its initial swift decay, is the same as that over which the single-particle velocity autocorrelation shows a reversal in sign (see fig. 4). Also, as already mentioned, ω_T^2 [or the downward curvature of $H(\kappa,t)/H(\kappa,0)$] tends to a constant value $\approx 2\Omega^2(0)$ as the wavelength increases; $\Omega^2(0)$, however, is a property of $\langle v(0)v(t)\rangle_{av}$ and is independent of the wavelength. Hence it

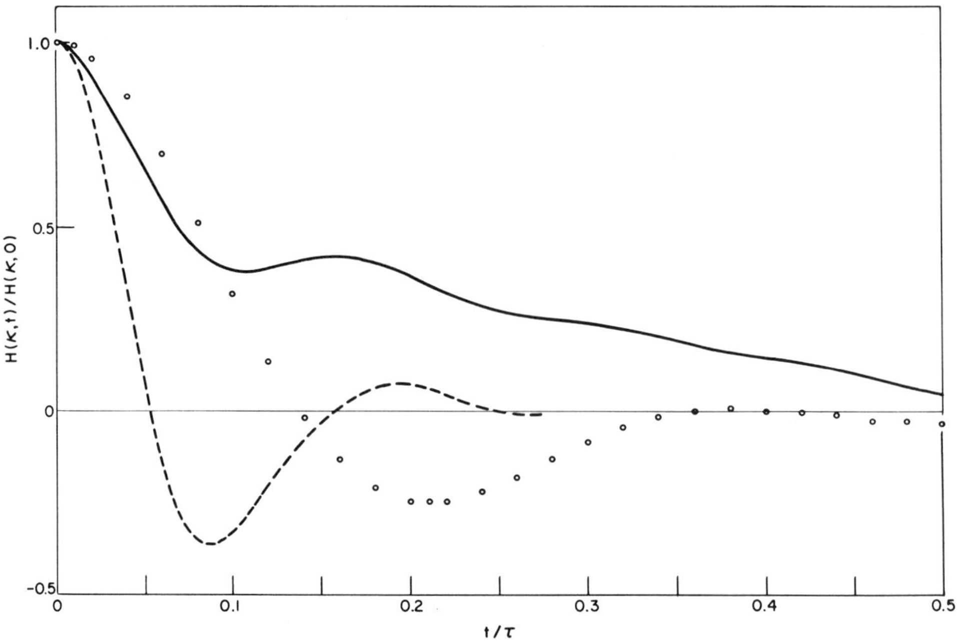

Fig. 4. Kinetic-energy density fluctuation H($\varkappa$,t) shown on enlarged time scale. The memory function $K_T(t)$ is shown as a dashed line (---); note that $K_T(t)$ is large and negative in the region where the decay of H($\varkappa$,t) suddenly changes character. o o o o indicate the single-particle velocity autocorrelation $\langle v(0)v(t)\rangle/\langle v^2\rangle$.

is to be expected that the short time behavior of H($\varkappa$,t)/H($\varkappa$,0) for very long wavelengths ($\varkappa \to 0$) will be similar to that seen in figure 1.

This property of H($\varkappa$,t) at short times gives rise to the long tail at high frequencies in T($\varkappa$, ω) which is seen in figure 2. It is to be expected that this property of T($\varkappa$, ω) will persist even in the limit $\varkappa \to 0$.

The peak in T($\varkappa$, ω) at $\omega\tau \sim 5$, though broad, is quite unmistakable, and there is a qualitative similarity between S($\varkappa$, ω) and T ($\varkappa$, ω). As already mentioned, the shape of S($\varkappa$, ω) is a reminder that the hydrodynamic regime is already in operation at $\varkappa = 0.23$ Å^{-1} for the system under consideration; it is shown in the Appendix that the usual solution of the linearized hydrodynamic equations leads to the well-known three-line phenomenon in the spectrum of density fluctuations and of the local thermodynamic temperature fluctuations; figure 2 indicates that the K.E.D. fluctuations also have the same qualitative behavior at $\lambda \sim 27$ Å.

8. DENSITY AND TEMPERATURE FLUCTUATIONS AT SHORT WAVELENGTHS

In the previous section we have seen that already at a wavelength $\lambda \sim 27$ Å ($= 2\pi/0.23$ Å^{-1}) the behavior of N.D. and K.E.D. fluctuations is qualitatively similar to the behavior expected at macroscopic values of the wavelength. We also saw in that section a remarkable feature in the short time behavior of the K.E.D. fluctuations arising out of the mean square force experienced by individual particles.

In this section we present data to show that at $\lambda \sim 4.5\,\text{Å}$ both, N.D. and K.E.D. fluctuations show a rather different behavior. We shall follow the same manner of presentation as before. At $\kappa = 1.38\,\text{Å}^{-1}$ we get

$$S(\kappa) = 0.129, \quad \tau\omega_o = 12.23, \quad \tau\omega_\ell = 19.21. \tag{38}$$

The functions $F(\kappa,t)$, $G(\kappa,t)$, $H(\kappa,t)$ are shown in figure 5, and the real part of the Laplace-Fourier transforms in figure 6. It is seen in figure 6 that apart from a slight shoulder at $\omega\tau \sim 14$ there is no side peak in $S(\kappa, \omega)$ as in figure 2.

Assuming

$$K_c(t) = \exp(-\mu^2 t^2) \tag{39}$$

(see equation [35] for comparison) now leads to $\pi S(\kappa, \omega = 0) = S(\kappa)\,\pi^{\frac{1}{2}}(\omega_\ell^2 - \omega_o^2)/\omega_o^2 2\mu$ which is to be compared with equation (37). Using $\pi S(\kappa, \omega = 0)/\tau S(\kappa) = 0.1$ (fig. 6), we get $\mu\tau = 13.0$. The same value of μ is obtained on using the equation analogous to equation (36) of the previous section.

Using this value of μ in $K_c(t)$ and substituting in equations (32a), (32b), one can calculate the autocorrelations F and G and their spectra. The calculated values are shown in figures 5 and 6 as circles. We conclude that a good representation of the data can be obtained with a frictional function $K_c(t)$ consisting of only one simple decaying function. This is very different from the stiuation at $\kappa = 0.23\,\text{Å}^{-1}$ (section 7) where $K_c(t)$ consists of two functions with very different decay characteristics.

To emphasize the remarkable change that occurs in $H(\kappa,t)$ on going from $\kappa = 0.23\,\text{Å}^{-1}$ to $1.38\,\text{Å}^{-1}$, we have drawn the two functions together in figure 7.

9. DENSITY FLUCTUATIONS IN AN ARGON-LIKE LIQUID

In the previous two sections we have shown the change that occurs in the time dependence of N.D. and K.E.D. fluctuations in going from a wavelength of about 6 times the interparticle separation in the liquid to a wavelength about equal to this separation.

Detailed analysis of molecular dynamics data on a liquid-argon-like system for wavelengths of the second category, i.e., comparable to and shorter than the interparticle distance, has already been published by various authors [10]. We shall briefly summarize the work of Ailawadi et. al. in which they used the method of section 8 to show that fluctuations in current density can be represented in terms of a Gaussian memory function with a time constant ($1/\mu$ of section 8) varying smoothly with κ. In this work, the molecular-dynamics data at each value of κ was fitted with $\pi C(\kappa, \omega)$ using μ as a least-squares fitting parameter. Other authors (Chung and Yip, and Akcasu and Daniels) have on the other hand suggested methods involving suitable approximations for constructing the memory function.

In figure 8 the values of μ obtained by Ailawadi et al. [10] are plotted in units of $(\omega_\ell^2 - \omega_o^2)^{\frac{1}{2}}$. It should be noticed that $\mu(\omega_\ell^2 - \omega_o^2)^{-\frac{1}{2}}$, though close to unity, shows a

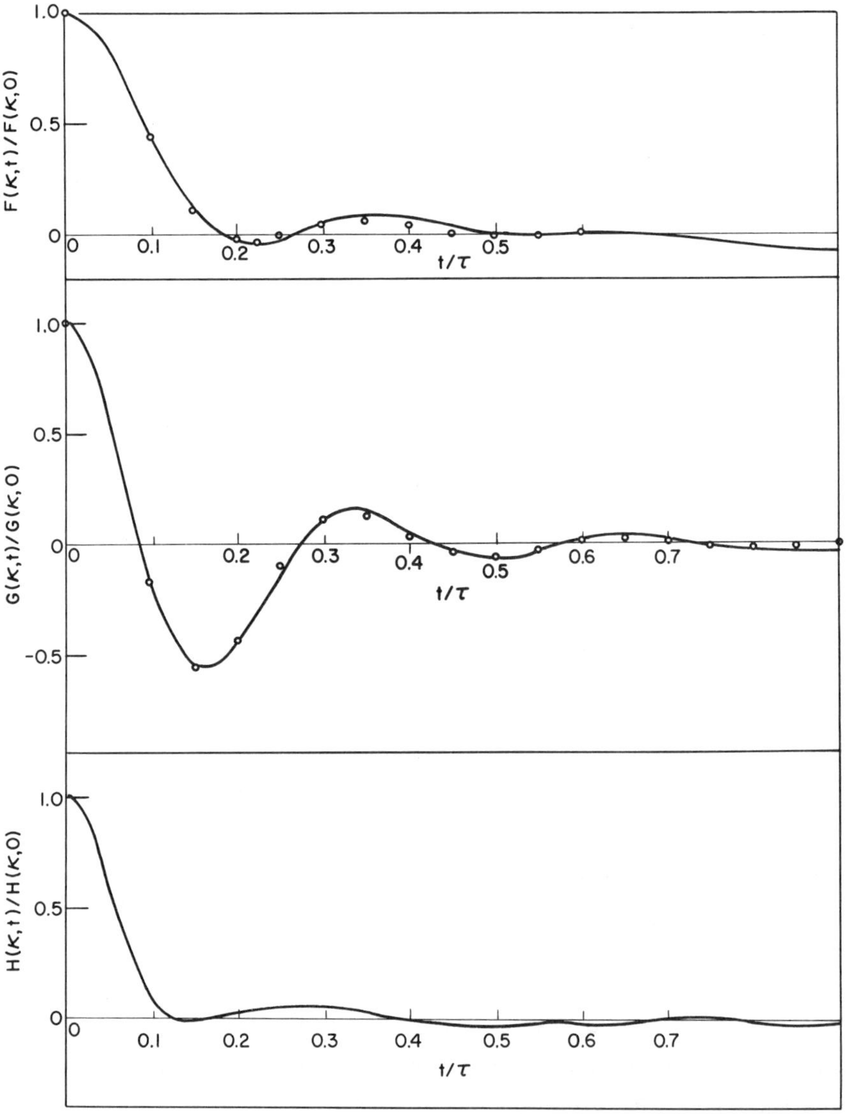

Fig. 5. Number, current and kinetic-energy density fluctuations F, G, H, respectively, at $\varkappa = 1.38$ Å^{-1}. Note that the scale of time is very different from that in fig. 1. H($\varkappa$,t) shows no abrupt change as at $t/\tau = 0.1$ in fig. 1. The values calculated for F($\varkappa$,t) and G($\varkappa$,t) with a one-Gaussian memory function, equation (39), are shown as o o o o.

significant departure from unity in the region ($\varkappa \sim 2$ Å^{-1}) of the maximum of S($\varkappa$) and for smaller values of $\varkappa$, i.e., for longer wavelengths.

The quantities ω_o^2 and ω_ℓ^2 are determined by the pair correlation function g(r) and the potential φ(r) of interaction between particles. We thus come to the interesting conclusion that g(r) and φ(r) contain enough information for us to predict--not exactly, of course, but with fair

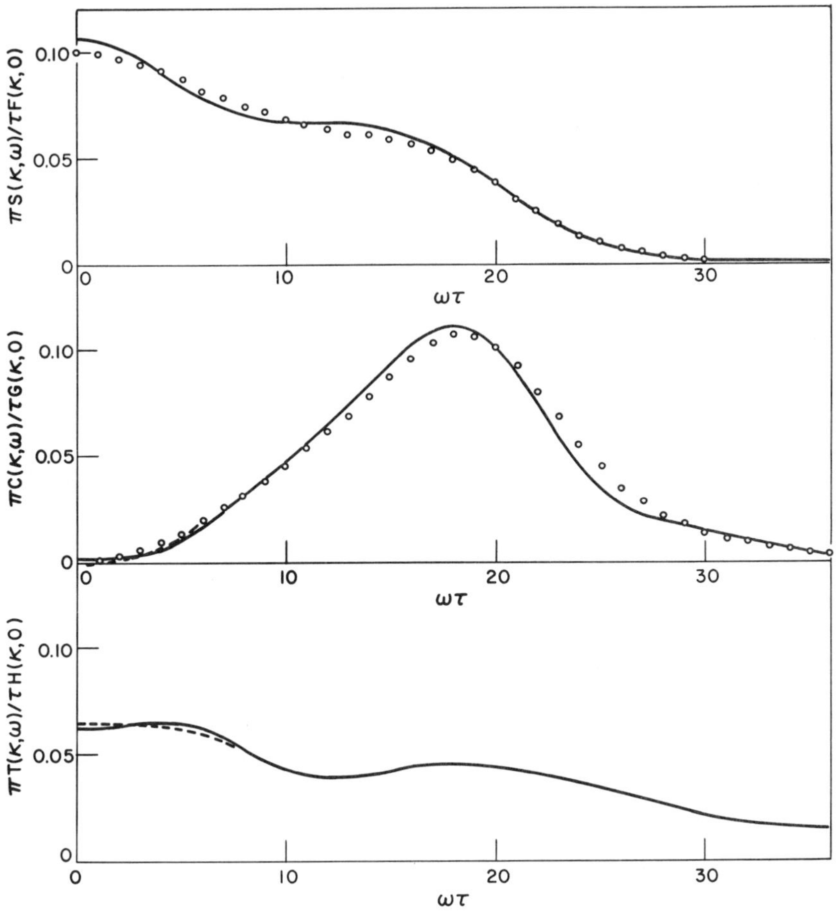

Fig. 6. Cosine transforms of data shown in fig. 5. o o o o indicates the calculated values of $S(\varkappa,\ \omega)$ and $C(\varkappa,\ \omega)$ with the one-Gaussian memory function (equation [39]).

accuracy--the whole time dependence of N.D. fluctuations provided the wavelength is not too large [13].

10. THE SCALING METHODS FOR CONSTRUCTING $S(\varkappa,\ \omega)$

Recently an interesting attempt has been made to use an ideal gas as the basis for constructing an $S(\varkappa,\ \omega)$ for the liquid for $\varkappa \gtrsim 1\ \text{Å}^{-1}$. Since this is in the nature of a scaled ideal-gas model, it will be appropriate first to summarize other attempts to construct $S(\varkappa,\ \omega)$ out of some known and (hopefully) simpler function of $\varkappa$ and ω.

(i) The first such prescription was that of Vineyard in which $S(\varkappa)S_s(\varkappa,\ \omega)$ was given as a prescription for $S(\varkappa,\ \omega)$ [$S_s(\varkappa,\ \omega)$ is analogous to $S(\varkappa,\ \omega)$ but contains information only about the variable $\underset{\sim}{r}_{i,t} - \underset{\sim}{r}_{i,o}$ whereas $S(\varkappa,\ \omega)$ analyzes $\underset{\sim}{r}_{i,t} - \underset{\sim}{r}_{j,o}$ where j runs over all particle

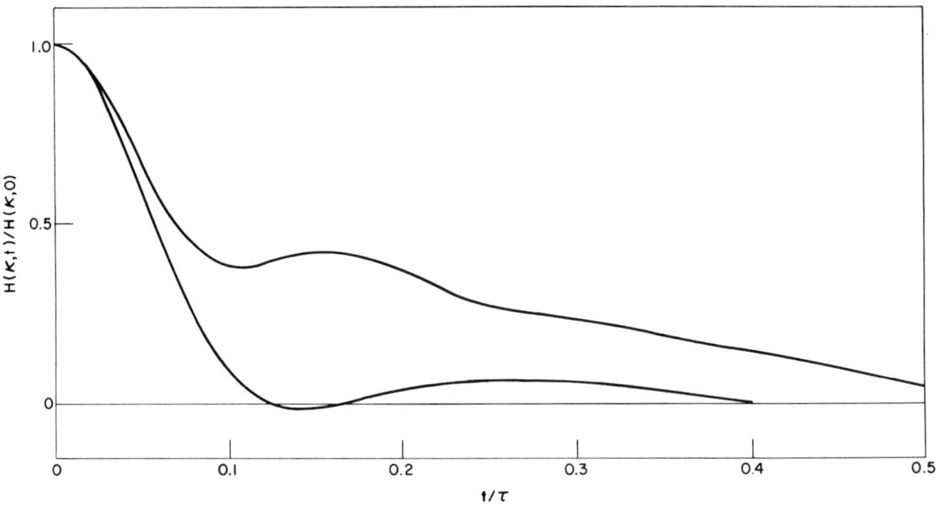

Fig. 7. Comparison of $H(\varkappa, t)$ at $\varkappa = 0.23$ Å^{-1} and 1.38 Å^{-1}.

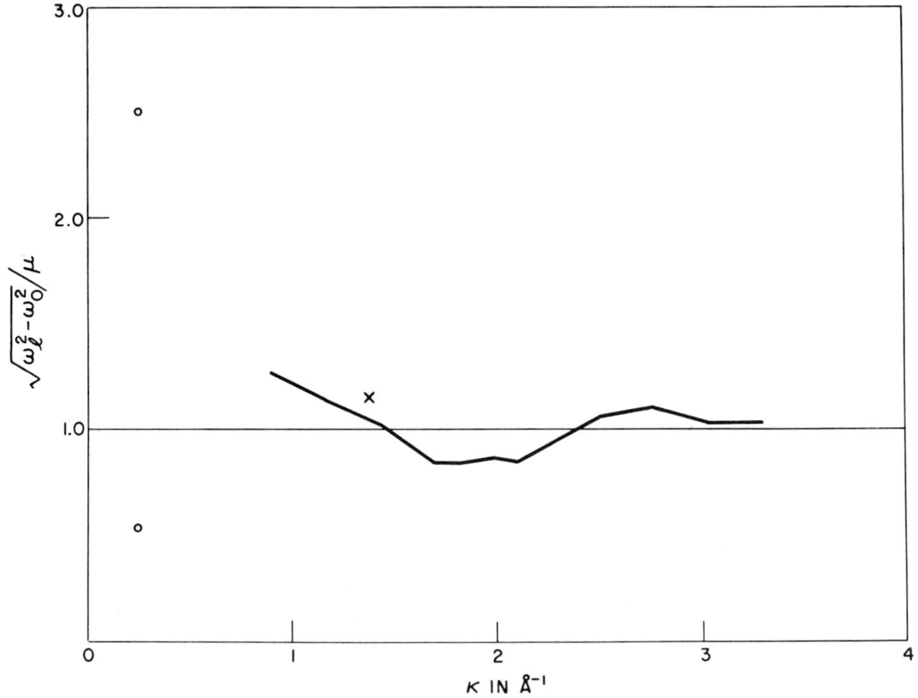

Fig. 8. Time constant $(1/\mu)$ of "best" Gaussian memory function $\exp(-\mu^2 t^2)$ for current fluctuations in liquid argon (Ailawadi et al. [10]). Circles, the two time constants in equation (35) for liquid sodium at $\varkappa = 0.23$ Å^{-1}; cross, the one time constant in equation (39) for liquid sodium at $\varkappa = 1.38$ Å^{-1}. Thus $\mu = (\omega_\ell^2 - \omega_o^2)^{\frac{1}{2}}$ fails below $\varkappa \sim 0.5$ Å^{-1} but is a good first approximation for higher values of $\varkappa$.

labels including i]. Assuming that the correct $S_s(\varkappa, \omega)$ is known, this prescription gives F_2 incorrectly, namely, $F_2 = S(\varkappa)\varkappa^2 k_B T/M$ (instead of $\varkappa^2 k_B T/M$). We remind the reader that the equation $F_2 = \varkappa^2 k_B T/M$ is a consequence of the conservation of the number of particles. It has now become standard practice in the field of neutron inelastic scattering to judge the accuracy of the data on $S(\varkappa, \omega)$ partly by requiring that this equality be satisfied.

(ii) The second model of this kind was to write $S(\varkappa, \omega) = S(\varkappa) S_s\{\varkappa/[S(\varkappa)]^{\frac{1}{2}}, \omega\}$; this prescription is equivalent to a scaling of $\varkappa$ (or wavelength) but not of time. This way of scaling the wavelength gives the correct value of F_o and F_2 and hence improves the Vineyard approximation but gives $F_4 = F_2[3\omega_o^2 + \Omega^2(0)]$ instead of the value in equation (26).

(iii) Alternatively we could have written $S(\varkappa, \omega) = S^{3/2} S_s[\varkappa, \omega S^{\frac{1}{2}}(\varkappa)]$ to get F_o and F_2 correctly, but $F_4 = F_2[3\omega_o^2 + \Omega^2(0)/S(\varkappa)]$, which again is incorrect. This prescription has not been tested and presumably will show no great virtue compared with (ii) above.

(iv) Finally, writing $S(\varkappa, \omega) = [S(\varkappa)/\lambda(\varkappa)] S_s[\varkappa/\lambda S^{\frac{1}{2}}(\varkappa), \omega/\lambda(\varkappa)]$, where $\lambda^2(\varkappa) = (\omega_\ell^2 - 3\omega_o^2)/\Omega^2(0)$, ensures that F_o, F_2, and F_4 are given correctly. This prescription was tested by the author on the data on liquid argon, and it was concluded that the function $S_s(\varkappa, \omega)$ in spite of this manipulation does not give a satisfactory representation of $S(\varkappa, \omega)$.

(v) The last such prescription we shall mention is to write the memory function of equation (32a) as

$$K_c(t) = K_c^{i.g}\left[\left(\frac{\omega_\ell^2 - \omega_o^2}{2\omega_o^2 S(\varkappa)}\right)^{\frac{1}{2}} t\right],$$

where $K_c^{i.g}$ is the memory function for an ideal gas, and is easy to calculate since for an ideal gas $F(\varkappa, t) = \exp(-\varkappa^2 k_B T t^2/2M)$. This leads to the expression for $S(\varkappa, \omega)$ given by Pathak and Singwi [10]. It appears to describe $S(\varkappa, \omega)$ with fair accuracy for $\varkappa \gtrsim 0.5\,\text{Å}^{-1}$; however, a numerical comparison between the results of this prescription and the simpler assumption

$$K_c(t) = \exp[-(\omega_\ell^2 - \omega_o^2)t^2]$$

leads to almost identical results (even for an ideal gas for which the first is rigorously correct and the second only an approximation). In fact, at $\omega = 0$, both give the same $S(\varkappa, \omega = 0)$. We have seen (fig. 8) that there is a significant difference between the time constant of the best Gaussian that fits the data and the quantity $(\omega_\ell^2 - \omega_o^2)^{-\frac{1}{2}}$. Hence the conclusion is that for liquid argon this scaling procedure to construct $K_c(t)$ out of $K_c^{i.g.}(t)$ will lead to values of $S(\varkappa, \omega = 0)$ which will be about 10% lower than the correct value around $\varkappa = 2\,\text{Å}^{-1}$ and about 10% higher around $\varkappa = 1\,\text{Å}^{-1}$. As seen from figure 8, $K_c(t) = \exp[-(\omega_\ell^2 - \omega_o^2)t^2]$ (and the scaled ideal-gas model) produces an extremely good representation of the data beyond $\varkappa = 2.5\,\text{Å}^{-1}$.

SUMMARY

Molecular dynamics data on a liquid-sodium-like system show that already at a wavelength $\lambda \sim 27\,\text{Å}$ the fluctuations of number density (N.D.) and the kinetic energy density (K.E.D.) in space and time have the same characteristics that N.D. fluctuations have in the long-wavelength, hydrodynamic, regime. However, at $\lambda \sim 4.5$ Å the data show quite a different qualitative behavior.

The well-known memory function formalism is presented, and it is suggested that N.D. fluctuations are most profitably approached via the memory function of the longitudinal current density fluctuations.

Following this approach, the N.D. fluctuations at $\lambda \sim 27$ Å can be interpreted fairly accurately in terms of a two-Gaussian memory function for the longitudinal current density fluctuations, the decay constants in the two Gaussians being in a ratio of about 5:1. At $\lambda \sim 4.5$ Å, however, a memory function decaying simply as one Gaussian is sufficient, as has been found previously for liquid-argon-like systems.

The short time behavior of K.E.D. fluctuations is shown to depend on the mean square force on a single particle even in the long-wavelength limit, and this dependence is clearly seen in the molecular-dynamics data at $\lambda \sim 27$ Å.

APPENDIX

Linearized hydrodynamic equations have been treated extensively in the literature (see, e.g., Mountain [11]) and need no detailed introduction. The Navier-Stokes equation giving the rate of change of current density is

$$\frac{\partial^2 N(\varkappa,t)}{\partial t^2} = -\varkappa^2 C_T^2 N(\varkappa,t) - b\varkappa^2 \frac{\partial N(\varkappa,t)}{\partial t} - \varkappa^2 C_T^2 M'(\varkappa,t), \qquad \text{(A1)}$$

where $N(\varkappa,t)$ is as defined in equation (22a), C_T is the isothermal sound velocity, and b is $\nu/m\rho_o$, ν being the kinematic viscosity, $m\rho_o$ the average mass density. $M'(\varkappa,t)$ has no dimensions and is equal to β times the transform of the variations in space of the local thermodynamic temperature (LTT), where β is the coefficient of thermal expansion.

Notice that in equation (A1) the rate of change of $\partial N/\partial t$ is expressed, in addition to two other quantities, in terms of $\partial N/\partial t$ itself with a constant factor bk^2, as in equation (4).

The energy conservation equation of hydrodynamics gives

$$\frac{\partial M'(\varkappa,t)}{\partial t} = (\gamma - 1)\frac{\partial N(\varkappa,t)}{\partial t} - a\varkappa^2 M'(\varkappa,t), \qquad \text{(A2)}$$

where $\gamma = C_p/C_v$, $a = \lambda/\rho_o C_v$, λ being the thermal conductivity and C_v the constant-volume heat capacity per particle.

Notice again that in equation (A2) the rate of change of M' is expressed, in addition to one other quantity, in terms of M' itself with a constant factor $a\varkappa^2$.

The solution of equations (A1) and (A2) is given in Mountain [11]. In obtaining this solution it is usual to assume further that LTT and N.D. fluctuations are statistically independent; in other words $\langle N^*(\varkappa,o)M'(\varkappa,o)\rangle_{av}$ is considered negligible. Using the notation in equations (32), this solution can be written as

$$\pi C(\varkappa,\omega) = \frac{G_o}{i\omega + (\omega_o^2/i\omega) + b\varkappa^2 + \omega_o^2(\gamma-1)/(i\omega + a\varkappa^2)} = i\omega F_o + \omega^2 \pi S(\varkappa,\omega), \tag{A3}$$

$$\pi T'(\varkappa,\omega) = \frac{H_o'}{i\omega + a\varkappa^2 + \omega_o^2(\gamma-1)/(i\omega + \omega_o^2/i\omega + b\varkappa^2)} \tag{A4}$$

where T' and H_o' are analogous to T and H_o (equations [29] and [31c]), but pertain to LTT fluctuations.

On calculating the real part of $S(\varkappa,\omega)$, $C(\varkappa,\omega)$, or $T'(\varkappa,\omega)$, the denominator will be seen to be identically the same (being a cubic polynomial in ω). It is this denominator which is responsible for giving the "three-line phenomenon" in $S(\varkappa,\omega)$ and in $T'(\varkappa,\omega)$; in the limit of $\varkappa \to 0$ the widths are proportional to $\varkappa^2$. We remind ourselves that the data presented in section 7 show a qualitative similarity between $S(\varkappa,\omega)$ and $T(\varkappa,\omega)$, the K.E.D. fluctuations, even at $\varkappa = 0.23$ Å^{-1}.

Comparison of equations (A3) and (A4) with equations (32) shows that the simple Langevin-type friction coefficients $b\varkappa^2$ and $a\varkappa^2$ in equation (A1) and (A2), respectively, lead to singular memory functions:

$$(\omega_\ell^2 - \omega_o^2)K_c(t) = b\varkappa^2\delta(t) + \omega_o^2(\gamma-1)\,\theta(t),$$

$$\omega_T^2 K_T'(t) = a\varkappa^2\delta(t) + \omega_o^2(\gamma-1)\,\lambda(t),$$

where $\theta(t)$ and $\lambda(t)$ satisfy the equations

$$\dot\theta(t) = -a\varkappa^2\,\theta(t),$$

$$\dot\lambda(t) = -\omega_o^2\int_0^t \lambda(s)ds - b\varkappa^2\lambda(t).$$

The solution, equation (A3) above, is known to be adequate to describe the long wavelength and long time behavior of density fluctuations. But it is obviously inadequate for describing the short-time behavior at all wavelengths.

A possible and simple extension of the solution which allows us to get the correct short-time behavior is to write

$$(\omega_\ell^2 - \omega_o^2)\,K_c(t) = (\omega_\ell^2 - \gamma\omega_o^2)\,\varphi(t) + \omega_o^2(\gamma-1)\,\theta(t), \tag{A5}$$

$$\omega_T^2 K_T'(t) = \eta^2\psi(t) + \omega_o^2(\gamma-1)\lambda(t), \tag{A6}$$

where $\varphi(t)$ will occur in equation (A1) as

$$-(\omega_\ell^2 - \gamma\omega_o^2)\int_0^t \varphi(t-s)\,\frac{\partial N(\varkappa,s)}{\partial s}\,ds$$

in place of $-b\varkappa^2\partial N(\varkappa,t)/\partial t$, and $\psi(t)$ will occur in equation (A2) as

$$-\eta^2\int_0^t \psi(t-s)M'(\varkappa,s)\,ds,$$

in place of $-a\varkappa^2 M'(\varkappa,t)$, with $\eta^2 = \omega'^2_T - \omega_o^2(\gamma - 1)$ where ω'^2_T is analogous to ω_T^2 of equation (30).

In all these definitions the dimensionless functions of time K_c, φ, θ, K_T', ψ, λ are unity at $t = 0$. They also depend on $\varkappa$ even though this dependence is not shown explicitly.

With this modification of equations (A1) and (A2), the solution of the linearized hydrodynamic equations can be written in terms of two unknown functions $\varphi(t)$, $\psi(t)$ [12]:

$$\pi C(\varkappa, \omega) = G_o \left[i\omega + \frac{\omega_o^2}{i\omega} + (\omega_\ell^2 - \gamma\omega_o^2)\,\tilde{\varphi}(\omega) + \frac{\omega_o^2(\gamma - 1)}{i\omega + \eta^2\tilde{\psi}(\omega)} \right]^{-1}, \tag{A7}$$

$$\pi T'(\varkappa, \omega) = H_o' \left[i\omega + \eta^2\tilde{\psi}(\omega) + \frac{\omega_o^2(\gamma - 1)}{i\omega + \omega_o^2/i\omega + (\omega_\ell^2 - \gamma\omega_o^2)\tilde{\varphi}(\omega)} \right]^{-1} \tag{A8}$$

or, equivalently, for $\theta(t)$ and $\lambda(t)$ of equations (A5), (A6)

$$\tilde{\theta}(\omega) = (i\omega + \eta^2\tilde{\psi}(\omega))^{-1}, \quad \tilde{\lambda}(\omega) = (i\omega + \omega_o^2/i\omega + (\omega_\ell^2 - \gamma\omega_o^2)\ddot{\varphi}(\omega))^{-1}. \tag{A9}$$

The data on $F(\varkappa,t)$ and $G(\varkappa,t)$ (section 7) has been interpreted successfully by assuming $\varphi(t)$ and $\theta(t)$ of equation (A5) to be Gaussian in time (compare equations [A5] and [35]).

However, as expected, the same $\theta(t)$ and $\varphi(t)$ do not reproduce the data on $T(\varkappa, \omega)$. This is because $T(\varkappa, \omega)$ of equation (31c) pertains to K.E.D. fluctuations and not to the LTT fluctuations. At $\omega = 0$, for example $\tilde{\theta}(\omega = 0) = \pi T'(\varkappa, o)/H_o'$. The function $\theta(t)$ used in equation (35) gives $\tilde{\theta}(\omega = 0) = \pi^{\frac{1}{2}}/2\nu \simeq 0.5\tau$ whereas $\pi T(\varkappa, o)/H_o = 0.22\tau$ (fig. 2).

We thus conclude that already at $\lambda \sim 27$ Å and presumably also for larger values of λ, K.E.D. fluctuations have a behavior quantitatively different from but qualitatively similar to that predicted by the hydrodynamic equations for the LTT fluctuations in the system.

Calculations on energy density fluctuations will be undertaken in the near future.

REFERENCES

1. L. P. Kadanoff and P. C. Martin, Ann. Phys. (N.Y.), 24, 419 (1963).

2. H. Mori, Prog. Theor. Phys. 33, 423 (1965).

3. R. Zwanzig, in Lectures in Theoretical Physics, ed. W. E. Brittin (New York: Wiley, 1961).

4. L. Van Hove, Phys. Rev. 95, 249 (1954).

5. B. R. A. Nijboer and A. Rahman, Physica, 32, 415 (1966).

6. R. Zwanzig, Phys. Rev. 156, 190 (1967).

7. W. Shyu, K. S. Singwi and M. P. Tosi, Phys. Rev. B3, 237 (1970).

8. J. Kurkijarvi, D. Levesque, and L. Verlet, unpublished.

9. A. Rahman, Rivista del Nuovo Cimento, Numero Speciale 1, 315 (1969).

10. C. H. Chung and S. Yip, Phys. Rev. 182, 323 (1969); A. Z. Akcasu and E. Daniels, Phys. Rev. A2, 962 (1970); K. N. Pathak and K. S. Singwi, Phys. Rev. A2, 2457

(1970); N. K. Ailawadi, A. Rahman, and R. Zwanzig, Phys. Rev. **A4**, 1616 (1971).

11. R. D. Mountain, Rev. Mod. Phys. **38**, 205 (1966).
12. P. Grey and C. A. Cooper (Proceedings of this conference).
13. S. W. Lovesey, Phys. Letters, **36A**, 413 (1971).

DISCUSSION

P. Selwyn: Since the kinetic energy density is not one of the conserved variables whereas the total energy density is, one would not expect its behavior to be described by a hydrodynamical formulation. Furthermore, the "driving mechanism" alluded to in the decay of kinetic-energy density fluctuations is to be expected, presumably reflecting exchange of kinetic and potential energy. One would not expect to see such behavior in the total energy density.

A. Rahman: The total energy density fluctuations are being calculated and will be reported in due course. The part of K.E.D. which is normal to the N.D., namely,

$$N^{-\frac{1}{2}} \sum_i (\underline{v}_i^2 - 3k_BT/M)\delta(\underline{r} - \underline{r}_i),$$

is also of some interest. Calculation for this is nearly complete.

M. G. Velarde: I would like to point out that if you consider the eigenvalue problem, referred to by Resibois this morning, for moderate nonlinear gradients (up to q^3) in a Boltzmann (dilute) model, one finds that the q^3 term is absent in the viscosity (η) and thermal diffusivity (λ) eigenvalues and that only the sound propagation eigenvalues have this cubic dependence [$\omega_\pm = \pm icq - \frac{2}{3}(\eta + \frac{1}{5}\lambda)q^2 \mp \frac{1}{6} \times 15^{-\frac{1}{2}} \eta^2 q^3 L$, where L is a linear combination of the Burnett transport coefficients (here $C_p/C_v = \frac{5}{2}$) which are not expressible straightforwardly in terms of simple thermodynamic quantities.] These q^3 results are indeed expected on the basis of symmetry arguments since the only vector eigenvalue is that corresponding to sound propagation. The interesting thing perhaps is that, though arising from a second-order perturbation calculation, these corrections point in the direction of the finite-q results presented graphically by Rahman, where the only affected peaks seems to be those of the Brillouin-Mandelshtam doublet. Indeed, taking into account weak nonlinearity, one expects a broadening and so a lowering of the doublet peaks.

SOME MODERN DEVELOPMENTS IN THE STATISTICAL THEORY OF TURBULENCE

Robert H. Kraichnan

1. NATURE OF TURBULENCE

Turbulent flow constitutes an unusual and difficult problem of statistical mechanics, characterized by extreme statistical disequilibrium, by anomalous transport processes, by strong dynamical nonlinearity, and by a perplexing interplay of chaos and order. The definition of turbulence is a matter of some controversy, and the term has been applied to a wide variety of disparate phenomena. We shall be concerned here with turbulence in an incompressible, Navier-Stokes or Boussinesq fluid, with principal emphasis on homogeneous, isotropic Navier-Stokes turbulence. We shall mean by turbulence a flow in which the velocity varies in a complicated way with position and time and which is subject to a manifold of instabilities so that the detailed evolution of the flow is not well determined by gross initial and boundary conditions [1].

In order to describe qualitatively some of the important physics of turbulence and illustrate its peculiar statistical mechanics, let us start by considering turbulence driven by the gravitational instability of a thin layer of Boussinesq fluid heated from below and cooled from above [2]. Here the fluid obeys the incompressible Navier-Stokes equation, with constant density, except that there is a buoyancy force, proportional to temperature (measured from some reference temperature), acting on each fluid element. The stability of the layer depends on the Rayleigh number

$$Ra = \alpha g \, \Delta T D^3/(\varkappa \nu) ,$$

where α is the bulk coefficient of thermal expansion, g is acceleration of gravity, ΔT is the impressed temperature difference across the layer, D is the thickness of the layer, $\varkappa$ is thermometric heat conductivity, and ν is kinematic viscosity. We shall suppose $\nu \sim \varkappa$ (for air, $\nu = 0.8\,\varkappa$).

If Ra is less than a critical value of order 1000 (the precise value depends on the boundary conditions at top and bottom of the layer), there is no motion, the fluid is stable to all disturbances, and the heat transport per unit area through the layer has the value $H = \varkappa \, \Delta T/D$.

This work was supported by the Fluid Dynamics Branch of the Office of Naval Research, under contract N00014-67-C-0284.

For Ra moderately above critical, the fluid executes a stable cellular or roll-like motion whose intensity increases with Ra. At still higher Ra, the primary cellular motion is unstable and the flow pattern looks irregular, and, finally, at Ra the order of 10^5, there is a rich variety of instabilities, yielding fully turbulent motion. As Ra rises above the critical value, convective heat transport supplants molecular conduction in the interior of the layer, with the result that H increases, becoming greater than $\varkappa \Delta T/D$ by an order of magnitude for Ra $\sim 10^6$. Close to the boundaries, however, vertical motions are inhibited because the fluid is incompressible, and here heat transport is principally molecular. The constant temperature gradient of the stable, subcritical regime is replaced at large Ra by a profile in which the mean temperature is nearly constant throughout the body of the layer while there are intense temperature gradients in the thin conduction-dominated boundary layers at top and bottom.

Although the high-Ra flow is chaotic, it is far from completely random. Observations indicate that the dominant mechanism by which hot fluid leaves the lower conduction boundary layer is through the rather abruptly intermittent formation of well-defined plumes whose boundaries with the surrounding cooler fluid become increasingly irregular and convoluted as the plumes rise [3]. (The formation of cumulus clouds is an example of this process on a large scale, the clouds serving as markers for the hot updraft plumes.) In the body of the fluid, the heat transport is dominated by motions whose scale, or domain of coherency, is the order of D.

The transport of momentum by turbulent shear flows is broadly analogous, in important aspects, to the turbulent heat transport just discussed. Consider the flow down a pipe, which is historically the first turbulent flow studied in the laboratory. If D now represents the diameter of the pipe, the parameter which determines stability of flow is the Reynolds number UD/ν, where U is the mean flow velocity on the pipe axis. If UD/ν sufficiently exceeds a critical value, the flow is turbulent, and the rate of momentum transport from the interior to the walls of the pipe can be many times that of a laminar flow at the same U [1]. Again, molecular transport, here by viscosity, takes over in a thin boundary layer at the walls. As in the thermal turbulence, the flow simultaneously exhibits chaos and order. Transport of momentum from the laminar sublayer at the wall into the fluid appears to be mediated to a substantial extent by well-organized horseshoe-like vortices which grow abruptly and then explosively detach from the wall and become more chaotic as they enter the main body of fluid [4]. Momentum transport in the central regions of the pipe is dominated by motions whose scale is of order D [5,6].

In both the thermal turbulence and the pipe-flow example, we can form a turbulent Reynolds number $u\ell/\nu$, where u is a typical value of turbulent velocity (defined as the difference between actual and mean velocity) and ℓ is a typical scale size of the large turbulent motions ($\ell \sim$ D in our examples). If $u\ell/\nu$ is large (> 1000, say), the turbulence appears to be subject to an hierarchy of instabilities, whereby the large-scale motions, or eddies, break down into successively smaller-scale motions. Once again, this does not happen in a purely chaotic fashion. The small-scale motions are not little eddying motions smoothly distributed over the large-scale motions. Instead, they are typically vortex sheets of filaments wherein shear is intense, separated by regions of low shear [7-9].

Much theoretical interest attaches to the limit $u\ell/\nu \to \infty$. Suppose we generate turbulence in such a way that u and ℓ are fixed and consider what happens as $\nu \to 0$. The body of experimental data suggests that the rate of dissipation of turbulent kinetic energy into heat by viscous friction in the shear layers approaches a constant of order u^3/ℓ per unit mass. That is, the decay time of the turbulent kinetic energy is of the order of the "large-scale eddy-circulation time" ℓ/u. This appears to be true for moderate as well as large turbulent Reynolds numbers [7].

A phenomenological theory of why the dissipation rate is independent of viscosity in the limit is provided by the Kolmogorov-Onsager-von Weizsacker theory of inertial transfer, which has profoundly influenced all modern work on turbulence [7]. The basic assumption is that the transformation of kinetic energy from large to small scales is a local cascade process in which appreciable energy transfer occurs only between scales of comparable size, or, in a spatial Fourier analysis of the velocity field, between wavenumbers of comparable size. An argument for this is that scales very much larger than a given one should simply convect the given scale without distorting it, while scales of comparable size should shear and stretch the given scale, thereby increasing its vorticity, according to Kelvin's circulation theorem, and producing a net transfer of kinetic energy to higher wavenumbers. A further assumption of the theory is that detailed statistical information about the large scales is degraded in the cascade, so that, if the latter is extensive enough, the small scales tend toward a universal, spatially isotropic statistical distribution determined by two parameters: ν and ϵ, the rate of energy dissipation by viscosity per unit mass.

Since the transfer of energy between scales is conservative, ϵ is also the rate of energy cascade. It follows from dimensional analysis that the wavenumber spectrum of the turbulent kinetic energy in the universal range must have the form

$$E(k) = \epsilon^{2/3} k^{-5/3} f(k/k_s) \ , \qquad k_s = (\epsilon/\nu^3)^{1/4} \ , \tag{1}$$

where k is wavenumber and the kinetic energy per unit mass is $\int_0^\infty E(k)\,dk$. The theory assumes that f is a universal function and that f(0), in particular, has a universal value. According to the Navier-Stokes equation, the rate of dissipation of kinetic energy by viscosity at all wavenumbers less than k is $2\nu \int_0^k E(p)p^2\,dp$. If we take $E(p) = f(0)\epsilon^{2/3}k^{-5/3}$, and $f(0) = O(1)$, this dissipation equals ϵ at a $k \sim k_s$, so that k_s is a characteristic dissipation-range wavenumber above which we expect $f(k/k_s)$ to fall off. The fall-off must be at least exponential if all velocity derivatives are to exist in mean square, and this is consistent with experiment [7-9].

The wavenumber range $k \ll k_s$ in equation (1) is called the inertial range. Here, according to the theory, smaller scales adjust themselves to the flow ϵ from larger scales, and ν plays no role. The $k^{-5/3}$ inertial range and the form (1) for the dissipation range appear consistent with experiment. However, criticisms of the theory have been raised by a number of authors, including Kolmogorov himself, and some experimental data on higher statistical properties may support these criticisms. Later in this paper we shall discuss this further.

Now let us return to the Boussinesq turbulence problem and look at it from the point of view of microscopic statistical mechanics. A fact of dominant importance is that the anomalous heat transport can exceed $\varkappa\, \Delta T/D$ many-fold while departure from local equilibrium is negligible. For air at standard conditions, $\alpha g/\varkappa \nu \sim 120$, so that $\Delta T = 1°K$ across $D = 1$ meter gives $Ra \sim 10^8$. The local disequilibrium measured by the fractional change of absolute temperature over a mean free path is minute. Clearly turbulent transport of heat (and, by a similar argument, momentum) represents a singular perturbation of absolute statistical equilibrium. An implication is that microscopic distribution functions are not an auspicious direct starting point for a theory of turbulence. Instead, it seems reasonable to do double averaging and describe turbulence by an ensemble of solutions of the Navier-Stokes equation, wherein the microscopic dynamics appear only in the parameters $\varkappa$ and ν which are relevant to macroscopic motion. We may note that air and water appear to give experimentally indistinguishable low-Mach-number turbulence, and they have only the Navier-Stokes equation in common. Generally, all turbulent scales significantly excited in laboratory and meteorological turbulence are large enough compared to a mean free path that the Navier-Stokes equation is valid [7]. When this is true, local parcels of fluid have no way of knowing whether the motion they are participating in is turbulent or laminar. The two kinds of flow are indistinguishable at the level of the Boltzmann equation. There has unfortunately been a substantial amount of mysticism about this point in discussions of turbulence.

We are led then to study turbulence in terms of the statistical mechanics of the continuous macroscopic velocity field $\hat{\underset{\sim}{u}}(\underset{\sim}{x},t)$. Suppose that $\hat{\underset{\sim}{u}}(\underset{\sim}{x},t)$ obeys cyclic boundary conditions on a large cubical box of side L. Setting

$$\hat{\underset{\sim}{u}}(\underset{\sim}{x},t) = \sum_{\underset{\sim}{k}} \underset{\sim}{u}(\underset{\sim}{k},t)\exp(i\underset{\sim}{k}\cdot\underset{\sim}{x}) , \tag{2}$$

we may write the Navier-Stokes equation as

$$(\partial/\partial t + \nu k^2)u_i(\underset{\sim}{k},t) = -i\lambda k_m P_{ij}(\underset{\sim}{k}) \sum_{\underset{\sim}{p}+\underset{\sim}{q}=\underset{\sim}{k}} u_j(\underset{\sim}{p},t)u_m(\underset{\sim}{q},t) + f_i(\underset{\sim}{k},t) , \tag{3}$$

where $P_{ij}(\underset{\sim}{k}) = \delta_{ij} - k_i k_j/k^2$ is a projection operator which expresses the effect of the pressure force in maintaining the incompressibility condition

$$\underset{\sim}{k}\cdot\underset{\sim}{u}(\underset{\sim}{k},t) = 0. \tag{4}$$

The quantity $f_i(\underset{\sim}{k},t)$ represents an external stirring force, which can excite or maintain the turbulence, and λ is a strength or ordering parameter ($\lambda = 1$ in the physical case) inserted for later use in discussing perturbation approaches. The kinetic energy per unit mass is $\frac{1}{2}\sum_{k} |\underset{\sim}{u}(\underset{\sim}{k},t)|^2$. It is conserved by the term in λ, which represents advection and pressure forces. Moreover, energy is conserved individually by each interacting triad of wavevectors $\pm\underset{\sim}{k}$, $\pm\underset{\sim}{p}$, $\pm\underset{\sim}{q}$.

If ν and f_i are zero, Liouville's theorem holds in the phase space whose coordinates are the real and imaginary parts of the $u_i(\underset{\sim}{k},t)$, and it follows that a canonical distribution, giving equipartition of energy among the $u_i(\underset{\sim}{k},t)$, is an absolute equilibrium distribution [10]. However, this distribution can be normalized only if equation (3) is truncated by removing all wavenumbers which exceed some cutoff value K, an operation which preserves the conservation properties. It is important to note that, no matter what value is taken for K, this distribution differs profoundly from equation (1). The absolute equilibrium distribution gives $E(k) \propto k^2$ $(k < K)$ while equation (1) gives $E(k) \propto k^{-5/3}$ in the limit $\nu \to 0$. Thus the inertial-range cascade represents a strong departure from absolute statistical equilibrium of the $u_i(\underline{k},t)$. A further manifestation of disequilibrium is the tendency of small scales to organize into well-defined shear layers, or filaments. This shows up in measured values of high-order moments of velocity-space derivatives which differ markedly from those of a normal distribution [7-9].

2. PERTURBATION EXPANSIONS AND RENORMALIZATION

The statistical problem associated with equation (3) may be formulated as the following initial-value problem: given the initial distribution of $\underline{u}(\underline{k},t)$, and the distribution of $\underline{f}(\underline{k},t)$, if $\underline{f}(\underline{k},t) \neq 0$, find interesting moments or distribution functions of $\underline{u}$ at later times. In particular, it is of fundamental interest to find $\langle u_i(\underline{k},t)u_i^*(\underline{k},t')\rangle$. It is fairly clear from the foregoing discussion that only minimal help can be expected from techniques of equilibrium statistical mechanics. We nevertheless hope that statistical quantities (averages over ensemble) will, in some respects, behave more simply and stably than the velocity amplitude of an individual realization, so that there remains a valid motivation for doing statistical dynamics [11].

The question of existence and uniqueness of solutions of the Navier-Stokes equation is still open, with the result that it is not possible to state precisely what conditions on initial and forcing-function distributions make the statistical initial-value problem well posed [12]. We shall sidestep this problem by truncating equation (3) to contain only wavenumbers below some K, as described above. Then the conservation property assures that bounded initial values and forcing amplitudes yield a unique solution bounded for all t, at least for finite L. The truncation is actually not as drastic as it may seem; if no K exists such that excitation at $k > K$ is negligible without truncation, then the Navier-Stokes equation is an invalid description. We must verify, of course, that any final statistical results are sensible for $K \to \infty$ and for $L \to \infty$, if the latter limit is taken.

Several authors have worked with the Liouville equation for the distribution of $\underline{u}(\underline{k},t)$ or $\hat{\underline{u}}(\underline{x},t)$, or with the corresponding equation of motion for the characteristic functional of the $\hat{\underline{u}}(\underline{x},t)$ distribution [13-17]. A generalization to the many-time distribution of $\hat{\underline{u}}(\underline{x},t)$ has also been described [18]. So far, little seems to have emerged from these studies that cannot be stated in addition by the less abstruse formulation in terms of moments. Powerful functional techniques that could be hoped for in the distribution approach have not emerged. Discussion here will therefore be confined to the moment equations which follow from equation (3). We should like to note, however, the recent formulation in terms of partial distributions, by Lundgren [19]. It is analogous to the Bogolubov, Born, Green, Kirkwood and Yvon hierarchy.

An obvious procedure with equation (3) is to treat the nonlinear term as a perturbation and thereby expand any desired statistical quantity as a power series in λ [20]. This can be done in a straightforward fashion by introducing the Green's function of the linearized ($\lambda = 0$) equation and iterating to develop $u_i(\underset{\sim}{k},t)$ as a power series in λ, the coefficients being functionals of the initial values and of $\underset{\sim}{f}$. Then, to evaluate any moment of the $\underset{\sim}{u}$ field, the factors which compose the moment are replaced by their series expansions, the series are multiplied out, and the ensemble average is taken, yielding the coefficients of the powers of λ as moments of the initial $\underset{\sim}{u}$ and the $\underset{\sim}{f}$ distributions. It is easy to see that, because equation (3) is bilinear, the coefficient of λ^n in the expansion of $u_i(\underset{\sim}{k},t)$ is a functional polynomial of degree $n+1$ in the initial $\underset{\sim}{u}$ field.

The expansion in λ amounts, in terms of physical parameters, to an expansion in $1/\nu$, or in powers of a characteristic turbulent Reynolds number [20]. Since the cases of greatest interest are large Reynolds numbers, this would appear to be a poor starting point. Nevertheless, most analytical treatments of turbulence statistics are based upon, or intimately related to the λ expansion, simply because more appropriate systematic procedures for stochastically nonlinear problems have not been devised.

The analyticity properties of the λ expansion for moments depend on those of the underlying Navier-Stokes equation, and upon the choice of initial distribution. With the K cutoff, $u_i(\underset{\sim}{k},t)$ has a nonzero radius of convergence in λ in any realization where the initial amplitudes and the $\underset{\sim}{f}$ amplitudes are bounded. It seems almost certain that the radius of convergence is finite in a typical realization, and that it typically tends to zero as $t \to \infty$ or as the velocity amplitude (Reynolds number) or the realization $\to \infty$. Such behavior is illustrated by the simple system [20]

$$\dot{y}_1 + \nu y_1 = \lambda A_1 y_2 y_3, \quad \dot{y}_2 + \nu y_2 = \lambda A_2 y_3 y_1, \quad \dot{y}_3 + \nu y_3 = \lambda A_3 y_1 y_2, \tag{5}$$

with $A_1 + A_2 + A_3 = 0$. This system has the same kind of nonlinearity as equation (3), and it conserves $y_1^2 + y_2^2 + y_3^2$ if $\nu = 0$.

If the radius of convergence in individual realizations behaves as just suggested, then the radius of convergence of a moment like $\langle u_i(\underset{\sim}{k},t)u_i^*(\underset{\sim}{k},t')\rangle$ is finite if the initial distribution admits no realizations with unbounded amplitudes, and the $\underset{\sim}{f}$ distribution is similarly bounded, but the radius of convergence is zero if the initial $\underset{\sim}{u}$ distribution is a normal distribution, or other distribution that is unbounded in our present sense. In the first case, the radius of convergence tends to zero as t and/or t' increases, even if the system is statistically stationary. If the turbulent Reynolds number is large, the radius of convergence becomes less than one when $t \ll \ell/u$, the characteristic eddy circulation time [20].

Padé approximants and other summation techniques may be used to obtain converging approximation sequences from the λ expansion where it is divergent or poorly convergent. Some promising results have recently been obtained in this way, particularly by a method of expanding the Fourier transform of the λ series in orthogonal functions [21-23]. Some basic restrictions should be noted, however. When the radius of convergence is zero, the unknown statistical function is not uniquely determined by its power series; it is ambiguous by any function whose power series vanishes. This means that unique results can be obtained by summation techniques only if the power series is supplemented by additional information about the function. This problem arises even when the initial distribution is uniquely determined by its moments. When the initial distribution falls off more slowly than exponentially at large amplitudes, then it is not uniquely determined by its moments, and the entire formulation in terms of moments is inadequate [24]. Fortunately, such distributions do not seem physically relevant.

When the initial $\underset{\sim}{u}$ distribution and the $\underset{\sim}{f}$ distribution are multivariate normal, the λ expansion has a diagram representation similar to those for quantum field theories. This is because any moment of a normal distribution can be evaluated in terms of the covariance by a pairing rule which is the classical analogue of Wick's theorem. It is thereby possible to carry out partial summations which express line and vertex renormalizations [20, 22, 25–28]. To describe this in its simplest case, suppose that $\underset{\sim}{f}$ vanishes, so that the turbulence decays freely from a normal initial state. Then the coefficient of λ^n in the expansion of $u_i(\underset{\sim}{k},t)$ is homogeneous of degree $n + 1$ in the quantities

$$u_i^{(0)}(\underset{\sim}{k},t) = u_i(\underset{\sim}{k},0)\exp(-\nu k^2 t) ,$$

which are the solutions of the linearized equation, and homogeneous of degree n in the linearized Green's functions

$$G^{(0)}(k,t,t') = \exp[-\nu k^2(t-t')] .$$

Each coefficient consists of a number of terms, which rises rapidly with n according to the branching properties of equation (3) under the iteration.

Now, let us specialize to homogeneous, isotropic ensembles (not an essential restriction) and consider the expansion of the covariance scalar defined by [29]

$$(L/2\pi)^3 \langle u_i(\underset{\sim}{k},t)\, u_i^*(\underset{\sim}{k},t')\rangle = \tfrac{1}{2} P_{ij}(\underset{\sim}{k}) U(k,t,t') , \qquad (6)$$

where the form of the right-hand side is required by isotropy. Only even powers of λ occur in the expansion of $U(k,t,t')$. The coefficient of λ^{2n} is homogeneous of degree $n + 1$ in the linearized scalars $U^{(0)}(k,t,t')$ and degree 2n in the $G^{(0)}(k,t,t')$. Each term, or diagram, in the coefficient arises from a particular branching in the expansions of the $\underset{\sim}{u}$ factors and from a particular one of the $(2n)!/(2^n n!)$ pairings by which a (2n)th-order moment of $\underset{\sim}{u}^{(0)}$ factors is reduced to covariances according to the normal-distribution rule. The line renormalization (self-energy renormalization) consists of discarding all diagrams with self-energy parts and, in the remaining ones, replacing all $U^{(0)}$ factors with U factors and all $G^{(0)}$ factors with G factors, where $G(k,t,t')$ is defined as the average response scalar of the full nonlinear system to infinitesimal perturbations [29]. That is,

$$\langle \delta u_i(\underset{\sim}{k},t)/\delta f_j(\underset{\sim}{k},t')\rangle = P_{ij}(\underset{\sim}{k}) G(k,t,t') . \qquad (7)$$

In absolute equilibrium we would have [30]

$$G(k,t,t') = U(k,t,t')/U(k,t,t) \quad (t \geq t') , \qquad (8)$$

which is the fluctuation-dissipation relation. In the actual case, equation (8) does not hold, and a consequence turns out to be that the renormalization cannot be carried out consistently on $U(k,t,t')$ directly. Instead, we form from equation (3) the equation of motion ($\underset{\sim}{f} = 0$)

$$(\partial/\partial t + \nu k^2)U(k,t,t') = S(k,t,t') \ , \tag{9}$$

$$S(k,t,t') = -(L/2\pi)^3 i\lambda k_m \sum_{\underset{\sim}{p}+\underset{\sim}{q}=\underset{\sim}{k}} \langle u_i(\underset{\sim}{p},t)u_m(\underset{\sim}{q},t)u_i^*(\underset{\sim}{k},t')\rangle \tag{10}$$

The coefficient of λ^{2n} in the expansion of $S(k,t,t')$, called the energy transfer function for $t = t'$, is homogeneous of degree $n + 1$ in the $U^{(0)}$ and degree $2n - 1$ in the $G^{(0)}$. A consistent line renormalization on the S expansion is accomplished by the algorithm stated above [25].

To complete the line renormalization program, we need a renormalized series for $G(k,t,t')$. The latter satisfies [25, 29]

$$(\partial/\partial t + \nu k^2)G(k,t,t') = H(k,t,t') \ , \tag{11}$$

$$H(k,t,t') = -\tfrac{1}{2} i\lambda k_m \sum_{\underset{\sim}{p}+\underset{\sim}{q}=\underset{\sim}{k}} \langle u_i(\underset{\sim}{p},t)\delta u_m(\underset{\sim}{q},t)/\delta f_i(\underset{\sim}{k},t') + u_m(\underset{\sim}{q},t)\delta u_i(\underset{\sim}{p},t)/\delta f_i(\underset{\sim}{k},t')\rangle \ . \tag{12}$$

The expansion of H again contains only even powers of λ, and the coefficient of λ^{2n} is homogeneous of degree n in the $U^{(0)}$ and degree 2n in the $G^{(0)}$. The line renormalization consists of discarding all diagrams with self-energy parts and, as before, replacing all $G^{(0)}$ factors with G factors and all $U^{(0)}$ factors with U factors in the surviving diagrams. The renormalized expansions of S and H are unchanged in form if $\underset{\sim}{f} \neq 0$.

The renormalized expansions have the consistency properties that truncation of the S expansion at any order gives energy conservation,

$$\int_0^\infty S(k,t,t)k^2 dk = 0, \tag{13}$$

while truncation of S and H at the same order gives equation (8) in absolute equilibrium (K cutoff and $\nu = 0$) [25]. A vertex renormalization which keeps these consistency properties can also be carried out, but it is very complicated. Away from absolute equilibrium it involves three distinct vertex functions, associated with the quantities $\langle u_i(\underset{\sim}{k},t)u_i(\underset{\sim}{p},t')u_m(\underset{\sim}{q},t'')\rangle$, $\langle u_i(\underset{\sim}{k},t)\delta u_i(\underset{\sim}{p},t')/\delta f_m^*(\underset{\sim}{q},t'')\rangle$, $\langle \delta^2 u_i(\underset{\sim}{k},t)/\delta f_i^*(\underset{\sim}{p},t')\delta f_m^*(\underset{\sim}{q},t'')\rangle$, where $\underset{\sim}{k} + \underset{\sim}{p} + \underset{\sim}{q} = 0$. Some special cases have been described [25, 31], but the full expansion seems never to have been published. A vertex-renormalized expansion which does not have the stated consistency properties has been given by Wyld and Lee [26, 27].

The physical motivation of the renormalized expansions is that the characteristic times of G and U should be more relevant to the high-Reynolds-number dynamics than those of the linearized quantities $G^{(0)}$ and $U^{(0)}$. However, it must be remembered that the underlying λ expansion is divergent so that no valid reason really exists for assuming that a partial infinite summation of terms will, of itself, give answers closer to reality than simple truncation of the λ series. In fact, it has been found that, in general, truncations of the renormalized series give even more spectacular nonsense (negative energies, divergencies) at large Reynolds number than truncations of the λ series [20, 25]. The one exception, which we shall discuss soon, is truncation at the lowest nontrivial level. It should also be pointed out that the partial summations implied by the line and vertex renormalizations are very ineffective in the sense that, in the limit of infinite order in λ, they eliminate the explicit appearance of only a finite fraction of the diagrams in each order of the primitive λ expansion [20, 22].

Padé approximants and other techniques for constructing convergents have given some interesting and promising results when applied to the renormalized expansions [21, 22]. However, the attempts are, so far, entirely heuristic, without mathematical justification. The analyticity properties of the renormalized expansions seem quite inaccessible. It cannot be asserted, for example, that S and H are single-valued functionals of U and G [22]. Thus, even if a sequence of approximants converges, it may converge to some unphysical branch.

3. EXPANSIONS ABOUT STOCHASTIC MODELS

The explicit form of the lowest-order nontrivial truncations of the line-renormalized perturbation expansions is [32]

$$S(k,t,t') = \pi\lambda^2 k \iint_\Delta pq\,dp\,dq \left[\int_0^{t'} a_{kpq} G(k,t',s)U(p,t,s)U(q,t,s)\,ds - \int_0^{t} b_{kpq} U(k,t',s)G(p,t,s)U(q,t,s)\,ds\right], \tag{14}$$

$$H(k,t,t') = -\pi\lambda^2 k \iint_\Delta pq\,dp\,dq\,b_{kpq} \int_{t'}^{t} G(k,s,t')G(p,t,s)U(q,t,s)\,ds, \tag{15}$$

where the wavenumber sums are expressed as integrals ($L \to \infty$). The integration $\iint_\Delta$ is over all p, q such that k, p, q form a triangle, and

$$b_{kpq} = (p/k)(xy + z^3), \quad 2a_{kpq} = b_{kpq} + b_{kqp} \geq 0, \tag{16}$$

where x, y, z are the cosines of the interior angles opposite k, p, q, respectively. These geometric factors arise from the $P_{ij}(\underset{\sim}{k})$ factors in equations (3) and (6).

These expressions, inserted into equations (9) and (11), have the property, uniquely among the truncations, of representing exactly the statistics of certain model dynamical systems. The model systems can be constructed in several ways, the original formulation being [33]

$$(\partial/\partial t + \nu k^2)u_i(\underset{\sim}{k},t) = -i\lambda k_m P_{ij}(\underset{\sim}{k}) \sum_{\underset{\sim}{p}+\underset{\sim}{q}=\underset{\sim}{k}} \phi_{\underset{\sim}{kpq}} u_j(\underset{\sim}{p},t)u_m(\underset{\sim}{q},t) + f_i(\underset{\sim}{k},t), \tag{17}$$

which differs from equation (3) only by the factor $\phi_{\underset{\sim}{kpq}}$. The latter is invariant to all index permutations, and to $\underset{\sim}{k} \to -\underset{\sim}{k}$, $\underset{\sim}{p} \to -\underset{\sim}{p}$, $\underset{\sim}{q} \to -\underset{\sim}{q}$, but, apart from that constraint, takes the value ± 1 at random for each choice of the triad $\underset{\sim}{k}, \underset{\sim}{p}, \underset{\sim}{q}$. The result is that, when the renormalized expansions are formed for the model S and H, only the lowest terms survive for $L \to \infty$.

Equations (14) and (15) have been called the direct-interaction approximation [29]. The model representation equation (17) assures equations (8) and (13), which are also true for other truncations and, in addition, uniquely assures that $U(k,t,t')$ is the covariance scalar of a realizable distribution of the vector process $u_i(\underset{\sim}{k},t)$, in particular, that $U(k,t,t) \geq 0$. We may note that the strength of interaction of each wave vector triad is the same in equation (3) and (17), the difference lying entirely in the random phasing of the individual interactions in equation (17).

When $\underset{\sim}{f} \neq 0$, the additional term [29]

$$\int_0^{t'} G(k,t',s)F(k,t,s)ds \tag{18}$$

appears on the right-hand side of equation (9), where F is the covariance scalar of the $\underset{\sim}{f}$ field. This expression is exact for equation (3) if $\underset{\sim}{f}$ is normally distributed and is exact for equation (17) for general $\underset{\sim}{f}$ statistics. Note that the kinetic energy spectrum is

$$E(k,t) = 2\pi \int_0^{\infty} U(k,t,t)k^2 dk.$$

An alternative model representation of the direct-interaction approximation is [23,28,34]

$$(\partial/\partial t + \nu k^2)u_i(\underset{\sim}{k},t) + \int_0^t \eta(k,t,s)u_i(\underset{\sim}{k},s)ds = q_i(\underset{\sim}{k},t) + f_i(\underset{\sim}{k},t), \tag{19}$$

where

$$q_i(\underset{\sim}{k},t) = -i\lambda k_m P_{ij}(\underset{\sim}{k}) \sum_{\underset{\sim}{p}+\underset{\sim}{q}=\underset{\sim}{k}} \xi_j(\underset{\sim}{p},t)\xi_m(\underset{\sim}{q},t) \tag{20}$$

and

$$\eta(k,t,s) = \pi\lambda^2 k \iint_\Delta b_{kpq} G(p,t,s)U(q,t,s)pqdpdq. \tag{21}$$

Here $\xi_i(\underset{\sim}{k},t)$ is a solenoidal vector field which has the same covariance scalar as $u_i(\underset{\sim}{k},t)$ but is statistically independent of the $\underset{\sim}{u}$ field and has quasi-normal statistics; that is, the fourth order moments of $\underset{\sim}{\xi}$ are related to second-order moments as in a normal distribution. This model portrays the nonlinear interaction as a balance between an extra random forcing term $q_i(\underset{\sim}{k},t)$ and a dynamical damping, with memory, the $\eta(k,t,s)$ term. In contrast to equations (3) and (17), we see that equation (19) is linear in the stochastic variable $\underset{\sim}{u}$.

The direct-interaction equations have been integrated numerically, and the results show good agreement with laboratory measurements of decaying isotropic turbulence at moderate turbulent Reynolds numbers [32,35,36]. The corresponding equations for diffusion of particles in isotropic turbulence, and for thermal turbulence at large $\nu/\varkappa$ agree well with computer simulation experiments [37,38]. At large Reynolds numbers, the approximation reproduces the empirical result that the decay time of turbulent energy is $\sim \ell/u$, as discussed in section 1, and it gives a local cascade of energy from lower to higher wavenumbers [29]. It yields an inertial range spectrum $k^{-3/2}$, off by only $k^{1/6}$ from equation (1), a result which will be discussed in detail in the following section. These facts, together with the internal consistency properties of the models, suggest that we abandon the renormalized expansions, whose analyticity properties seem inaccessible, and expand, instead, about the stochastic models equations (17) and (19). It should be stressed here that these models are meaningful without reference to the renormalized expansions or to the underlying perturbation expansion [39]. Equation (17) is obtained by a direct modification of equation (3). Equation (19) can be obtained from equation (3) by assuming that the right-hand side is purely random and then requiring that energy conservation be restored in the mean by a dynamical damping term [40].

Expansion of the exact dynamics about the direct-interaction models can be carried out in a number of ways [28]. The most elementary is to expand moments in the form

$$U(k,t,t') = U_{DI}(k,t,t') + \sum_{n=2} \lambda^{2n} U_{(2n)}(k,t,t'), \tag{22}$$

where the $U_{(2n)}$ are found by subtracting the λ expansion of the direct-interaction solution U_{DI} from the exact expansion. A procedure which exploits more fully the properties of the models is to modify equation (17) by taking [23]

$$\phi_{\underset{\sim}{kpq}} = \pm(1-\mu^2)^{1/2} + \mu \, , \tag{23}$$

where the $\pm$ sign is taken at random, as before. Then $\mu = 1$ gives the exact dynamics while $\mu = 0$ is the direct-interaction model. $G(k,t,t')$ and $U(k,t,t')$ can then be developed into power series in μ, by an iteration procedure. This expansion is independent of λ, and we can take $\lambda = 1$ at the outset. Similarly, we can set $\lambda = 1$ and modify equation (19) to the form [23]

$$\begin{aligned}(\partial/\partial t + \nu k^2)u_i(\underset{\sim}{k},t) + (1-\mu^2)\int_0^t \eta(k,t,s)u_i(\underset{\sim}{k},s)ds &= (1-\mu^2)^{1/2} q_i(\underset{\sim}{k},t) \\ - i\mu k_m P_{ij}(\underset{\sim}{k}) \sum_{\underset{\sim}{p}+\underset{\sim}{q}=\underset{\sim}{k}} u_j(\underset{\sim}{p},t)u_m(\underset{\sim}{q},t) &+ f_i(\underset{\sim}{k},t) \, .\end{aligned} \tag{24}$$

Again, $\mu = 0$ is the direct-interaction model, $\mu = 1$ is the exact dynamics, and the exact $U(k,t,t')$ can be developed as a power series in μ by iteration. Equation (23) gives exact energy conservation. With equation (24), conservation is exact at all μ values only if $\xi_i(\underset{\sim}{k},t)$ has fully normal statistics. With quasi-normal $\xi_i(\underset{\sim}{k},t)$, conservation is exact only at the endpoints $\mu = 0$ and $\mu = 1$. A modified, easier-to-iterate form of equation (24) can be constructed also, in which $\eta(k,t,t')$ always takes its direct-interaction value and so is μ-independent.

These expansions about the direct-interaction models differ profoundly from the renormalized diagram expansions in that they are expansions in an actual dynamical parameter μ. As a result, their analyticity properties can be investigated fairly straightforwardly. The numerical successes of the direct-interaction approximation give the hope that the lowest few terms of these expansions may exhibit better convergence than the primitive λ expansion, and this is supported by some examples that have been studied already [21,23]. However, it is not difficult to see that the convergence in high orders cannot be better than that of the λ expansion; it is controlled by the radius of convergence of the Navier-Stokes equation in the individual realizations. Therefore, we must still appeal to Padé approximants and other similar methods for constructing systematically improving higher approximations. The advantage over the renormalized diagram expansions is that we have here much more hope of being able to work out the convergence theory [23].

For a case where the μ expansions have advantages over equation (22), consider free decay ($\underline{f} = 0$) turbulence initially confined to $k < k_0$. The turbulence energy will cascade to higher wavenumbers, and this is qualitatively well described by the direct-interaction approximation. However, it is easy to see from the convolution structure of equation (3) that if $k/k_0 \gg 1$ the λ expansion of $U(k,t,t)$ starts with a high power of λ^2. Therefore equation (22) will give corrections to the direct-interaction value only in high orders, which in practice will be infeasible to compute. On the other hand, the μ expansions give corrections already in the lowest orders. In contrast to the renormalized λ expansions, the μ expansions are readily adaptable to nonnormal initial $\underline{u}$.

4. DYNAMICS OF THE SMALL-SCALE MOTIONS

The chaotic nature of turbulence suggests that the small-scale motions, which arise from the large-scale motions through an hierarchy of instabilities, may have some universal characteristics at large Reynolds numbers, largely independent of the structure of the large scales. The Kolmogorov theory, described in section 1, is perhaps the most direct embodiment of this expectation. In recent years, the structure of the small-scale motions in both geophysical and laboratory flows has been examined in detail, and this has led to substantial doubts about Kolmogorov's theory and to renewed interest in theoretical attacks on the small-scale dynamics [8,9,41-52].

Kolmogorov's hypotheses imply not only equation (1) but also that other statistics of the small scales have universal values when nondimensionalized with ϵ and ν. In the inertial range, there should be universal scaling with ϵ alone. In particular, the skewness $\langle |\Delta\hat{\underline{u}}(\underline{r})|^3\rangle / \langle |\Delta\hat{\underline{u}}(\underline{r})|^2\rangle^{3/2}$ and kurtosis $\langle |\Delta\hat{\underline{u}}(\underline{r})|^4\rangle / \langle |\Delta\hat{\underline{u}}(\underline{r})|^2\rangle^2$, where $\Delta\hat{\underline{u}}(\underline{r}) = \hat{\underline{u}}(\underline{x} + \underline{r}) - \hat{\underline{u}}(\underline{x})$, should be universal numbers in homogeneous turbulence, if $2\pi/r$ is an inertial-range wavenumber. The recent measurements at high Reynolds numbers support equation (1) rather well, but suggest that the skewness and kurtosis rise with Reynolds number and with $1/r$ [8,9,51]. The latter effects are evidence that the velocity-field structure becomes increasingly intermittent as the scale size r decreases, in accord with an intuitive picture of the smallest scales as very stretched out shear layers or filaments of vorticity.

One way of obtaining Kolmogorov's inertial-range spectrum is to assume that the time constant for decay of the energy in scales of order r by cascade into smaller scales is r/u_r, where u_r is some typical value of $|\Delta \hat{u}(r)|$. This assumption is suggested by the idea of localness of cascade, together with the observation (section 1) that the time constant for decay of total energy is ℓ/u. Constancy of energy cascade as r decreases (energy conservation) then gives $u_r \sim \epsilon^{1/3} r^{1/3}$, whence a Fourier transformation yields the inertial-range limit ($k/k_s \ll 1$) of equation (1). Now suppose this is so in a homogeneous turbulence. If we increase the intermittency by making the fluid into quiescent regions with negligible velocity and action regions, of equal extent, where u_r increases by $\sqrt{2}$, then the mean kinetic energy in scales of order r is unchanged but the time constant decreases, and hence ϵ increases, by $\sqrt{2}$. This example suggests, first, that if Kolmogorov's theory holds in subregions of the fluid, then the constant f(0) in the inertial-range law can be universal only if intermittency in the local dissipation ϵ_r, defined as averaged dissipation over a domain of size r, somehow tends to a universal distribution. Second, if intermittency increases as scale size decreases, and Kolmogorov's basic ideas hold in local regions, then the cascade becomes more efficient as r decreases and E(k) must fall off more rapidly than $k^{-5/3}$ if, according to conservation of energy, the overall cascade rate is r independent.

Kolmorogov and others have recently assumed that ϵ_r has a log normal distribution with variance that increases logarithmically with 1/r [41,42]. There is some experimental support [9,51]. Some authors have proposed models with nested intermittency, with each active region subdivided into active and quiescent regions, and so forth [45,47]. It should be emphasized that these hypotheses are purely conjectural, with no support from analysis using the Navier-Stokes equation. The experimental picture is rather indecisive. The highest Reynolds numbers attained are apparently about the highest attainable in the future, and they barely give a clean inertial range of about a decade in k; above this, dissipation effects alter the spectrum [53]. Presumably, higher statistical properties, like skewness and kurtosis, would require still higher Reynolds numbers to attain asymptotic forms. The very concept of strong spatial intermittency implies statistical dependence through extensive ranges of wavenumbers, to give the requisite phase correlation, and this suggests that the dynamical interactions would be only weakly local in wavenumber. It therefore does not seem possible to conclude from the existing experimental data whether Kolmogorov's original theory, the recently suggested modifications, or none of them, are asymptotically correct. Nor is there much hope for importantly better data in the future.

The basic questions about small-scale structure are highly resistant to resolution by perturbation-related approaches like those discussed in sections 2 and 3. The description of intermittency first of all requires high-order moments, which means high orders of the expansions. But there is also another, almost unfair difficulty peculiar to the self-convection mechanism. The basic physical idea underlying Kolmogorov's theory is that sufficiently large scales should simply convect small scales, without distorting them and so without affecting the energy cascade process within the small scales. We do not know whether this really is so, but an idealized version certainly is true. Suppose that each flow system in the ensemble is subjected to a spatially constant translation velocity which, however, changes randomly from realization to realization. This can produce no distortion and therefore cannot affect the energy transfer, spectrum, or other statistics of a homogeneous ensemble. We may call this stochastic Galilean invariance. The λ expansion of section 2 is stochastically Galilean invariant in each order of λ. However, the renormalized expansions, which sum some, but not all, terms from all orders of the λ expansion, are not invariant in any order. In particular, the direct-interaction approximation is not invariant [31]. The noninvariance shows up in the following rather subtle way. Suppose a very strong uniform, random convecting velocity of rms value v_0 is imposed. The decay time of $G(k,t,t')$ or $U(k,t,t')$ as a function of $t - t'$ is $\sim(v_0 k)^{-1}$, which is the convective de-phasing time for simple translation of a sinusoidal disturbance of wavenumber k. This is true in the exact dynamics, and is easily verified for the direct-interaction approximation also. However, in the latter, it can be seen from equation (14) that this convective de-phasing time controls the rate of energy transfer among the wavenumbers. As a result the direct-interaction approximation gives a local energy cascade in k, but with the spectrum

$$E(k) = C_{DI}(\epsilon v_0)^{1/2} k^{-3/2} , \tag{25}$$

where C_{DI} is a number of order one, instead of equation (1). It is clear that this effect of convective de-phasing time on energy transfer is spurious, regardless of what the correct inertial-range law may turn out to be.

The failure of stochastic Galilean invariance is a very basic problem in the analytical treatment of turbulence and probably affects any scheme which seeks to replace the irrelevant viscous-decay time-constants of $G^{(0)}$ and $U^{(0)}$ with physically relevant dynamical times, within the framework of moments of the Eulerian velocity field [31]. In that framework, the characteristic times for distortion of velocity structures are washed out by the convective de-phasing times; Eulerian moments do not distinguish between them. An enlarged framework capable of expressing this distinction is provided by the generalized velocity field $\hat{\underset{\sim}{u}}(\underset{\sim}{x},t\,|\,t')$, defined as the velocity at time t' in that fluid element which arrives at $\underset{\sim}{x}$ at time t. This field includes the Eulerian and Lagrangian velocity fields as special cases. It is determined by the Eulerian field through [54]

$$\partial\hat{u}(\underset{\sim}{x},t\,|\,t')/\partial t = -\hat{\underset{\sim}{u}}(\underset{\sim}{x},t)\cdot\nabla\hat{\underset{\sim}{u}}(\underset{\sim}{x},t\,|\,t') , \quad \hat{\underset{\sim}{u}}(\underset{\sim}{x},t\,|\,t) = \hat{\underset{\sim}{u}}(\underset{\sim}{x},t) . \tag{26}$$

If the ordering parameter λ is inserted in the nonlinear term of equation (26), and the Navier-Stokes equation is adjoined, a perturbation analysis of moments of $\hat{u}(\underset{\sim}{x},t\,|\,t')$ may be carried out as in section 2, and a generalized direct-interaction approximation, with model representations, can be constructed. Now, however, the generalized framework permits a modification of the final statistical relation (14) in which the integration back in time is carried out along particle trajectories instead of at fixed points in the fluid. The form of the modified equations appears to be uniquely determined by the requirement that the final result exhibit stochastic Galilean invariance and retain, at the same time, the consistency properties of the unmodified direct-interaction approximation [54].

The result, which has been called the Lagrangian-history direct-interaction approximation, gives answers which differ little either qualitatively or quantitatively from the direct-interaction approximation at low Reynolds numbers or for the large scales at large Reynolds numbers [36]. However, it gives equation (1) for the small scales at high Reynolds number, and yields numerical results for $f(k/k_s)$ which agree well with experiment [53].

The model representation of the direct-interaction representation appears to be lost in the Lagrangian modification, so that the equivalent of the μ expansions of section 3 cannot be constructed. However, model amplitude equations, suitable for μ expansions, can be constructed after a fashion. Instead of equation (19), consider the class of model equations [34,35]

$$[\partial/\partial t + \nu k^2 + \eta(k,t)]u_i(\underset{\sim}{k},t) = q_i(\underset{\sim}{k},t) + f_i(\underset{\sim}{k},t) , \tag{27}$$

where

$$\eta(k,t) = \pi\lambda^2 k \int\int_\Delta b_{kpq}\, \theta_{pqk}(t) U(q,t,t) pq dp dq, \tag{28}$$

$$q_i(\underset{\sim}{k},t) = i\lambda k_m P_{ij}(\underset{\sim}{k}) w(t) \sum_{\underset{\sim}{p}+\underset{\sim}{q}=\underset{\sim}{k}} [\theta_{kpq}(t)]^{1/2} \xi_j(\underset{\sim}{p},t)\xi_m(\underset{\sim}{q},t) . \tag{29}$$

Here b_{kpq} and $\xi_i(\underset{\sim}{k},t)$ are defined as before, $w(t)$ is a white-noise process,

$$\langle w(t)w(t')\rangle = 2\delta(t-t') ,$$

and $\theta_{kpq}(t)$ is an effective memory time for energetic interaction of k, p, q, which takes the place of the integration over time in equation (19). The flexibility of choice of $\theta_{kpq}(t)$ gives this class of Langevin equations wider possibilities than equation (19), at the expense of less realistic representation of the dependence of $U(k,t,t')$ on $t-t'$ for small $t-t'$ [55]. The choice

$$\theta_{kpq}(t) = \int_0^t G(k,t,s)G(p,t,s)G(q,t,s)ds \tag{30}$$

gives results for energy transfer which are close to direct-interaction results and identical with Edwards' approximation for turbulence [14]. Thus equations (27)-(30) give a model representation of Edwards' theory. It turns out that, when equation (27) is generalized to corresponding equations for $\hat{u}(\underset{\sim}{x},t\,|\,t')$, choices of $\theta_{kpq}(t)$ can be made which yield stochastic Galilean invariance, and provide the base for a stochastically Galilean-invariant μ expansion. It is also possible to construct models of the class of equation (27) which give stochastic Galilean invariance without explicitly going outside the Eulerian framework [55]. The principal objection to equation (27) is that the white-noise process in time is foreign to the nature of turbulence, which offers no intrinsic separation of time scales, in contrast to, say, kinetic theory.

The application of appropriate convergent-forming techniques, such as Padé approximants and related things, to the λ expansion or the stochastically Galilean-invariant μ expansions should yield systematically improving approximations to $E(k)$ at high k and, in principle, indicate whether, or under what conditions, equation (1) is a correct asymptotic form [23]. The same thing, in fact, is possible with the non-Galilean invariant μ expansion of section 3, if the fact of noninvariance is taken into account in the method of forming convergents. In practice, however, only the lowest few terms of any of these expansions are feasible to compute; the asymptotic behavior of the high-order terms does not seem within present grasp. Although low-order approximants formed from the expansions may turn out to yield good quantitative representations of the actual E(k) at high Reynolds numbers accessible to experiment, there is not much hope of finding for certain from them what the correct asymptotic behavior is and, in particular, what the asymptotic intermittency properties of the small scales are.

It seems likely that novel analytical approaches are needed to settle the asymptotic behavior of the small scales. Although the small scales are unimportant in the gross transport properties of turbulence, they have great intrinsic interest. In particular, it is interesting to ask whether intermittency does increase as scale size decreases so that asymptotically, as $\nu \to 0$, the dissipation of kinetic energy into heat takes place in a vanishingly small fraction of the fluid, somewhat in analogy to dissipation of compressive energy in shock waves.

5. BOUNDS ON STATISTICAL FUNCTIONS

During the past 15 years, Malkus and his co-workers have explored an analytical approach to turbulence quite distinct from the perturbation-theoretic methods discussed in sections 2 and 3. Malkus hypothesized that the turbulent flow in Boussinesq convection or shear flow tended to be close to that solution of the flow equations which maximized the gross transport of heat or momentum, subject to the boundary constraints of the problem [56]. This hypothesis leads to a program of attempting to find upper bounds to heat transport using such partial constraints from the equations of motion as yield a tractable problem.

Howard and Busse have in this way obtained upper bounds in the thin-layer Boussinesq convection problem discussed in section 1 [57, 58]. The equations of motion here are

$$(\partial/\partial t - \nu\nabla^2)\underset{\sim}{u} + \rho^{-1}\underset{\sim}{\nabla} p = \alpha g T \underset{\sim}{n} - \underset{\sim}{u}\cdot\underset{\sim}{\nabla}\underset{\sim}{u} \quad , \tag{31}$$

$$\underset{\sim}{\nabla}\cdot\underset{\sim}{u} = 0 \;, \tag{32}$$

$$(\partial/\partial t - \varkappa\nabla^2)T = -\underset{\sim}{u}\cdot\underset{\sim}{\nabla} T \;, \tag{33}$$

where $\underset{\sim}{u}$, T, p are velocity, temperature, and pressure fields in $\underset{\sim}{x}$-space, ρ is density, and the other symbols are given in section 1. The boundary conditions are vanishing velocity and constant, uniform T on the boundaries at top and bottom. The unit vector $\underset{\sim}{n}$ points up. These equations yield the integral relations [57]

$$\alpha g\{wT\} = \nu\{|\underset{\sim}{\nabla}\underset{\sim}{u}|^2\} \;, \tag{34}$$

$$\varkappa^{-1}[\{wT\}^2 - \{(\overline{wT})^2\}] + (\Delta T/D)\{wT\} = \varkappa\{|\underset{\sim}{\nabla} T|^2\} \;, \tag{35}$$

where $w = \underset{\sim}{n}\cdot\underset{\sim}{u}$, the overbar represents an average over the (infinite) horizontal at fixed verbal coordinate, and { } denotes an average over the entire fluid layer, while $|\underset{\sim}{\nabla}\underset{\sim}{u}|^2 \equiv |\underset{\sim}{\nabla} u|^2 + |\underset{\sim}{\nabla} v|^2 + |\underset{\sim}{\nabla} w|^2$, where u and v are the horizontal components of $\underset{\sim}{u}$. These are conservation equations, equation (34) representing the balance between work by gravity and viscous dissipation of kinetic energy, and equation (35) representing the balance between entropy production in the layer and the net flux of entropy through the boundaries.

Using only equations (34) and (35) and the boundary conditions, as constraints, Howard sought the $\underset{\sim}{u}$, T fields which maximized heat transport H. In this way he found the rigorous upper bound

$$H/(\kappa \Delta T/D) \leq N(Ra), \quad N(Ra) \to (3Ra/64)^{\frac{1}{2}}, \quad Ra \to \infty, \tag{36}$$

where N(Ra) is a numerically computed monotonically increasing function of Ra. Busse included equation (32) as a third constraint in addition to equations (34) and (35). He obtained a bound lower by an order of magnitude at large Ra,

$$N(Ra) \to (Ra/1035)^{\frac{1}{2}}, \quad Ra \to \infty, \tag{37}$$

but this is not yet a rigorous bound because Busse was unable to examine all the solutions of the Euler equations for the variational problem. He does, however, give plausible arguments that the solutions examined are the ones with the largest H.

The bounds found in this way are not actually close to experimental values, expression (36) lying roughly two orders of magnitude above the observations at large Ra and the nonrigorous closer bound, given asymptotically in expression (37), still lying above the measurements by a factor of 3 or more [58]. However, they are of extreme interest for two reasons. First, the rigorous upper bound, together with companion results for turbulent shear flow to be described below, constitutes the only rigorous quantitative dynamical result so far obtained in turbulence dynamics. Second, if Busse's arguments for expression (37) are accepted, the addition of an additional constraint in the variational problem produces a large reduction of the bound, suggesting a rapid convergence as still more conditions are imposed.

Busse has obtained parallel results for the mathematically similar problem of momentum transport in plane Couette flow, and for turbulent pipe flow [59,60]. He finds a rigorous bound for the momentum transport using only the equation of energy balance for the fluid as a whole, together with the boundary conditions, and a putative improved bound using also the continuity equation. The agreement with experimental data is roughly similar to that in the Boussinesq convection case.

It is not clear, at the present early stage of this work, how broad a class of statistical functions can be bounded by similar procedures, and what the nature of the convergence is as more moment relations obtained from the equations of motion are added as constraints. In the cases of the heat and momentum transport, if we assume that the best possible bound, fully using the equations of motion, can be found, there is still the matter of Malkus' original hypothesis. How close to the extremalizing solution is the actual turbulent flow? This is essentially a stability problem. Perhaps it could best be resolved by adding to the problem a small random forcing which would spoil unstable solutions, and then seeking both upper and lower bounds.

In the work of Howard and Busse, an actual realization of the flow field is sought, subject to integral constraints. It may be possible, instead, to work solely with moments by adjoining a suitable set of realizability inequalities to a set of moment relations obtained from the equations of motion.

As yet, no extension of this work to isotropic turbulence has been published. It is possible that similar methods may yield bounds on the inertial-range energy cascade and so contribute to understanding of the small-scale dynamics and statistics discussed in section 4. More directly analogous to the work of Howard and Busse would be a search for bounds on aspects of the gross energy dynamics, such as the rate of decay of total kinetic energy. In two-dimensional turbulence, which we have not discussed, the vorticity of each fluid element is an inviscid constant of motion, in addition to kinetic energy [34]. This highly constrained system, which is of meteorological importance, may prove a more fruitful starting point for bounding techniques than the three-dimensional case.

6. CONCLUDING REMARKS

We have not attempted to give a complete account of all the lines of investigation pursued in the analytical study of incompressible turbulence in recent years. We wish, in closing, to call attention to the work by Edwards [14,61] and Herring [15] on expansion of the distribution function about products of univariate $u_i(\underset{\sim}{k},t)$ distributions, to that of Balescu and Senatorski using the methods of the Brussels school [16], and to that of Meecham and Siegel and others on the expansion of $\hat{\underset{\sim}{u}}(\underset{\sim}{x},t)$ in Wiener's series of Hermite functionals of a white-noise process [62-66]. The usable final statistical approximations which have emerged from these studies are close in form and content to those obtained from moment expansions.

The embryonic theory of bounds on turbulence statistics, described in section 5, differs greatly from all the other approaches. So far, no bridges have been built. A possible link to perturbation approaches would be the isolation of statistical functions whose expansions are series of Stieltjes so that they are bounded by Padé approximants. It is to be hoped that substantial effort will be devoted to the theory of bounds in the next few years.

Note should also be made of recent rapid advances in the numerical simulation of turbulence by high-speed computers. The use of fast transforms and representation of the turbulent field by optimal orthogonal function sets make feasible computer experiments on isotropic turbulence, turbulent convection, and shear-excited turbulence [67-69]. This in turn makes possible a confrontation of analytical theory with experiment under controlled initial conditions, which are not attainable in the laboratory. The analytical approximations described for isotropic turbulence in sections 2-4 all extend also to inhomogeneous flows with mean fields, such as the Boussinesq thermal turbulence problem, and turbulent channel, Couette, and pipe flow [38, 70]. Little integration of the approximations has yet been carried out for such cases because of computational demands. The advances in numerical techniques promise substantial progress here also.

In line with remarks made in section 1, not much work on incompressible turbulence theory has been based on microscopic distribution functions. The general difficulties associated with microscopic description of turbulence correlations are discussed by Frisch [73]. These difficulties are illustrated by the attempt of Prigogine, Balescu, and Krieger, who use a closure at the two-particle-distribution level but lose completely the nonlinear transfer of turbulent energy [71, 72]. The latter is described in $\underset{\sim}{x}$-space by $\langle \hat{\underset{\sim}{u}}(\underset{\sim}{x}',t) \cdot [\hat{\underset{\sim}{u}}(\underset{\sim}{x},t) \cdot \underset{\sim}{\nabla}\hat{\underset{\sim}{u}}(\underset{\sim}{x},t)] \rangle$, which, because of the $\underset{\sim}{\nabla}$, involves three-body distributions with macroscopic separations of all three particles. The dynamical effects of such intrinsic three-point correlations, with incompressibility correctly described, must survive in any successful closure.

There has not yet been sufficient confrontation with laboratory and computer experiments to permit a definitive judgment of the performance of the various perturbation-theory-related schemes discussed in section 2, 3, and 4. What results there are, suggest that relatively simple approximations, such as the direct-interaction approximation or approximants formed from the first few terms of the primitive λ expansion, do perhaps surprisingly well at predicting spectra and covariances.

A test situation simpler than turbulence dynamics itself is the diffusion of fluid particles in a random velocity field with prescribed statistics. Figure 1 compares computer-experiment and direct-interaction results [37] for the eddy diffusivity when the Eulerian velocity field is frozen in time; is multivariate-normal, isotropic, and homogeneous; and has the two choices of wave-number spectrum

$$E_1(k) = \frac{3}{2} v_0^2 \delta(k - k_0) , \quad E_2(k) = 16(2/\pi)^{\frac{1}{2}} v_0^2 k^4 k_0^{-5} \exp(-2k^2/k_0^2). \qquad (38)$$

Both choices peak at $k = k_0$ and are normalized to yield $\langle |\underset{\sim}{u}(\underset{\sim}{x})|^2 \rangle = 3v_0^2$. The trajectory of a fluid element starting at $\underset{\sim}{x}$ at time $t = 0$ is given by

$$d\underset{\sim}{y}(t)/dt = \underset{\sim}{v}(t) , \quad \underset{\sim}{v}(t) = \underset{\sim}{u}(\underset{\sim}{y}) , \quad \underset{\sim}{y}(0) = \underset{\sim}{x} . \qquad (39)$$

The Lagrangian velocity covariance and eddy diffusivity are defined, respectively, by

$$U_L(t) = \frac{1}{3} \langle \underset{\sim}{v}(0) \cdot \underset{\sim}{v}(t) \rangle , \quad \varkappa(t) = \frac{1}{3} \langle \underset{\sim}{y}(y) \cdot \underset{\sim}{v}(t) \rangle . \qquad (40)$$

They are related in the present case by $\varkappa(t) = \int_0^t U_L(s)ds$. The expansions discussed in sections 2 and 3 can be formed by using the concept of a density of marked fluid particles $\psi(\underset{\sim}{x},t)$, which is convected according to

$$(\partial/\partial t + \lambda \underset{\sim}{u}(\underset{\sim}{x}) \cdot \underset{\sim}{\nabla}) \psi(\underset{\sim}{x},t) = 0 . \qquad (41)$$

Then

$$U_L(t) = \frac{1}{3} \int \langle \underset{\sim}{u}(\underset{\sim}{x}) \cdot \underset{\sim}{u}(\underset{\sim}{x}') \psi(\underset{\sim}{x}',t) \rangle d\underset{\sim}{x}' , \quad \psi(\underset{\sim}{x}',0) = \delta^3(\underset{\sim}{x} - \underset{\sim}{x}') . \qquad (42)$$

The physical case, as before, is $\lambda = 1$.

The primitive λ expansion for $U_L(t)$ with spectrum E_2 is

$$U_L(t)/v_0^2 = 1 - \frac{5}{4} v_0^2 k_0^2 (\lambda t)^2/2! + \frac{235}{32} v_0^4 k_0^4 (\lambda t)^4/4! - \ldots , \qquad (43)$$

a series with zero radius of convergence in λt. Figure 2 shows approximants to $U_L(t)$ constructed by using the coefficients of expansion (43) to determine the coefficients of a Hermite-function

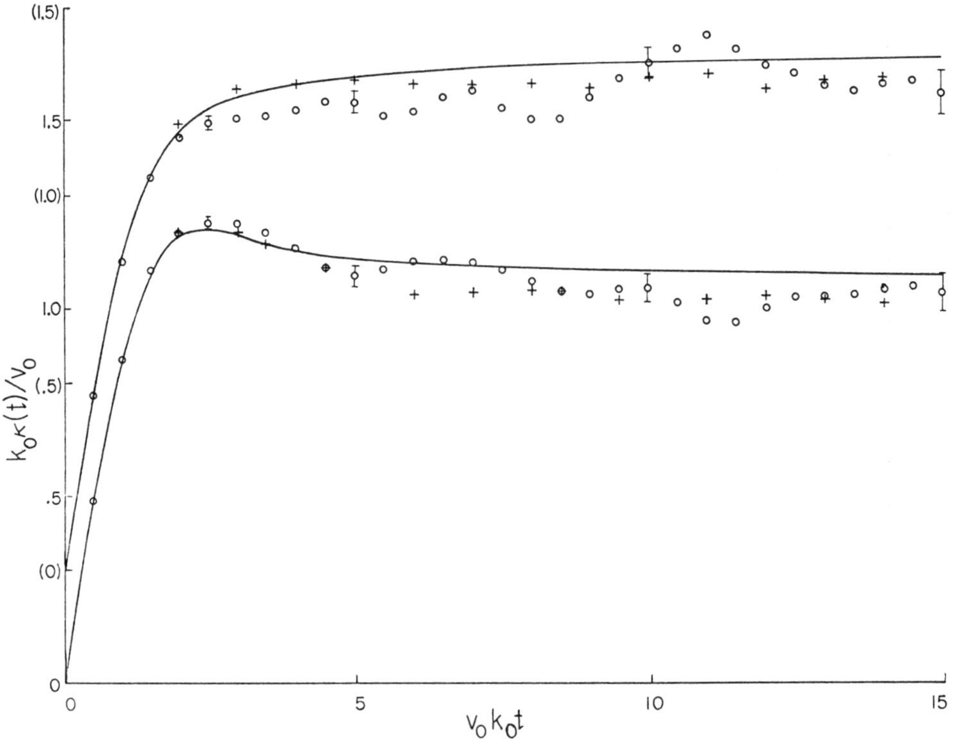

Fig. 1. Eddy diffusivity versus time. Upper data and scale ( ) are for spectrum E_2; lower data, for spectrum E_1. Curves are direct-interaction approximation. Circles are computer-experiment data for ensembles of 2000 marked particles, plus signs for ensembles of ~ 20,000 particles. Bars indicate standard deviation of data mean.

expansion of the Fourier transform of $U_L(t)$ with respect to λ, an asymptotic expansion in the present case [23]. Approximant 1 is based on the first two terms of expansion (43); approximant 2, on the first three terms.

Figures 3 and 4 compare computer experiment and direct-interaction approximation for the thermal turbulence problem discussed in section 1, but for the case of high Prandtl number ($\nu \gg \varkappa$) rather than air [38]. The high-Prandtl-number case is the easiest to compute. The computer experiment imposed slippery top and bottom boundaries and equations of motion which were truncated in a spatial Fourier representation so as to include the lowest three vertical wavenumbers and all horizontal wavevectors (some 76 in all) that fell in the annulus $\alpha_0 < |\underset{\sim}{\alpha}| < 2\alpha_0$. The evolution was followed for an ensemble of solutions with an indented initial mean temperature profile and random initial fluctuations. A similar truncation, and equivalent statistical initial conditions, were used with the direct-interaction equations. Figure 3 shows the evolution of the normalized heat transport $Nu = H/(\varkappa \Delta T/D)$ for computer experiment and direct-interaction approximation, together with a measure of the statistical error of the experiment. Figure 4

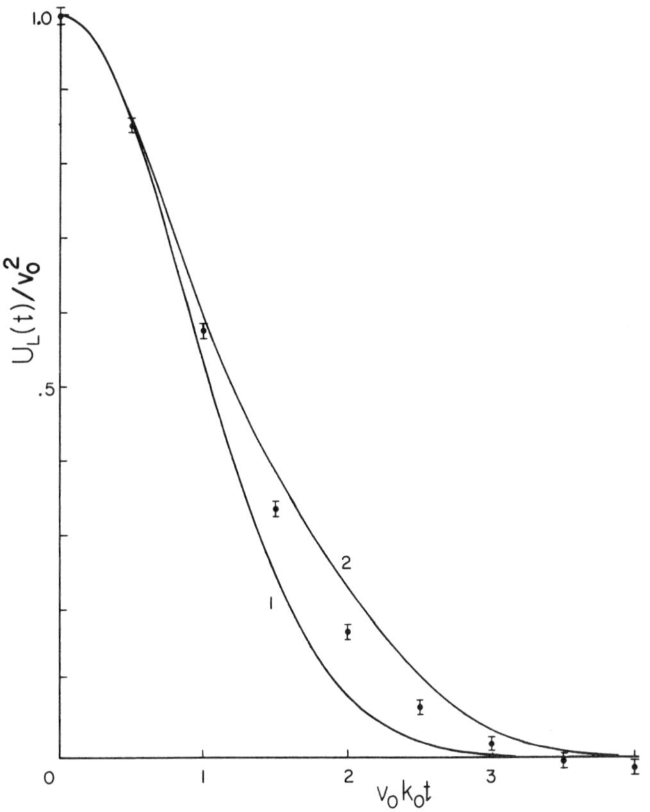

Fig. 2. Lagrangian velocity covariance $U_L(t)$, for spectrum E_2. Points are computer experiment, with bars indicating standard deviation of mean. Curves 1 and 2 are approximants constructed from primitive λ expansion.

compares the spectrum of the temperature fluctuations in the gravest vertical mode as a function of horizontal wavenumber. The Rayleigh number was 3000.

The accuracy of the direct-interaction approximation here is notable because the computer experiment shows a high degree of order in the flow, a phenomenon also observed in laboratory experiments at the same parameter values.

Comparison of turbulence evolution according to the statistical approximations and to laboratory and field experiments is handicapped by an inability to assure identical initial conditions. This ambiguity is hopefully minimized by comparing spectral regions, principally the higher wavenumbers, where both statistical approximation and experiment show self-preserving or nearly universal behavior. Figure 5 shows a comparison of the dissipation spectrum of decaying isotropic turbulence at moderate Reynolds number as determined by wind-tunnel experiments and by the direct-interaction approximation (section 3 [32]) and an abridged form of the Lagrangian-

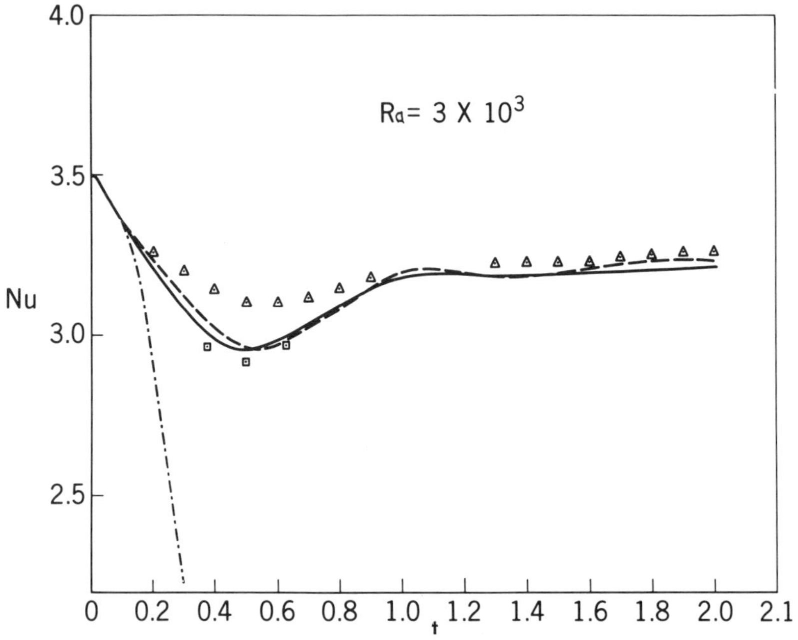

Fig. 3. Heat transport versus time for high-Prandtl-number turbulent convection. Solid curve, computer experiment. Squares are a more refined computer experiment and measure the statistical error. Dashed curve, direct-interaction approximation. Triangles and dot-dashed curve are, respectively, the quasi-linear and quasi-normal approximations, two additional perturbation-related approximations [38]. Ra is the Rayleigh number.

history direct-interaction approximation (section 4 [53]). In all three cases, the initial conditions are such that the spectrum is nearly self-preserving in time when normalized as shown. Figure 6 shows a comparison of the universal high-Reynolds-number inertial- and dissipation-range spectrum predicted by the abridged Lagrangian-history direct-interaction approximation with high-Reynolds-number measurements in a tidal channel [53]. It is to be hoped that comparisons such as these can soon be supplemented with more definitive tests against computer experiments, at least at small and moderate Reynolds numbers. Full-scale computer simulations of isotropic turbulence decay at moderate Reynolds numbers have now been achieved. The results agree well with the direct-interaction approximation [74].

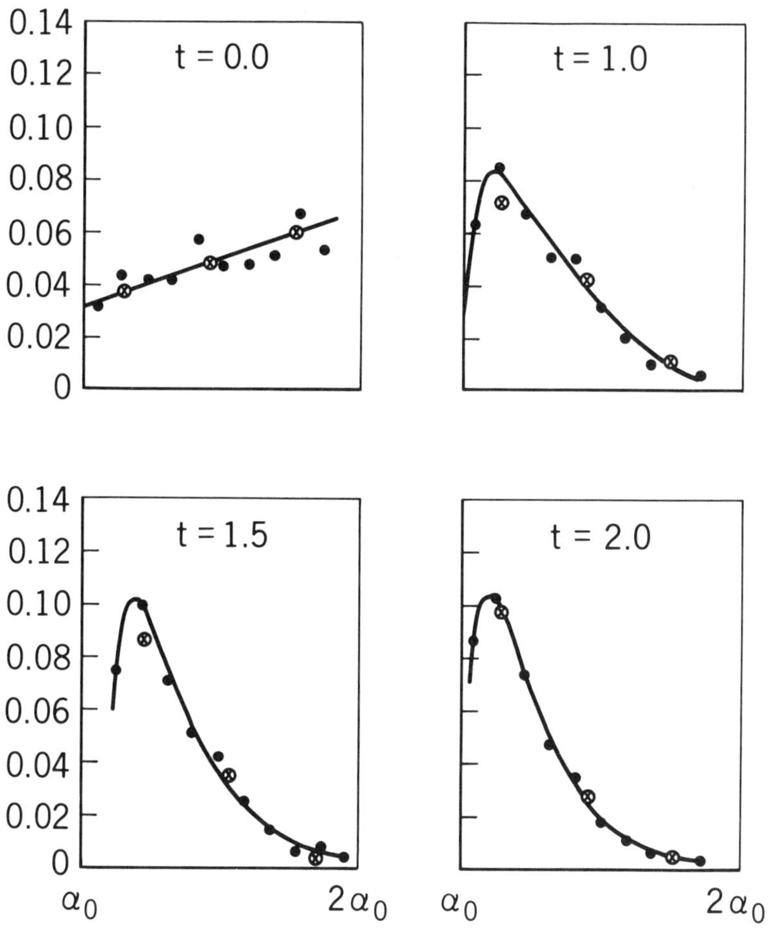

Fig. 4. Temperature-fluctuation spectrum level in gravest vertical mode versus horizontal wavenumber α. Dots, computer experiment; curves, fits to the dots. Crosses in circles, direct-interaction approximation.

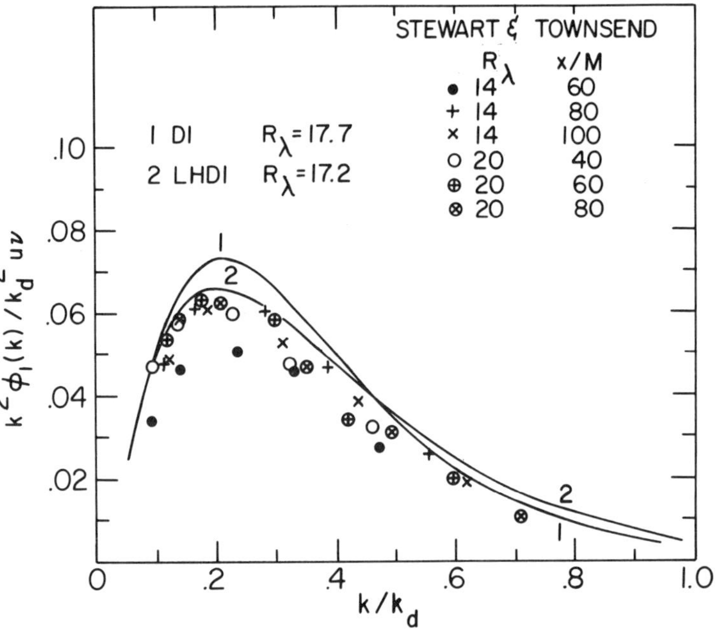

Fig. 5. One-dimensional dissipation spectrum of decaying isotropic turbulence. (See ref. [7] or [32] for definition of one-dimensional spectrum $\phi_1(k)$.) Curve 1 is direct-interaction approximation, curve 2 is abridged Lagrangian-history direct-interaction approximation. Data-point symbols are wind-tunnel experiments. R_λ is Reynolds number based on Taylor microscale, k_d is a characteristic dissipation wavenumber of the direct-interaction approximation [32].

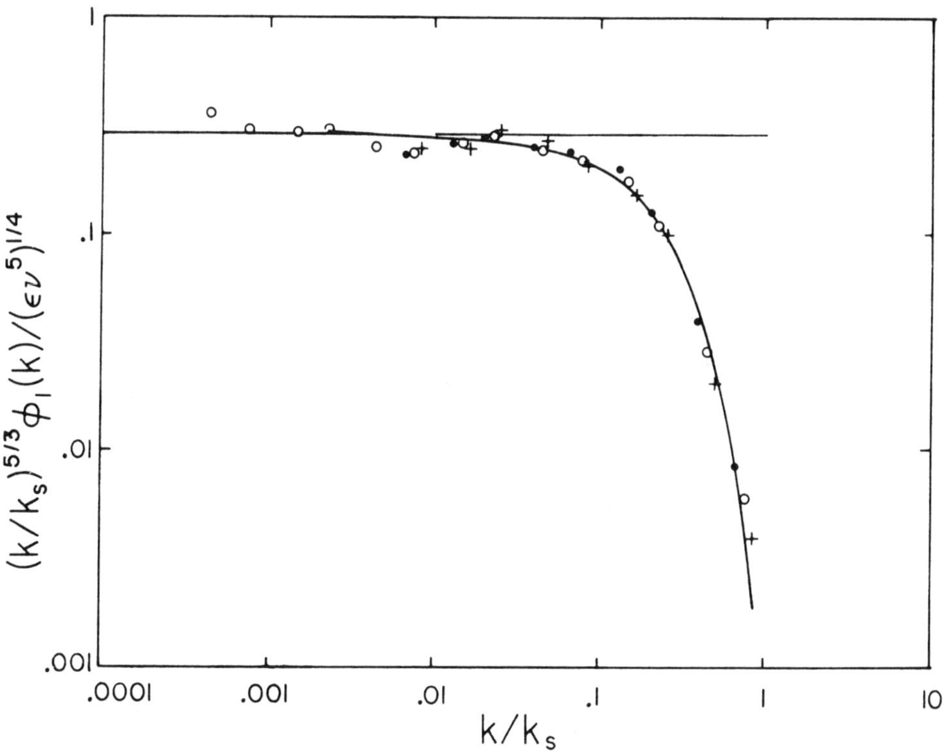

Fig. 6. One-dimensional energy spectrum in dissipation and inertial range at high Reynolds number according to abridged Lagrangian-history direct-interaction approximation (curve) and tidal-channel experiment (data-point symbols). The horizontal line is the inertial-range spectrum level given by the statistical approximation. The quantity k_s is defined in section 1.

REFERENCES

1. L. D. Landau and E. M. Lifshitz, Fluid Mechanics (Reading, Mass.: Addison-Wesley Publishing Company, Inc., 1959).

2. M. Jakob, Heat Transfer (New York: John Wiley and Sons, Inc., 1949), Vol. 1.

3. A. A. Townsend, J. Fluid Mech. **5**, 209 (1959).

4. S. J. Kline et al., J. Fluid Mech. **30**, 741 (1967).

5. J. Laufer, Report 1174, N.A.C.A., 1954.

6. A. A. Townsend, The Structure of Turbulent Shear Flow (Cambridge University Press, 1956).

7. G. K. Batchelor, Theory of Homogeneous Turbulence (Cambridge University Press, 1953).

8. R. W. Stewart, J. R. Wilson, and R. W. Burling, J. Fluid Mech. **41**, 141 (1970).

9. C. H. Gibson, G. R. Stegen, and R. B. Williams, J. Fluid Mech. **41**, 153 (1970).

10. T. D. Lee, Quart. J. Appl. Math. **10**, 69 (1952).

11. R. H. Kraichnan, in Proceedings of Symposia in Applied Mathematics, Vol. **13** (American Mathematical Society, 1962).

12. O. Ladyzhenskaya, Mathematical Theory of Viscous Incompressible Flow (New York: Gordon and Breach, 1963).

13. E. Hopf, J. Rational Mech. Anal. **1**, 87 (1952).

14. S. F. Edwards, J. Fluid Mech. **18**, 239 (1964).

15. J. R. Herring, Phys. Fluids, **8**, 2219 (1965); **9**, 2106 (1966).

16. R. Balescu and A. Senatorski, Ann. Phys. **58**, 587 (1970).

17. G. Rosen, Phys. Fluids, **3**, 519 (1960).

18. R. M. Lewis and R. H. Kraichnan, Comm. Pure and Appl. Math. **15**, 397 (1962).

19. T. S. Lundgren, Phys. Fluids, **10**, 969 (1967).

20. R. H. Kraichnan, in Dynamics of Fluids and Plasmas, ed. S. I. Pai (London: Academic Press, 1966).

21. R. H. Kraichnan, Phys. Rev. **174**, 240 (1968).

22. R. H. Kraichnan, in The Padé Approximant in Theoretical Physics, eds. G. Baker and J. Gammel (London: Academic Press, 1970).

23. R. H. Kraichnan, J. Fluid Mech. **41**, 189 (1970).

24. H. S. Wall, Continued Fractions (Chelsea Publishing Company, 1967).

25. R. H. Kraichnan, J. Math. Phys. **2**, 124 (1961); **3**, 205 (1962).

26. H. W. Wyld, Ann. Phys. **14**, 143 (1961).

27. L. L. Lee, Ann. Phys. **32**, 292 (1965).

28. R. Phythian, J. Phys. A (Gen. Phys.), **2**, 181 (1969).

29. R. H. Kraichnan, J. Fluid Mech. **5**, 497 (1959).

30. R. H. Kraichnan, Phys. Rev. **113**, 1181 (1959).

31. R. H. Kraichnan, Phys. Fluids, **7**, 1723 (1964).

32. R. H. Kraichnan, Phys. Fluids, **7**, 1030 (1964).

33. R. H. Kraichnan, in Second Symposium on Naval Hydrodynamics, ed. R. Cooper (Office of Naval Research, Publication ACR-38, 1958).

34. C. E. Leith, J. Atmos. Sci. **28**, 145 (1971).

35. C. W. Van Atta and W. Y. Chen, J. Fluid Mech. **38**, 743 (1969).

36. S. A. Orszag, J. Fluid Mech. **41**, 363 (1970).

37. R. H. Kraichnan, Phys. Fluids, **13**, 22 (1970).

38. J. R. Herring, Phys. Fluids, **12**, 39 (1969).

39. U. Frisch and R. Bourret, J. Math. Phys. **11**, 364 (1970).

40. B. B. Kadomtsev, Plamsm Turbulence (New York: Academic Press, 1965).

41. A. N. Kolmogorov, J. Fluid Mech. **13**, 82 (1962).

42. A. M. Oboukhov, J. Fluid Mech. **13**, 77 (1962).

43. S. Pond, R. W. Stewart, and R. W. Burling, J. Atmos. Sci. **20**, 319 (1963).

44. E. A. Novikov, Soviet Physics-Doklady, 11, 497 (1966).

45. A. S. Gurvich and A. M. Yaglom, Phys. Fluids, 10, S59 (1967).

46. E. A. Novikov, Atmospheric and Oceanic Phys. 1, 455 (1965).

47. E. A. Novikov and R. W. Stewart, Izv. Akad. Nauk., Ser Geofiz., No. 3, p. 408 (1964).

48. F. N. Frenkiel and P. S. Klebanoff, Phys. Fluids, 10, 507 (1967).

49. J. C. Wyngaard and H. Tennekes, Phys. Fluids, 13, 1962 (1970).

50. P. G. Saffman, Phys. Fluids, 13, 2193 (1970).

51. C. H. Gibson, G. R. Stegen, and S. McConnell, Phys. Fluids, 13, 2448 (1970).

52. R. H. Kraichnan, Phys. Fluids, 10, 2081 (1967).

53. R. H. Kraichnan, Phys. Fluids, 9, 1728 (1966).

54. R. H. Kraichnan, Phys. Fluids, 8, 575 (1965).

55. R. H. Kraichnan, J. Fluid Mech. 47, 513 (1971).

56. W. V. R. Malkus, Proc. Roy. Soc. (London), A225, 196 (1954).

57. L. N. Howard, J. Fluid Mech. 17, 405 (1963).

58. F. H. Busse, J. Fluid Mech. 37, 457 (1969).

59. F. H. Busse, J. Appl. Math. Phys. (ZAMP), 20, 1 (1969).

60. F. H. Busse, J. Fluid Mech. 41, 219 (1970).

61. S. F. Edwards and W. D. McComb, J. Phys. A (Gen. Phys.), 2, 157 (1969).

62. W. C. Meecham, J. Fluid Mech. 41, 179 (1970).

63. S. A. Orszag and L. R. Bissonnette, Phys. Fluids, 10, 2603 (1967).

64. G. H. Canavan, J. Fluid Mech. 41, 405 (1970).

65. S. C. Crow and G. H. Canavan, J. Fluid Mech. 41, 387 (1970).

66. S. E. Bodner, Phys. Fluids, 12, 33 (1969).

67. S. A. Orszag, Phys. Fluids, 12, II-250 (1969).

68. S. A. Orszag, J. Atmos. Sci. 27, 890 (1970).

69. S. A. Orszag, J. Fluid Mech. 71, 689 (1971).

70. R. H. Kraichnan, Phys. Fluids, 7, 1048 (1964).

71. I. Prigogine, R. Balescu, and I. M. Krieger, Physica, 26, 529 (1960).

72. I. M. Krieger, Phys. Fluids, 4, 649 (1961).

73. U. Frisch, in Geophysical Fluid Dynamics Notes (Woods Hole Oceanographic Institution, 1964), Vol. 2.

74. S. A. Orszag and G. S. Patterson, Jr., Phys. Rev. Letters, 28, 76 (1972).

DISCUSSION

M. G. Velarde: (1) Is it true that the maximum-heat-transport criterion of Malkus is not entirely applicable in the context of the Benard-Rayleigh instability (Malkus and Veronis, J. Fluid Mech. **4**, 225 [1958])? That, for example, this criterion fails if the amplitudes of the nonlinear solutions are not very small. See for instance, Schlüter, Lortz, and Busse, J. Fluid Mech. **23**, 129 [1965], and some recent numerical computations reported by T. Foster in the same journal. In the latter reference the inapplicability of such a criterion for turbulence is suggested.

(2) Is it not true that such a physical criterion was introduced to avoid having to solve an initial (and boundary condition) value problem?

R. H. Kraichnan: In the context of a systematic bounding program, maximum heat transport is not appealed to as a physical hypothesis. Instead, we seek to find how the maximum possible transport permitted by the integral (or other) constraints is reduced as the constraints (derived from the equations of motion) are made stronger.

COLLECTIVE MODES IN CLASSICAL LIQUIDS

Robert Zwanzig

INTRODUCTION

In recent years, considerable work in liquid state physics has made use of concepts described loosely by the term "phonon," "collective mode," and "elementary excitation." I propose to discuss here the utility of these concepts, and the interpretation that might be given to these terms, in the context of the physics of simple classical liquids.

In solid state and superfluid physics, there seems to be no question about the utility and meaning of these concepts. For example, phonons arise because the Hamiltonian of a solid can be approximated reasonably well as a quadratic form in momenta and displacements. By a normal mode transformation, this Hamiltonian can be reduced to a sum of independent harmonic oscillator Hamiltonians, one for each normal mode. By going over to a representation using creation and destruction operators, the phonon language is seen to be very natural. Further, this language is a useful one, because, in real solids, anharmonic interactions are often weak enough to be treated by low order perturbation theory. Phonons are approximate eigenstates of the Hamiltonian, and have long lifetimes.

The same situation holds in superfluid physics. Quantum excitations from the ground state of helium are described reasonably well by density fluctuations having a well defined (Landau) dispersion, by vortices, etc. These excitations have long lifetimes and interact weakly.

In both cases, we assume that there is some kind of vacuum or background, either a rigid perfect lattice or the quantum ground state of a superfluid, and we are concerned with excitations out of this vacuum.

It is well known that quantities somewhat like the phonons of solid state physics can be observed in classical liquids. For example, longitudinal sound waves can be excited in liquids, and at long wavelengths they have long lifetimes. In supercooled liquids it is even possible to excite transverse sound waves having long lifetimes. This suggests that there may be a point to using the concept of "phonon" in liquid state physics.

Research supported by the National Science Foundation, grant GP-12591.

Another example arises in the study of anomalous transport processes near thermodynamic critical points. Theories of such processes, e.g., the mode-mode coupling theory, are based on the use of perturbation theory to describe nonlinear interactions between certain collective hydrodynamic modes of the fluid. The existence of these collective modes is an essential ingredient of the theories.

A number of methods have been proposed to define phonons, collective modes, and elementary excitations in classical liquids. Before discussing these methods, it may be useful to distinguish between two different interpretations of the quantities that are being defined.

One interpretation, and the most common one, is that they are collective coordinates of a many body system which can be used, by a canonical transformation, to reduce a Hamiltonian to a sum of almost independent single-mode Hamiltonians. This is precisely how they are used in solid state and superfluid physics. Attempts to find collective variables like this have not been particularly successful in liquid state physics. Some attempts along these lines will be discussed shortly.

The other interpretation is that they are collective properties of a many body system that obey simple dynamical laws (e.g., hydrodynamic equations of motion). This is the interpretation that seems to be most useful in discussing classical liquids.

In the first interpretation, collective variables are used to describe excitations out of a vacuum state, taken to be the state of the system at zero temperature. This means that they may be used to calculate thermodynamic properties of the system at a nonzero temperature. In the second interpretation, collective variables are used to describe deviations from an equilibrium state at some given nonzero temperature. This equilibrium state is the "vacuum" here; its properties are assumed to be fully known. In the second interpretation, collective variables cannot be used to calculate properties of an equilibrium state. Nevertheless, they can be useful in dynamical problems. For example, they provide a tool for calculating the structure factor for inelastic scattering of light or neutrons.

In this second interpretation, one has gone quite far from the way that collective variables are used in solid state physics. The liquid state variables are <u>not</u> used to simplify a many body Hamiltonian. Perhaps one should not even use the terms "phonon" and "elementary excitation" in liquid state physics; but since they are used, we should remember how they are to be interpreted.

With this distinction in mind, let us survey some methods that have been used to define collective quantities in classical liquids. These methods fall into two classes; in one, emphasis is put on the search for collective coordinates or elementary excitations analogous to those appearing in solid state or superfluid physics; and in the other, emphasis is put on the search for collective variables having a simple dynamical significance.

ELEMENTARY EXCITATIONS

Two quite different kinds of collective variables have been proposed to describe elementary excitations in classical liquids. These are (1) density fluctuations, and (2) approximate eigenfunctions of the Liouville operator.

Density fluctuations are described by Fourier components of the number density,

$$\rho_{\underset{\sim}{k}} = \sum_{j=1}^{N} \exp(i\underset{\sim}{k} \cdot \underset{\sim}{R}_j), \tag{1}$$

where $\underset{\sim}{k}$ is a wave vector, and $\underset{\sim}{R}_j$ is the position of the j(th) particle. Suppose that these are regarded as new collective coordinates of the many body system. Then two problems arise. First, one must decide on the appropriate set of wave vectors to include--with too many, one has overdetermined the state of the system. The other problem is to find the momenta that are canonically conjugate to the coordinates. Until this is done, one cannot express the Hamiltonian in terms of the new variables. This has turned out to be extremely difficult to do, and not much progress has been made.

One may question whether density fluctuations are a reasonable set of quantities to be used as collective coordinates. In a solid state problem, they are not enough, because they do not take account of the important transverse degrees of freedom. Why should transverse motions be ignored in liquids? They can surely be observed in supercooled liquids.

Another scheme for defining elementary excitations was proposed a few years ago by this author [1]. It was based on the idea that collective variables associated with elementary excitations should be nearly periodic in time. Let (p, q) denote a complete set of momenta and coordinates, and let A(p, q) be some complex function of the dynamical state of the system. Then if A is a good collective variable, its time dependence should be given approximately by

$$\frac{d}{dt} A \cong i\omega A , \tag{2a}$$

where ω is some appropriate frequency. But the rate of change of any dynamical quantity can be calculated with the Liouville operator L, defined in terms of the Hamiltonian H in the following way:

$$iL = \frac{\partial H}{\partial p}\frac{\partial}{\partial q} - \frac{\partial H}{\partial q}\frac{\partial}{\partial p} = \frac{d}{dt} \tag{2b}$$

By comparing the preceding two equations, we see that a good collective variable ought to be an approximate eigenfunction of the Liouville operator, and its frequency ought to be the corresponding eigenvalue,

$$LA \cong \omega A . \tag{3}$$

Approximate eigenfunctions can be found by a variational method. Let angular brackets, $\langle\ \rangle$, denote an average over some stationary ensemble distribution in phase space (i.e., a distribution that is invariant to L). Then the variational condition

$$\delta \langle A^* L A \rangle , \tag{4}$$

together with the constraint

$$\langle A^* A \rangle = 1 \ , \tag{5}$$

leads to the desired eigenfunction-eigenvalue equation. Note that if A can be varied in an entirely arbitrary way, then the weighting function (ensemble distribution) used in the averaging, $\langle\ \rangle$, drops out. But if only certain restricted kinds of variations are considered, then the result of the calculation may depend on the weighting function. In particular, we may use a canonical ensemble distribution function, in which case the approximate eigenfunctions and eigenvalues may be temperature dependent. This is an example of the use of a thermal equilibrium "vacuum" state, as mentioned in the introduction.

A number of different trial eigenfunctions have been investigated. One very simple class of functions, motivated by the structure of phonon variables in solids, has the general form

$$A = B + b\dot{B} \ , \tag{6}$$

where B(p,q) is some observable and b is a variational parameter. When the variational condition is applied, one finds that this parameter is related to the eigenvalue ω by

$$b = 1/i\omega \ . \tag{7}$$

The frequency is given by

$$\omega^2 = \langle \dot{B}^* \dot{B} \rangle / \langle B^* B \rangle \ . \tag{8}$$

The quantities appearing here are equilibrium properties of the fluid.

An obvious candidate for the function B is a Fourier component of number density, ρ_k. Then the eigenfunction is a linear combination of number density and longitudinal current density, and the frequency of this eigenfunction is

$$\omega_o^2(k) = k^2 \frac{k_B T}{m\, S(k)} \tag{9}$$

where k_B is Boltzmann's constant, T is the temperature, m is the mass of a single particle, and S(k) is the equilibrium structure factor of the fluid. This frequency has been calculated from Rahman's computer experiments on liquid argon [2]; it is identified as ω_o in figure 1. Note that this dispersion curve has a strong similarity to the kind of phonon dispersion curve obtained for solids: it "almost" has a Brillouin zone structure.

Another candidate for the function B(p,q), motivated by analogy with solid state theory, was proposed in Ref. [3]. This is a Fourier component of the current density,

$$J_{k\lambda} = \underset{\sim}{\epsilon}_{k\lambda} \cdot \sum_{j=1} \dot{\underset{\sim}{R}}_j \exp(i\underset{\sim}{k} \cdot \underset{\sim}{R}_j) \ , \tag{10}$$

where $\dot{\underset{\sim}{R}}_j$ is the velocity of the j(th) particle, and $\underset{\sim}{\epsilon}_{k\lambda}$ is a polarization vector, used to select either longitudinal or transverse components of the current density. The resulting expressions for the frequencies are given in Ref. [1]; we do not write them down here. These frequencies have also been computed from Rahman's experiments. They are identified as $\omega_\ell(k)$ (for the longitudinal case) in figure 1, and as $\omega_t(k)$ (for the transverse case) in figure 2.

Further calculations along the same lines have been discussed by Nossal and Zwanzig [4].

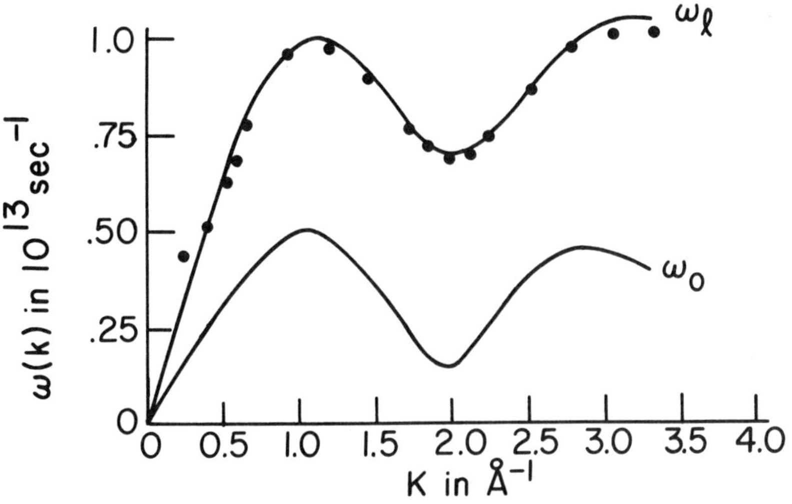

Fig. 1. Frequencies of two kinds of longitudinal excitations in liquid argon, as determined from computer experiments.

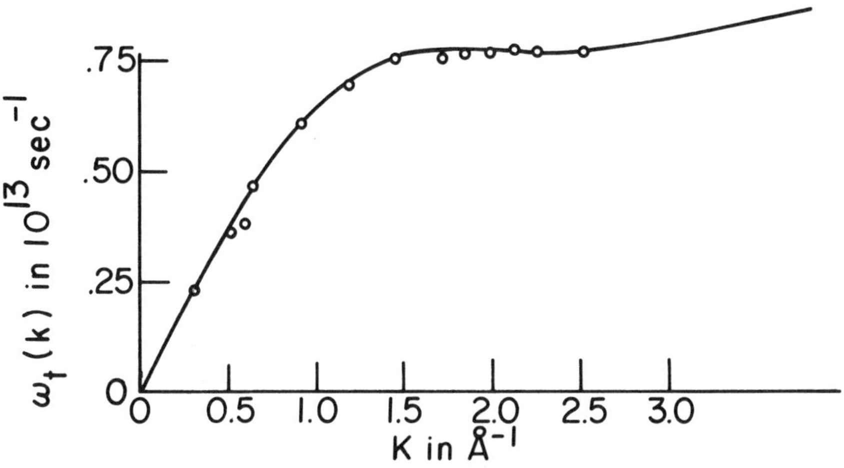

Fig. 2. The frequency of a transverse excitation in liquid argon, as determined from computer experiments.

In this article, linear combinations of various "hydrodynamic" variables and their time derivatives were considered. The resulting eigenfunctions and eigenvalues were discussed in connection with the low frequency and high frequency limits of nondissipative hydrodynamics.

Can these approximate eigenfunctions be regarded as good collective variables for classical liquids? The answer of course depends on what one means by "good." They have a certain physical plausibility, and they bear a close relationship to the variables that appear in hydrodynamic equations. For example, the eigenfunction obtained from ρ_k and its time derivative looks like a normal mode for longitudinal sound propagation in a nonviscous fluid. If one carries the analogy further, and assumes that the actual time dependence of this quantity is given by hydrodynamic laws, then one can estimate its lifetime. The result is that the lifetime is roughly $(\nu_\ell k^2)^{-1}$, where ν_ℓ is the longitudinal kinematic viscosity. At long wavelengths, these eigenfunctions will have long lifetimes.

The other eigenfunctions, based on the current density and its time derivative, can be viewed also as hydrodynamic quantities. This was done in Ref. [3], with the conclusion that the lifetimes of these eigenfunctions are expected to be very short--of the order of viscoelastic relaxation times. The lifetimes do not vary as $1/k^2$ for small k, but approach a finite limit as k goes to zero. The lifetimes are never long compared with a period of oscillation. One must conclude that if these eigenfunctions are to be used as collective variables, then interactions between them are likely to be very strong. This suggests that they will not be useful collective variables.

Another class of approximate eigenfunctions was investigated in Ref. [1]. They have the following structure:

$$A = \sum_{j=1}^{N} \psi(\underset{\sim}{R}_j, \dot{\underset{\sim}{R}}_j) , \tag{11}$$

where the specific function $\psi(\underset{\sim}{R}, \dot{\underset{\sim}{R}})$ is to be determined by the variational condition. The result will then be the best approximate eigenfunction that can be represented as a sum of single particle functions.

When the variational condition is applied, one gets an integral equation for the single particle function ψ. This integral equation has exactly the same structure as the linearized Vlasov equation in plasma physics; the only difference is that the Coulomb force law is replaced by an effective force, derived from an effective potential

$$U_{eff}(k) = - k_B T\, C(k) , \tag{12}$$

where C(k) is the direct correlation function of the fluid. The approximate eigenfunctions have a structure similar to the eigenfunctions of the Vlasov equation, and these have been studied extensively. As in the plasma case, the eigenvalue spectrum is continuous, and the eigenfunctions are singular. Also as in the plasma case, these eigenfunctions and eigenvalues can be used to solve initial value problems, and one finds a phenomenon analogous to Landau damping. In particular, a longitudinal sound wave will be damped in time; but the decay is quite different from what one expects from hydrodynamic considerations.

The classical-liquid version of the linearized Vlasov equation has been used in several applications. For example, Nelkin and Ranganathan [5] and Chihara [6] have discussed collisionless or "zero(th)" sound in classical liquids.

In my own opinion, this general approach to the search for collective variables in classical liquids has not been successful. It is easy to guess plausible forms for approximate eigenfunctions, and to apply the variational condition. But it is not easy to decide whether or not the results have any real significance. And it is hard to see how the results can be used. For example, one would like to construct a kind of "mode-mode coupling" theory to calculate the lifetimes of the approximate eigenfunctions; with the exception of an attempt by Ailawadi [7], little has been done along these lines.

The most useful result to come out of these investigations has been, in my opinion, the indication that it might be more appropriate to focus attention directly on hydrodynamic variables and their equations of motion.

DYNAMICS OF COLLECTIVE VARIABLES

If we give up the urge to imitate solid state and superfluid physicists, and abandon the search for elementary excitations that (hopefully) interact with each other only weakly, we are left with the investigation of collective variables that have a simple dynamical behavior. The existence of such variables is well known; in particular, hydrodynamic variables are collective and obey known simple equations of motion. While other collective variables may be important, we focus attention on these only.

The standard hydrodynamic variables are (1) mass density, (2) momentum density (or the corresponding velocity field), and (3) energy density (or entropy density). In simple classical liquids, the equations of motion of these quantities are the conservation law connecting mass and momentum density, the Navier-Stokes equation for the momentum density, and the Fourier heat law for the energy density.

Normally, these equations of motion are said to be concerned only with variables that have been averaged over some statistical ensemble. But it is easy to generalize the equations so that they apply to the actual values of the variables in any single member of the ensemble. In a fundamental paper [8], Mori has shown how to do this.

The idea is contained in Langevin's form of the theory of Brownian motion. The average velocity of a particle suspended in a viscous fluid obeys the equation of motion

$$m \frac{d}{dt} \langle v \rangle = - \zeta \langle v \rangle , \tag{13}$$

where ζ is a friction coefficient. The actual velocity obeys the Langevin equation

$$m \frac{d}{dt} v = - \zeta v + F(t) , \tag{14}$$

where F(t) is a fluctuating force, determined (in principle) by the motions of the environment of the particle. The ensemble average of the fluctuating force is zero, and its second moment is given by the fluctuation-dissipation theorem

$$\langle F(t)F(t') \rangle = 2k_B T \zeta \delta(t - t') . \tag{15}$$

In the standard theory of Brownian motion, the distribution of the fluctuating force is assumed to be Gaussian, so that it is fully determined by the mean value and second moment.

It is evident that if the actual motion is given by equation (14), then the average motion is given by equation (13). But equation (14) contains much more information; it can be used also to calculate the average behavior of any function of the actual velocity.

Mori generalized these ideas in two ways. First, he discussed the motion of any given set $a_1, a_2, a_3, \ldots = \underset{\sim}{a}$ of dynamical properties of the system. Second, he took into account the possibility of non-Markovian effects (i.e., frequency dependent transport coefficients). His results are as follows.

Let us construct an ensemble of repeated experiments, in which the mean values of the variables $\underset{\sim}{a}$ are given at some initial time $t = 0$. At this time, the state of the system is assumed to be one of thermal equilibrium subject to constraints giving rise to the specified initial mean values. The variables $\underset{\sim}{a}$ are defined so that they vanish in the state of unconstrained thermal equilibrium. Let us suppose further that the initial mean values are small enough to allow linearization of the equations of motion. Mori showed that the actual motion of this set of variables is given by generalized Langevin equations,

$$\frac{d}{dt}\underset{\sim}{a}(t) = i\underset{\sim}{\Omega}\cdot\underset{\sim}{a}(t) - \int_0^t ds \ \underset{\sim}{\varphi}(t-s)\cdot\underset{\sim}{a}(s) + \underset{\sim}{f}(t) . \tag{16}$$

The matrix $\underset{\sim}{\Omega}$ is determined by certain equilibrium properties of the system, and the matrix kernel $\underset{\sim}{\varphi}(t)$ is determined by certain equilibrium time correlation functions of the system. Explicit formulae are given by Mori. The quantity $\underset{\sim}{f}(t)$ is a vector composed of fluctuating generalized forces (analogous to the fluctuating force in the ordinary Langevin equation). The equilibrium mean value of the fluctuating force vanishes, and the second moment is related to the kernel $\underset{\sim}{\varphi}(t)$ by a generalization of the familiar fluctuation-dissipation theorem. Equation (16) is exact to the first order in $\underset{\sim}{a}(t)$; it fails only if the initial mean values are large enough that higher orders in $\underset{\sim}{a}(t)$ must be considered.

It is interesting to observe that the frequency matrix $\underset{\sim}{\Omega}$ is closely related to the approximate eigenfrequencies obtained by applying a variational condition to observables of the form of equation (6). These approximate eigenfrequencies are now seen to have a definite physical significance. Mori's theory goes much further, however, because it provides expressions for the lifetimes of these observables.

On the other hand, one might say that Mori's theory goes too far. It does not contain any internal criteria for the choice of the observables; any set of variables $\underset{\sim}{a}$ will be described by his equations, within the limitations on initial values that we have referred to already. In this sense the theory is entirely formal.

The freedom of choice of the variables $\underset{\sim}{a}$ allows us to explore different kinds of collective variables. Let us select, in particular, the set of hydrodynamic variables: Fourier components (in space) of the mass density, current density, and energy density. Then the resulting equations of motion correspond to the linearized continuity equation, the Navier-Stokes equation, and the heat conduction equation. The application of Mori's theory to these variables has been worked out independently by Ackasu and co-workers [9] and by Ailawadi, Rahman, and Zwanzig [10]. Some further extensions, including more than the standard five hydrodynamic variables, have been studied also; see Ref. [11].

While inclusion of the energy density does not pose any special problems, it is simpler, for the purpose of illustration, to leave it out. Then there are four dynamical variables for any given $\underset{\sim}{k}$, the density ρ_k defined in equation (1), and the three components of momentum density $J_{k\lambda}$, defined in equation (10). In a translationally invariant fluid, different Fourier components do not mix. For a given $\underset{\sim}{k}$, these four variables obey four equations:

$$\frac{d}{dt} \rho_k = i\, k\, J_{k1} \; ; \tag{17}$$

$$\frac{d}{dt} J_{k1} = i\, k\, C^2(k)\, \rho_k - \int_0^t ds\, k^2\, \Phi_1(k, t-s)\, J_{k1}(s) + f_{k1}(t) \; ; \tag{18}$$

$$\frac{d}{dt} J_{k\lambda} = - \int_0^t ds\, k^2\, \Phi_\lambda(k, t-s)\, J_{k\lambda}(s) + f_{k\lambda}(t) \; , \tag{19}$$

where the last equation applies for $\lambda = 2$ and 3.

The first equation is the linearized continuity equation, and the coefficient ik comes from the $\underset{\sim}{\Omega}$ matrix. There is no dissipative term or fluctuating force here. In the second equation, for the longitudinal current density, the coefficient $i\, k\, C^2(k)$ comes from the $\underset{\sim}{\Omega}$ matrix. The quantity C(k) is a k-dependent sound velocity, and is given explicitly by

$$\omega_o^2(k) = C^2(k)\, k^2 \; , \tag{20}$$

where $\omega_o(k)$ is the same frequency as in equation (9). The equations for longitudinal and transverse current densities both contain kernels $\Phi_\lambda(k; t)$, and fluctuating forces related to these kernels by generalized fluctuation-dissipation theorems. Aside from factors k^2 which have been separated out for later convenience, these kernels are just elements of the matrix $\underset{\sim}{\varphi}$ in equation (16).

Let us compare these equations for the dynamical variables with the familiar hydrodynamic laws (linearized and expressed in Fourier components). The continuity equation is exactly the same. The equations for the averaged current density are

$$\frac{d}{dt} \langle J_{k1}(t) \rangle = \left(\frac{1}{m} \nabla P\right)_k - \nu_\ell\, k^2 \langle J_{k1}(t) \rangle \; ; \tag{18'}$$

$$\frac{d}{dt} \langle J_{k\lambda}(t) \rangle = - \nu_t\, k^2 \langle J_{k\lambda}(t) \rangle \qquad (\lambda = 2,3) \; . \tag{19'}$$

In these equations, P is the thermodynamic pressure, and ν_ℓ and ν_t are the longitudinal and transverse kinematic viscosities. Aside from the fluctuating force, which vanishes upon averaging, the hydrodynamic laws can be put into one-to-one correspondence with the exact Mori-Langevin equations. For example, the gradient of pressure becomes

$$-\left(\frac{1}{m} \nabla P\right)_k = - \frac{1}{m} \frac{\partial p}{\partial \rho} (\nabla \rho)_k = i\, k\, C_o^2\, \rho_k \; , \tag{21}$$

where C_o is the ordinary sound velocity. In the exact Langevin equation, this is generalized to a k-dependent sound velocity. The viscosity coefficients correspond to the kernels $\Phi_\lambda(k,t)$. If we define generalized frequency- and k-dependent viscosity coefficients by

$$\nu_{\ell}(k,\omega) = \int_0^\infty dt\ e^{-i\omega t}\ \Phi_1(k,t)\ ,$$
$$\nu_t(k,\omega) = \int_0^\infty dt\ e^{-i\omega t}\ \Phi_2(k,t)\ , \tag{22}$$

than the ordinary viscosity coefficients are just $\nu_{\ell}(0,0)$ and $\nu_t(0,0)$.

If we had included the energy density, then the equation for the longitudinal current density would contain another term from the $\underset{\sim}{\Omega}$ matrix, and there would be a fifth equation containing a generalized heat conductivity. These also could be put into one-to-one correspondence with familiar hydrodynamic quantities.

Therefore, Mori's theory provides explicit equations of motion for the actual hydrodynamic variables, and these equations are seen to be plausible generalizations of standard hydrodynamic laws. Because they contain fluctuating force terms, they can be used to describe the actual motion of hydrodynamic variables, and not just the mean values. They can be used to discuss phenomena at frequencies and wavelengths of a molecular order of magnitude. Their only serious limitation is that they are linear approximations, and cannot be used when initial deviations from equilibrium are too large.

It must be mentioned that much of the information that is contained in Mori's Langevin theory, in particular the $\underset{\sim}{\Omega}$ and $\underset{\sim}{\varphi}$ and matrices, is contained in apparently quite different theories. For example, the correlation function formalism of Kadanoff and Martin [12] has been applied to hydrodynamics by Chung and Yip [13]. The essential ingredient of this formalism is the introduction of a "damping function" depending on k and ω. It is possible to establish an exact correspondence between this damping function and the $\underset{\sim}{\Omega}$ and $\underset{\sim}{\varphi}$ of Mori's theory, so that in a sense there is no difference between the two points of view. In the Kadanoff-Martin theory, emphasis is put on analytic properties of a correlation function or Green's function; in the Mori theory, emphasis is put on equations of motion of the Langevin form. But both theories are concerned with the same phenomena, and lead to equivalent results.

As Rahman has pointed out [14], the generalized mean field theory of collective variables put forth by Pathak and Singwi [15] can be shown to correspond to a special choice of the memory kernel $\underset{\sim}{\varphi}(t)$ in the Mori theory.

I believe that this is a general situation. Any theory of collective motions in classical liquids which invokes at some stage an assumption that the equations of motion of the collective variables are linear, must be a special case of Mori's theory, perhaps with special hypotheses about the structure of the matrices $\underset{\sim}{\Omega}$ and $\underset{\sim}{\varphi}(t)$.

CONCLUSIONS

Two general conclusions can be drawn from this survey. First, it appears that not much progress has been made in trying to force a similarity between the techniques of solid state or superfluid physics and techniques of the physics of classical liquids. One can define collective variables in various ways, but so far no one has been able to use them to reduce the Hamiltonian of some realistic liquid (with strong interactions between particles) to a sum of almost independent parts. Estimates of lifetimes indicate that the proposed collective variables must interact very strongly; thus there is no reason to prefer them to the standard momentum and position variables of interacting particles.

Second, it is not hard to construct collective variables that obey simple dynamical equations. In particular, the familiar (unaveraged) hydrodynamic variables have this property: they obey generalized Langevin equations having the same structure as hydrodynamic equations. These collective variables, and their associated dynamical equations, can be used in a variety of calculations concerned with fluctuation and relaxation processes.

REFERENCES

1. R. Zwanzig, Phys. Rev. **144**, 170 (1966).
2. N. K. Ailawadi, A. Rahman, and R. Zwanzig, Phys. Rev. **A4**, 1616 (1971).
3. R. Zwanzig, Phys. Rev. **156**, 190 (1967).
4. R. Nossal and R. Zwanzig, Phys. Rev. **157**, 120 (1967).
5. M. Nelkin and S. Ranganathan, Phys. Rev. **164**, 222 (1967).
6. J. Chihara, Progr. Theor. Phys. **41**, 285 (1969).
7. N. K. Ailawadi, Physica, **49**, 345 (1970).
8. H. Mori, Progr. Theor. Phys. **33**, 423 (1965).
9. A. Z. Akcasu and J. J. Duderstadt, Phys. Rev. **188**, 479 (1969); A. Z. Akcasu and E. Daniels, Phys. Rev. **A2**, 962 (1970).
10. N. K. Ailawadi, Ph.D. thesis, University of Maryland (1969).
11. G. D. Harp and B. J. Berne, Phys. Rev. **A2**, 975 (1970); E. Tong and R. C. Desai, Phys. Rev. **A2**, 2129 (1970).
12. L. P. Kadanoff and P. C. Martin, Ann. Phys. (N.Y.), **24**, 419 (1963).
13. C.-H. Chung and S. Yip, Phys. Rev. **182**, 323 (1969).
14. A. Rahman, private communication.
15. K. N. Pathak and K. S. Singwi, Phys. Rev. **A2**, 2457 (1970).

NONLINEAR DYNAMICS OF COLLECTIVE MODES

Robert Zwanzig

INTRODUCTION

This article is concerned with the dynamics of collective modes in liquids. In particular, it presents a method for treating nonlinear interactions between collective modes, and an application of this method to a problem of current interest.

As background, let us consider first the linear theory of dynamics of collective modes, as set forth by Mori [1]. First, one selects some set of dynamical variables $\{a_k\}$. The choice of variables to be included is somewhat arbitrary: more will be said about this later. Second, one restricts attention to a particular class of experiments, in which the initial state of the system is described by a constrained thermal equilibrium ensemble. The constraints are set by specifying initial average values of the $\{a_k\}$. Then the evolution of these variables is given by a generalized Langevin equation of the form

$$\frac{d}{dt} a_k = \sum_{\ell} i\, \Omega_{k\ell}\, a_\ell - \int_0^t ds \sum_{\ell} \phi_{k\ell}(s)\, a_\ell(t-s) + F_k(t) \; . \tag{1}$$

The matrix Ω is determined by certain equilibrium average values, the matrix $\phi(t)$ is constructed from certain equilibrium time correlation functions, and the vector $F_k(t)$ is a Langevin fluctuating force. The mean value of the force (averaged over the initial distribution) is zero, and the second moments of the forces are related to the memory kernels $\phi(t)$ by a fluctuation-dissipation theorem. In Mori's derivation of equation (1), the initial average values are assumed to be small enough that a linear theory is valid. The motion of the <u>average</u> values at later times is given by equation (1) without the fluctuating force.

The memory kernels correspond to transport coefficients. In particular, we may define frequency dependent transport coefficients by

---

Research supported by the National Science Foundation, grant GP-12591.

$$\gamma_{k\ell}(\omega) = \int_0^\infty dt \ \exp(-i\omega t) \ \phi_{k\ell}(t) \tag{2}$$

When the memory of the kernels can be treated as short, or

$$\phi_{k\ell}(t) \simeq 2\,\gamma_{k\ell}\,\delta(t)\ , \tag{3}$$

then the resulting transport coefficients are approximately independent of frequency, and the equations of motion take on a "Markovian" form,

$$\frac{d}{dt}\,a_k \simeq \sum_\ell \ (i\Omega_{k\ell} - \gamma_{k\ell})\ a_\ell + F_k(t)\ . \tag{4}$$

These equations contain all of conventional linear transport theory, e.g., the linearized hydrodynamic equations. The quantities $\gamma_{k\ell}$ contain familiar transport coefficients, e.g., the viscosity.

Intuition suggests that the memory of the kernels is probably important only for times of a molecular order of magnitude, so that the transport coefficients are probably independent of frequency except for frequencies of a molecular order of magnitude. If this is so, then there is a wide range of frequency for which the above Markovian approximation is valid.

Occasionally we are surprised. Consider, for example, sound attenuation in a gas of diatomic molecules. According to standard (Markovian) hydrodynamics, the sound attenuation depends on frequency in a simple and well defined way, determined by the low frequency limits of viscosity and thermal conductivity. But experiments, e.g., on molecular nitrogen, do not fit this standard theory; there is extra sound attenuation at frequencies in the megahertz range. This extra attenuation can be attributed to a frequency dependent bulk viscosity [2]; but frequency dependent transport coefficients correspond to non-Markovian equations of motion. In order to recover the desired Markovian behavior, we introduce an extra variable--the energy contained in internal molecular vibrations. We couple this variable, linearly, to the standard hydrodynamic variables, thereby introducing a new transport coefficient for the collisionally induced relaxation of vibrational states. When the new variable is included, the new set of transport equations is again approximately Markovian.

This example shows that the transport coefficients needed to describe an experiment may depend on the level of description. We may use <u>either</u> a non-Markovian (frequency dependent) theory <u>or</u> a Markovian (frequency independent) theory, depending on what variables are taken into account. Both procedures are correct.

Further, this illustrates how one usually tries to select the variables $\{a_k\}$. We look for variables that lead to a theory that is approximately Markovian; this is the most economical description of an experiment because it involves only parameters $\gamma_{k\ell}$, instead of functions $\gamma_{k\ell}(\omega)$. It is clear that this cannot be done in a precise way, because the transport equations are ultimately non-Markovian at very high frequencies. But we believe that there is usually a low frequency "hydrodynamic" regime where the Markovian approximation ought to be valid. Whenever we find non-Markovian behavior in this hydrodynamic regime, we try to eliminate it by introducing new variables.

Recently some situations have come up where this procedure does not seem to work. For example, transport coefficients near critical points show striking anomalies and non-Markovian behavior in the hydrodynamic regime; nonlinear interactions between collective modes seem to be involved. The theory of transport processes near critical points has been studied intensively for several years; see the review by Kawasaki in this volume for discussion and references.

Another example is suggested by the computer experiments of Alder and Wainwright [3] on the asymptotic behavior of the velocity correlation function in two dimensions. They found that the velocity correlation function for hard disks decays with time approximately as $1/t$. If this behavior is asymptotically correct, then the frequency dependent self-diffusion coefficient varies at low frequency as in $1/\omega$, and actually diverges at zero frequency. Then there is no Markovian approximation at all.

This computer experiment led to a number of theoretical investigations on the asymptotic time dependence of various correlation functions, by Dorfman and Cohen [4], Ernst, Hauge, and van Leeuwen [5], Pomeau [6], Kawasaki [7], and Wainwright, Alder, and Gass [8]. The theories predict an asymptotic decay as $t^{-d/2}$, where d is the dimensionality of the system, for a variety of correlation functions including those connected with viscosity, diffusion, and thermal conductivity.

In the examples just mentioned, no one has been able to account for the observed non-Markovian behavior of transport coefficients by adding new variables that are coupled linearly to the standard hydrodynamic variables. This suggests that we should look at the possibility of nonlinear interactions between collective modes. The theory to be described in this article is in fact a paraphrase of Kawasaki's mode-mode coupling theory, which seems to work well near the critical point.

NONLINEAR DYNAMICS OF COLLECTIVE MODES

I present here a model theory of the nonlinear interaction of collective modes in fluids. The theory is a model one in the sense that it starts with a set of rules that are intuitively plausible but not proved to be valid. These rules are fully equivalent to Kawasaki's formulation of mode-mode coupling theory; he has given them a heuristic justification on the basis of a generalized Fokker-Planck theory of transport processes [9, 10]. For the present, however, the rules are to be regarded only as a means of doing certain calculations. If the results of the calculations are good, then this provides some validation of the rules.

We begin by selecting some set of dynamical variables, just as in Mori's linear theory. Also, as in the linear theory, the variables are chosen so that a certain Langevin equation is approximately Markovian. The subscript k on a_k may denote a Fourier component in a plane-wave expansion. We allow complex values for a_k, and we impose the condition

$$a_k^* = a_{-k} \quad . \tag{5}$$

Let us assume that the thermal equilibrium distribution of the $\{a_k\}$ is Gaussian. The variables can always be chosen so that their equilibrium mean values vanish,

$$\langle a_k \rangle_{eq} = 0 \ . \tag{6}$$

By choosing appropriate linear combinations, we can also satisfy the condition

$$\langle a_k^* \ a_\ell \rangle_{eq} = \delta_{k\ell} \tag{7}$$

on the second moments. (Note that this requires a great deal of information about the equilibrium state of the fluid; one must know all equilibrium second moments of some initially chosen set of variables before linear combinations can be constructed to satisfy equation (7).) Then, aside from a constant of normalization, the equilibrium distribution of the $\{a_k\}$ is

$$g_o(a) \sim \exp(-\sum_k a_k^* a_k) \ . \tag{8}$$

Because this distribution is Gaussian, and because it is related to the entropy of a nonequilibrium state, we see that the thermodynamic force driving the system toward equilibrium is linear in the displacement from equilibrium.

The second rule of this theory is that the equations of motion of the $\{a_k\}$ are nonlinear Langevin equations,

$$\frac{d}{dt} \ a_k = v_k(a) - \sum_\ell \gamma_{k\ell} \, a_\ell + F_k(t) \ . \tag{9}$$

The quantities $\gamma_{k\ell}$ are transport coefficients. They are not the same as the transport coefficients appearing in linear transport laws, e.g., equation (4). We will call them "bare" transport coefficients, for a reason to be given shortly. The vector $F_k(t)$ is a Langevin fluctuating force. We assume that it has a Gaussian distribution, with zero mean value, and with second moments given by the fluctuation-dissipation theorem

$$\langle F_k^*(t) \ F_\ell(t') \rangle_{eq} = 2 \, \gamma_{k\ell} \ \delta(t - t') \ . \tag{10}$$

The vector $v_k(a)$ is called a streaming velocity.

The third rule of this theory concerns the definition of the streaming velocity. The variables a_k are numerical values of certain functions $A_k(p,q)$ of the dynamical state of the system in phase space (p,q). Then the streaming velocity is just the component of the time derivative of $A_k(p,q)$ that lies on the surface of constant $\{a\}$ in phase space,

$$v_k(a) = (dA_k/dt)_{A=a} \ , \tag{11}$$

or more explicitly

$$v_k(a) = \frac{\int\int dp \ dq \ f_o(p,q) \ \delta[A(p,q) - a] \ \dot{A}(p,q)}{\int\int dp \ dq \ f_o(p,q) \ \delta[A(p,q) - a]} \ , \tag{12}$$

where $f_o(p,q)$ is the thermal equilibrium distribution in phase space. The denominator in equation (12) is just the distribution $g_o(a)$.

This definition has two important consequences. First, the equilibrium average of the streaming velocity vanishes,

$$\langle v_k(a)\rangle_{eq} = \int da \; g_o(a) \; v_k(a) = o \; . \tag{13}$$

Second, the streaming current density $v_k(a)g_o(a)$ at equilibrium obeys a divergence condition in a-space,

$$\sum_k \frac{\partial}{\partial a_k} \left[v_k(a) \; g_o(a) \right] = o \; . \tag{14}$$

Approximations to $v_k(a)$ can be found by expanding the δ-function in equation (12) in some complete orthonormal set. A logical choice, suggested by the Gaussian nature of the equilibrium $g_o(a)$, is the set of multidimensional Hermite polynomials. Thus we expect that $v_k(a)$ can be approximated by finite sums of these polynomials. Such approximations automatically satisfy the conditions of equations (13) and (14).

This concludes our formulation of rules for the analysis of nonlinear dynamics of collective modes. Only the "bare" transport coefficients are left undetermined; for the present these must be regarded as adjustable parameters in the theory.

While a Langevin equation gives a graphic description of dynamical processes, it is often more convenient to do calculations with the corresponding Fokker-Planck equation. This equation can be found by standard methods. The result is

$$\frac{\partial g}{\partial t} + \sum_k \frac{\partial}{\partial a_k} \left[v_k g - \sum_\ell \gamma_{k\ell} \, a_\ell \, g \right] = \sum_k \sum_\ell \gamma_{k\ell} \frac{\partial}{\partial a_k} \frac{\partial}{\partial a_\ell^*} \, g \; . \tag{15}$$

Note that this is exactly the Fokker-Planck equation used by Kawasaki in his most recent version of mode-mode coupling theory [9].

It is easy to show that the equilibrium distribution function $g_o(a)$ is a time-independent solution of the Fokker-Planck equation. The terms containing transport coefficients cancel exactly, and the streaming term vanishes because of equation (14).

The Fokker-Planck equation takes on a much more interesting and suggestive form when the following transformation is made:

$$g(a;t) = (g_o)^{\frac{1}{2}} f(a;t) \; . \tag{16}$$

The new distribution function $f(a;t)$ obeys the equation

$$\frac{\partial f}{\partial t} + \sum_k \frac{\partial}{\partial a_k} (v_k f) - \tfrac{1}{2} \sum_k a_k^* v_k f = \sum_k \sum_\ell \gamma_{k\ell} \left[\frac{\partial^2 f}{\partial a_k \partial a_\ell^*} + \left(\tfrac{1}{2} \delta_{k\ell} - \tfrac{1}{4} a_k a_\ell^* \right) f \right] . \tag{17}$$

Note that the right hand side resembles the quantum mechanical Hamiltonian operator for coupled harmonic oscillators. This suggests another transformation, to an operator representation analogous to use of creation-destruction operators in quantum field theory. We define these operators by

$$a_k = \alpha_k + \alpha_{-k}^\dagger \; , \quad \frac{\partial}{\partial a_k} = \tfrac{1}{2} (\alpha_{-k} - \alpha_k^\dagger) \; . \tag{18}$$

They obey the standard commutation rules

$$[\alpha_j, \alpha_k] = [\alpha_j^\dagger, \alpha_k^\dagger] = o, \quad [\alpha_j, \alpha_k^\dagger] = \delta_{jk} \quad . \tag{19}$$

In this representation the Fokker-Planck equation becomes

$$\frac{\partial f}{\partial t} = - H f \tag{20}$$

where the operator H is

$$H = \sum_k \sum_\ell \gamma_{k\ell} \alpha_k^\dagger \alpha_\ell - \sum_k \alpha_k^\dagger v_k(\alpha, \alpha^\dagger) \quad . \tag{21}$$

To complete the analogy with field theory, the "vacuum" state is f_o, or

$$|0\rangle = f_o(a) = [g_o(a)]^{\frac{1}{2}} \quad , \tag{22}$$

and is evidently a stationary solution of the transformed quation.

This transformation to oscillator variables has the following general effect. Let us define the Fokker-Planck operator by

$$\frac{\partial g}{\partial t} = \mathcal{D} g \tag{23}$$

Then, for arbitrary f, we have the identity

$$\mathcal{D}[f\, g_o^{\frac{1}{2}}] = - g_o^{\frac{1}{2}} H f \quad . \tag{24}$$

In particular, this leads to the useful formula

$$[\exp(t\,\mathcal{D})] [f\, g_o^{\frac{1}{2}}] = g_o^{\frac{1}{2}} [\exp(-t H)] f \quad . \tag{25}$$

The field-theoretic apparatus that we have just introduced can be used in a variety of practical calculations. An example is given next.

CORRELATION FUNCTIONS AND LINEAR RESPONSE TRANSPORT COEFFICIENTS

The purpose of the following calculation is to show that the linear response (Green-Kubo) transport coefficients arising from the nonlinear Langevin equation are not the same as the "bare" transport coefficients; they are renormalized as a result of nonlinear interactions between collective modes. This renormalization is precisely why some transport coefficients turn out to be non-Markovian in the hydrodynamic regime.

In the linear response theory of transport coefficients, as set forth, for example, by Kadanoff and Martin [11], one starts by constructing equilibrium time correlation functions of dynamical variables. Therefore, let us consider the time correlation function of the variable a_k,

$$C_k(t) = \langle a_k^*(o)\; a_k(t)\rangle_{eq} \quad . \tag{26}$$

On using the Fokker-Planck operator $\mathcal{D}$, this may be written more explicitly as

$$C_k(t) = \int da \quad a_k^* \, [\exp(t\,\mathcal{D})]\, a_k\, g_o(a) \quad . \tag{27}$$

Now we use equation (25) to move a factor $g_o^{\frac{1}{2}}$ through the exponential operator,

$$C_k(t) = \int da \; [a_k^* \, g_o^{\frac{1}{2}}] \; [\exp - (t H)] \; [a_k\, g_o^{\frac{1}{2}}] \quad , \tag{28}$$

and we recognize that the combination $a_k g_o^{\frac{1}{2}}$ has the operator representation

$$a_k \, g_o^{\frac{1}{2}} = \alpha_k \, g_o^{\frac{1}{2}} = |k\rangle \, . \tag{29}$$

So the correlation function becomes

$$C_k(t) = \langle k| \, \exp(-t\,H)\, |k\rangle \, . \tag{30}$$

The Laplace transform of the correlation function is

$$C_k(\epsilon) = \int_0^\infty dt \, \exp(-\epsilon t)\, C_k(t) = \langle k| \frac{1}{\epsilon + H} |k\rangle \, . \tag{31}$$

This looks like the diagonal matrix element of a resolvent operator in quantum many body theory, and the same kind of analysis is suggested.

As an illustration, let us consider a special case. First, we suppose that the transport coefficient matrix $\gamma_{k\ell}$ is diagonal,

$$\gamma_{k\ell} = \gamma_k \, \delta_{k\ell} \, . \tag{32}$$

This can generally be arranged by taking appropriate linear combinations of the variables. Next, we suppose that the streaming velocity contains only a quadratic nonlinearity:

$$v_k(a) = \sum_\ell \sum_m V_{-k\ell m} \left[a_\ell \, a_m - \langle a_\ell \, a_m \rangle_{eq} \right] , \tag{33}$$

where the $V_{-k\ell m}$ are coupling constants. This particular streaming velocity is a linear combination of second-order Hermite polynomials, and satisfies the conditions (13) and (14) if the coupling constants satisfy

$$V_{k\ell m} = V_{km\ell} \, , \quad V_{k\ell m} + V_{\ell mk} + V_{mk\ell} = 0 \, . \tag{34}$$

We separate the operator H into an unperturbed part H_o and a perturbation H_1,

$$H = H_o + H_1 \tag{35}$$

where

$$H_o = \sum_k \gamma_k \, \alpha_k^\dagger \, \alpha_k \, , \tag{36}$$

$$H_1 = \sum_k \sum_\ell \sum_m V_{-k\ell m} \, \alpha_k^\dagger \left[\alpha_\ell \, \alpha_m + \alpha_{-\ell}^\dagger \, \alpha_m + \alpha_\ell \alpha_{-m}^\dagger + \alpha_{-\ell}^\dagger \, \alpha_{-m}^\dagger \right] .$$

Note that H_o, representing dissipative effects, is Hermitian and positive definite, while H_1, representing energy conserving interactions, is anti-Hermitian.

By analogy with quantum many-body theory, the diagonal element of the resolvent operator can be written in a "self-energy" form.

$$\hat{C}_k(\epsilon) = \langle k| \frac{1}{\epsilon + H} |k\rangle = -[\epsilon + \gamma_k + \hat{E}_k(\epsilon)]^{-1} \, . \tag{37}$$

The self-energy contribution can be calculated by a variety of field theoretic methods--for example, perturbation theory. If we use perturbation theory, and stop at the second order in H_1, the result is

$$\hat{E}_k(\epsilon) = 2 \sum_\ell \sum_m V_{-k\ell m} \left[\frac{1}{\epsilon + \gamma_\ell + \gamma_m} \right] V_{k-\ell-m} + O(V^3) \, . \tag{38}$$

This expression has a simple interpretation as a time correlation function of the streaming velocity. It is easy to see by direct calculation that

$$\hat{E}_k(\epsilon) = \int_0^\infty dt\ e^{-\epsilon t} \langle v_{-k}(a)\, v_k[a^0(t)] \rangle_{eq} + O(V^3), \tag{39}$$

where the time dependence of $a^0(t)$ is to be found from the unperturbed equation of motion

$$\frac{d}{dt} a_k^0 = - \gamma_k a_k^0 . \tag{40}$$

The average is taken with the equilibrium distribution function $g_o(a)$.

According to linear response theory, the denominator in equation (37) determines a non-Markovian transport coefficient $\Gamma_k(\epsilon)$,

$$\Gamma_k(\epsilon) = \gamma_k + \hat{E}_k(\epsilon) \tag{41}$$

This illustrates the renormalization of the "bare" transport coefficient associated with the k(th) collective mode. The new transport coefficient is evidently non-Markovian in the sense we discussed earlier; the parameters γ_k are decay rates for collective modes in a "hydrodynamic" regime, so that $\hat{E}_k$ may have important frequency or time dependence in this regime.

The above transport coefficient is the Laplace transform of a memory kernel $\phi_k(t)$. On using equation (39) for the extra contribution $E_k(t)$, this memory kernel is

$$\phi_k(t) = 2\gamma_k\ \delta(t) + \langle v_{-k}(a)\ v_k[a^0(t)]\rangle_{eq} + O(V^3) . \tag{42}$$

The renormalized linear transport law for the average $\langle a_k(t)\rangle$ is

$$\frac{d}{dt} \langle a_k(t)\rangle = \int_0^t ds\ \phi_k(s)\ \langle a_k(t-s)\rangle . \tag{43}$$

It gives exactly the same correlation function as in equation (37).

What this calculation has accomplished is to separate the linear response memory kernel of Mori's theory into two parts: one, containing the δ-function, relaxes rapidly compared with hydrodynamic times, and the other part relaxes more slowly. This separation is the basis of the work in References [4-8].

A specific example of renormalization will be given soon, in a discussion of the velocity of a two-dimensional fluid.

TRANSPORT EQUATIONS FOR AVERAGED VARIABLES

The Langevin equation that we started with gives the motion of dynamical variables including the effects of thermal noise,

$$\frac{da_k}{dt} = -\gamma_k\ a_k + \sum_\ell \sum_m V_{-k\ell m} \left[a_\ell a_m - \langle a_\ell a_m\rangle_{eq}\right] + F_k(t) . \tag{44}$$

It is interesting to find the corresponding equation of motion for the averaged variables, defined by

$$\langle a_k(t)\rangle = \int da\ a_k\ g(a;t) . \tag{45}$$

This has been done (to second order in nonlinearity) in collaboration with William C. Mitchell; details will be presented elsewhere. The result is given here.

In order to avoid irrelevant complications, we restrict attention to a special class of experiments, characterized by an initial condition on the distribution function $g(a; 0)$. We assume that this distribution is Gaussian, with nonequilibrium mean values $\langle a_k(0)\rangle$, but with equilibrium second moments. This distribution can be realized experimentally as follows. Suppose that external forces are imposed on the system, so that it comes to equilibrium with the mean values $\langle a_k(0)\rangle$. The second moments will then approach their equilibrium values. At time $t = 0$, the external forces are turned off, so that the system will begin to relax. The initial distribution for this experiment has the desired properties.

The resulting equations of motion for average values are

$$\frac{d}{dt}\langle a_k(t)\rangle = -\gamma_k \langle a_k(t)\rangle - \int_0^t ds\; E_k(t-s)\,\langle a_k(s)\rangle + \sum_{\ell}\sum_{m} V_{-k\ell m}\langle a_\ell(t)\rangle\,\langle a_m(t)\rangle + O(V^3)\;. \tag{46}$$

The extra memory kernel $E_k(t)$ is exactly the same as we found from linear response theory. This transport equation is valid to second order in the nonlinear coupling constants, and contains a streaming term that is first order in the coupling constants and second order in deviations from equilibrium.

The effects of averaging over thermal noise are accounted for in two ways. First, the average of the function $v_k(a)$ is replaced by the corresponding function of the average. This neglects certain dynamical correlations. Second, extra dissipation described by $E_k(t)$ is introduced. This compensates for the neglect of these dynamical correlations. Additional renormalization, and further nonlinear terms in $\langle a(t)\rangle$, will be introduced by carrying out the calculations to higher order in the coupling constants.

So we see that the renormalized transport coefficient is obtained not only from linear response theory but also by investigating the nonlinear motion of average values. The renormalized coefficient is the one that is observed in conventional experiments.

APPLICATION TO NONLINEAR HYDRODYNAMICS

Here we apply the preceding formalism to a problem of current interest, the renormalization of the viscosity of a fluid. For simplicity we consider only a two-dimensional incompressible fluid.

We begin with an <u>assumption</u>, that the instantaneous or unaveraged fluid viscosity $\underset{\sim}{v}(\underset{\sim}{R}; t)$ obeys a nonlinear Langevin equation having the same general structure as the Navier-Stokes equation,

$$\frac{\partial \underset{\sim}{v}}{\partial t} + \underset{\sim}{v}\cdot\nabla\underset{\sim}{v} = -\nabla P/\rho + \nu_0\,\nabla^2\,\underset{\sim}{v} + \text{noise}\;. \tag{47}$$

Here, P is the thermodynamic pressure, ρ is the fluid density, and ν_o is the "bare" kinematic viscosity. This is not the same as the viscosity observed in ordinary experiments. The fluid is incompressible, so we impose the divergence condition $\nabla \cdot \underset{\sim}{v} = 0$.

It is convenient to introduce the stream function ψ, such that

$$v_x = \partial\psi/\partial y \ , \ v_y = - \partial\psi/\partial x \ . \tag{48}$$

The divergence condition is satisfied automatically. Let us expand the stream function in Fourier components in space,

$$\psi(\underset{\sim}{R}) = \sum_k \psi_k \exp(i\underset{\sim}{k} \cdot \underset{\sim}{R}) ; \quad \psi_{-k} = \psi_k^* \ . \tag{49}$$

Subscripts k, ℓ, m, . . . are used to denote wave vectors in the two-dimensional Fourier space.

When equations (47)-(49) are combined, and the pressure is eliminated, we get the equivalent equation

$$\frac{\partial}{\partial t} \psi_k = - \nu_o k^2 \psi_k + \sum_\ell \sum_m{}' G_{k\ell m} \psi_\ell \psi_m + F_k(t) \ . \tag{50}$$

The first term comes from linear viscosity. The second term comes from nonlinearities in the original equation, and contains the coupling constants

$$G_{k\ell m} = \delta_{k,\ell+m} [(m^2 - \ell^2)/2k^2] (\underset{\sim}{m} \times \underset{\sim}{\ell})_z \ . \tag{51}$$

The third term contains effects of thermal noise.

We assume that the equilibrium distribution of fluid velocity is a Boltzmann distribution at temperature T, determined by the total kinetic energy of the fluid. (Boltzmann's constant is set equal to unity.) After changing variables,

$$a_k = (\rho/2T)^{\frac{1}{2}} |\underset{\sim}{k}| \ \psi_k \ , \tag{52}$$

the equilibrium distribution of the $\{a_k\}$ is Gaussian,

$$g_o(a) \sim \exp \left(-\sum_k a_k^* a_k\right) \ . \tag{53}$$

Because we use a continuum description of the fluid, which does not contain any information about molecular size, we may have to impose a cutoff on sums over wave vectors k; the cutoff will be denoted by k_o, approximately the inverse of some molecular length.

When the $\{a_k\}$ variables are used, we have a nonlinear Langevin equation of the same structure that was discussed earlier. To complete the correspondence, we note that the "bare" transport coefficients are

$$\gamma_k = \nu_o k^2 \ , \tag{54}$$

and the nonlinear coupling constants are

$$V_{-k\ell m} = (2T/\rho)^{\frac{1}{2}} (k/\ell m) \ G_{k\ell m} \ . \tag{55}$$

These satisfy the requirements of equation (34).

According to the previous discussion, the "bare" transport coefficients γ_k are renormalized to $\Gamma_k(\epsilon)$. When the calculation is carried out to second order in the coupling constants, we may use equation (38) to get the correction $\hat{E}_k(\epsilon)$. Because the calculation is a bit messy for arbitrary k, we present results only in the long wavelength limit $k \to 0$. Then we may define a long wavelength non-Markovian viscosity by

$$v(\epsilon) = \lim_{k \to 0} \Gamma_k(\epsilon)/k^2 . \tag{56}$$

The result of the calculation is

$$\nu(\epsilon) = \nu_o + \frac{T}{32\,\pi\rho\nu_o} \log \frac{\epsilon + 2\nu_o k_o^2}{\epsilon} , \tag{57}$$

where k_o is the cutoff mentioned earlier. The second term comes from $\hat{E}_k(\epsilon)$.

At small ϵ (or low frequency), $\nu(\epsilon)$ diverges logarithmically. The inverse Laplace transform is the memory kernel

$$\phi(t) = 2\,\nu_o\,\delta(t) + \frac{T}{32\pi\rho\nu_o} \left[1 - \exp(-2\nu_o k_o^2 t)\right]/t . \tag{58}$$

This decays for long times as $1/t$. This asymptotic behavior, and the coefficient of the renormalized term, are in essential agreement with results found by earlier investigators. One significant difference is that our results contain the "bare" viscosity ν_o, while the earlier results do not distinguish explicitly between this and the observed viscosity $\nu(\epsilon)$.

Our calculation was done only to second order in perturbation theory. One obvious extension is to do a self-consistent calculation, in which the zero(th) order (bare) viscosity in the denominator of equation (38) is replaced by the renormalized viscosity. This changes the analytic behavior of $\nu(\epsilon)$ from $\log(1/\epsilon)$ to $[\log(1/\epsilon)]^{\frac{1}{2}}$, a result found also by Wainwright, Alder, and Gass [8]. It is not known whether further changes will develop from a more complete calculation. In any case, the $1/t$ result from second order perturbation theory matches the behavior found so far from computer experiments.

There is an interesting connection between the calculation and one performed by Dorfman and Cohen [4]. They used the linearized Boltzmann equation to find the asymptotic decay of Green-Kubo expressions for transport coefficients. Their calculation does not make any explicit reference to nonlinearities. But the time correlation functions involved in the Green-Kubo formulae are just the memory kernels $\phi_k(t)$ that we referred to earlier. We found that application of second order perturbation theory to the nonlinear Langevin or Fokker-Planck equation led to a separation of $\phi_k(t)$ into two parts, one decaying rapidly and the other decaying slowly. The slow part, as in equation (42), corresponds to the asymptotic part of the Green-Kubo formula that was investigated by Dorfman and Cohen. Their calculation of the time dependence of this part using the linearized Boltzmann equation corresponds to our calculation of this part using the linear transport law in equation (40). And, in fact, they used the linearized Boltzmann equation in the long time hydrodynamic regime, so that the correspondence is even closer.

SUMMARY

The results of this discussion are summarized in table 1.

Table 1

Type of Theory	Fluctuations	Average Values
Nonlinear	Nonlinear Langevin equation with "bare" transport coefficients (Markovian)	Nonlinear equation of motion with renormalized transport coefficients corresponding to Green-Kubo transport coefficients (non-Markovian)
Linear	Mori theory--linear Langevin equation with Green-Kubo transport coefficients (non-Markovian)	Linear equation of motion with Green-Kubo transport coefficients (non-Markovian)

We are concerned with both linear and nonlinear dynamical theories, and with both fluctuation phenomena and the motion of average values. Linear fluctuations are described correctly by Mori's theory; this contains non-Markovian transport coefficients of the Green-Kubo type. In Mori's theory, the linear motion of average values is governed by the same transport coefficients.

In our theory, the dynamics of fluctuations is described by a nonlinear Langevin equation; this contains "bare" Markovian transport coefficients. From this we derive the linear motion of average values, governed by renormalized transport coefficients corresponding to those of Mori. We also derive nonlinear equations of motion for average values; to second order, these equations contain the same renormalized (Green-Kubo) transport coefficients. The nonlinear equations of motion for average values can be linearized safely by merely omitting the nonlinear terms. This is not true for the nonlinear Langevin equation--we do not get Mori's linear Langevin equation by dropping the nonlinear terms. A more elaborate calculation, which takes into account the nonlinear mixing of thermal noise, must be made. This has not yet been done. The arrows in table 1 indicate what is known so far.

Our theory shows how transport coefficients that are non-Markovian in the "hydrodynamic" regime can be related to Markovian transport coefficients describing the nonlinear dynamics of fluctuations. In particular, it shows how one may understand the asymptotic long time decay of various Green-Kubo transport coefficients. This asymptotic decay in the hydrodynamic regime comes from nonlinear interactions between collective modes.

Finally, this theory--which is a simplified form of the one used by Kawasaki [9]--may be of some value in understanding the renormalization of transport coefficients near critical points.

REFERENCES

1. H. Mori, Progr. Theor. Phys. **33**, 423 (1965).

2. R. Zwanzig, J. Chem. Phys. **43**, 714 (1965).

3. B. J. Alder and T. E. Wainwright, Phys. Rev. **A1**, 18 (1970).

4. J. R. Dorfman and E. G. D. Cohen, Phys. Rev. Letters, **25**, 1257 (1970).

5. M. H. Ernst, E. H. Hauge, and J. M. J. van Leeuwen, Phys. Rev. Letters, **25**, 1254 (1971).

6. Y. Pomeau, Phys. Rev. **A3**, 1174 (1971).

7. K. Kawasaki, Phys. Letters, **32A**, 379 (1970); **34A**, 12 (1971).

8. T. E. Wainwright, B. J. Alder, and D. M. Gass, preprint (1971).

9. K. Kawasaki, lectures to be published in the proceedings of the International School of Physics "Enrico Fermi," summer 1970.

10. R. Zwanzig, Phys. Rev. **124**, 983 (1961).

11. L. P. Kadanoff and P. C. Martin, Ann. Phys. **24**, 419 (1963).

DISCUSSION

N. Van Kampen: From your Langevin equation it follows that $\langle v(a) \rangle$ in equilibrium vanishes. That also follows from the explicit expression you wrote for v(a). However, from that same expression it also follows that $\langle a^n v(a) \rangle$ in equilibrium vanishes. That seems to leave very little room for the function v.

R. Zwanzig: I disagree. It does not follow that $\langle a^n v(a) \rangle = 0$. For example, suppose we calculate $\langle a^* v(a) \rangle$. This in the same as $\langle A^* dA/dt \rangle$, which is the $i\Omega$ of Mori's theory.

R. Kubo: The concept of bare and renormalized transport coefficients has been discussed often in the literature. For example, it is the same as what I called internal (or local) mobility and external mobility in one of my review articles (Rept. Progr. Phys. 29 [1961]). Definition of transport coefficients, or rather admittance functions, depends on the choice of the set of macro-variables to describe the physical process, as clearly explained by Professor Zwanzig in this lecture. This choice may be made intuitively or by experience or for some dynamical reasons. It should also be decided by a careful analysis of the structure of the external response function because this structure itself reflects the hydrodynamics, for instance in the low-frequency, long wavelength limit. Bare transport coefficients are calculated from the correlation function formalism by carefully projecting out the hydrodynamic part of motion.

R. Kraichnan: I think the logarithmic divergence is real, not an artifact of the perturbation method. It occurs only in two dimensions, and can be found also from simple considerations based on random-walk models.

B. J. Berne: There is some tentative experimental evidence that the long time tail of the velocity correlation function (v.c.f.) exists in real fluids. It should be noted that

a. Alder and Wainwright have determined D/D_E from computer experiments for a hard-sphere model of CH_4 and have compared this with measurements of Trappiners and Oosting as a function of density. It can be concluded from this comparison that if the long time tail in the v.c.f. is omitted, the density dependence of D/D_E does not even qualitatively agree with the experiment; whereas if the long time tail is included, there is qualitative agreement between the computer experiment and laboratory experiment. Both the computer experiment and the laboratory experiment have a maximum in D/D_E as a function of density at approximately the same density; but they differ in the magnitude of the maximum. Differences of this sort may arise from the fact that methane molecules are not hard spheres. D_E is the Enskog diffusion coefficient.

b. According to the theory presented by Ernst et al., the long time tail in the v.c.f. goes as

$$\frac{2kT}{mn} \; [4\pi(\nu + D)t]^{-3/2} \; ,$$

where ν and D are respectively the kinematic viscosity and the self-diffusion coefficient. We find that the addition of this term leads to an enhancement of the diffusion coefficient over the Enskog diffusion coefficient D_E, of a form such that D/D_E has a maximum as a function of density with the same qualitative behavior mentioned in section a. We are presently trying to make a detailed comparison.

I. Prigogine: Some caution is necessary when one speaks of non-Markovian processes, long tails in time, and so on. As a simple example let us consider the decay of an unstable atom as discussed, for example, in the textbook on scattering by Goldberger and Watson. We have then the exponential decay due to the pole on the second Riemann sheet and the "non-Markovian" $t^{-3/2}$ due to the branch cut. This class of contributions may be eliminated by using "natural" initial conditions. An interesting example is discussed by Schieve and Hawker in the contributed papers to this conference.

A beautiful and general theorem has been proved by Cl. George (it is even a "rigorous result!"). Let us write the distribution function (velocity distribution function or in general the vacuum part of ρ; see for definitions Prigogine, George, and Henin, Physica **45**, 418 [1969]):

$$\rho_o(+) = \frac{1}{2\pi i} \int dz \; e^{-izt} \; \frac{1}{z - \Psi(z)} \; [\rho_o(0) + \sum_k \mathcal{D}_k(z)\, \rho_k(0)] \tag{1}$$

If the natural initial conditions for the correlations are used,

$$\rho_k(0) = C_k \rho_o(0) \; , \tag{2}$$

then

$$\frac{1}{z - \Psi(z)} \; [1 + \sum_k \mathcal{D}_k(z)\, C_k] = \frac{1}{z - (\Omega\Psi)_{z=0}} \; . \tag{3}$$

The <u>only</u> contribution to the resolvent then comes from the pole at $\Omega\Psi$. We have a strictly

Markovian description.

This of course does <u>not</u> mean that the time behavior will be exponential (or that the transport coefficients will be analytic functions of the concentrations). As is well known, the operator $\Psi(z)$ acts on the velocities (in the classical case), and the velocities are themselves distributed in equilibrium according to the Maxwellian distribution. As a consequence, no general statement can be made about the time dependence of functions such as $\langle v(t)v(0)\rangle$ over an equilibrium ensemble. But it should be possible to clarify the situation studying specific examples. Only then would it be possible to decide if one should turn to a phenomenological description in which the connection with Hamiltonian dynamics is lost.

IV. Phase Transitions

DYNAMICAL BEHAVIOR NEAR CRITICAL POINTS

Kyozi Kawasaki

I. INTRODUCTION

Real understanding of the dynamical behavior of a system near a critical point started only several years ago, despite the fact that experimental studies have been pursued for more than half a century [1]. The principal reason for this delay was the lack of a general theoretical framework to describe transport phenomena in dense systems. Thus, the early theoretical ideas were based on the thermodynamics of irreversible phenomena and the assumption that all the critical anomalies ought to be attributable to those of the thermodynamic driving forces (the so-called conventional theory [2]). The theory met little success.

The first breakthrough in the theoretical development is due to Marshall Fixman [3], who in 1962 demonstrated the importance of nonlinear coupling among long-wavelength fluctuations in determining critical anomalies in the transport coefficients. This idea was followed up by several authors [4,5,6,7] and has now been developed into the mode-mode coupling theory of transport phenomena.

Another approach to the description of the dynamical behavior was stimulated by the success of the scaling law idea [8], which assumes that all the critical anomalies in equilibrium properties are expressible in terms of a unique characteristic length which grows indefinitely as the critical point is approached. The scaling law idea was successfully extended to include dynamical properties by Ferrell et al. [9] and by Halperin and Hohenberg [10].

The above two approaches have provided the basis for our present theoretical understanding of dynamical behavior near critical points. However, both approaches are phenomenological in nature, and are at most semimicroscopic. Attempts to formulate a completely microscopic theory [11], though quite interesting and promising, are plagued with enormous complexity, and also with unclear approximations, just as are the phenomenological approaches. Complementing these approaches are the studies on artificial but precisely defined mathematical models, now

The work was supported by the National Science Foundation under grant 545-782-46.

known as kinetic Ising models.[1]

In the next two sections we describe the mode-mode coupling approach in a somewhat different and heuristic manner. Since the specific predictions of the dynamical scaling and mode-mode coupling theories coincide, we shall discuss the experimental tests of these predictions in section IV. In this paper we have tried to clarify the assumptions and approximations involved in the theory and also the remaining difficulties and problems.

II. MODE-MODE COUPLING THEORY OF CRITICAL DYNAMICS

In this section a heuristic derivation is given of the mode-mode coupling theory. Consider a set of gross variables a_j for the macroscopic modes.[2] These variables are normally the densities (or their spatial Fourier components) of the conserved quantities and the order parameter of the phase transition if it is not conserved [7, 14]. Here we choose these variables in such a way that their equilibrium averages always vanish. When such a set $\{a\}$ completely determines a macroscopic state, the set $\{a\}$ obeys the following equations of motion (a dot stands for the time derivative):

$$\dot{a}_j = v_j(a) - \sum_{\ell} L_{j\ell} X_\ell(a) , \tag{2.1}$$

where $v_j(a)$ describes a nondissipative instantaneous rate of change of a_j, while the second term gives the dissipative change familiar in the thermodynamics of irreversible processes [15]. Here a set of the variables $\{a\}$ is simply denoted as a in the arguments of v_j and X_ℓ. The function $X_\ell(a)$ is the thermodynamic driving force related to the entropy $S(a)$ by

$$X_\ell(a) = - \partial S(a)/\partial a_\ell^* ,$$

and $L_{j\ell}$ are the Onsager kinetic coefficients which are assumed to be independent of a and to obey the Onsager relation $L_{j\ell} = L_{\ell j}$. An example of equation (2.1) is provided by the classical hydrodynamic equations.

Equation (2.1) describes, however, only the average behavior of the gross variables, whereas thermal fluctuations introduce randomly fluctuating components in $\{a\}$. Such randomly fluctuating components can be included in equation (2.1) by adding the random force $f_j(t)$ on the right-hand side of the equation. Thus the equation of motion for a_j, including fluctuations, is

$$\dot{a}_j = v_j(a) - \sum_{\ell} L_{j\ell} X_\ell(a) + f_j . \tag{2.2}$$

The fluctuation-dissipation theorem [16] then connects f_j and $L_{j\ell}$ by

---

[1] The properties of kinetic Ising models have been recently reviewed by the author [12]. The most interesting unsolved problem here is the failure of the conventional theory in the opposite direction to what the mode-mode coupling theory suggests, namely, the Onsager kinetic coefficient associated with spin relaxation appears to vanish at the critical point [13].

[2] The suffix j specifies, for example, the types of variables as well as the spatial Fourier component of the density variables.

$$\langle f_i(t) f_\ell^*(t') \rangle = 2 k_B L_{i\ell} \delta(t - t') , \tag{2.3}$$

where $\langle \ . \ . \ . \ \rangle$ denotes an equilibrium average, and k_B is the Boltzmann constant.

The form of $v_i(a)$ is taken from the usual nonlinear macroscopic equations of motion, such as the hydrodynamic equations. If such macroscopic equations are not known, or if the validity of such macroscopic equations in the vicinity of the critical point is in doubt, $v_i(a)$ can still be found in the following way. Suppose that the system is in thermal equilibrium except that the set of variables $\{a_i\}$ are constrained to take on some nonequilibrium values which we again denote by $\{a_i\}$ or simply by a. At the instant when the constraints are removed, the $\{a_i\}$ begin to change. At that instant, however, any dissipative process has not yet had time to develop. Thus, only the first term of equation (2.1) remains after averaging over the states with given $\{a_i\}$, that is, $v_i(a)$ is identical to the average of $\dot{a}_i$ in this situation. This is formally expressed by introducing the classical phase-space function or the quantum-mechanical operator $\hat{a}_i$ corresponding to a_i, as follows:

$$v_i(a) = \langle \hat{a}_i ; a \rangle , \tag{2.4}$$

where $\langle \ . \ . \ . \ ; a \rangle$ denotes an average over the constrained equilibrium state with given $\{a_i\}$, $\hat{X}$ denotes the variable X expressed in molecular variables.

For small values of the a's, that is near equilibrium, $v_i(a)$ can be expanded in powers of the a's:

$$v_i(a) = i \sum_\ell \omega_{i\ell} a_\ell - i \sum_{\ell m} V_{i\ell m} (a_\ell a_m - \langle a_\ell a_m \rangle) + \ . \ . \ . \ . \tag{2.5}$$

The expansion coefficients are expressed in terms of equilibrium averages of Poisson brackets and various products of a's (here classical mechanics is assumed). In particular, if different a's are statistically independent and the factorization approximation is used for equal time correlations of the several a's, we have

$$\omega_{i\ell} = - i k_B T \langle \{\hat{a}_i, \hat{a}_\ell^*\} \rangle / \chi_i , \tag{2.6}$$

$$V_{i\ell m} = \frac{i}{2} k_B T \langle \{\hat{a}_i, \hat{a}_\ell^* \hat{a}_m^*\} \rangle / \chi_\ell \chi_m , \tag{2.7}$$

where $\{ , \}$ denotes a Poisson bracket and

$$\chi_i \equiv \langle |\hat{a}_i|^2 \rangle . \tag{2.8}$$

The terms in the expansion (2.5) have a rather simple physical meaning. The first term gives rise to oscillatory motions, such as sound waves, with characteristic frequencies given by the eigenvalues of the matrix $\omega_{i\ell}$. The second term then describes nonlinear coupling among three hydrodynamic modes i, ℓ, m. The subsequent terms not explicitly shown in equation (2.5) produce coupling among four, five, . . . , hydrodynamic modes.

In order to use equation (2.2) near the critical point we must make one basic assumption, that the Onsager kinetic coefficients $L_{i\ell}$ remain finite at the critical point. This assumption is based on the intuitive idea that the random forces f_i describe purely microscopic molecular motions, and hence should maintain their random character even at the critical point. For the moment we do not know of any way to test this assumption except by comparing its consequences with experiments, or by the numerical study of possible critical anomalies in $L_{i\ell}$ for special models such as the kinetic Ising models [13].

If we accept this basic assumption, then all the remaining critical anomalies in equation (2.2) appear in the thermodynamic driving forces X_ℓ and in $v_i(a)$ (equation [2.4]), which are then problems associated with equilibrium critical phenomena. Thus the problem of critical dynamics reduces to the investigation of the consequences of equations (2.2)-(2.4). In this sense it is legitimate to regard equation (2.2) as the fundamental kinetic equation of critical dynamics in the same sense that the Boltzmann equation is the fundamental kinetic equation in gas dynamics.

We should point out that equations of the type of (2.2) are not new. In fact, Landau and Lifshitz derived equations of the same type in their discussion of fluctuations in fluid dynamics [17]. Such equations have been used recently by Zwanzig in studying the long-time behavior of correlations of fluctuations [18]. Questions related to this problem were also discussed by Kubo recently [19].

Since the derivation of the fundamental kinetic equations presented here is entirely heuristic, it is appropriate to add a few words about derivations of these equations from first principles. These derivations make use of the recent developments in the Brownian-motion theory of gross variables. One can either start from the microscopic Langevin equation for the gross variables, due to Mori, which is suitably extended; or one may utilize the Fokker-Planck equation for the probability distribution function of gross variables, due to Green and Zwanzig [14]. With the assumption that the random forces acting upon the gross variables remain random at the critical point (that is, memory effects can always be ignored), both approaches lead to the following Langevin-type equation:[3]

$$\dot{a}_i = v_i(a) - \sum_\ell L_{i\ell}(a) X_\ell(a) + k_B \sum_\ell \partial L_{i\ell}(a)/\partial a_\ell^* + f_i \, , \tag{2.9}$$

where $L_{i\ell}(a)$ is given by

$$L_{i\ell}(a) = \frac{1}{k_B} \int_0^\infty dt \, \langle f_i(o) f_\ell^*(t); a \rangle \, . \tag{2.10}$$

Here $L_{i\ell}(a)$ is in fact not the Onsager kinetic coefficient which we normally observe, but is a "bare" kinetic coefficient in the sense used by Zwanzig [18]. We emphasize again that this result is obtained from the exact generalized Langevin or Fokker-Planck equations which include memory effects by assuming that all the motions occurring in the system can be classified into

---

[3] Derivation of this result is greatly facilitated by the use of the operator identity: $\frac{d}{dt} e^{it\mathcal{L}} = e^{it\mathcal{L}} Pi\mathcal{L} + \int_0^t ds\, e^{i(t-s)\mathcal{L}} Pi\mathcal{L}\, e^{is(1-P)\mathcal{L}} (1-P) i\mathcal{L} + e^{it(1-P)\mathcal{L}}(1-P)i\mathcal{L}$. In our case $\mathcal{L}$ is the time-displacement Liouville operator and P is a certain projection operator. This will be discussed elsewhere.

rapid motions associated with f_i and slower motions associated with the gross variables $\{a\}$ in a clear-cut manner. This is an idealization which puts some restrictions upon the region of validity of this type of equation. We believe that strictly speaking equation (2.9) is meaningful only for those physical quantities that are independent of the "bare" $L_{i\ell}$ at the end of calculations. If we assume that $L_{i\ell}$ does not depend upon a, equation (2.9) reduces to equation (2.2).

Let us now turn to the consequences of equation (2.2).

The simplest thing to do is to take an average, that is, to return to equation (2.1). It is still a nonlinear equation, and we linearize it assuming a small deviation from equilibrium, where the thermodynamic driving force is now given by

$$X_\ell(a) = (k_B/\chi_\ell)a_\ell \; . \tag{2.11}$$

In the absence of oscillatory motions, equation (2.1) becomes

$$\dot{a}_i = -\sum_\ell (k_B L_{i\ell}/\chi_\ell)a_\ell \; . \tag{2.12}$$

Thus, in this treatment all the critical anomalies appear through "susceptibilities" χ_ℓ in the thermodynamic driving forces. This is in fact the conventional theory referred to in section I. For instance, in a binary solution near the critical consolute point, the theory predicts that the diffusion constant is proportional to $(\partial\mu/\partial c)_{p,T}$, where μ and c are the relative chemical potential and concentration, respectively.

Unfortunately, however, for most cases where detailed studies have been made, the predictions of the conventional theory turn out to be wholly inadequate; some of the Onsager kinetic coefficients in equation (2.12) diverge near the critical point [20]. Thus, if our starting equation (2.2) is to be valid, the procedures leading to equation (2.12), in particular the linearization of equation (2.2), are inadequate. Indeed, in such cases, the nonlinear coupling among fluctuating hydrodynamic modes in $v_i(a)$ (equation [2.5]) plays a crucial role (the mode-mode coupling). Fixman considered the influence of mode-mode coupling on the shear viscosity in critical mixtures by calculating the rate of entropy production [3]. The mode-mode coupling can be taken into account more readily in the Green-Kubo time correlation function formulae for transport coefficients, where the flux entering the correlation function is expanded in powers of the gross variables whose time development is assumed to obey macroscopic laws [4,5]. Combining the static scaling law ideas [8], the theory is capable of predicting the nature of the critical anomalies in the transport coefficients [6]; these predictions are receiving increasing support from experiment [21].

The theory briefly described above, which is widely known as the mode-mode coupling theory, is, however, limited primarily to the so-called hydrodynamic regime where wavelengths associated with macroscopic disturbances are much greater than the range of correlation of critical fluctuations. Here we briefly describe a different approach which is free from this restriction [7,12], however, limiting ourselves to the mode-mode coupling explicitly shown in equation (2.5) and assuming equation (2.11) for $X_\ell(a)$. We consider a set of propagators of gross variables, defined by

$$G_{j\ell}(t) \equiv \langle a_j(t)\, a_\ell^*(o)\rangle / \chi_\ell \ , \tag{2.13}$$

which contain all the information on linearized macroscopic laws and (in general, nonlocal) transport coefficients. Regarding the mode-mode coupling term in equation (2.2) as a perturbation, one can readily develop a formal perturbation theory for $G_{j\ell}(t)$. However, whenever the mode-mode coupling dominates, terminating the perturbation expansion at some finite order makes no sense. The difficulty is circumvented by the use of a standard renormalization technique, which leads to the following self-consistent set of equations for $G_{j\ell}(t)$:

$$\frac{\partial}{\partial t} G_{j\ell}(t) = \sum_m (i\,\omega_{jm} - k_B L_{jm} \chi_m^{-1})\, G_{m\ell}(t) - 2 \sum_{mnfgh} \frac{\chi_f \chi_g}{\chi_h} v_{jmn} v_{hfg}^* \int_0^t ds\, G_{mf}(s)\, G_{ng}(s)\, G_{h\ell}(t-s) \tag{2.14}$$

with the initial condition $G_{j\ell}(o) = \delta_{j\ell}$.

The second term of equation (2.14) is schematically represented in figure 1.

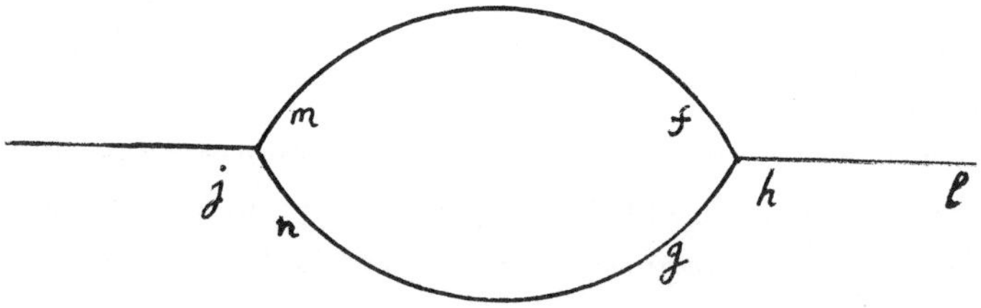

Fig. 1. Graphical representation of the simplest self-energy term.

In obtaining equation (2.14) all the vertex corrections have been omitted. Transport coefficients are readily deduced by taking the hydrodynamic limit of the "self-energy" part of the propagators.

In the next section we describe applications of the present general theory to some specific problems. Here we repeat the important approximations or assumptions made in obtaining the result (2.14) apart from the basic assumption that the $L_{j\ell}$ are finite:

(AI) The probability distribution of gross variables $\{a_j\}$ is Gaussian.

(AII) Terms containing products of more than three a's in the expansion of v_j are ignored.

(AIII) The vertex corrections are entirely dropped.

Furthermore, the choice of the gross variables is not always self-evident, and occasionally, for specific problems, certain subsets of the gross variables are chosen in place of the entire set on plausible arguments. Therefore, we include:

(AIV) The specific choice of the gross variables.

(AV) Clear-cut separation of the time-scales discussed in connection with equation (2.9).

Equations having structure similar to that of equation (2.14) but with varying degrees of accuracy have been derived by a number of authors [22].

III. APPLICATIONS

In this section we describe in some detail an application of the general theory to binary critical liquid mixtures and the results obtained. Applications to other systems will also be described briefly.

In binary liquid critical mixtures there are six different kinds of gross variables, namely, the relative concentration c, the local pressure p, the entropy density s, and three components of the local velocity $\underset{\sim}{u}$. Among these, the local pressure and the logitudinal component of the local velocity are associated with sound propagation, and change very rapidly in time compared to the diffusive motions associated with the relative concentration, the entropy density, and transverse components of the local velocity. Here we will be concerned only with slow diffusive motions, and thus only the gross variables associated with them will be taken into account.

We consider the dynamics of concentration fluctuation, and for simplicity, we will also drop the local entropy, since the conclusion is not affected by this omission. Therefore, the gross variables we use consist of the relative concentration and the transverse components of the local velocity.[4]

The hydrodynamic equation for the relative concentration gives for $v_{c(\underset{\sim}{r})}$ that is, $v_i(a)$ for $c(\underset{\sim}{r})$, the following:

$$v_{c(\underset{\sim}{r})} = -\underset{\sim}{u}(\underset{\sim}{r}) \cdot \underset{\sim}{\nabla}\, c(\underset{\sim}{r}) \ . \tag{3.1}$$

Hence the kinetic equation for the Fourier component of the local concentration $c_{\underset{\sim}{q}}$ becomes

$$[\partial/\partial t + q^2 D_o(q)]c_{\underset{\sim}{q}} = -\frac{i}{V^{\frac{1}{2}}} \sum_{\underset{\sim}{k}} \underset{\sim}{u}_{\underset{\sim}{q}-\underset{\sim}{k}} \cdot \underset{\sim}{k}\, c_{\underset{\sim}{k}} + f_{\underset{\sim}{q}c} \ , \tag{3.2}$$

where $D_o(q)$ is the $\underset{\sim}{q}$-dependent diffusion constant of the conventional theory, and is inversely proportional to $\chi_{\underset{\sim}{q}c} \equiv \langle |c_q|^2 \rangle$.

Equation (3.1) can be also derived by using the definition (2.4) if we define the $\hat{a}$'s by dividing the system into macroscopically infinitesimal cells. The fact that there is no other term in $v_{c(\underset{\sim}{r})}$ is one important simplifying feature of the binary fluid, a point emphasized by Ferrell [24]. The same is expected to hold for the local entropy associated with a unit mass of fluid because the local entropy does not change in adiabatic fluid flow.

The corresponding kinetic equation for $\underset{\sim}{u}_{\underset{\sim}{q}}$, the Fourier component of the local velocity $\underset{\sim}{u}(\underset{\sim}{r})$, is not so simple. A part of $v_{\underset{\sim}{u}(\underset{\sim}{r})}$ which involves only $\underset{\sim}{u}(\underset{\sim}{r})$ is obtained from hydrodynamics, which, however, is not important near the critical point and will be ignored. A more important part of $v_{\underset{\sim}{u}(\underset{\sim}{r})}$ near the critical point involves the concentration, and is not contained in the hydrodynamic equations. This arises from the concentration dependence of local stress tensor.

---

[4] If we include the local entropy, then the thermal conductivity and thermal diffusion enter also. This problem was considered by Swift [23] as well as by the author (unpublished). The thermal conductivity and the thermal diffusion coefficient remain finite, and the critical anomaly of the diffusion constant is unaffected.

Thermodynamics gives only the average of the diagonal elements of the stress tensor, that is, the pressure. On the other hand, the off-diagonal part of the stress tensor is required for the transverse components of $\underset{\sim}{u}$. Thus, we must use the expansion formulae (2.5)-(2.8). The resulting kinetic equation for $\underset{\sim}{u}_{\underset{\sim}{q}}$ (transverse components only) is [7,12]

$$\frac{\partial}{\partial t} \underset{\sim}{u}_{\underset{\sim}{q}} = - q^2 \nu_o \underset{\sim}{u}_{\underset{\sim}{q}} - i \frac{k_B T}{2\rho V^{\frac{1}{2}}} \sum_{\underset{\sim}{k}} \left(\frac{1}{\chi_{\underset{\sim}{k}c}} - \frac{1}{\chi_{\underset{\sim}{q}-\underset{\sim}{k}c}} \right) [\underset{\sim}{k} - \frac{\underset{\sim}{q}}{q^2} (\underset{\sim}{q} \cdot \underset{\sim}{k})] c_{\underset{\sim}{k}} c_{\underset{\sim}{q}+\underset{\sim}{k}} + f_{\underset{\sim}{q}u} \tag{3.3}$$

where ν_o is the kinetic viscosity (which does not include mode-coupling contributions).

The equation for $G_{\underset{\sim}{q}c}(t) = \langle c_{\underset{\sim}{q}}(t) c^*_{\underset{\sim}{q}}(o)\rangle / \chi_{\underset{\sim}{q}c}$ which corresponds to equation (2.14) is then

$$\left(\frac{\partial}{\partial t} + q^2 D_o\right) G_{\underset{\sim}{q}c}(t) = - \frac{k_B T}{\rho} \frac{1}{V} \sum_{\underset{\sim}{k}} \left[q^2 - \left(\frac{\underset{\sim}{q} \cdot \underset{\sim}{k}}{k}\right)^2\right] \frac{\chi_{\underset{\sim}{q}-\underset{\sim}{k}c}}{\chi_{\underset{\sim}{q}c}} \int_0^t ds\, G_{\underset{\sim}{k}u}(s) G_{\underset{\sim}{q}-\underset{\sim}{k}c}(s)$$

$$\times \; G_{\underset{\sim}{q}c}(t - s) \,, \tag{3.4}$$

where $G_{\underset{\sim}{k}u}$ is the propagator of the transverse Fourier component of the local velocity. A similar equation can be written for $G_{\underset{\sim}{k}u}(t)$ to complete the self-consistent scheme for the calculation of $G_{\underset{\sim}{q}c}$ and $G_{\underset{\sim}{q}u}$ [7,12].

Examination of this self-consistent set of equations shows that the kinetic shear viscosity ν remains finite at the critical point [7,12], which allows us to replace $G_{\underset{\sim}{k}u}(t)$ in equation (3.4) by $\exp(- k^2 \nu t)$. In this way we obtain $G_{\underset{\sim}{q}c}(t)$ which is valid in both the hydrodynamical and critical regimes. That is, for times t greater than $(\nu q^2)^{-1}$ or $(\nu \kappa^2)^{-1}$, whichever is smaller, where κ is the inverse range of correlation of critical fluctuations, $G_{\underset{\sim}{q}c}(t)$ exhibits a simple exponential decay,

$$G_{\underset{\sim}{q}c}(t) = \exp(- \Gamma_q t) \,, \tag{3.5}$$

with the decay rate given by

$$\Gamma_q = (k_B T / 6\pi\eta)\, \kappa^3 F(q/\kappa) \tag{3.6}$$

$$F(x) \equiv \frac{3}{4} [1 + x^2 + (x^3 - x^{-1}) \tan^{-1} x] \tag{3.7}$$

where η is the shear viscosity and the Ornstein-Zernike form is assumed for $\chi_{\underset{\sim}{k}c}$. The results (3.5)-(3.7) were also derived by Ferrell [24] using the Green-Kubo formulae for transport coefficients.

If we adopt the Onsager hypothesis that the regression of fluctuations is described by macroscopic laws, Γ_q in the hydrodynamic regime $q \ll \kappa$ is given by $q^2 D$ with the diffusion constant D given by $(k_B T / 6\pi\eta)\kappa$ which can be interpreted in terms of Stokes law for a sphere of radius κ^{-1} [12,25].

Let us now review the approximations entering equations (3.5)-(3.7) in addition to the use of the Ornstein-Zernike form (which can be readily improved). At the moment, it is difficult to examine the approximation (AI) assuming a Gaussian distribution of fluctuations $\{a_i\}$, due to lack of knowledge of equal-time correlations of more than three a's. On the other hand, in the approximation where the vertex corrections are dropped, the approximation (AII) mentioned in the preceding section becomes exact owing to equation (3.2). That is, there is no contribution represented by the diagram  . Thus, the next corrections to equation (3.6) come from including the vertex corrections where in general the approximation (AII) ceases to be exact. The simplest vertex corrections are represented by  , a part of which scales as κ^3 and must be included in Γ_q. This problem is currently under investigation in collaboration with Mr. S. Lo. Preliminary results indicate that corrections to equation (3.6) amount to 2.4% and 0.4% for $q \ll \kappa$ and $q \gg \kappa$, respectively.

Because equations (3.6) and (3.7) involve approximations which have not been fully justified, it is worthwhile to give a somewhat intuitive derivation of these results. Take a non-equilibrium average of the diffusion current density

$$\mathbf{J}_d(\mathbf{r}) = \langle \mathbf{u}(\mathbf{r})\, \delta c(\mathbf{r})\rangle_{av} \tag{3.8}$$

where $\mathbf{u}(\mathbf{r})$ is the transverse component of the local velocity and $\delta c(\mathbf{r})$ is the local concentration fluctuation. Now, when the concentration is not uniform, a fluid element $d\mathbf{r}$ exerts a force $\mathbf{f}(\mathbf{r})d\mathbf{r}$ upon the surrounding fluid elements. Such a force is found by considering a change of the free energy due to a small local virtual displacement of the fluid.[5] Thus, the Fourier transform $\mathbf{f}_k$ of $\mathbf{f}(\mathbf{r})$ is given by

$$\mathbf{f}_k = \frac{k_B T}{2V^{\frac{1}{2}}}\, i \sum_{\boldsymbol{\ell}}{}' \left(\frac{\boldsymbol{\ell}}{\chi_{\boldsymbol{\ell} c}} + \frac{\mathbf{k}-\boldsymbol{\ell}}{\chi_{\mathbf{k}-\boldsymbol{\ell} c}} \right) c_{\mathbf{k}-\boldsymbol{\ell}}\, c_{\boldsymbol{\ell}}\, , \tag{3.9}$$

where the sum excludes the terms with $\boldsymbol{\ell} = o$ or $\mathbf{k} = \boldsymbol{\ell}$.[6] If there is a periodic average concentration $\langle c_{\mathbf{q}}\rangle_{av}$, we have $c_{\mathbf{k}} = \delta c_{\mathbf{k}} + \delta_{\mathbf{q},\mathbf{k}} \langle c_{\mathbf{q}}\rangle_{av}$, where $\delta c_{\mathbf{k}}$ is the spontaneous concentration fluctuation. Since we are concerned with the average diffusion current induced by $\langle c_{\mathbf{q}}\rangle_{av}$, we replace equation (3.9) by

$$\mathbf{f}_k = \frac{k_B T}{2V^{\frac{1}{2}}}\, i \left(\frac{\mathbf{q}}{\chi_{\mathbf{q} c}} + \frac{\mathbf{k}-\mathbf{q}}{\chi_{\mathbf{k}-\mathbf{q} c}} \right) \delta c_{\mathbf{k}-\mathbf{q}} \langle c_{\mathbf{q}}\rangle_{av}\, . \tag{3.10}$$

This fluctuating force produces a fluctuating local velocity field $\mathbf{u}_{\mathbf{k}}$ given by

$$\mathbf{u}_{\mathbf{k}} = T(\mathbf{k}) \cdot \mathbf{f}_{\mathbf{k}} \tag{3.11}$$

where $T(\mathbf{k})$ is the Fourier transform of the Oseen tensor [26] given by

$$T(\mathbf{k}) = -\frac{1}{\eta k^2} \left(\mathbf{1} - \frac{\mathbf{k}\mathbf{k}}{k^2}\right). \tag{3.12}$$

Substitution of equation (3.11) into equation (3.8), or its Fourier transform

$$\mathbf{J}_d(\mathbf{q}) = V^{-\frac{1}{2}} \sum_{\mathbf{k}} \langle \mathbf{u}_{\mathbf{k}}\, c_{\mathbf{q}-\mathbf{k}}\rangle_{av} \tag{3.13}$$

---

[5] A similar idea has been used by M. Fixman, J. Chem. Phys. **47**, 2808 (1967).

[6] In fact, reaction to this force is responsible for the second term of equation (3.3).

gives one part of the diffusion current,

$$\underset{\sim}{J}_d^{(1)}(\underset{\sim}{q}) = -i \frac{k_B T}{\eta} \frac{1}{V} \sum_{\underset{\sim}{k}} \frac{1}{k^2} (\underset{\sim}{q} - \frac{\underset{\sim}{q} \cdot \underset{\sim}{k}}{k^2} \underset{\sim}{k}) (\frac{\chi_{\underset{\sim}{q}-\underset{\sim}{k}c}}{\chi_{\underset{\sim}{q}c}} - 1) \langle c_{\underset{\sim}{q}} \rangle_{av} . \quad (3.14)$$

Another contribution to the diffusion current arises from considering a change in local concentration $\Delta c(\underset{\sim}{r})$ due to a small displacement $\underset{\sim}{\Delta}(\underset{\sim}{r})$ in the fluid in the presence of an average nonuniform concentration $\langle c(\underset{\sim}{r}) \rangle_{av}$. That is,

$$\Delta c(\underset{\sim}{r}) = - \underset{\sim}{\Delta}(\underset{\sim}{r}) \cdot \nabla \langle c(\underset{\sim}{r}) \rangle_{av} ,$$

or, in Fourier space,

$$\Delta c_{\underset{\sim}{k}} = -i V^{-\frac{1}{2}} \underset{\sim}{q} \cdot \underset{\sim}{\Delta}_{\underset{\sim}{k}-\underset{\sim}{q}} \langle c_{\underset{\sim}{q}} \rangle_{av} . \quad (3.15)$$

Substituting equation (3.15) into equation (3.13), we need to find the correlation $\langle \underset{\sim}{u}_{\underset{\sim}{k}} \underset{\sim}{\Delta}_{-\underset{\sim}{k}} \rangle$. Since $\underset{\sim}{u}_{\underset{\sim}{k}} = \dot{\underset{\sim}{\Delta}}_{\underset{\sim}{k}}$, we have

$$\langle \underset{\sim}{u}_{\underset{\sim}{k}} \underset{\sim}{\Delta}_{-\underset{\sim}{k}} \rangle = \int_{-\infty}^{0} dt \, \langle \underset{\sim}{u}_{\underset{\sim}{k}}(o) \underset{\sim}{u}_{-\underset{\sim}{k}}(t) \rangle = \frac{\rho k^2}{\eta} \langle \underset{\sim}{u}_{\underset{\sim}{k}} \underset{\sim}{u}_{-\underset{\sim}{k}} \rangle , \quad (3.16)$$

where we have assumed that no correlation exists between $\underset{\sim}{u}_{\underset{\sim}{k}}(o)$ and $\underset{\sim}{\Delta}_{-\underset{\sim}{k}}(-\infty)$ and have used the fact that $\langle \underset{\sim}{u}_{\underset{\sim}{k}}(o) \underset{\sim}{u}_{-\underset{\sim}{k}}(t) \rangle = \langle \underset{\sim}{u}_{\underset{\sim}{k}} \underset{\sim}{u}_{-\underset{\sim}{k}} \rangle \exp(-\eta \rho^{-1} k^2 |t|)$. Thus we find for the remaining part of the diffusion current,

$$\underset{\sim}{J}_d^{(2)}(\underset{\sim}{q}) = -i \frac{k_B T}{\eta} \frac{1}{V} \sum_{\underset{\sim}{k}}{}' \frac{1}{k^2} (\underset{\sim}{q} - \frac{\underset{\sim}{q} \cdot \underset{\sim}{k}}{k^2} \underset{\sim}{k}) \langle c_{\underset{\sim}{q}} \rangle_{av} . \quad (3.17)$$

Hence, adding equation (3.14) and equation (3.17), the total diffusion current now becomes

$$\underset{\sim}{J}_d(q) = -i \frac{k_B T}{\eta} \frac{1}{V} \sum_{\underset{\sim}{k}}{}' \frac{1}{k^2} (\underset{\sim}{q} - \frac{\underset{\sim}{q} \cdot \underset{\sim}{k}}{k^2} \underset{\sim}{k}) \frac{\chi_{\underset{\sim}{q}-\underset{\sim}{k}c}}{\chi_{\underset{\sim}{q}c}} \langle c_{\underset{\sim}{q}} \rangle_{av} . \quad (3.18)$$

This result, together with the continuity equation $\langle \dot{c}_{\underset{\sim}{q}} \rangle_{av} = -i \underset{\sim}{q} \cdot \underset{\sim}{J}_d(\underset{\sim}{q})$ leads to equation (3.6).

Let us now turn to other systems. First let us consider an isotropic ferromagnet in the disordered phase, where the gross variable is chosen to be the local spin density $\underset{\sim}{S}(\underset{\sim}{r})$. The instantaneous rate of change of $S(\underset{\sim}{r})$ is then

$$v_{\underset{\sim}{S}(\underset{\sim}{r})} = \underset{\sim}{S}(\underset{\sim}{r}) \times \underset{\sim}{H}(\underset{\sim}{r}) , \quad (3.19)$$

where $\underset{\sim}{H}(\underset{\sim}{r})$ is the local magnetic field ($g\mu_B$ is chosen to be unity for simplicity) which is obtained from the thermodynamic potential $\Phi(\{\underset{\sim}{S}\}, T)$ in the state of nonuniform magnetization by

$$\underset{\sim}{H}(\underset{\sim}{r}) = \delta \Phi(\{\underset{\sim}{S}\}, T) / \delta \underset{\sim}{S}(\underset{\sim}{r}) . \quad (3.20)$$

Furthermore, if we assume the expansion,

$$\Phi(\{\underset{\sim}{S}\}, T) = \tfrac{1}{2} \iint A(\underset{\sim}{r}, \underset{\sim}{r}') \underset{\sim}{S}(\underset{\sim}{r}) \cdot \underset{\sim}{S}(\underset{\sim}{r}') d\underset{\sim}{r} \, d\underset{\sim}{r}' , \quad (3.21)$$

where the Fourier transform of $A(\underset{\sim}{r}, \underset{\sim}{r}')$ is the inverse of the wave-vector-dependent magnetic susceptibility, we find that

$$v_{\underset{\sim}{S}(\underset{\sim}{r})} = \int A(\underset{\sim}{r}, \underset{\sim}{r}') \underset{\sim}{S}(\underset{\sim}{r}) \times \underset{\sim}{S}(\underset{\sim}{r}') d\underset{\sim}{r}' . \quad (3.22)$$

The kinetic equation for $\underset{\sim}{S}(\underset{\sim}{r})$ is then obtained by assuming the second term of equation (2.2) to be $D_o \nabla^2 \underset{\sim}{S}(\underset{\sim}{r})$, where D_o is the spin diffusion constant of the conventional theory, which is inversely proportional to the magnetic susceptibility. The resulting self-consistent equation for the time-displaced spin correlation function is identical to that of references [7] and [12], and similar self-consistent equations have been discussed by a number of authors and were compared favorably with experiments [22].

A similar treatment can be made for the liquid helium near the λ-transition. Here the relevant gross variables are the local order parameter $\psi(\underset{\sim}{r})$ and the local entropy. In units with $\hbar = 1$, we assume the following Poisson brackets:[7]

$$\{\hat{\psi}(\underset{\sim}{r}), \hat{\psi}^{+}(\underset{\sim}{r}')\} = \frac{1}{i}\,\delta(\underset{\sim}{r}-\underset{\sim}{r}'),\quad \{\hat{\psi}(\underset{\sim}{r}), \hat{S}(\underset{\sim}{r}')\} = i\,s\,\hat{\psi}(\underset{\sim}{r})\,\delta(\underset{\sim}{r}-\underset{\sim}{r}')$$

$$\{\hat{\psi}^{+}(\underset{\sim}{r}), \hat{S}(\underset{\sim}{r}')\} = -i\,s\,\hat{\psi}^{+}(\underset{\sim}{r})\,\delta(\underset{\sim}{r}-\underset{\sim}{r}')\ , \tag{3.23}$$

where s is the entropy per helium atom.

The instantaneous rate of change of the gross variables is then obtained from

$$v_{\psi(\underset{\sim}{r})} = \langle\{\hat{\psi}(\underset{\sim}{r}),\ \Phi(\{\hat{\psi}\},\{\hat{\psi}^{+}\},\ \{\hat{S}\};\ T)\};\ \{a\}\rangle \tag{3.24}$$

$$= \int \langle\{\hat{\psi}(\underset{\sim}{r}),\ \hat{\psi}^{+}(\underset{\sim}{r}')\}\rangle \frac{\delta\Phi}{\delta\psi^{+}(\underset{\sim}{r}')}\,d\underset{\sim}{r}' + \int \langle\{\hat{\psi}(\underset{\sim}{r}),\ S(\underset{\sim}{r}')\};\ a\rangle \frac{\delta\Phi}{\delta S(\underset{\sim}{r}')}\,d\underset{\sim}{r}' \tag{3.25}$$

with similar expressions for $v_{\psi^{+}(\underset{\sim}{r})}$ and $v_{S(\underset{\sim}{r})}$, where Φ is the thermodynamic potential and $\{a\}$ denotes a set of values for the gross variables.

If we can expand Φ in powers of the Fourier components of ψ, ψ^{+} and S as

$$\Phi = \Phi_o + \tfrac{1}{2}\sum_{\underset{\sim}{k}} \frac{k_B T}{\chi_{\underset{\sim}{k}}}\,|\delta\psi_{k}|^2 + \tfrac{1}{2}\sum_{\underset{\sim}{k}} \frac{k_B T}{T\rho C_k}\,|\delta S_{\underset{\sim}{k}}|^2\ ,\quad (T > T_\lambda), \tag{3.26}$$

where $\chi_{\underset{\sim}{k}} = \langle|\delta\psi_{\underset{\sim}{k}}|^2\rangle$ and C_k is the k-dependent specific heat per unit mass at constant pressure, and Φ_o is the thermodynamic potential in the absence of nonuniform fluctuations, the resulting kinetic equation reduces to that of reference [7].[8]

A useful feature of the present treatment, as given above for a ferromagnet and for liquid helium, is that the validity of the approximation AII in these examples is seen to be intimately connected with the validity of taking only the quadratic terms in the expansions of Φ, equation (3.21) and equation (3.26), which is also the condition for the validity of AI since the probability distribution of the gross variables is proportional to $\exp(-\Phi/k_B T)$.

A final example to be discussed is the planar ferromagnet which is considered to be a magnetic analogue of liquid helium [27]. Indeed, if the (x,y)-plane is chosen to be the plane

---

[7] The Poisson brackets involving the entropy density are discussed in reference [7], Appendix B. This is also consistent with the Poisson bracket between the entropy density and the phase of the order parameter used by Ferrell et al. [9].

[8] This set of equations has been used recently by D. A. Kruger and D. L. Huber, Phys. Letters, 33A, 149 (1970), to find the thermal conductivity above T_λ. Theoretical value found is one-seventh of the experimental value. However, the value of κ entering the theory is not known, and must be inferred.

in which spontaneous magnetization can occur, and if no external magnetic field is present, the critical dynamics of the planar ferromagnet in the disordered phase is readily obtained from that of liquid helium by replacing $S(\underline{r})$, $\psi(\underline{r})$, and $\psi^{\dagger}(\underline{r})$ by the spin densities $S^z(\underline{r})$, $S^+(\underline{r})$, and $S^-(\underline{r})$, respectively. However, if an external field in the z-direction is present, then $S^z(\underline{r})$ can couple with the energy density; and the spin diffusion coefficient, the thermal conductivity and the thermal spin diffusion coefficient diverge as $\kappa^{-\frac{1}{2}}$ near the critical point.[9] In the absence of an external field, of course, the thermal spin diffusion coefficient disappears and the thermal conductivity no longer diverges.

The mode-mode coupling approach described above has been applied to many other systems, and in each case specific predictions have been made for the transport anomalies near the critical point [21]. These predictions coincide with those of dynamical scaling whenever the latter applies [21]. The relationship between the two approaches has been discussed elsewhere [12].

Finally, it is worth noting that the mode-mode coupling mechanism can become important for transport properties even far away from the critical point. This is manifested in the recent discovery [28, 18, 29] that time correlation functions entering transport coefficients exhibit unexpected long time tails which lead to divergent transport coefficients in two-dimensional fluids.

IV. COMPARISON WITH EXPERIMENTS

Experimental tests of the predictions of both the mode-mode coupling and dynamical scaling theories as of the summer of 1968 were reviewed by Kadanoff [21] (see table 1 of this reference). Here we summarize the subsequent developments which are significant for experimental tests of the theories.

We start with the most controversial case, namely, SF_6. The difficulty here is the apparent asymmetry observed in the critical exponents of the thermal diffusivity above and below the liquid-gas critical point. According to the latest report of Benedek et al. [30]. It is quite possible that in the experiments that have been done so far the thermal conductivity, as well as the specific heat at constant pressure, has not yet reached the asymptotic region, in which case one is not able to deduce the critical exponents.

Next we take up magnetic critical scattering of neutrons. Earlier, the spin diffusion coefficients of iron and nickel above T_c were claimed to have the critical exponents 0.14 and 0.51, respectively, in disagreement with the theoretical prediction of $\frac{1}{2}(1-\eta)\nu \approx 1/3$ [31]. Subsequently, it was realized that these experiments were done still outside the hydrodynamic

---

[9] Here the thermal conductivity was defined in terms of the heat current in the absence of a spin density gradient. If, however, the thermal conductivity is defined in terms of the heat current in the absence of a spin diffusion current, the divergent contributions in the thermal conductivity cancel. This system is more closely analogous to liquid helium in fine pores than in free space [27]. Further details together with analogous transport properties for liquid helium will be discussed elsewhere.

regime. Reanalyses of these and more recent data now seem to indicate the correctness of the predicted exponent in iron [32]. In nickel, data seem to be consistent with the prediction, but the situation is less clear than in the case of iron [33].

We must also mention the beautiful experimental tests by the Brookhaven group of the dynamical scaling prediction for the cubic antiferromagnet $RbMnF_3$ above, at, and below the Neél temperature [34].

Let us now consider the first sound attenuation coefficient in liquid helium near the λ-point. In this case the controversy centered on the apparent asymmetry of the critical exponents characterizing the critical anomaly of the attenuation below and above T_λ. Here again it was recently realized that the exponent $\sim 1/2$ above T_λ was obtained by including data from outside the hydrodynamic region $\omega\tau \ll 1$, where ω is the sound-wave frequency and τ is the order parameter relaxation time. If the data in the nonhydrodynamical region are excluded, the predicted behavior $\omega^2|T - T_\lambda|^{-1}$ is found for the attenuation both above and below T_λ [35]. Furthermore, the theoretical prediction [36] that the attenuation should behave as $\omega|T - T_\lambda|^0$ for $\omega\tau \gg 1$ seems to be supported by these experiments.

On the other hand, the sound attenuation in magnetic insulators still continues to be controversial. The observed critical exponents are consistently smaller than the theoretical predictions [37]. Sound attenuation via coupling to the energy of the spin system has been suggested; however, the controversy has not been settled [37,38].[10]

Interesting developments have taken place in the study of critical dynamics in binary liquid critical mixtures and in simple liquids near the liquid-gas critical points. By light scattering techniques, Berge et al. [39] and, subsequently, Cummins and coworkers [40] have succeeded in determining the decay rate of order parameter fluctuations from the hydrodynamic through the critical region. The results generally confirm the dynamical scaling idea [10] as well as the mode-mode coupling theory described in the preceding section, though details are still controversial [30]. The sound attenuation and dispersion in these systems have also been studied over wider ranges of temperature and frequency [41,42], and the results compare favorably with the theoretical predictions [43]. However, the high-frequency sound attenuation in nitrobenzene-n-hexane and aniline-cyclohexane critical mixtures does not show the predicted asymptotic behavior $\omega|T - T_c|^0$ [42].

Neutron critical scattering experiments on the anisotropic antiferromagnets MnF_2 and FeF_2 have been carried out at Brookhaven [44]. An unexpected feature of the results is the fact that damping rate of the staggered magnetization with a small wavenumber remains very small below the Neél point for MnF_2. Heller [45] suggested that owing to mixing of the staggered spin mode with the spin energy density, the observed small damping rate is that of the spin energy

---

[10] After this paper was written, the author received a preprint of T. J. Moran and B. Luthi's paper which gives some experimental evidence for the mechanism of spin-lattice relaxation proposed here for $RbMnF_3$ and MnF_2. [Now published in Phys. Rev. B4, 122 (1971)].

density. If this is the case, the neutron scattering will provide the only known method to measure the spin thermal conductivity at elevated temperatures. The behavior above the Neél point appears to be well accounted for by the mode-mode coupling theory of anisotropic magnets due to Riedel and Wegner [46].

Finally, we must mention the very precise measurements of the shear viscosity of mixtures near the critical consolute points [25,47]. The shear viscosity exhibits a very weak logarithmic divergence. It was suggested that the self-consistent theory described in section III may give an apparent logarithmic divergence which turns into a sharp cusp at the critical point.

Reviewing these experiments, we note that extreme care is required in deducing the critical exponents of the transport coefficients from critical scattering experiments. In the first place, one must be absolutely sure that only data in the hydrodynamic regime are included; and in the second place, possible large background contributions must be taken into account. These points are now finally realized after a long period of confusion.

V. CONCLUDING REMARKS

In the preceding sections, dynamical behavior near critical points has been reviewed from the point of view of the mode-mode coupling theory. The problem is somewhat similar to that of hydrodynamical turbulence [48] in that nonlinear coupling among hydrodynamic modes plays a dominant role, and it is thus beset with the same type of difficulties. Here the approximation AIII of section II ignores the basic difficulty. It is hoped that further theoretical studies will shed light on this kind of nonlinear statistical problem. One advantage of studying critical dynamics instead of turbulence is the availability of very precise experiments under well-defined conditions.

The difficulties encountered in determining the critical exponents of transport coefficients discussed in section III suggest the usefulness and also the intrinsic interest in further studies of transport properties in nonhydrodynamical regions, including possible nonlinear effects [49,28].

On the theoretical side, the region of validity and the degree of accuracy of the dynamical scaling and mode-mode coupling theories are not yet precisely known, and any successful attempt at microscopic theory must be able to answer these questions.

REFERENCES

1. J. V. Sengers, in Critical Phenomena, ed. M. S. Green and J. V. Sengers, N.B.S. Miscellaneous Publication No. 273 (1966); also lectures delivered at the International Summer School "Enrico Fermi," Varenna, Italy (to be published).

2. See, for instance, a brief review in reference [7].

3. M. Fixman, J. Chem. Phys. **36**, 310 (1962).

4. K. Kawasaki, Phys. Rev. **150**, 291 (1966).

5. J. M. Deutch and R. Zwanzig, J. Chem. Phys. **46**, 1612 (1967); R. Mountain

and R. Zwanzig, J. Chem. Phys. **48**, 1451 (1968).

6. L. P. Kadanoff and J. Swift, Phys. Rev. **166**, 89 (1968).

7. K. Kawasaki, Ann. Phys. **61**, 1 (1970).

8. L. P. Kadanoff et al., Rev. Mod. Phys. **39**, 395 (1967), and the references cited therein.

9. R. A. Ferrell et al., Phys. Rev. Letters, **18**, 891 (1967); Phys. Letters, **24A**, 493 (1967); Ann. Phys. (N.Y.), **47**, 565 (1968).

10. B. I. Halperin and P. C. Hohenberg, Phys. Rev. Letters, **19**, 700 (1967); Phys. Rev. **177**, 952 (1969).

11. P. Résibois and M. de Leener, Phys. Rev. **178**, 806, 819 (1969); B. L. Clarke and S. A. Rice, Phys. Fluids, **13**, 271 (1970); A. M. Polyakov, Soviet Phys. JETP, **30**, 1164 (1970); F. Wegner, Z. Phys. **216**, 433 (1968); **218**, 260 (1969); G. Reiter, J. Appl. Phys. **41**, 1368 (1970), and to be published.

12. K. Kawasaki, in Phase Transitions and Critical Phenomena, ed. C. Domb and M. S. Green (New York: Academic Press) (in press).

13. M. Suzuki, H. Ikari, and R. Kubo, J. Phys. Soc. (Japan), **26**, Suppl. 153 (1969); A. Sadiq, Ph. D. thesis, University of Illinois (1971).

14. K. Kawasaki, Lectures at International Summer School of Physics "Enrico Fermi" Varenna, Italy, 1970 (to be published).

15. S. R. de Groot and P. Mazur, Nonequilibrium Thermodynamics (Amsterdam: North-Holland Publishing Co., 1962).

16. R. Kubo, Rept. Progr. Phys. **29**, 255 (1966).

17. L. D. Landau and E. M. Lifshitz, Fluid Mechanics (London: Pergamon Press, 1959), chap. 17.

18. R. Zwanzig, paper presented at this conference.

19. R. Kubo, J. Phys. Soc. (Japan), **26**, Suppl. 1 (1969).

20. This was briefly reviewed in references [7] and [12].

21. L. P. Kadanoff, J. Phys. Soc. (Japan), **26**, Suppl. 122 (1969).

22. Reviewed in K. Kawasaki, J. Appl. Phys. **41**, 1311 (1970). See also M. Blume and J. Hubbard, preprint; and J. Hubbard, preprint; P. Résibois and C. Piette, Phys. Rev. Letters, **24**, 514 (1970); E. Riedel and F. Wegner, Phys. Rev. Letters, **24**, 730, 930 (1970).

23. J. Swift, Phys. Rev. **173**, 257 (1968).

24. R. Ferrell, Phys. Rev. Letters, **24**, 1169 (1970).

25. G. Arcovito et al., Phys. Rev. Letters, **22**, 1040 (1969).

26. R. Zwanzig, in Stochastic Processes in Chemical Physics, ed. K. E. Shuler (New York: Interscience, 1969).

27. B. I. Halperin and P. C. Hohenberg, Phys. Rev. **188**, 898 (1969), and the references quoted therein.

28. T. Yamada and K. Kawasaki, Progr. Theor. Phys. (Kyoto), 38, 1031 (1967); T. E. Wainwright, B. J. Alder, and D. M. Gass, "Decay of Time Correlations in Two Dimensions," UCRL Report No. 72906 (1971) and the references cited therein; K. Kawasaki, Phys. Letters, 34A, 12 (1971); M. Ernst, private communication.

29. W. C. Mitchell, preprint.

30. G. B. Benedek et al., paper presented at Battelle Memorial Colloquium on Critical Phenomena, Geneva and Gstaad, 1970 (to be published).

31. M. F. Collins et al., Phys. Rev. 172, 417 (1969); V. J. Minkiewicz et al., Phys. Rev. 182, 608 (1969).

32. J. Als-Nielsen, Phys. Rev. Letters, 25, 730 (1970); G. Parette and R. Kahn, preprint (1970).

33. V. J. Minkiewicz, Fordham Conference on Dynamical Aspects of Critical Phenomena, June, 1970 (to be published).

34. H.-Y. Lau et al., J. Appl. Phys. 41, 1384 (1970).

35. R. D. Williams and I. Rudnick, Phys. Rev. Letters, 25, 276 (1970).

36. K. Kawasaki, Phys. Letters, 31A, 165 (1970), and Phys. Rev. A3, 1097 (1971).

37. B. Luthi, T. J. Moran, and R. J. Pollina, J. Phys. Chem. Solids, 31, 1741 (1970).

38. K. Kawasaki, Phys. Letters, 29A, 406 (1969).

39. P. Berge et al., Phys. Rev. Letters, 23, 693 (1969); 24, 1223 (1970); P. Berge and M. Dubois, Phys. Rev. Letters (to be published).

40. D. L. Henry, H. L. Swinney, and H. Z. Cummins, Phys. Rev. Letters, 25, 1170 (1970).

41. M. Barmatz, Phys. Rev. Letters, 24, 641 (1970); C. W. Garland, D. Eden, and L. Mistura, Phys. Rev. Letters, 25, 1161 (1970).

42. D. Sette and L. Mistura, Lectures at International Summer School of Physics "Enrico Fermi," Varenna, Italy, 1970 (to be published).

43. K. Kawasaki, Phys. Rev. A1, 1750 (1970).

44. M. P. Schulhof, paper presented at the Fordham Conference on Dynamical Aspects of Critical Phenomena (1970), and the references cited therein. Results on FeF_2 are quoted by Riedel, reference [46].

45. P. Heller, paper presented at the Fordham Conference on Dynamical Aspects of Critical Phenomena (1970)

46. E. K. Riedel, J. Appl. Phys. (March 1971) and the references cited therein. Qualitative difference in the spin dynamics of isotropic and anisotropic magnets of the type described by Riedel and Wegner was found somewhat earlier. See K. Kawasaki, Progr. Theor. Phys. (Kyoto), 39, 285 (1968).

47. Further details on the latest status of transport anomalies in fluids are given in the Varenna Summer Lectures by J. V. Sengers, reference [1].

48. R. Kraichnan, paper presented at this conference.

49. W. Botch and M. Fixman, J. Chem. Phys. 36, 3100 (1962).

DISCUSSION

R. S. Wilson: Professor Zwanzig was asked yesterday how one knows a **priori** which variables should be chosen as "good" critical dynamical variables. One might ask the same question here about the choice of Kawasaki's generalized macroscopic dynamical variable (GMDV's). We know that for the case of diffusion in critical binary mixtures, the critical fluctuations in composition and the transverse local fluid velocity were sufficient within Kawasaki's mode-mode coupling theory to give an Onsager coefficient L which diverged as κ^{-1} at critical composition as one approached the critical separation temperature T_c. Perhaps it is worthwhile to give a simple physical explanation or picture for an a priori choice of the local fluid velocity as a good candidate for a GMDV. Fixman and I worked out via his general variational method a general formula for the anomalous diffusion constant: $\sim \lim_{r \to 0} \langle u(r) \rangle_\theta$, where $u(r)$ is a conditional fluid velocity at r when a molecule is fixed at the origin. The analogy was made with the electrophoretic effect in simple electrolytes where the molecules which appear about a particular molecule due to critical fluctuations are analogous to the ionic atmosphere in electrolytes and there under an electric field cause ions to move upstream and here cause molecules to be dragged along (Stokes-Einstein) with the local diffusion flow. $G^o(r) \sim ae^{-\kappa r}/r$, we found $L \sim \kappa^{-1}$ as already mentioned. No use was made of static or dynamic scaling. One could also calculate the concentration dependence of L or equivalently D. This is in agreement with Professor Kawasaki's result in the hydrodynamic regime.

K. Kawasaki: In order to choose relevant variables, one first writes down all the conceivable slowly varying variables (see references [1] and [14] of the text for the choice of such variables), and then drops the variables which are less slowly varying. For instance, in critical mixtures there are altogether six kinds of such variables, i.e., the local entropy density, the local concentration, the two transverse components of the local velocity, the longitudinal local velocity, and the local pressure. The last two of them vary like sound waves and are hence less slowly varying compared to the other four which obey diffusive-type equations. Here, for simplicity, we also omitted the local entropy density which can be included without changing the critical anomaly of the diffusion constant.

M. S. Green: In response to the question of Wilson's about the practical way to choose gross variables and to Kawasaki's answer, I would like to make the comment that in all instances one must choose the approximate integrals of motion. In the critical region one must choose the usual macroscopic observables which are also approximate integrals of motion out of the critical region and the order parameters which are approximate integrals of motion only in the critical region.

B. U. Felderhof: I would like to come back to Professor Fisher's question and say a word to the defense of the kinetic Ising model and Suzuki's work on the model. Although it is true that the dynamics of this model is not described in terms of a Hamiltonian, it nevertheless has a good physical basis. The kinetics may be found from the interaction of the Ising system with a heat bath of phonons. The microscopic kinetics contains nothing about the critical point, and the transition probabilities are regular functions of temperature and magnetic field. The critical slowing down which is found is therefore a true expression of the statistics of the magnetic part of the system as given by the Ising-Hamiltonian.

K. Matsuno: Can you predict the functional form of the scaling function which has been observed in the antiferromagnet MnF_2 by Schulhof et al.? If not, is it possible to extend the "extended mode-mode coupling theory" so that such a scaling function may be predicted?

K. Kawasaki: For magnets with a uniaxial anisotropy and a unique direction of spontaneous magnetization, the mode-mode coupling theory in the strict sense does not work because there are no slowly varying variables that can couple with the order parameter. See K. Kawasaki, Progr. Theor. Phys. (Kyoto), 39, 285 (1968). Nevertheless, if the anisotropy is not too large, one can still find the approximately slowly varying variables that couple to the order parameter and develop the mode-mode coupling theory, as has been done recently by Riedel and Wegner. See E. Riedel, paper presented at the 16th Annual Conference on Magnetism and Magnetic Materials, Miami, November, 1970, to be published in J. Appl. Phys.

STATIC SCALING IN THE CRITICAL REGION

Robert Brout

1. INTRODUCTION

We present here a resume of the hypotheses of static scaling and their consequences. We then argue for their exact or approximate validity in terms of the cluster expansion. In addition, an argument is offered which explains the value of the critical parameter in three-dimensional systems ($\delta = 5$).

At the outset, I wish to make clear that I have more confidence in some arguments than others. Thus I believe that the argumentation [1] which leads to thermodynamic scaling (Domb and Hunter [2], Widom [3], Fisher [4]) very probably is the basis of an exact theorem. The same argumentation leads to approximate length scaling (Kadanoff [5,6], Patashinskii and Pokrovskii [7]). I see no reason why this part of the theory should be exact and every reason why it should be a very good approximation. Still a third argument yields $\delta = 5$, and again there is no compelling reason for its exactitude (in fact, there is even less than for length scaling; the argument fails in two dimensions). I mention these considerations right at the beginning because I am rather disturbed that there are strong indications from the King's College group [8] that $\delta = 5$ is an exact result both for the Ising model and the self-avoiding walk approximation [9] to it. If this turns out to be the case, it is very probable that the analysis presented below for $\delta = 5$ is lacking in some essential ingredient, presumably concerned with analyticity of the free energy in magnetization.

Section 2 of this paper contains the definition of the critical parameters along with the motivations for these definitions. The reader is also referred to Fisher's review [10] in this regard. The thermodynamic-scaling hypothesis is stated and the consequences drawn. The length scaling hypothesis is then presented and further consequences deduced. Finally, the question of symmetry of critical parameters above and below T_c is broached.

This work is partially supported by Instituts Internationaux de Solvay de Physique et Chimie.

Section 3 begins with a very brief review of the linked cluster expansion of the Ising model. Some approximations (random phase approximation [RPA] and self-avoiding walk [SAW]) are introduced and analyzed in this context. In particular it is shown how scaling arises in the SAW. The general analysis of all graphs ensues, and the lessons drawn from SAW are exploited. The argument for exact thermodynamic scaling and approximate length scaling follows.

Section 4 takes up the question of $\delta = 5$ and the dilemma presented. Things are left where they began: with a question.

2. SCALING HYPOTHESES

The beginning of this section contains a bunch of motivated definitions in order to keep the lecture self-contained. The uninitiated are nevertheless well advised to refer to the reviews of Fisher [10] and Kadanoff et al. [6].

The definition of the critical parameters is predicated on the supposition that all thermodynamic properties are power law in character (interpreted as a single power of a logarithm or a finite discontinuity). Evidence for the truth of this proposition is based on (1) an exact two-dimensional Ising model, (2) good fits to series expansions, and (3) experimental evidence. Throughout we use the three-dimensional Ising model with the notation T = temperature; T_c = critical temperature; R = magnetic moment per spin in dimensionless units $[-1 \le R \le 1]$; $\mathcal{H}$ = magnetic field in energy units; $\epsilon = (T - T_c)/T_c$ (we are concerned with $\epsilon \ll 1$); χ = isothermal susceptibility $= (\partial R/\partial \mathcal{H})_{T,\mathcal{H}=0}$; C_v = specific heat; and R_o = zero field magnetization $\epsilon < 0$, i.e., $R_o = \lim_{\mathcal{H} \to 0^+} R(\mathcal{H}, T)$. The quantity χ diverges in the neighborhood of $\epsilon = 0$, and one defines γ, γ' according to

$$\begin{aligned} \chi &\sim \epsilon^{-\gamma} \quad (\epsilon > 0) \\ &\sim \epsilon^{-\gamma'} \quad (\epsilon < 0) . \end{aligned} \tag{2.1}$$

(Note that for $\epsilon < 0$, $\chi = \lim_{\mathcal{H} \to 0} \dfrac{R - R_o}{\mathcal{H}}$.)

The quantity χ is of course equal to the spin fluctuation. This is seen from the Hamiltonian

$$H = -\tfrac{1}{2} \sum v_{ij} \mu_i \mu_j - \sum \mu_i \mathcal{H} , \tag{2.2}$$

where v_{ij} is the potential of interaction (>0 for ferromagnets) and μ_i is the two-valued spin variable ($\mu_i = \pm 1$). The free energy is

$$-\beta F = \log \operatorname{tr} e^{-\beta H} \tag{2.3}$$

whence $R = -N^{-1} \partial F/\partial \mathcal{H}$ and $\chi = -N^{-1} \partial^2 F/\partial \mathcal{H}^2$ yielding

$$\begin{aligned} \chi &= \sum_j \langle \mu_i \mu_j \rangle - \langle \mu_i \rangle \langle \mu_j \rangle \\ &= \sum_j \langle \mu_i \mu_j \rangle \quad (\epsilon > 0). \end{aligned} \tag{2.4}$$

Thus the divergence in χ is indicative of a divergence in the spin-density fluctuation. We should then expect a similar divergence in the energy-density fluctuation which by the same reasoning as for χ is equal to C_v. We therefore define indices

$$C_v \sim \epsilon^{-\alpha} \quad (\epsilon > 0)$$

$$\sim \epsilon^{-\alpha'} \quad (\epsilon < 0) . \qquad (2.5)$$

To have χ diverge as $\epsilon \to 0$, the correlation function $\langle \mu_i \mu_j \rangle \equiv g(R_{ij})$ must be such as to cause the divergence of the integral $\int g(R) d^3R \sim \chi$. One parametrizes g(R) according to

$$\lim_{R \to \infty} g(R) \sim e^{-\kappa R}/R^{1+\eta}$$

$$\kappa = \epsilon^{\nu} = [\text{correlation length}]^{-1} . \qquad (2.6)$$

The quantity η is not an independent parameter but is arranged to make the integral of g(R) proportional to $\chi(=\epsilon^{-\gamma})$. Thus $\gamma = (2-\eta)\nu$.[1]

For $\epsilon < 0$, R_o traces out a singular curve which is parametrized by the parameter β; R_o vanishes according to

$$R_o = [-\epsilon]^{\beta} ; \quad \epsilon < 0 . \qquad (2.7)$$

Finally for $\epsilon = 0$ the isotherm $R(\mathcal{H}, T_c)$ in the vicinity of $\mathcal{H} = 0$ has inflection behavior characterized by the parameter δ. R vanishes according to

$$R = \mathcal{H}^{1/\delta} ; \quad \epsilon = 0 . \qquad (2.8)$$

For a general experimental rundown of the values of these parameters (α, β, γ, δ, ν) see Kadanoff et al. [6]. We quote here a typical set containing recent calculational data based on series computation of the three-dimensional Ising model. My references are the King's College group (many private communications from Domb, Sykes, and Gaunt) and the Illinois group [11].

$\gamma = 5/4$ (very precise)

γ' (presently in doubt, but there are indications for $\gamma' > \gamma$ [12])

$\alpha \simeq 0.1$ (Professor Domb and others like 1/8)

α' (rather uncertain with a tendency for $0 < \alpha' < 0.1$)

$\beta = 0.31$ (again a favorite value is 5/16)

$\delta = 5.0$ (and not more than 5.05 [8])

$\nu = 0.638$ (and hence with great precision $\eta = 0.04$)

Apart from the general disorder, two things stand out: $\delta = 5$, and η is very small. Any theory must explain these facts either quantitatively or qualitatively; we will attempt the latter in section 4. A much more subtle ordering of these numbers has been achieved through the beautiful inductive scaling hypotheses to which we now turn.

The general notion of thermodynamic scaling is that one can make the spin system more "rigid" either by lowering the temperature or by increasing the strength of the magnetic field. Thus one can efface the effect of raising $\mathcal{H}$, by raising ϵ in proportion to some power of $\mathcal{H}$. One then postulates that F depends on $\mathcal{H}$ through $\mathcal{H}/\epsilon^{\Delta}$ ($\Delta > 0$) (Δ is called the gap parameter). We first handle $\epsilon > 0$ where F is a function of $\mathcal{H}^2$ alone, the spin Hamiltonian having up and down

---

[1] Eq. 2.6 is called strong scaling. This is now known to be incorrect. Rather $\lim_{R \to \infty} g(R) = \kappa^{\eta} e^{-\kappa R}/R$. What is true is that when $\epsilon = 0$, $g(R) \to 1/R^{1+\eta}$. It is easy to show that with this latter used as definition for η, one still maintains the equality $\gamma = (2-\eta)\nu$.

spin symmetry. When $\mathcal{H} = 0$, we have $F \sim \epsilon^{2-\alpha}$ since $C_v \sim \partial^2 F/\partial T^2$ for its most divergent part. One therefore postulates

$$F \sim \epsilon^{2-\alpha} f(\mathcal{H}^2/\epsilon^{2\Delta}) . \tag{2.9}$$

For $\epsilon > 0$, f may be expanded. The coefficient of $\mathcal{H}^2$ is $\chi/2$ since $R = \partial F/\partial \mathcal{H}$. Thus $\epsilon^{2-\alpha} f'(0)(\mathcal{H}^2/\epsilon^{2\Delta}) = \epsilon^{-\gamma}\mathcal{H}^2$, and

$$2\Delta = 2 - \alpha + \gamma \;\; (= \tfrac{25}{8} \text{ for } \gamma = \tfrac{5}{4}, \; \alpha = \tfrac{1}{8}). \tag{2.10}$$

Consider now the limit $\epsilon \to 0$, $\mathcal{H} \neq 0$. Under such conditions F is analytic in ϵ (e.g., R versus T at $\mathcal{H}$ fixed is a smooth curve with an inflection in the vicinity of $T = T_c$). Therefore the function f(x) of equation (2.9) must be such as to cancel the ϵ singularity, i.e.,

$$\lim_{x \to \infty} f(x) \sim x^{2-\alpha/2\Delta} \tag{2.11}$$

whence

$$F \sim \mathcal{H}^{2-\alpha/\Delta} \qquad \epsilon = 0;\; \mathcal{H} \neq 0 . \tag{2.12}$$

Recalling that at $\epsilon = 0$, $R = \partial F/\partial\mathcal{H} \sim \mathcal{H}^{1/\delta}$, we get

$$1/\delta = [(2-\alpha)/\Delta] - 1$$

$$\delta = \frac{2-\alpha+\gamma}{2-\alpha-\gamma} \quad (= 5 \text{ with } \gamma = \tfrac{5}{4};\; \alpha = \tfrac{1}{8}). \tag{2.13}$$

We now turn to $\epsilon < 0$. Scaling generally supposes that because of the possibility of smooth passage through $\epsilon = 0$ at $\mathcal{H} \neq 0$, it is then possible to take a subsequent limit of $\mathcal{H} = 0$ with impunity. We discuss this point in more detail subsequently and for the moment simply go along with this reasonable assumption. Rather than adhere to the form (2.9) we find it more perspicuous to change over to R as variable and to write the free energy in the form

$$F = G(\epsilon, R) - R\mathcal{H};\quad \partial F/\partial R = 0 , \tag{2.14}$$

where $G(\epsilon, R)$ is a "trial" free energy. In statistical mechanics it is calculated from a restricted trace ($\Sigma \mu_i$ = NR is fixed). R is then centered on its correct value according to equation (2.14). Now observe that

$$R = \epsilon^{-\gamma}\mathcal{H} f'(\mathcal{H}^2/\epsilon^{2\Delta}) \tag{2.15}$$

may be inverted by multiplication by $\epsilon^{\gamma-\Delta} \equiv \epsilon^{-\beta}$. We get

$$\mathcal{H}/\epsilon^{\Delta} = h(R/\epsilon^{\beta}) ,$$

$$\beta = \Delta - \gamma = \tfrac{1}{2}(2 - \alpha - \gamma) . \tag{2.16}$$

Observe further that $R\mathcal{H}$ appearing in equation (2.15) is then of the form $\epsilon^{2-\alpha}\xi(R/\epsilon^{\beta})$ (ξ is some function) and expression (2.9) is also of this form. Therefore $G(\epsilon, R)$ of equation (2.14) is again of the same form, and we write

$$F = \epsilon^{2-\alpha} g(R^2/\epsilon^{2\beta}) - R\mathcal{H} . \tag{2.17}$$

This second form of scaling can be used with equal ease to derive the previous results (left to the reader). What is of interest now is the equation of state

$$\epsilon^{\gamma} R\, g'(R^2/\epsilon^{2\beta}) - \mathcal{H} = 0. \tag{2.18}$$

This equation makes perfect sense at $\epsilon = 0$, and one recovers $R = \mathcal{H}^{1/\delta}$. We now extrapolate its solution smoothly to $\epsilon < 0$ and then go to the limit $\mathcal{H} \to 0$. We have

$$R_o g'(R_o^2/\epsilon^{2\beta}) = 0 \ , \tag{2.19}$$

which must be such as to yield $R_o = 0$ ($\epsilon > 0$) and $R_o \sim |\epsilon|^\beta$ for $\epsilon < 0$. The smoothness requirement then also forces us to replace ϵ everywhere by $|\epsilon|$. Clearly a good bit of smoothness is being asked in the limit $\{\mathcal{H} \to 0, \ \epsilon \to 0\}$. Notwithstanding, the expression (2.16) is compatible with present series calculations. The result $\alpha = \alpha'$ is deduced from equation (2.17) and is not terribly well tested. The result $\gamma = \gamma'$ follows from one derivative equation (2.18) and is presently in the throes of anguish [12]; indications of its failure certainly exist, but an unexpected singular behavior in the series analysis has arisen which has yet to be sorted out [13]. We emphasize then the wisdom in keeping apart the scaling assumption for $\epsilon \geq 0$ leading to δ (equation [2.13]) and the further smoothness assumption down to the coexistence curve at $\epsilon < 0$, leading to β, α', γ'.

Stell [14] has made this point clear. Define $Z \equiv R^{1/\beta}$ and take any thermodynamic function, say χ. Then

$$\chi \sim \epsilon^{-\gamma} \mathcal{F}(\epsilon/z).$$

The contours of constant χ are similar curves in the (ϵ, Z)-plane since $\chi(a\epsilon, aZ) = a^{-\gamma}\chi(\epsilon, Z)$. Suppose scaling to be true for $\epsilon > 0$ and also continuable into the $\epsilon < 0$ region for $\mathcal{H} \neq 0$ and $\epsilon \neq 0$. We want to characterize what horrors might arise as we let $\mathcal{H}$ and ϵ approach zero on the $\epsilon < 0$ side.

The usual (respectable) state of affairs is that the coexistence curves $\pm Z_c = -A\epsilon$ are straight lines joining at the origin of ϵ, Z. The spinodal lines are enclosed within these. In this case the contours of constant χ take a vertical jump (parallel to the Z-axis) when they hit the coexistence lines. Similarity requires that the intersection points between the contours and the coexistence always have ϵ and Z in the same proportion, thereby verifying $Z_c = \pm A|\epsilon|$ or $R_o = \pm A|\epsilon|^\beta$.

However, the coexistence curves may not be linear, and the contours in the limit $\epsilon \to 0$ may not obligingly jump between them. Stell suggests two possibilities: (1) Z_c becomes tangent to the ϵ-axis at $\epsilon = 0$. In that case the whole argument is lost; and while scaling is maintained for $\epsilon > 0$, the limit $\epsilon = 0^-$ has lost all predictive power. (2) The coexistence lines become parallel to the spinodal lines at the origin. Consider the example of a coexistence curve

$$Z_c = -\epsilon + c(-\epsilon)^p \quad (p > 1, \ c > 0)$$

and suppose $\chi \sim (\epsilon + Z)^{-\gamma}$ $[= \epsilon^{-\gamma}(1 + Z/\epsilon)^{-\gamma}]$. Then on the coexistence curve

$$\chi \sim (-\epsilon)^{-\gamma'} \sim (\epsilon + z_c)^{-\gamma} \sim (-\epsilon)^{p\gamma},$$

$$\gamma' = p\gamma .$$

We conclude that it is well to keep scaling and smoothness assumptions as independent postulates. In the latter sections of the paper we will be concerned only with scaling at $\epsilon > 0$ and we now drop (for lack of knowledge) any further discussion of this fascinating point.

Turning now to Kadanoff's formulation, the fundamental notion here is that near the critical region the correlation length K^{-1} is extremely large, so within a finite distance small compared to K^{-1} and large compared to interparticle distance one has a large number of spins correlated together. Clearly there is a large leeway on how to pick such distances as $\epsilon \to 0$. Since all the spins within a cell of such dimensions act in consort, one may consider an Ising model of cells, but then with an effective temperature and effective magnetic field. One assumes that in going from one system of cells to another a scaling occurs according to

$$\epsilon \to L^y \epsilon \ , \qquad \mathcal{H} \to L^x \mathcal{H} \ . \tag{2.20}$$

Finally, since the number of cells is reduced by a factor of L^{-3}, we get a new cell free energy which is the same as the original:

$$F(\epsilon, \mathcal{H}) = L^{-3} \ F(L^y \epsilon, \ L^x \mathcal{H}) \ , \tag{2.21}$$

thereby implying

$$F = \epsilon^{3/y} \ f(\mathcal{H}/\epsilon^{x/y}) \tag{2.22}$$

Thermodynamic scaling (equation [2.9]) is thus implied by length scaling with the identification

$$2 - \alpha = 3/y \ , \quad \Delta = x/y \ . \tag{2.23}$$

What is new in this formulation is that under length scaling the distance between cells scales like R/L. However, the spin correlation must be independent of L. Since ϵ scales like L^y, we must then have at $\mathcal{H} = 0$ that g(R) depends on R only through $R\epsilon^{1/y}$. [This function is to be multiplied by the power of ϵ which makes $\chi = \int g(R)\, d^3R$ come out to be $\epsilon^{-\gamma}$. If $\mathcal{H} \neq 0$, then g(R) depends on $\mathcal{H}$ through $\mathcal{H}/\epsilon^{\Delta}$.] The important result ensues

$$\kappa = \epsilon^{/y} = \epsilon^{\nu} \ , \ 2 - \alpha = 3\nu , \tag{2.24}$$

where we have used equation (2.23). One should also notice that the same correlation length K^{-1} appears in all correlation functions (energy density, etc.) and this appears to be an a priori weakness in length scaling. This point will be made clear in section 3. Equation (2.24) is fulfilled in very good approximation; the error is about 1.5% (though α is not known that well, δ and γ are; thermodynamic scaling then fixes α).

3. APPROXIMATIONS AND ANALYSIS

Thermodynamic scaling is easily explained (and, as mentioned, probably provable) through the cluster expansion. The form which suits me best is the original one announced in 1959 [15] and is found in more digestible form in the Appendix of reference [16]. [This article is based on Coopersmith's 1961 Cornell thesis and is relevant in the present context in that it is shown there that what holds for spin systems is also true for the liquid-gas condensation; the graphs have the same structure when the potential is separated into a core (unperturbed piece) and the attractive tail (treated as a perturbation). As an example, for long-range forces, just as Weiss theory is the infinite range limit, so the van der Waals theory is for the fluid. It was also stated in this work that the dense liquid should be described in terms of the hard-core distribution alone since the rigidity of the core system holds down the fluctuations due to attractions. This prediction has been amply borne out. Thus, almost everything that goes for spins is expected to go for fluids (see, however, the comments in section 4).] The expansion for $\epsilon > 0$, $\mathcal{H} = 0$ is summarized by: the n(th) order in β for the energy is the sum over linked irreducible graphs containing n bonds. With a bond goes a factor of βv_{ij}; an even number of bonds emanates from each vertex. In addition, a graph contains a statistical weighting factor (unity for cycles and calculable otherwise according to a simple rule irrelevant to our present purpose). What is important is the exclusion; no two spins are on the same site. Unlike some more elegant versions [17] which are rearrangements of the above, the above clusters are physical. Other versions contain unphysical configurations which mask the points I wish to bring out. We also need the rules for open graphs. For $\epsilon > 0$ expand F in powers for $\mathcal{H}^2$:

$$F = F_o(\epsilon) + \sum_{r=1}^{\infty} A_{2r}(\epsilon)\, \mathcal{H}^{2r} \quad (A_2 = \chi/2) \;. \tag{3.1}$$

Then A_{2r} contains a sum over the positions of 2r fixed points. One draws connected graphs between these fixed points from which emanates an odd number of bonds. Pieces which do not contain these fixed points are irreducible. Finally for correlation functions one adopts the same rule as above but without summing on the fixed points (e.g., $\langle \mu_1 \mu_2 \rangle$ is the sum of connected graphs between 1 and 2, and χ is the sum over points 2).

Of all the graphs, the cycles (for E) or chains (for A_{2r}, g(R) . . .) are the most important for phase transitions. It is they that transmit essentially the correlations between spins. A spin, known to be up at the origin, polarizes its neighbors which in turn polarizes its neighbors, etc. This chain of correlation is in fact the chain in the graph. The reason why a phase transition occurs at all is the enormous diversity of paths between two points (in more than one dimension). If a spin is "wrong" on one path, the message can get through on another; as T is lowered, there are enough "good" paths so that the message travels over infinite distance and long-range order sets in. Other graphs are best looked upon as refinements of the single chain.

A more technical point concerning chains is the following: Let the force have a range covering Z points. For the first $\sim$Z graphs (i.e., containing n < Z bonds), chains dominate and their numerical value is approximated with precision in a simple manner to yield the Ornstein-Zernike and Curie-Weiss laws. Since the series converges in Z terms for $\epsilon > 1/Z$, one has an extremely simple theory outside the critical region as defined by the above inequality. Since Z is never very small in physical systems, one then explains the early success of molecular field theory quite simply.

The description of the critical region is to be sought in the graphs of very high order since we are interested in characterizing the nature of divergences. We first analyze chains (cycles for closed graphs), for a near-neighbor interaction of strength J. We refer to this set as the self-avoiding walk [SAW] [9]. Following Domb we have

$$\frac{2\beta E}{N} = \sum_n U_n (\beta J)^n \quad (\epsilon > 0) \;, \tag{3.2}$$

$$\chi = \sum C_n (\beta J)^n \tag{3.3}$$

where U_n is the number of walks that return to the origin in n steps and C_n is the total number of walks in n steps with the proviso that no site is visited more than once in a promenade.

As a first orientation, we drop the proviso of exclusion. Then

$$C_n = Z^n , \tag{3.4}$$

since there are Z positions available around each site. Equation (3.3) gives

$$\chi_{RPA} = \sum (\beta Z J)^n = \frac{1}{1 - \beta Z J} , \tag{3.5}$$

and we find the Curie-Weiss law $\chi = (T - T_c)^{-1}$; $kT_c = ZJ$. Observe that the exclusions are not important for the first Z terms and that these dominate for $(\beta Z J) < (1 - 1/Z)$ $[\epsilon \geq 1/Z]$, thereby instating the Curie-Weiss law for $\epsilon \geq 1/Z$.

To count the number of cycles without exclusion, write the n(th) term of equation (3.2)

$$U_n(\beta J)^n = \sum_{i_2 \ldots i_n} \beta^n [v_{i i_2} \ldots v_{i_n i}] = \frac{1}{N} \mathrm{tr}\, [\beta v]^n , \tag{3.6}$$

where we have written v_{ij} as a matrix. Since v_{ij} has translational symmetry, its eigenvalues are

$$v(q) = \sum_j v_{ij} \exp[iq \cdot (R_i - R_j)] \quad (q \text{ in Brillouin zone}) ,$$

so that equation (3.6) becomes

$$\frac{1}{N} \mathrm{tr}\, [\beta v]^n = \frac{1}{N} \sum_q [\beta v(q)]^n . \tag{3.7}$$

Substitution into equation (3.2) gives

$$2\beta E_{RPA} = \sum_{n=2}^{\infty} \sum_q [\beta v(q)]^n = \sum \frac{\beta v(q)}{1 - \beta v(q)} . \tag{3.8}$$

Equation (3.8) is the RPA form, again valid for $\beta\epsilon > (1/Z)$. The Ornstein-Zernike formula follows from expanding about $q = 0$ $[v(0) = \mathrm{Max}\ v(q)]$. Equation (3.8) then yields [with $v(q) = v(0) - Cq^2$] $\langle |\mu_q|^2 \rangle \simeq [q^2 + K^2]^{-1}$, small q, with $K^2 \sim \epsilon$. [One uses $\langle |\mu_q|^2 \rangle = \partial \beta F / \partial v(q)$ to obtain this result.] Therefore, $g(R) \sim e^{-KR}/R$, large R (the length scale is $Z^{1/3}$). To extract U_n for large n, we approximate the series in the neighborhood of $\beta v(0) = 1$. Then $v(q) \sim v(0) - Cq^2 \simeq v(0) \exp(-Cq^2)$, and

$$(\beta J)^n (U_n)_{RPA} = \int d^3q\, [\beta v(0)]^n e^{-Cnq^2} = \frac{[\beta v(0)]^n}{n^{3/2}} ,$$

$$(U_n)_{RPA} \sim \frac{Z^n}{n^{3/2}} . \tag{3.9}$$

This result may be understood as follows. In a random walk $\langle R_n^2 \rangle \sim n$; hence $n^{3/2} = \langle R_n^2 \rangle^{3/2} \equiv V_n$, the volume swept out:

$$(U_n)_{RPA} = \frac{Z^n}{V_n} = Z^{n-1} \left(\frac{Z}{V_n}\right) , \tag{3.10}$$

where Z/V_n is the number of sites neighboring the origin over the volume of the walk and thus is the appropriate factor for the probability of return.

We get a first glimpse into Kadanoff's [5] scaling. Equations (3.2) and (3.9) give

$$E \sim \sum (\beta ZJ)^n/V_n \sim \epsilon^{1-\alpha} , \tag{3.11}$$

whence

$$V_n \sim n^{2-\alpha} . \tag{3.12}$$

Contact with the correlation function is made through the second moment $\langle R_n^2 \rangle$:

$$K^{-2} = \sum \langle R_n^2 \rangle (\beta J)^n C_n / \sum (\beta J)^n C_n$$

$$= \sum n(\beta ZJ)^n / \sum (\beta ZJ)^n = \epsilon^{-1} . \tag{3.13}$$

More generally $K^{-2} = \epsilon^{-2\nu}$ so that

$$\langle R_n^2 \rangle = n^{2\nu} . \tag{3.14}$$

If in fact the V_n of expression (3.12) is equal to $\langle R_n^2 \rangle^{3/2}$ (as it is in the present case), we get

$$V_n = n^{2-\alpha} = \langle R_n^2 \rangle^{3/2} = n^{3\nu} ,$$

$$2 - \alpha = 3\nu \text{ (RPA)} . \tag{3.15}$$

We now record Domb's numerical results for the SAW:

$$C_n = \mu^n n^{1/6} , \quad U_n = \mu^n / n^{7/4} , \quad \langle R_n^2 \rangle = n^{6/5} , \tag{3.16}$$

where $\mu \simeq Z - 2$ for the f.c.c. ($\mu = 10.035$). We then get $kT_c = \mu J$ or a reduction from the Weiss value of $O(2/Z)$. This value approximates closely to the true one; the error is $\sim 2\%$. We see that the SAW is quite successful in catching the essentials, though the critical indices are not quite as good. For the SAW, $\alpha = 1/4$, $\gamma = 7/6$, $\nu = 3/5$; for the complete Ising model, $\alpha \simeq 1/8$, $\gamma = 5/4$, $\nu = 0.638$. Length scaling ($2 - \alpha = 3\nu$) fails to the same extent in both ($\sim 2\%$). In addition, Domb has calculated Δ for the SAW and verified that equation (2.10) holds in this approximation. Thus thermodynamic scaling holds in SAW, and one may calculate δ from equation (2.13) to find $\delta_{SAW} = 5$ just as $\delta_{true} = 5$.

To understand why scaling arises [18] in SAW we first interpret the results (3.16). The factor $\mu < Z$ is a short-range effect due to local exclusion of sites (in fact, $\mu < Z - 1$ is obvious). The factor $n^{1/6}$ in C_n is more subtle. It arises because of the expanded surface of the SAW as compared to the free walk. A typical endpoint rides around the outside of the cluster, being excluded from the middle. This blows out the cluster somewhat, and the increased phase space is reflected in the correction factor $n^{1/6}$. We are then led to conjecture

$$\frac{(C_n)_{SAW}}{(C_n)_{free}} = \left(\frac{\mu}{Z}\right)^n \frac{(S_n)_{SAW}}{(S_n)_{free}} = \left(\frac{\mu}{Z}\right)^n \frac{(S_n)_{SAW}}{n} , \tag{3.17}$$

where S_n is some measure of the surface. From equation (3.3) we get

$$S_n \sim n^{\gamma} . \tag{3.18}$$

The factor $n^{-7/4}$ in U_n comes from arguments related to expressions (3.9)-(3.12), namely, U_n is proportional to the density in the center of the cluster $\sim V_n^{-1}$. For the free walk we were able to "measure" this volume in terms of $\langle R_n^2 \rangle$, but in general there is no reason to expect the exact relation $V_n = \langle R_n^2 \rangle^{3/2}$. However, wo do expect that (1) this is a good approximation, (2) $(V_n)_{SAW} > (V_n)_{free}$. Point 2 is obvious ($n^{7/4} > n^{3/2}$). Point 1 is tested by Kadanoff's scaling relation (equation [3.15], $7/4 \cong 9/5$). We see that for length scaling to be exact, $P(R_n)$, the probability of an endpoint at R_n, would have to be a one-parameter function (a single length parameter).

To understand thermodynamic scaling we turn to equation (3.1). In the SAW, the coefficient of $(\beta J)^n$ in A_4 is the number of ways to lay down two intersecting chains of length n_1, $n - n_1$ which have one intersection. A_6 contains three intersecting chains of length n_1, n_2, $n - n_1 - n_2$, and two intersections, etc. Again as first orientation we drop the exclusion proviso. Then A_4 is the number of ways to lay down the two chains times the number of intersections times the number of splits of n into n_1 and $n - n_1$ $= Z^{n_1} Z^{n-n_1} \times n^2 \times n = Z^n n^3$.

$$(A_4)_{free} = \sum (\beta Z J)^n \, n^3 \sim \epsilon^{-4} \sim \epsilon^{-\gamma} \epsilon^{-2\Delta} \quad (2\Delta = 3). \tag{3.19}$$

$(A_6)_{free}$ is calculated from three chains with two intersections to yield a factor $Z^n n^6$. $(A_{2r})_{free} \sim Z^n n^{3(r-1)}$. Thus

$$F_{free} = \epsilon^{3/2} + \epsilon^{-1} \mathcal{H}^2 \sum_{r=0} \alpha_r \sum_{n=0}^{\infty}{}' (\beta Z J)^n \, n^{3r} , \tag{3.20}$$

where α_r is a set of numerical coefficients and we have used $F(\epsilon, \mathcal{H} = 0) = \epsilon^{2-\alpha}$ with $\alpha_{free} = \frac{1}{2}$. One sees that equation (3.20) cannot be reduced to the form (2.9) and the "free" walk approximation cannot scale correctly (actually for infinite-range forces, the term in $\epsilon^{3/2}$ vanishes and is replaced by ϵ^2; the above calculation is then the Weiss approximation and becomes rigorous. Scaling is then reinstated).

We now calculate with exclusions beginning with A_4. The number of ways to lay down two chains is $\sim \mu^n (n^{1/6})^2$. The number of splits is n. The number of intersection is $n^2 \times$ probability of intersection. This last factor occurs because spins away from the surface are blocked out and are less likely to occur as intersection points. To estimate this effect one notes that $U_n (= n^{-7/4})$ already has such a blocking factor in it; namely, the suppression factor $n^{-1/4}$ as compared to the free walk factor $n^{-3/2}$ is precisely due to the blocking out of the origin by surrounding spins. We may then conjecture that this self-blocking factor also can be used for blocking between spins on different clusters and try for the number of intersections $\equiv I_n$ through $I_n \sim n^2/n^{1/4}$. Then

$$(A_4)_{SAW} \simeq \sum (\beta \mu J)^n \, n^3 (n^{1/6})^2 / n^{1/4} = \epsilon^{-\gamma - 2\Delta} ,$$

$$2\Delta = 2 + 7/6 - 1/4 = 2 + \gamma - \alpha , \tag{3.21}$$

and equation (2.10) is verified. This estimate has been confirmed by direct calculation.

A_{2r} follows the same pattern: r chains, r - 1 intersections give a coefficient of $(\beta\mu J)^n$ equal to $n^{3(r-1)}(n^{1/6})^r/(n^{1/4})^{r-1}$. Thus $A_{2r} \sim \epsilon^{-\gamma}\epsilon^{-2r\Delta}$ and

$$F_{SAW} = \epsilon^{7/4} + \epsilon^{7/6}\mathcal{K}^2\mathfrak{F}(\mathcal{K}^2/\epsilon^{3\frac{1}{12}}) \tag{3.22}$$

which can now be written in the form (2.9) by virtue of equation (3.21).

An alternative, equally instructive argument due to Mackenzie [19] is to observe that I_n is precisely the number that occurs in the calculation of the two-body virial coefficient of two polymers of length n where the only interaction is the proviso of exclusion. If the polymers are tightly coiled up, then they act as hard cores and the virial coefficient will be a measure of the volume V_n. Thus $I_n \sim V_n$. For general r we then get

$$A_{2r} \sim \sum (\beta\mu J)^n (n^{1/6})^r n^{r-1} V_n^{r-1} . \tag{3.23}$$

However, we have argued that $U_n \sim \mu^n/V_n$, thereby identifying V_n with $n^{2-\alpha}$. Since $n^{7/6} = n^{\gamma}$, equation (3.23) becomes

$$A_{2r} = \sum (\beta\mu J)^n n^{\gamma} n^{(2-\alpha+\gamma)(r-1)} = \epsilon^{-\gamma}\epsilon^{-2(r-1)\Delta} , \tag{3.24}$$

where $2\Delta = 2 - \alpha - \gamma$, and thermodynamic scaling follows. We will see below that this argument is at the crux of the affair and is based on thermodynamic consistency.

We now try to generalize to all graphs. Assume C_n known and consider A_4. This is the sum of all graphs connecting four points and then summed freely on these points. Of all these graphs we focus attention on those with two strands connecting, say, (12) and (34), respectively, containing n_1 and $(n - n_1)$ spins. The strands must intersect since the graph is connected, and we consider those graphs in which the region of intersection is far from the ends. Such a graph can be calculated on exact analogy to the SAW but with two free parameters, one to describe the chain length γ and the other to describe the intersection α. Before doing so, we compare the chosen class to other classes. Suppose, for example, there were a strand connecting the ends 1 and 3. Then the ratio of the two graphs in question is

$$\sim \int g(R_1)g(R_3)g(|R_1 - R_3|)d^3R_1 d^3R^3 \Big/ \int g(R_1)g(R_3)d^3R_1 d^3R^3$$

$$\sim K^{1+\eta} = \epsilon^{(1+\eta)\nu} . \tag{3.25}$$

The more connected graph is asymptotically negligible. Clearly the more the ends are tied together the less will be the divergence. This is almost certainly the graphical reason for the scaling phenomenon, and we feel that the heuristic argument presented above should provide the basis of a complete proof. Of course, the basic difficulty is to prove that the most divergent graphs are not outnumbered significantly as $\epsilon \to 0$.

In summary: the most divergent part of A_{2r} is that which is minimally connected, i.e., the trees containing r strands and r - 1 intersections.[2] In accordance with our considerations in SAW, we call the intersection factor V_n and the number of ways to lay down a strand (S_n/n).

---

[2] This argument is not valid. Thus in A_6 one may have two separate intersections each of two strands or one intersection of three strands together. The latter is expressible in terms of a three-body virial C_3. It is easy to show that if $C_3 \sim V_3^2$, as one expects on dimensional grounds, then these graphs scale as well.

We expect V_n to be a measure of volume and S_n to be a measure of surface. Then

$$A_{2r} \sim \sum \frac{(\beta\mu J)^n}{n} \quad S_n \; (S_n V_n)^{r-1} \quad (r = 1, 2, \ldots) \;, \tag{3.26}$$

where we have used the number of splits $\sim n^{r-1}$. Calling $V_n = n^l$ and using $A_2 = \chi$ to identify S_n with n^γ, we get

$$A_{2r} \sim \epsilon^{-\gamma} \, \epsilon^{-2(r-1)\Delta} \,,$$

$$2\,\Delta = \gamma + l \;. \tag{3.27}$$

F is given by

$$F \sim \epsilon^{2-\alpha} + \sum A_{2r}(\epsilon)\, \mathcal{H}^{2r}$$

$$\sim \epsilon^{2-\alpha} + \epsilon^{-\gamma}\mathcal{H}^2 \; g\;(\mathcal{H}^2/\epsilon^{2\Delta}). \tag{3.28}$$

We now require thermodynamic consistency for this maximally divergent portion of F [20]. Thus as $\epsilon \to 0$, $\mathcal{H} \neq 0$, equation (3.28) must have a finite limit.[3] Therefore, $g(x) \to x^{-1}$ and $2\Delta = 2 - \alpha + \gamma$. Referring to equation (3.27) gives us $l = 2 - \alpha$, and thermodynamic scaling is established. This double identification of V_n gives us confidence in the interpretation of V_n as a geometric volume, and we would be very surprised if V_n should differ radically from $\langle R_n^2\rangle^{3/2}$. By the same token there is no compelling reason for an exact equality of these two quantities.

We then conclude

$2 - \alpha \approx 3\nu$ (very good approximation)

$F = \epsilon^{2-\alpha} f(\mathcal{H}^2/\epsilon^{2\Delta})$ (presumably exact limiting behavior).

We remind the reader that we have always held $\epsilon > 0$ and we are still at the mercy of the smoothness assumption to go over to the limit $\mathcal{H} \to 0$, $\epsilon \to 0^-$.

4. $\delta = 5$

We have seen that length scaling ($V_n \sim \langle R_n^2\rangle^{3/2}$ or equivalently $2 - \alpha = 3\nu$) is approximately true to about 2%. To the same extent the arguments of expressions (3.17) and (3.18) suggest $S_n \sim \langle R_n^2\rangle$ (i.e., $\gamma = 2\nu$). In this manner one explains why η is small, but again there is no compelling reason for it to vanish [21].

However, something astonishing seems to occur in the Ising model [8] (also the SAW [9], and quite possibly the Heisenberg model); namely, though $V_n \sim \langle R_n^2\rangle^{3/2}$ and $S_n \sim \langle R_n^2\rangle$ are both merely good approximations, it appears that $V_n \sim S_n^{3/2}$ is exact, i.e., $2 - \alpha = 3/2\,\gamma$. Recall that

$$\delta = (2 - \alpha + \gamma)/(2 - \alpha - \gamma) \;, \tag{4.1}$$

so that one gets a model independent value of $\delta (= 5)$ on dimensional grounds alone. This is perfectly consistent with Gaunt's best estimates which he holds to be uncertain to $< 1\%$. The

---

[3] This point has been criticized in the discussion by Professor R. Griffiths. See the discussion remarks at the end of this paper.

errors in length scaling are $\sim 2\%$, and we would have expected a corresponding error in δ. We conclude that if $\delta = 5$, we are missing an essential element in our argument.

One may appreciate more fully this point by returning to the form (2.14). Direct comparison of this form with equation (3.1) yields:

$$F = F_o(\epsilon) + \sum B_{2r} R^{2r} - R\mathcal{H},$$

$$B_2 = \tfrac{1}{2}\chi^{-1},$$

$$B_4 = (1/4)\, A_4/\chi^4,$$

$$B_6 = \frac{1}{6}[A_6/\chi^6 - 3A_4^2/\chi^7]. \qquad (4.2)$$

We now write the coefficient of $(\beta\mu J)^n$ which appears in the various series for the numerators and denominators of B_{2r} $[A_{2r} \sim S_n(S_n V_n)^{r-1}]$:

$$B_2 \sim 1/S_n,$$

$$B_4 \sim V_n S_n^2/S_n^4,$$

$$B_6 \sim V_n^2 S_n^3/S_n^6. \qquad (4.3)$$

To find out the behavior of B_{2r} one may formally cancel common terms in numerator and denominator, and one of course finds $B_{2r}/B_{2r-2} \sim V_n/S_n \sim n^{2\beta}$ or $\epsilon^{-2\beta}$ with $2\beta = 2 - \alpha - \gamma$. Observe that whereas B_2, $B_4 \to 0$ as $\epsilon \to 0$, $B_6 \to V_n^2/S_n^3 \to$ constant if $V_n \sim S_n^{3/2}$. Thus if the exact dimensional relationship is maintained, the free energy appears analytic in R at $\epsilon = 0$ (only near $R = 0$):

$$F \sim R^6 - R\mathcal{H}.$$

Our reasoning would seem model independent, and one would expect $\delta = 5$ to be true for all saturating forces in three dimensions. Calculations indicate this to be true not only for the Ising model and the SAW but also for the Heisenberg model and dilute ferromagnets [22], though only the Ising model and the SAW are very precise in this regard. There is also experimental information on the Heisenberg model [23]. Both calculation and experiment seem to agree that $\gamma = 1.4$ (Als-Nielsen quotes a provisional value of 1.38 before a complete error analysis). There are also indications that $\alpha \simeq -0.1$. The first is calculational [22], and the second is Als-Nielsen's result [23] $\nu = 0.7$. If we use $2 - \alpha = 3\nu$, this gives $\alpha = -0.1$. It is absolutely remarkable that if $\gamma = 1.40$ and $\alpha = -0.10$, one gets $\delta = 5$. Thus magnetic systems are certainly pushing us toward the acceptance of $\delta = 5$ as an exact result and hence $V_n \sim S_n^{3/2}$.

On the other hand, the value of δ in fluids (experiments on Xe and CO_2) are internally consistent and are consistent with $\delta = 4.4 \pm 0.4$. If we assume $\gamma = 5/4$ for these systems, then thermodynamic scaling would yield $\alpha \simeq 0$ which is consistent with experiment. We are then inclined at the present stage to accept a difference between the spin and fluid systems, and if this state of affairs persists, some explanation must be forthcoming. The first thing that comes to mind, of course, is that the apparent analyticity in R and not in $[\rho_{liq} - \rho_{gas}]$ is tied up with the up-down symmetry in spin systems, absent in fluids [24].

We close then with three questions; all within reach and all nontrivial. (1) $\delta = 5$: exact or approximate? distinction between fluids and magnetic systems? (2) lim $\epsilon = 0^-$, $\mathcal{H} = 0$: are Stell's analytic horrors real or will it turn out nice and simply with $\gamma = \gamma'$? (3) Will it be possible to solve the Ising model from the geometric information we now have on SAW? I think so.

It is a pleasure to express my gratitude to Professors Domb and Sykes and other members of the King's College group for stimulating exchanges on all aspects of this work.[4]

REFERENCES

1. R. Brout, Phys. Letters, 34A, 115 (1971).

2. C. Domb and D. L. Hunter, Proc. Phys. Soc. 86, 1147 (1965).

3. B. Widom, J. Chem. Phys. 43, 3898 (1965).

4. M. Fisher, J. Appl. Phys. 38, 981 (1967); Proc. U. of Kentucky Conf. on Phase Transitions (1965).

5. L. Kadanoff, Physics, 2, 263 (1966).

6. L. Kadanoff, W. Gotze, D. Hamblen, R. Hecht, E. Lewis, V. Palciauskas, M. Rayl, J. Swift, D. Aspnes, and J. Kane, Rev. Mod. Phys. 39, 395 (1967).

7. A. Z. Patashinskii and V. L. Pokrovskii, Zh. Exp. i Theor. Fiz. 50, 439 (1966); JETP, 23, 292 (1966).

8. D. S. Gaunt, private communication.

9. C. Domb, J. Phys. C3, 256 (1970).

10. M. Fisher, Prog. Repts. in Physics (1967), p. 615.

---

[4] In the meantime, I have examined the problem of scaling using Migdal's field theoretic formalism, and I have proved the relation

$$A_{2n} = (2n-1)\,\epsilon^{-2\Delta}\,\epsilon\,\frac{\partial}{\partial\epsilon}\,A_{2n-2}$$

for the most divergent graphs. This class has the external legs of A_{2n} passing into the interaction region through an infinite series of vacuum polarization loops. The neglected class is smaller by $O(\epsilon^{\gamma+\alpha-1})$. The recurrence relation then leads to $A_{2n} = \frac{(2n)!}{2^n n!}\left[\epsilon^{-2\Delta}\,\epsilon\,\frac{\partial}{\partial\epsilon}\right]^n A_o$. Hence one finds an explicit form for F which has the correct properties of analytic continuation to $\epsilon = 0$:

$$F = \sum_{n=0}^{\infty} \frac{1}{2^n\, n!}\left[(\mathcal{H}^2/\epsilon^{2\Delta})(\epsilon\,\frac{\partial}{\partial\epsilon})\right]^n \epsilon^{2-\alpha}$$

$$= \epsilon^{2-\alpha}\left[1+\mathcal{H}^2/\epsilon^{2\Delta}\right]^{\frac{2-\alpha}{2\Delta}}.$$

It indeed does appear that thermodynamic scaling is a rigorous result, though there are still loose ends to be tidied up in making rigorous demonstrations. One must now investigate how this function continues to $\epsilon < 0$.

This work has been submitted to Physica for publication.

11. D. Jasnow and M. Wortis, Phys. Rev. 176, 739 (1968).

12. M. F. Sykes, private communication.

13. L. Guttman, private communication.

14. G. Stell (Orsay preprint March 1968, Th/68/11).

15. R. Brout, Phys. Rev. 115, 824 (1959); 118, 1009 (1960).

16. M. Coopersmith and R. Brout, Phys. Rev. 130, 2539 (1963).

17. G. Horwitz and H. Callen, Phys. Rev. 124, 1757 (1961); F. Englert, Phys. Rev. 129, 567 (1963).

18. We follow R. Brout, Physica, 50, 149 (1970).

19. Quoted by Domb (reference [9]).

20. It is Professor D. J. Thouless who called to my attention this use of thermodynamic consistency. Professor J. L. Lebowitz has called my attention to a proof of this point (J. L. Lebowitz and O. Penrose, Comm. Math. Phys. 11, 99 [1968]).

21. We may mention that in two dimensions this is a rather poor approximation (at the level of refinement which is being discussed here). The conjecture that $S_n \sim \langle R_n^2 \rangle^{1/2}$ along with $C_n \sim S_n / \langle R_n^2 \rangle_{\text{free}}$ yields $\gamma = 3/2$ whereas $\gamma_{\text{exact}} = 7/4$. All geometric reasoning in two dimensions is very suspect since the clusters are so open $\langle R_n^2 \rangle_{\text{exact}} = n^2$ and $\langle R_n^2 \rangle_{\text{SAW}} = n^{3/2}$. We note also that the SAW and exact results are very far removed. It is the compactness of the clusters in three dimensions that makes the analysis amenable to geometric reasoning.

22. C. Domb, private communication.

23. Dr. Als-Nielsen, private communication.

24. Professor Resibois has told me of an article by Migdal which is now circulating as a preprint in English. He believes that this work can shed light on this problem.

DISCUSSION

R. B. Griffiths: It seems to me that the following problem arises in using analyticity in ϵ for $\mathcal{H} > 0$. You have looked at the series expansion in $\mathcal{H}^2$ for $\epsilon > 0$, and the most divergent term in each coefficient as $\epsilon \to 0$. However, if we fix $\mathcal{H} > 0$, and let $\epsilon \to 0$, the free energy could be determined by the "less divergent" terms you have neglected. It seems to me, therefore, that to get from the vicinity of $\mathcal{H} = 0$, $\epsilon > 0$ to the critical isotherm, you have had to include scaling as one of your assumptions.

M. E. Fisher: Perhaps Professor Brout would let me respond. The ideas Brout has outlined can be interpreted as a program for proving "scaling" or, more precisely, aspects of scaling. If one wishes to be mathematically precise, one must specify the exact sense or senses in which the scaling function is supposed to approximate the true free energy. For example, one might merely require that the divergence as $T \to T_c$ of all the field derivatives evaluated at $\mathcal{H} = 0$ be given correctly. (This would be "gap exponent scaling," i.e., $\Delta_k = \Delta$ for all k.) With only this limited objective in mind Professor Griffiths' objections do not apply. However, as he correctly emphasizes, if one wishes to take the further step in the argument and relate Δ to α (as indicated in the lecture), it is essential to establish some uniformity for the smallness of the sum of the neglected diagrams. While this might prove possible by studying all the excluded-volume diagrams, it is certainly a much harder step than the first one suggested above, and I, personally, do not feel hopeful at this point. Generally, however, it is worth reemphasizing the point that there are very many ways in which "scaling" may be correct or may partially fail. In trying to construct proofs one will have to enter into the detailed criteria.

R. Brout: Professor Griffiths has raised a deep and interesting point which I can only hope to answer in speculative fashion. I will try to indicate why I believe it likely that the less divergent terms in fact do not affect the nature of the critical isotherm in the vicinity of $\mathcal{H} = 0$;

more particularly, the value of δ will not change due to these corrections.

What I have done is to choose fixed ϵ finite and positive and taken H sufficiently small so that the series in it (hopefully) converges. As $\epsilon \to 0$, this radius of convergence shrinks to nothingness and the only avenue open to the critical isotherm is through analytic continuation (as an example, think of an oscillating geometric series in $\mathcal{H}^2/\epsilon^{2\Delta}$). It is in this continued form that I appeal to analyticity at $\epsilon = 0$ and $\mathcal{H} \neq 0$ (but so small that corrections to $R = \mathcal{H}^{1/\delta}$ are negligible). In particular, the most divergent portion of the Cayley trees is continued to a form $\epsilon^{-\gamma}\mathcal{H}^2 \mathfrak{F}(\mathcal{H}^2/\epsilon^{2\Delta})$, which I use to cancel the most divergent contributions from closed graphs ($\epsilon^{2-\alpha}$) on the critical isotherm, through the identification $2 - \alpha = 1$. This latter has been made plausible through the volume-density identifications presented in my paper. Griffiths questions the mathematical legitimacy of this procedure, not necessarily the physical picture of equating the volumes.

Consider now less divergent terms, either corrections to trees or other classes, for example trees with a bridge across two ends. Most of the graph will still be in the form of a tree, but each one of them will be multiplied by a certain power of ϵ. Thus a given correction is expected to sum to a form $\epsilon^p \epsilon^{-\gamma} \mathcal{H}^2 \mathfrak{F}(\mathcal{H}^2/\epsilon^{2\Delta})$ ($p > 0$, and we take the nontrivial case of p noninteger). It is then impossible that such a form cancels the singularities at $\mathcal{H} \neq 0$, both in most divergent closed graphs, $\epsilon^{2-\alpha}$, and in the Cayley trees $\epsilon^{-\gamma}\mathcal{H}^2\mathfrak{F}(\mathcal{H}^2/\epsilon^{2\Delta})$. The only possibility is that there is a less divergent contribution to the closed graphs $= \epsilon^{2-\alpha+p}$. I am therefore suggesting that analyticity on the critical isotherm is maintained through class by class cancellation, where a class is characterized by its singularity in ϵ.

Now in the above case, once the cancellation is indeed affected on the critical isotherm, one is left with a form $\epsilon^{2-\gamma+p}\mathfrak{F}_p(\mathcal{H}^2/\epsilon^{2\Delta})$. Analyticity then further requires that $\mathfrak{F}_p(x) \xrightarrow[x \to \infty]{} x^{(2-\alpha+p)/2\Delta}$ and hence a term in the free energy $F = \mathcal{H}^{(2-\alpha+p)/\Delta}$ ($\epsilon = 0$). Such a term does not contribute to the leading behavior at small $\mathcal{H}$ which is $\mathcal{H}^{(2-\alpha)/\Delta}$. The above then suggests that the neighborhood of $\mathcal{H} = 0$ is indeed properly described by the most divergent contributions.

I again emphasize that the main mathematical point to prove in all of this is that the Cayley trees are indeed the dominant set in a certain domain $\epsilon > 0$, $\mathcal{H} > 0$. I believe that Professor Griffiths' point can then be countered in the manner given above. Added note: There appears to be some misunderstanding about what I mean by a tree. A tree is composed of chains connecting fixed points with a minimum number of intersections (e.g., in A_{2r} there are r chains and (r - 1) intersections). A chain is the totality of all graphs connecting two points. (An intersection has no articulation points; a chain has an asymptotically large number of them in the critical region.)

One more point is that if the above is correct, it predicts deviations from scaling like $O(\epsilon^p)$ for $\mathcal{H} = 0$ and $O(\mathcal{H}^p/\Delta)$ for $\epsilon = 0$. Since p comes from bridge-type corrections, we might expect $\epsilon^p \sim K^{-1} \sim \epsilon^{\nu}$ or $p = \nu$. The result seems to be consistent with experiment

(T. Schneider, private communication).

R. Thom: The scaling hypothesis amounts to saying that around the critical point all thermodynamical functions are analytic functions not in the temperature $T - T_c$, but in some rational desingularizing parameter $(T - T_c)^{1/\nu}$. What can be said a priori (before looking at molecular mechanisms) of the desingularizing exponent ν, if we require the differentiability of thermodynamical functions up to second or third order? Has somebody looked at this problem?

M. Coopersmith: (1) I have examined various specific algebraic functions (which are the functions for rational ν) in M. Coopersmith, Phys. Rev. (in press), and M. Coopersmith, Adv. Chem. Phys. **17**, 43 (1970). The results are given in terms of the topology of the function as exemplified by the behavior of the Riemann surface.

I have proved the following theorem: If S = S(E) has a continuous second derivation and if the free energy F(J, E) is analytic on the real axes except for the phase boundary, then F cannot be single-valued.

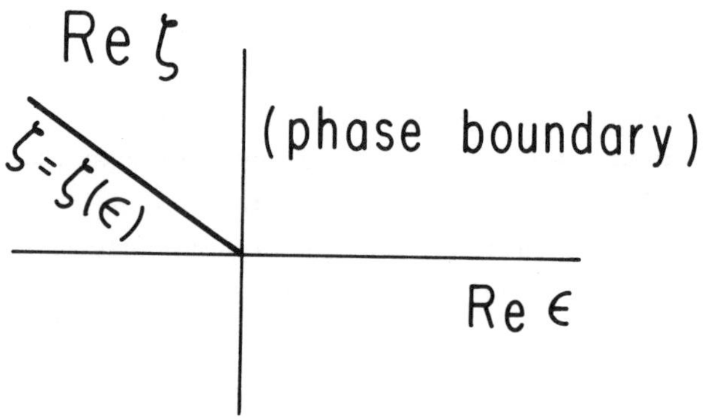

(2) The analytic continuation of your free energy, defined above the critical point by $F = \sum_{R=0}^{\infty} A_{2R}(E)\mathcal{H}^{2R}$ can be obtained by guessing at the multiple-valued function. For $\delta = 5$,

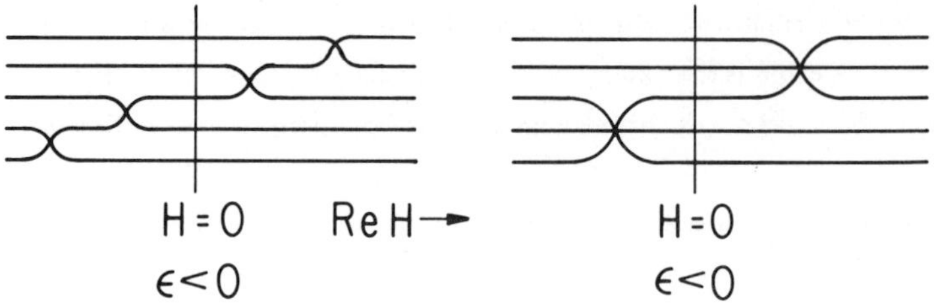

there are two possibilities for the Riemann surface. The only catch is that the coefficients, α_{2R}, be known.

M. N. Barber: For $\alpha > 3$, it seems likely that clusters would be more spherical. Hence the identification $V_r = \langle R_n^2 \rangle^{d/2}$ should be better, leading to $2 - \alpha = d\nu$, which with conventional definition of α does not hold. Have you considered the extension to higher dimensionality and the breakdown of the d-dependent scaling?

B. Widom: If α refers not to the leading part of C_v, but to the singular part, so that it can become negative, then $2 - \alpha = d\nu$ may continue to hold for higher d. Something like this is found by K. Wilson in his renormalization theory.

R. Brout: It is clear in all of this that I am dealing with an asymptotic characterization of the series in question. Thus those terms coming from early orders of perturbation theory are not within my scheme (nor within scaling a posteriori). For example, when I write the energy as $E \sim \epsilon^{1-\alpha}$, this refers to its singular part alone. The numerically more important contribution to E comes from the first Z terms of its expansion in βJ. (Z = coordination number) and is left out of my considerations. Similarly in higher dimensions, C_v is regular in its most dominant part, but this is omitted. These considerations explain Widom's answer.

H. E. Stanley: Additional support to Brout's arguments arises from calculations for spin systems for which the dimensionality of the <u>spin</u> space (D) is greater than 1. Specifically, for the spherical model ($D = \infty$) we have $\delta = 5$ rigorously, and for $1 < D < \infty$ numerical calculations suggest $\delta \cong 5$ (though the error bounds are not nearly as small as they are for $D = 1$). Of course, if one believes (as many workers do) that all critical point exponents are <u>monotonic</u> functions of D, then $\delta(D = 1) = \delta(D = \infty) = 5$ implies that $\delta(D) = 5$ for <u>all</u> D.

J. Percus: There is some ancient work of George Stell and myself which touches upon several of these points. We made a functional expansion of the density evoked by fixing a particle and found that it could be summed completely by assuming that correlations at the critical point are sufficiently long range. Although we thought we were dealing with a fluid, it turns out that the argument for the expansion is far stronger when the three-point correlation function vanishes identically at critical, which is precisely the case for the Ising model and not at all for a fluid. Our result was: the direct correlation function went asymptotically as the δ power of the indirect correlation. When we attempted to solve this, coupled with the Ornstein-Zernike relation, it behaved analytically and numerically like a nonlinear eigenvalue problem, with $\delta = 5$. Furthermore, there was no correlation exponent in the usual sense, but rather a $1/r\,(\ln r)^{1/6}$ dependence for h(r). This means that a measured exponent will appear as $\eta = 1/6 \ln r$, sinking to zero only for very large r. It also suggests that a simple branch-point behavior for thermodynamic quantities may not be mirrored in the spatial dependence of correlations.

THE STEADY STATE FAR FROM EQUILIBRIUM PHASE CHANGES AND ENTROPY OF FLUCTUATIONS

Rolf Landauer and James W. F. Woo

I. INTRODUCTION

The last decade has seen the increasingly detailed development of a number of analogies between the distribution functions for some highly excited, dissipative steady-state systems on the one hand, and the distribution functions for thermal equilibrium systems on the other hand. Analogies in this general vein are not entirely new. Closely related concepts are in fact basic to much of the theory of nonlinear oscillations [1, 2], which stress the similarities in stability and instability, between active systems on the one hand and particles moving in potentials on the other hand. At this level, however, where fluctuations are not taken into account, we cannot differentiate between stability and metastability. A particle trapped at the bottom of a metastable potential well will just stay there. Thermal fluctuations are needed to permit a jump over the barrier and relaxation to the state of ultimate stability. Recognition of analogies including fluctuations has also been developed within the statistical mechanics community, as recorded in a symposium [3] held in Chicago in 1965. In the present discussion we want to review and expand upon a series of particularly detailed analogies which have emerged from the consideration of electronic and quantum electronic systems. These are analogies which apply to the distribution function but which go beyond that and consider changes in the distribution function of the nonequilibrium system as some parameter is varied. The nonequilibrium system is then seen to have phenomena which are entirely analogous to first- and second-order phase transitions, and in the latter case are accompanied by soft modes and critical fluctuations. The analogy is rooted in the fact that the analogous systems have identical Fokker-Planck or master equations for the time development of their distribution functions. We shall, however, point out subsequently that the analogies seem to go beyond this purely statistical mechanics base, and can in a sense be given a thermodynamic significance.

The analogies we shall discuss seem to have a number of independent origins. First of all, we should mention the pioneering two-volume treatise of Stratonovich [4], to which the physics community, including the present authors, seems to have awakened belatedly. Stratonovich discusses a great many systems of electronic interest, and has also provided the basis for subsequent applications to control theory [5]. Another early version of the analogy was a discussion of the distribution function for some tunnel diode circuits [6]. This has subsequently been elaborated [7,8]. A completely separate source for these analogies is the quantum electronics community, which has shown a growing appreciation of the similarity between a laser passing through its threshold and a ferromagnet treated via the molecular field theory. An early version of such a discussion was given by Martin [9], for example. A much more detailed treatment, providing the basis for much of our subsequent discussion, has since been supplied by Graham and Haken [10], Haken [11], and DeGiorgio and Scully [12]. We shall not attempt to trace out in detail the relationship of these analogies to work which has been more in the mainstream of statistical mechanics, such as the work by M. S. Green [13] on the evolution of closed systems toward equilibrium. An excellent discussion of these relationships has been provided by Graham and Haken [14]. Instead, we shall in the subsequent discussion concentrate on the degenerate parametric oscillator as a sample vehicle, since it most clearly emphasizes the common denominator of the various authors who have contributed to this field.

We wish to stress here that the present paper concentrates on the steady state, or states very close to the steady state. More complex time-dependent phenomena discussing the relaxation to a distant steady state have been treated in the same spirit [6,7,8], but will not be reviewed here.

II. PARAMETRIC OSCILLATORS

Since this paper is addressed to an audience centered in statistical mechanics, a brief reminder about the behavior of parametric systems may be in order. The parametrically excited oscillator, in its simplest version, can be considered to be a harmonic oscillator, say an LC circuit, with a periodically time-modulated degree of freedom, say a capacitance C(t). This leads to the equation $d^2q/dt^2 + \omega^2(t)q = 0$, with $\omega^2 = 1/LC$. This is Hill's equation and, as is well known from the analysis of electrons in one-dimensional periodic potentials, can lead to exponential growth or decay of q(t). In the one-dimensional periodic Schrodinger equation the exponential behavior is connected with forbidden bands, or Bragg reflections, and this exponential growth typically is most pronounced in the lowest energy gap. This corresponds to the case where the frequency of the variation of $\omega(t)$ is twice the circuit's typical natural oscillatory frequency.

We can present the same phenomenon more physically by assuming that C(t) is varied as shown in figure 1a, say by a little man who at fixed intervals pushes capacitor plates together and pulls them apart. When he pulls them apart, he is reducing capacitance and increasing the stored energy, $q^2/2C$, doing work against the electrostatic attraction of the plates. Later, when he lets the plates come together, he is extracting energy from the circuit. If the signal q(t) is phased as shown in figure 1b, then charge is present only when he pulls the plates apart; hence

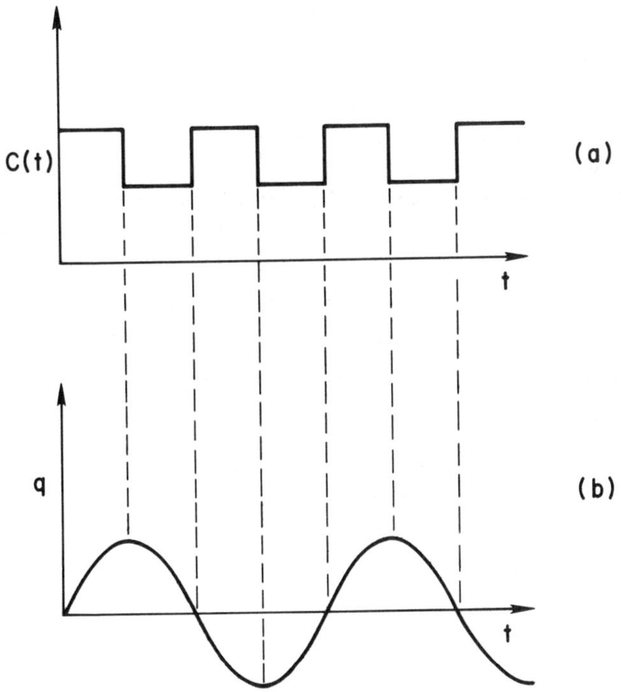

Fig. 1. (a) Capacitance variation with time, as plates are periodically brought together and apart. (b) Signal phased so that plates are maximally charged when capacitance decreases.

he is always doing work on the signal, and we can expect exponential signal growth. If the sign of the signal had been reversed, we would have come to the same conclusion; if, however, it had been shifted by 90° (at the signal frequency), we would have come to a situation in which energy is continually extracted from the circuit, and signal attenuation results. Thus we have a situation in which two phases, 180° apart, are particularly and equally favored. This bistability, recognized by von Neumann [15] in a patent filed in 1954, is the basis of the use of parametric devices in computers. The same point was also recognized independently by a Japanese group [16]. In the late fifties there were major attempts, in several laboratories, to develop these inventions. Parametric computing devices have, however, ceased to be an item of major technical interest.

Now instead of a linear time-dependent capacitance we can utilize a nonlinear capacitance. In this case, an exciting or "pump" voltage sweeps us back and forth along the nonlinear Q-V relationship, and thus modulates the effective capacitance seen by an additional perturbing signal. We can envision a circuit with two modes, one of them resonant to the externally supplied pump voltage, and the other mode resonant to half that frequency, with the two modes coupled through a common nonlinear capacitance, as illustrated in figure 2. In this case the resistance in series with the pump voltage sets a maximum for the power which can be extracted

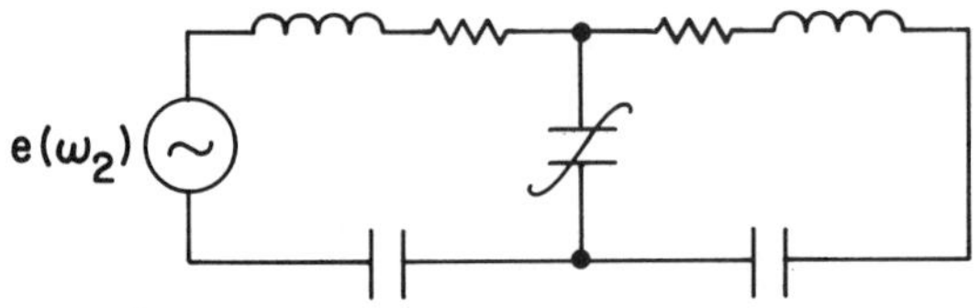

Fig. 2. Circuit with two resonant modes, with inductances and capacitances chosen so that one mode has twice the frequency of the other. External driving force is supplied only at the higher frequency. Capacitor in central branch is nonlinear.

from that voltage source. It is therefore immediately apparent that the growth of the subharmonic can only initially be exponential, but that subsequently the subharmonic amplitude cannot exceed some limit. The behavior of this system, without regard to fluctuations, has been analyzed by Lasher [17]. An analysis of a parametric oscillator in which there are no internal noise sources, but stochastic irregularities in the pump, has been given by Stratonovich (Vol. II) [4]. We shall here, instead, give an abbreviated account of the parametric oscillator with a perfect pump, and with noise sources which agree with fluctuation-dissipation theory, i.e., reside in the circuit's resistances. This is a sketch and simplification of a more elaborate treatment [18].

III. DISTRIBUTION FUNCTION FOR THE PARAMETRIC OSCILLATOR

The equations of motion for the two modes of the parametric oscillator are [18]:

$$i\dot{q}_1 = \omega q_1 - 2Tq_1^* q_2 - \frac{i\omega}{Q_1} q_1 + iL_1(t) , \tag{3.1}$$

$$i\dot{q}_2 = 2\omega q_2 - Tq_1^2 - i\frac{2\omega}{Q_2} q_2 + iL_2(t) - Ve^{-2i\omega t} . \tag{3.2}$$

The first equation represents the lower-frequency mode. Its right-hand terms, taken up in succession, are: (a) ωq_1 gives the resonant behavior of the mode, in the absence of nonlinearities, losses, and fluctuations; (b) the term involving T gives the nonlinear coupling effects to the higher-frequency mode, and represents the effect of the subharmonic beating with the pump frequency signal; (c) the term in $1/Q_1$ gives damping due to losses; and (d) the term $L_1(t)$ gives the fluctuating forces which in equilibrium balance the dissipation. Equation (3.2) describes the behavior of the pump frequency and as a final right-hand term includes a driving force. Terms which do not contribute to the two frequencies under consideration have been omitted. We have assumed (in arriving at this set of <u>first-order</u> differential equations) that the modes change amplitude slowly compared to their natural frequencies. The dissipation-fluctuation relation in this notation is

$$\langle L_i(t) L_j(t') \rangle = \frac{2}{Q_i} k\theta \delta_{ij} \delta(t - t') , \tag{3.3}$$

where θ is the temperature of the resistances and k is Boltzmann's constant.

If we now introduce a transformation to rotating coordinates, suitably normalized,

$$C_1 = \omega^{-1} T (Q_1 Q_2)^{1/2} a_1 e^{i\omega t} e^{i\pi/4} ,$$
$$C_2 = \omega^{-1} 2 Q_1 T a_2 e^{2i\omega t} , \tag{3.4}$$
$$v = \omega^{-2} Q_1 Q_2 T V ,$$

we obtain

$$\dot{C}_1 = -\frac{\omega}{Q_1} [-C_1^* C_2 + C_1] + \frac{T(Q_1 Q_2)^{1/2}}{\omega} e^{i\pi/4} L_1 e^{i\omega t} , \tag{3.5a}$$

$$\dot{C}_2 = \frac{2\omega}{Q_2} [-C_1{}^2 - C_2 + v] + \frac{2Q_1 T}{\omega} e^{2i\omega t} L_2 . \tag{3.5b}$$

Let us now take $Q_2 << Q_1$, so that mode 2 adjusts quickly to mode 1. Then in following the time variation of C_1 we can set $\dot{C}_2$, above, equal to zero and can use

$$C_2 = v - C_1{}^2 - \frac{Q_2}{2\omega} \tilde{L}_2 , \tag{3.6}$$

where $\tilde{L}_2$ represents the entire final term in equation (3.5b). This takes equation (3.5a) into the form

$$\dot{C}_1 = \frac{\omega}{Q_1} [C_1^* \; v - |C_1|^2 C_1 - C_1] \; \frac{Q_2}{2Q_1} C_1^* \tilde{L}_2 - \tilde{L}_1 , \tag{3.7}$$

where $\tilde{L}_1$ is given by the last term on the right-hand side of equation (3.5a). The first right-hand term in brackets gives the ensemble-average behavior; the remaining terms, the random or diffusive behavior. C_1 has both a real and an imaginary part. The imaginary part is driven to zero by the first right-hand term in equation (3.7), but of course does not quite stay there, since it is also driven by fluctuations. It contrasts with the real part of C_1, where the restoring forces to Re $C_1 = 0$ become weaker with growing v, and eventually, at $v = 1$, change sign. We shall therefore assume that C_1 always stays very close to the real axis. This approximation permits us to replace equation (3.7) by a one-dimensional Fokker-Planck equation, which for the steady state yields

$$\frac{\partial}{\partial C_1} [- \frac{\omega}{Q_1} (v - 1 - C_1{}^2) C_1 \rho + D_o \frac{\partial}{\partial C_1} (1 + C_1{}^2)\rho] = 0 , \tag{3.8}$$

where $D_o = Q_2 T^2 k\theta/\omega$, as the diffusion coefficient at $C_1 = 0$. Note that if we disregard the diffusive term, we find points of stability or instability when $(v - 1 - C_1{}^2)C_1$ vanishes, i.e., at $C_1 = 0$, and at $C_1 = \pm (v-1)^{1/2}$, for $v > 1$. Note also that for very large C_1 the drift velocity, $\omega Q_1^{-1} (v - 1 - C_1{}^2)C_1$, always tends to reduce the magnitude of C_1, and thus reduces the amplitude of oscillation. The amplitude of oscillation as a function of pump strength follows the curve shown in figure 3, and is very reminiscent of magnetization as a function of temperature.

Instead of continuing to belabor tha analytical details of the parametric oscillator, we shall now turn to a somewhat more general mode of discussion. Consider a stochastic variable q, which is a generalization of the variable C_1 of the preceding section. Assume that the probability distribution function ρ obeys a Fokker-Planck equation. (This is not an essential step--at least one case with a master equation which cannot be simplified to a Fokker-Planck equation has been treated [6]--but the analytical complexities become burdensome in that case.)

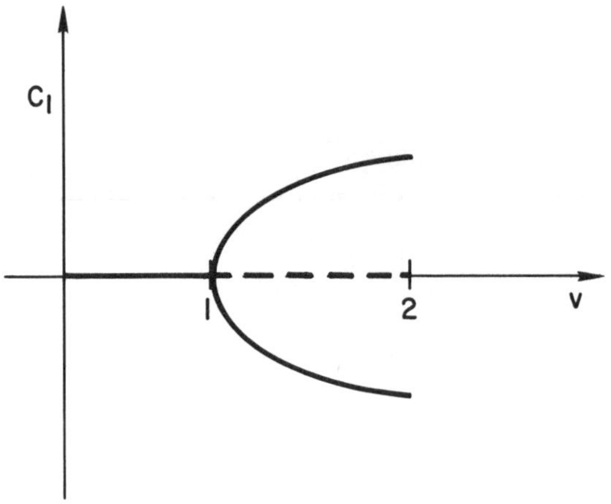

Fig. 3. Amplitude of oscillations as a function of pump strength. Below threshold ($v < 1$) the quiescent state is stable. Above threshold ($v > 1$) the quiescent state (dotted line) is a state of unstable equilibrium and the preferred states lie along the parabola.

We then have, for the probability flux in q space

$$j_q = \rho(q)\langle \dot{q} \rangle - \frac{\partial}{\partial q}(D\rho) , \tag{3.9}$$

with $\langle \dot{q} \rangle = \lim_{\Delta t \to 0} \langle \Delta q \rangle / \Delta t$ and $D = \frac{1}{2} \lim_{\Delta t \to 0} \langle \Delta q^2 \rangle / \Delta t$. In these equations Δq represents displacements away from the initial position, as discussed for example by Chandrasekhar [19]. Thus $\langle \dot{q} \rangle$ represents the average velocity away from a given point. Equation (3.9) can be rewritten

$$j_q = \rho(q) v_q - D \frac{\partial \rho}{\partial q} , \tag{3.10}$$

where $v_q = \langle \dot{q} \rangle - \partial D / \partial q$. As discussed in detail elsewhere [7], v_q measures the imbalance between left-to-right jumps and right-to-left jumps, i.e., the terms tending to create a nonuniform distribution. In the one-dimensional steady-state case, if j_q vanishes at $q = \pm \infty$, then $j_q = 0$ for all q. In the multidimensional case div $j_q = 0$ does not necessarily imply $j_q = 0$, and the extent of the applicability of the analogies in that case is still incompletely understood. (We will return to a discussion of the multidimensional case in a subsequent section.) If j_q vanishes, equation (3.10) yields

$$\rho(q) = \exp\left[\int v_q \, dq / D\right] . \tag{3.11}$$

Figure 4 shows log ρ (qualitatively) for the distribution function of equation (3.8), for various values of pumping strength. These curves are, of course, similar to the free energy near a second-order transition, with q as the order parameter, e.g., magnetization in a uniaxial magnet. We can instead draw the analogy to the distribution function for a modulated potential well going from a narrow monostable state to a deeply bistable state. In the latter case $\rho(q) = \exp(-U_{eff}/kT)$, where U_{eff} is the well energy. The potential U_{eff}, however, can be rewritten as an integral of

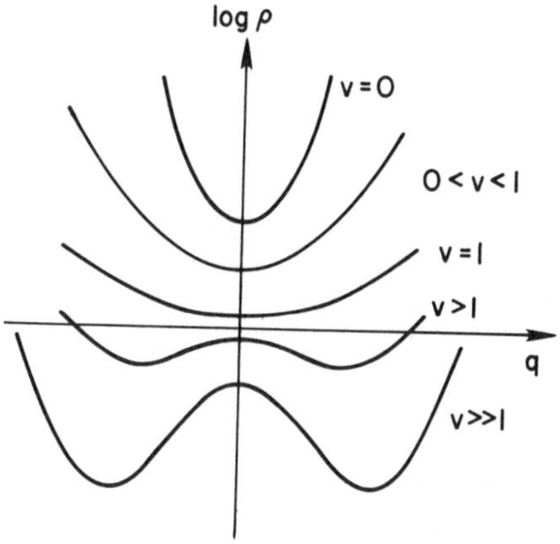

Fig. 4. Log of distribution function for parametric oscillator, shown (qualitatively) for various pumping strengths v. The vertical placement of the curves has no significance, and has been chosen so as to keep the curves from intersecting and obscuring each other. The curve $v = 1$ represents the transition from monostability to bistability, and has no quadratic term at $q = C_1 = 0$.

the force, hence $\rho(q) = \exp(-\int \nabla U_{eff} dq/kT)$. Multiplying numerator and denominator by the mobility μ associated with the well, we find $\rho(q) = \exp(-\int \mu \nabla U_{eff} dq/\mu kT) = \exp(\int (v/D) dq)$, where we have used the Einstein relation, $D = \mu kT$, and also $v = -\mu \nabla U_{eff}$. This is, however, identical to the expression given by equation (3.11).

The analogy also holds in the presence of an external field. For the ferromagnet in an externally applied magnetic field, the free energy curves of figure 4 are tilted so that in the bistable case one well is favored over the other. The same thing happens in the parametric oscillator when an external signal at the subharmonic frequency ω and with one of the favored phases is applied. This has been discussed in detail by DeGiorgio and Scully [12] for the laser, and the parametric oscillator is entirely analogous. A theme which runs through some of the cited papers [7,8,15] concerns the responsiveness of a system, near its transition, to external influences. Thus as we pass through threshold ($v = 1$), the parametric oscillator phase is more easily influenced by external signals than at high pump fields ($v >> 1$). This corresponds to the flatness of the U_{eff} curve for $v = 1$ in figure 4; for $v >> 1$ we are trapped in deep potential wells and it takes a very strong external signal to force a transition to the other well. This is the same viewpoint which has been used in proposals for magnetic memories to be addressed by a laser beam. Before the laser beam hits a spot, the temperature is low enough, and the coercive force high enough, to stabilize the information regardless of the externally applied magnetic field. After the laser beam impinges on the selected spot, it heats it up through, or close to, a transition temperature. At this elevated temperature the external field sets the information; upon cooling, the information becomes locked in. This concept of modulating a system to make it susceptible and then again "locking" it seems such a basic process that one cannot help but suspect that it may be quite widespread and perhaps occurs in biological systems. Haken has already made a suggestion in this direction [11].

It should be emphasized that the potential U_{eff} for the analogous potential wells is not related to the energy content of the steady-state dissipative system. For instance, in the case of the parametric oscillator, the energy content is approximately proportional to $|C_1|^2$ and does not exhibit the transition from a single to a double well, as the pump level is varied.

We will here briefly discuss a second dissipative system. This is the tunnel diode circuit illustrated in figure 5. The N-shaped curve gives the i-V characteristic of the tunnel diode,

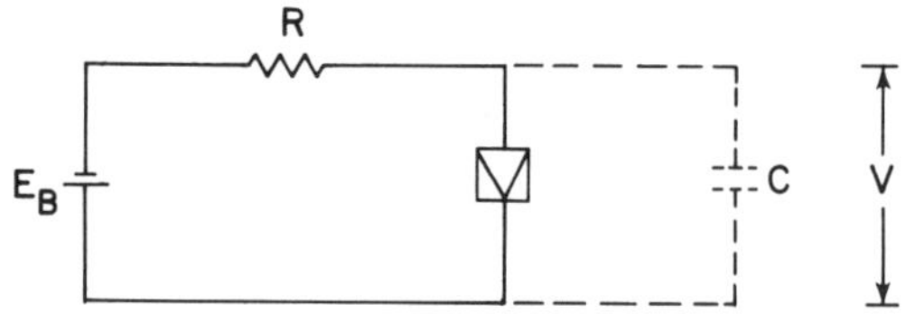

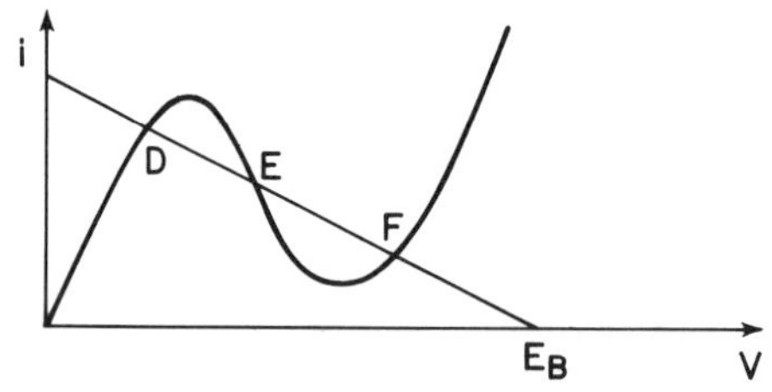

Fig. 5. Tunnel diode circuit at top, with device capacitance. Graph below plots nonlinear i-V characteristic of tunnel diode and also shows straight line $i_B = (E_B - V)R^{-1}$ for current through resistor as a function of the voltage V at the junction of diode and resistor.

i.e., the current through the diode, as a function of the voltage across it. The straight line intersecting it gives the current through the resistor R, as a function of the voltage at the point between the resistor and the diode. At the intersection of the curves we have continuity of current, and therefore solutions of the macroscopic steady-state circuit equations, with D and F as points of stability and E as a point of instability. The master equation for this circuit cannot, in fact, be validly approximated by a Fokker-Planck equation [6,7]. The Fokker-Planck equation is, however, valid [6] in the vicinity of the favored steady states, and this is our only concern in this section. Tunnel diode circuits, depending on the details, can exhibit analogies to both first- and second-order phase transitions [8], with the circuit shown in figure 5 exhibiting a first-order transition. Let us raise the battery voltage E_B, and move toward the situation shown in figure 6, where points D and E have almost coalesced. As is clear from the more detailed discussions elsewhere [6,7], D is now a metastable state, with F the stable state. Raising the

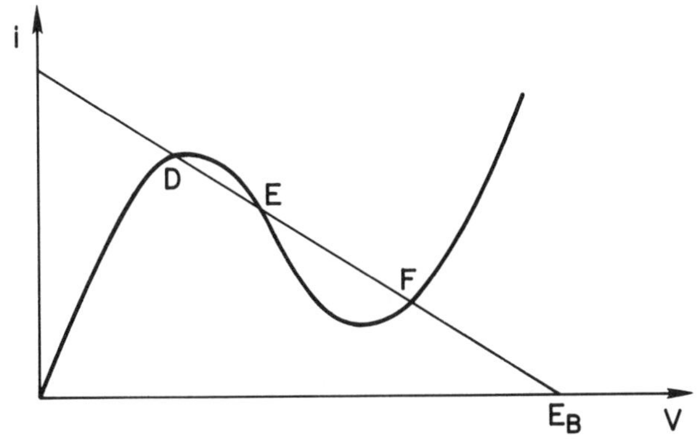

Fig. 6. Tunnel diode characteristic. E_B has been shifted (relative to fig. 5) so that the steady states D and E are close to merging, and the two curves are almost tangent in the vicinity of D and E.

battery voltage further will force a discontinuous transition to F. As D approaches this limit of metastability, the macroscopic restoration velocity v_q to the preferred state D goes smoothly to zero. This can be seen from the fact that the restoration process here consists of a charging or discharging of a capacitor. This charging process results from the difference, shown in figure 6, between the two currents, i.e., between the curved and the straight line. These current differences become small as the intersection, at point D, approaches a point of tangency. Since v_q goes to zero at the limit of metastability, the relaxation time to the preferred state goes to infinity, as it does at the second-order transition; i.e., in both cases we find "soft modes." (We use the term "soft mode" somewhat differently from some other recent authors. They have used the word "hard mode" to represent a disturbance which becomes unstable in a continuous fashion without passing through $\omega = 0$. In their terminology the parametric oscillator has a "hard mode." We prefer to use the expression "soft mode" for any instability pattern with a continuous onset. This is in keeping with the established usage of the words "hard" and "soft" excitation as used by the nonlinear oscillation community [2], in discussing the onset of oscillations.) Furthermore, in both cases (i.e., fig. 6 and fig. 4) the distribution $\exp[\int v_q dq/D]$ becomes broad as v_q approaches zero with parameter variation. Figure 6 thus illustrates an analogy with the "spinodal" behavior found at stability limits in first-order phase transitions and as recently emphasized in connection with the liquid-solid transition [20].

Let us here briefly discuss whether energy or entropy generation are useful guides to the identification of the preferred state. Figure 7 shows a tunnel diode with E_{B1} and E_{B3} representing limits of metastability. E_{B2} is the battery voltage below which the low-voltage tunnel diode state is preferred and above which the high-voltage state is preferred. The stored energy, $\frac{1}{2}CV^2$, is <u>always</u> less for the low-voltage state, and is therefore not a satisfactory criterion for absolute stability. The rate of entropy generation, $i_B E_B$, is <u>always</u> a minimum for the low-current state, and a maximum for the high-current state. For a situation as sketched in

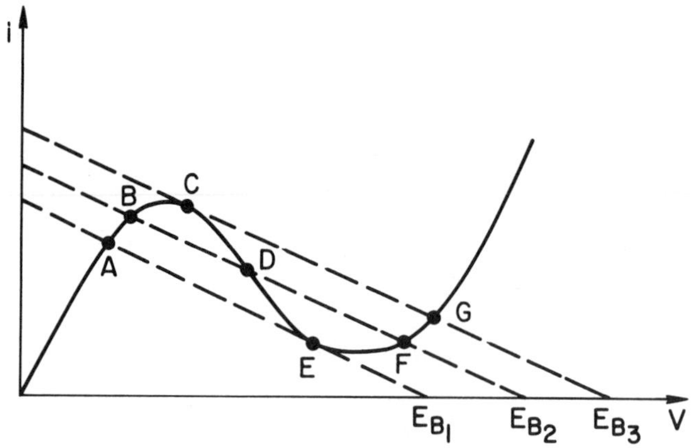

Fig. 7. E_{B2} is selected so that B and F are equally probable. E and C are metastable states at the limits of stability.

the diagram, E, F, and G all are associated with less entropy generation than their respective alternates A, B, and C, and therefore entropy generation cannot be used to predict a transition at E_{B2}.

IV. CRITICAL FLUCTUATIONS

We have already pointed to the fact that parametric oscillator fluctuations in the neighborhood of threshold are similar to those found near a second-order phase transition. Here we will explore this in more detail.

Consider an equilibrium system near its second-order phase transition, restricting ourselves to the region where the mean field theory is applicable. Let ψ be the order parameter defined so that its average value is zero in the high-temperature phase and nonzero in the low-temperature phase. The mean field expression for the free energy difference between the ordered and disordered phases is [21]

$$\Delta F = \alpha|\psi|^2 + \frac{1}{2}\beta|\psi|^4 + \frac{1}{2m}(\nabla\psi)^2 \tag{4.1}$$

where α is proportional to $T - T_c$, and changes sign at the critical temperature T_c. The quantities β and m depend weakly on T and can be treated as constants. For T close to, but slightly greater than, T_c, we can neglect the $|\psi|^4$ term. It is then convenient to consider, instead of $\psi(r,t)$, its Fourier transform $\psi_\kappa(t)$. We then have

$$\Delta F = \frac{1}{V}\sum_\kappa (\alpha + \frac{\kappa^2}{2m})|\psi_\kappa|^2, \tag{4.2}$$

where V is the volume of the system. The thermal averages are computed in the usual way. The result for the (equal time) correlation function is [19]

$$\langle \psi_\kappa^*, \psi_\kappa \rangle = \delta_{\kappa\kappa'} \frac{1}{2V} \frac{kT}{\alpha + \kappa^2/2m} \tag{4.3}$$

The expression for the frequency spectrum of the fluctuations is [21]

$$\langle|\psi_{\kappa\omega}|^2\rangle = 2\,\langle|\psi_\kappa|^2\rangle \frac{\Gamma_\kappa}{\Gamma_\kappa^{\,2} + \omega^2} , \tag{4.4}$$

where Γ_κ is the inverse relaxation time of a fluctuation of ψ_κ. The temperature dependence of Γ_κ can be found by using the equation of motion for the average value of ψ for small deviations from equilibrium value:

$$\dot{\psi}_\kappa = - \gamma_\kappa \frac{\partial F}{\partial \psi_\kappa{}^*} , \tag{4.5}$$

where γ_κ is a temperature-insensitive constant which depends on the microscopic details of the system. (Equation [4.5] is applicable below T_c, if we replace ψ and ψ^* by their respective deviations from equilibrium.) Equation (4.2) then gives the temperature dependence of $\partial F/\partial \psi_\kappa{}^*$ showing that the relaxation time is proportional to $(\alpha + \frac{\kappa^2}{2m})^{-1}$. Thus, for $\kappa = 0$, the relaxation time diverges as T approaches T_c. Note that as temperature is varied, $\langle|\psi_\kappa|^2\rangle$ is proportional to $\Gamma_\kappa^{\,-1}$. Thus for $\kappa = 0$ the spectral density of equation (4.4) assumes the form $C/(\epsilon^2 + \omega^2)$, where C is temperature insensitive and ϵ is proportional to α. Similar points can be made for $T < T_c$, leading to the same general results.

The above discussion can also be carried out for systems whose steady-state distribution function can be written as $\rho = A \exp(- CV)$, where V has the form of ΔF in equation (4.1). For instance, for the subharmonic oscillator, $\Delta F/kT$ is replaced by $U_{eff}/kT = \int v_q\, dq/D$. The order parameter is C_1; and instead of going through the transition via temperature changes, we change the pump amplitude. Equation (4.5) is replaced by the equation of motion for C_1, as given by the bracketed portion on the right-hand side of equation (3.7).

Our discussion of the subharmonic oscillator is valid throughout the transition region. In this case, and for lumped circuits in general, the mean field theory is exact, because the system has no spatial extent, and there is only <u>one</u> field involved.

It should be stressed that systems which are extended in space and which exhibit the kinds of analogies we have been discussing do exist. One example is the laser, which has been treated with regard to its spatial details [10,11]. In this case, the order parameter is the electromagnetic field which can be a function of position. In the existing treatments only the mean field approximation has been explored. It is not yet clear whether the analogies can be extended beyond the mean field description.

The distribution function, $\exp[\int v_q dq/D]$, measures deviations from the favored state, toward which we are returned with an average velocity v_q. We can therefore expect to find a term in the exponential expression for the distribution function of the spatially extended case, which measures the deviation from the laws of propagation, and is a minimum when this is satisfied. Indeed, such a term of the form

$$\int \gamma \left| \left(\frac{d}{dx} - \frac{i\omega_o}{c}\right) E(x) \right|^2 dx , \tag{4.6}$$

turns up in the treatment of that case [11]. The function E(x) is the propagating electric field, and the equation of propagation is

$$\left(\frac{d}{dx} - \frac{i\omega_o}{c}\right) E(x) = 0. \tag{4.7}$$

V. MANY-DIMENSIONAL CASES

Thus far, we have considered only systems for which the steady state corresponds to detailed balancing; i.e., in the steady state there is no net exchange of ensemble members between two nearby states. Since $\underline{j}_q$ vanishes, div $\underline{j}_q$ also vanishes and $\partial\rho/\partial t = 0$. In this section we will first generalize our earlier results within the assumption $\underline{j}_q = 0$, but then go beyond this to indicate that there are at least some cases in which $\underline{j}_q$ does not vanish, and the analogy can still be used. These points are due, respectively, to Stratonovich [4] and Haken [23]. We have considered systems whose Fokker-Planck equation has the form

$$\frac{\partial\rho}{\partial t} = \sum_i \frac{\partial}{\partial q_i} \left[k_i\rho + D\frac{\partial\rho}{\partial q_i}\right], \tag{5.1}$$

where D is a constant and $k_i = \partial u/\partial q_i$. The steady-state solution to this equation for zero current flow is

$$\rho = A e^{-u/D}, \tag{5.2}$$

where A is a normalization constant.

In our first generalization, the diffusion constant is allowed to be space dependent and nondiagonal, but current flow is still zero. Equation (5.1) becomes

$$\frac{\partial\rho}{\partial t} = \sum_i \frac{\partial}{\partial q_i} \left[k_i\rho + D_{ij}\frac{\partial\rho}{\partial q_j}\right]. \tag{5.3}$$

The steady-state solution still has the form $\rho = A e^{-u}$ provided that there exists a function u such that $k_i = \sum D_{ij}\, \partial u/\partial q_j$. If D has neither zero nor singular eigenvalues in the neighborhood of the transition, our previous conclusions concerning the nature of the fluctuations still hold.

The second generalization to be discussed is used by Haken in his treatment of lasers [23]. If we have a solution for the case $\underline{j}_q = 0$, and then add terms to $\underline{j}_q$ which do not disturb the density found for $\underline{j}_q = 0$, the distribution function remains unaltered. Consider the situation of equation (5.3), where the k_i permit the existence of a potential function u, leading to a steady-state distribution ρ. Then consider the restoration velocities of equation (5.1) supplemented by additional terms:

$$k_i' = k_i + v_i\ . \tag{5.4}$$

The original solution remains unaffected if

$$\sum_i \frac{\partial}{\partial q_i} (v_i\rho) = 0. \tag{5.5}$$

Alternatively we can write this

$$\rho \operatorname{div} \underline{v} + \underline{v} \cdot \nabla \log \rho = 0. \tag{5.6}$$

Thus, for instance, if $\underline{v}$ represents a rotation about an axis (div $\underline{v} = 0$) and $\nabla\rho$ is radial

($\underset{\sim}{v} \cdot \nabla\rho = 0$), the above condition is satisfied.

VI. ENTROPY OF FLUCTUATIONS ABOUT THE STEADY STATE

This section shows that our analogies can be given a thermodynamic significance that goes beyond the statistical mechanical features discussed in the cited literature. Once we have defined a distribution function $\rho(q)$ for the nonequilibrium system, we can formally define an entropy $S = -k\int \rho \log \rho \, dq$. As we change the operating point, say D in figure 5, we will be changing the entropy associated with the steady-state distribution function, and can ask if this entropy change corresponds to some identifiable heat flow dQ, divided by a physically meaningful temperature T. We shall, in fact, show that for the entropy of the steady state $dS = dQ/T$ as the system is slowly changed, remaining very close, however, to a sequence of nearly steady states, and where dQ and T are defined in ways which are natural extensions of their equilibrium role. T will turn out to be the noise temperature, i.e., the temperature characterizing the magnitude of the stored energy fluctuations about the favored point. The differential dQ is the (reversible) heat exchange with the reservoir in excess of the dissipation calculated from the macroscopic theory which describes the behavior of the ensemble average. The argument will be limited to the case where the energy storage associated with the stochastic variable is quadratic in that variable. In the case of figure 5, this means that the capacitance is independent of q, or V.

Consider small fluctuations in the vicinity of a single stability point in conjecture with our expression $\rho = \exp[\int v_q dq/D]$. Near the favored value of q we can write $v_q = -\delta/\tau$, where δ is the deviation in q from the point at which $v_q = 0$. Assuming D sensibly constant in the vicinity of the stable point then yields

$$\rho(\delta) = A \exp(-\delta^2/2D\tau), \tag{6.1}$$

with A a normalization constant. Now view the capacitive degree of freedom only through its fluctuations, as we might if determining temperature via electromagnetic radiation. For a capacitor we expect $\rho(\delta) = \exp(-\delta^2/2kTC)$, at a temperature T, so that the "noise" temperature corresponding to the fluctuations of equation (6.1) is $T_N = D\tau/kC$. As we approach a critical point or stability limit, the relaxation time τ, and with it T_N, become infinite. (Points of unstable equilibrium can be characterized by T_N negative.)

Now consider what happens as we change some parameter, such as the battery voltage in figure 5, shifting the steady state. As a simpler case consider first a resistance R at equilibrium in parallel with a capacitor C. Then $\langle i_R V_R \rangle = 0$. If we raise the temperature, power must flow out of the resistor into the capacitor to supply the increase in $\frac{1}{2}$ kT required by the capacitor. Therefore,

$$\int \langle i_R V_R \rangle dt = \frac{1}{2} kT' - \frac{1}{2} kT. \tag{6.2}$$

The quantity in equation (6.2) is just the change in stored energy. Hence

$$\int \langle i_R V_R \rangle dt = \langle \frac{(q')^2}{2C} \rangle - \langle \frac{(q)^2}{2C} \rangle, \tag{6.3}$$

where q and q' are, respectively, the fluctuating charges in the capacitor before and after the temperature changes. Now consider, instead, a resistor in parallel with a capacitor, but with a direct current fed to the combination. Assume that the resistor has not been disturbed enough by the d.c. flow to change its noise spectrum. Then if we let δ and δ' represent respective charge fluctuations about ensemble-average charge values before and after the temperature change, equations (6.2) and (6.3) become instead

$$\int \langle (i_R - \langle i_R \rangle)(V_R - \langle V_R \rangle) \rangle dt$$

$$= \int (\langle i_R V_R \rangle - \langle i_R \rangle \langle V_R \rangle) dt = \frac{1}{2} kT' - \frac{1}{2} kT , \tag{6.4}$$

$$\int (\langle i_R V_R \rangle - \langle i_R \rangle \langle V_R \rangle) dt = \langle \frac{(\delta')^2}{2C} \rangle - \langle \frac{\delta^2}{2C} \rangle . \tag{6.5}$$

The left-hand side of equation (6.5) has subtracted out of it the power flow into the resistor, as calculated from the macroscopic theory, applied to the ensemble average. Equations (6.2)-(6.5) are not a formal part of our reasoning here, but are presented only to make the appearance of the equations which follow more plausible and motivated.

Now return to our circuit of figure 5 with series resistance R, tunneling current i_t, and capacitor current i_C. The ensemble average of the energy supplied by the nonfluctuating battery is

$$\langle E_B \int i_R dt \rangle = \int (E_B - \phi) \langle i_R \rangle dt + \int \phi \langle i_t \rangle dt + \int \phi \langle i_C \rangle dt . \tag{6.6}$$

This is a purely mathematical identity, utilizing $i_R = i_t + i_C$ and valid for all functions $\phi(t)$. We shall take $\phi(t)$ to be $\langle V \rangle$, the ensemble average of the voltage at the junction of resistor and diode. The instantaneous energy flow before ensemble-averaging also obeys the law of conservation of energy

$$E_B i_R = (E_B - V) i_R + V i_t + V i_C . \tag{6.7}$$

Take the ensemble average of the time integral of equation (6.7) and subtract it from equation (6.6). We are then left with

$$\int (\langle V_C i_C \rangle - \langle V_C \rangle \langle i_C \rangle) dt = - \sum_{R,t} \int (\langle V_i i_i \rangle - \langle V_i \rangle \langle i_i \rangle) dt . \tag{6.8}$$

The left-hand integral in equation (6.8) represents the difference between the ensemble average of the capacitive energy, $\langle U \rangle$, and the energy associated with the ensemble average, or macroscopic behavior. If we assume that the capacitance C (or more generally, energy storage coefficient) is independent of q, then we have

$$\int (\langle V_C i_C \rangle - \langle V_C \rangle \langle i_C \rangle) dt = \int \langle V_C dq \rangle - \langle V_C \rangle \langle dq \rangle = \frac{1}{2C} (\langle q^2 \rangle - \langle q \rangle^2). \tag{6.9}$$

We shall call this difference the fluctuational energy, U_F. The right-hand side of equation (6.8) similarly represents, in its first term, the negative of the ensemble average of the heat dissipation and then in its second term subtracts from it the negative of the heat dissipation associated with the ensemble average. Thus $dU_F = dQ$, where dQ represents heat flow from the reservoir, above and beyond the heat flow associated with the macroscopic circuit equations. The remaining discussion must address the relationship $T_N dS = dU_F$, then it will follow that $T_N dS = dQ$.

As we slowly modulate parameters in figure 5, the distribution function changes in two ways. First of all, the average value of q changes; e.g., if the battery voltage is increased, the charge on the capacitor goes up. This effect is described by the macroscopic (or ensemble average) equations describing the system. Furthermore, the extent of the fluctuations can change. For distributions of the form given by equation (6.1) the quantities $\langle i_C \rangle dt = \langle dq \rangle$ and $\langle V_C \rangle$ characterize the center of the distribution, at $\delta = 0$. Let the capacitive energy associated with this point be U_o. Then the quantity U_F of equation (6.9) can be represented in the form

$$U_F = \int \rho(q) U(q) dq - U_o = \int \rho(q) [U(q) - U_o] dq \; , \tag{6.10}$$

or

$$U_F = \int \rho(\delta) \frac{\delta^2}{2C} d\delta \; . \tag{6.11}$$

For the fluctuational entropy we have

$$S = -k \int \rho(\delta) \log \rho(\delta) d\delta \; . \tag{6.12}$$

The relationships (6.11), (6.12), combined with $\rho(\delta) = \exp(-\delta^2/2kT_{NC})$, are a set of relationships in which we have only one remaining variable parameter as the point of stability is slowly shifted, and that is T_N. Since these relationships result in $T_N dS = dU_F$ when applied to the ordinary equilibrium case, they must do so here also.

T_N is typically (but not necessarily) higher than the ambient temperature. Thus a fluctuational entropy gain $dS = dQ/T_N$ can be less than the entropy taken from the reservoir at a lower temperature. This is, however, not a violation of the second law, since these entropy changes take place against a background of steady-state entropy generation, associated with the dissipation predicted from the macroscopic laws of motion for the system.

The many-dimensional case, in which U_F, D, and τ all have different principal axes, looks complex. If the principal axes agree, however, then we have $dQ = \sum_i T_{Ni} dS_i$.

REFERENCES

1. R. K. Brayton and J. K. Moser, Quart. J. Appl. Math. 22, 1 (1964); 22, 81 (1964).

2. N. Minorsky, Introduction to Nonlinear Mechanics (Ann Arbor: J. W. Edwards, 1947).

3. R. J. Donnelly, R. Herman, and I. Prigogine, Nonequilibrium Thermodynamics Variational Techniques and Stability (Chicago: University of Chicago Press, 1966).

4. R. L. Stratonovich, Topics in the Theory of Random Noise (New York: Gordon and Breach, 1963), Vols. 1 and 2.

5. W. C. Lindsey, Proc. IEEE, 57, 1705 (1969).

6. R. Landauer, J. Appl. Phys. 33, 2209 (1962).

7. R. Landauer in Proceedings of the Conference on Fluctuation Phenomena in Classical and Quantum Systems, Chania, Crete, Greece, 1969, ed. E. D. Haidemenakis. The publication fate of this volume remains unclear. R. Landauer and James W. F. Woo will, however, have a closely related paper appearing in the Proceedings of the International Symposium on Synergetics, Schloss Elmau, Germany, April 30-May 6, 1972, ed. H. Haken.

8. R. Landauer, Ferroelectrics, 2, 47 (1971).

9. P. C. Martin in Low Temperature Physics, ed. J. G. Daunt, D. O. Edwards, F. J. Milford, and M. Yagub (New York: Plenum, 1965).

10. R. Graham and H. Haken, Z. Physik, 213, 420 (1968); 237, 37 (1970); Phys. Letters, 33A, 335 (1970).

11. H. Haken, Festkörper Probleme X, ed. O. Madelung (Braunschweig: Pergamon Vieweg, 1970), p. 351.

12. V. DeGiorgio and M. O. Scully, Phys. Rev., A2, 1170 (1970).

13. M. S. Green, J. Chem. Phys., 20, 1281 (1952).

14. R. Graham and H. Haken, Z. Physik, 243, 289 (1971).

15. J. von Neumann, U. S. Patent #2,815,488.

16. E. Goto, Proc. IRE, 47, 1304 (1959).

17. G. J. Lasher, IBM J. Res. Develop., 5, 151 (1961).

18. J. W. F. Woo and R. Landauer, "Fluctuations in a Parametrically Excited Subharmonic Oscillator," submitted for publication. See also A.P.S. Bull., 15, 1638 (1970).

19. S. Chandrasekhar, Rev. Mod. Phys., 15, 1 (1943).

20. T. Schneider, R. Brout, H. Thomas, J. Feder, Phys. Rev. Letters, 25, 1423 (1970).

21. L. Landau and E. M. Lifshitz, Statistical Physics (Reading, Mass.: Addison-Wesley, 1958).

22. L. D. Landau and I. M. Khalatnikov, Collected Papers of Landau, ed. D. ter Haar (New York: Gordon and Breach, 1965), p. 626.

23. H. Haken, Z. Physik, 219, 246 (1969).

DISCUSSION

M. E. Fisher: I am not clear about the status of the expression $\int \rho \log \rho \, dq$ for the entropy. It seems to me that by inserting dq (rather than, say, $q^2 dq$) one is making a physical assumption--essentially that q is a "good" phase coordinate in the sense that one has some Liouville theorem and an ergodic hypothesis. It doesn't seem obvious that the charge, q, is such a coordinate.

R. Landauer: Admittedly we don't yet understand in a completely general way what a "good" coordinate is. Our derivation of $dS = dQ/T$ does, of course, require that energy be a strictly quadratic function of q. That prevents us from making some mistakes. Consider, for example, an electrostrictive capacitor, i.e., one where the plates are held apart by an elastic spring. This is a nonlinear capacitor, and the requirement that U must be a quadratic function of q now prevents us from viewing this simply as a system with one degree of freedom, q, and erroneously ignoring the hidden degree of freedom, i.e., the distance between the plates.

But there is a much more basic point. Our derivation $dS = dQ/T$ can hardly be in error. It is the identification of the quantities involved in this equation with their equilibrium equivalents which is open to question. If, however, we choose as our coordinate one which is "good" in the equilibrium case (as is certainly true for the charge on a capacitor), then there doesn't seem to be much question about the fact that we have a real generalization of the equilibrium relationship.

N. van Kampen: In principle Dr. Fisher is of course quite right, but I feel that it should not be hard to verify that the whole range of q-values covered by this Gaussian distribution is so narrow that any phase space factor or weight function for q may be constant. This feeling is corroborated by the fact that on maximizing the expression which Dr. Landauer gave for the entropy, the correct Gaussian distribution for q is found.

J. Rothstein: In answer to Fisher's comment, perhaps an easy way to justify the choice of entropy function is to note that (a) it has the form of Boltzmann's entropy, and (b) q is the analogue of a coordinate in the circuit equation, the electromechanical analogy being complete even to the quadratic dependence of the energy on coordinates and velocities. The standard discussion should then apply.

The possibility of having T_N become infinite while the temperature of the rest of the system is finite suggests an analogy with spin systems exhibiting "negative temperatures." The discussion of two regimes (depending on the phase of a driving force) in which work is done either on or by the system in the two cases supports the aptness of the analogy. In the spin case, positive, negative, and the intermediate infinite temperatures are properly sorted out via the relation $1/T = \partial S/\partial E$. Can one do that here? If one can, I would consider this strong justification for the correctness of the choice of entropy function actually made; and if not, I would suspect the existence of a gap or error in the discussion.

R. Landauer: As you point out, there are strong analogies between our discussion and discussions of spin temperature. In our case the temperature goes through infinity to negative temperatures as a point of stability becomes a point of instability, whereas in the spin system this happens at the point where the population is inverted. The transition to a state of inverted population is a transition to an emitting state from an absorbing state, and is therefore really also a transition to a state of instability.

We wouldn't expect a relationship here quite as simple as $1/T = \partial S/\partial E$. We have to separate the quantities associated with the fluctuations and the quantities which simply define the shift in operating point of the steady state. But once we have done that, we certainly do have $1/T = \partial S/\partial U_F$.

N. van Kampen: I would like to comment that the use of a nonlinear Langevin equation, or the corresponding Fokker-Planck equation, is dangerous. More specifically, it is erroneous when computing the effect of nonlinearity on the fluctuations. However, it can be used for two restricted purposes: either to derive from it the macroscopic, phenomenological equation neglecting all fluctuations, or to describe fluctuations near equilibrium in the linear approximation. I feel that Dr. Landauer has managed to stay within these restrictions, and his use of the nonlinear Fokker-Planck equation is therefore justified. I am not so sure that the question whether to put D in front or behind $\partial/\partial q$ is meaningful within the framework of the Fokker-Planck approximation, since it involves the dependence of D on q, that is, of nonlinear terms.

R. Landauer: Professor van Kampen is completely correct in a case like that of the tunnel diode. That's a situation where the capacitive charge can change by one electron at a time. That can be a transition to a state which can have an appreciably different probability of occupation, and the difference terms in the master equation cannot be accurately replaced by derivatives. Therefore, the Fokker-Planck equation is valid only where the probabilities vary particularly slowly. That leaves it valid only in the immediate vicinity of the points of stability and instability, as has been discussed in detail in reference [6] and amplified in reference [7]. Luckily that's where we need it most!

I do not believe, however, that Professor van Kampen's point is relevant to a system like the parametric oscillator where the nonlinearities are in the reactances, and where the noise sources are the thermal-equilibrium noise sources residing in the linear resistors. Here the statistically independent event is the motion of one electron, through a mean free path, and that can be a very small total charge transport. I furthermore believe that my comments in relation to equations (3.9) and (3.10) do apply to ordinary spatial diffusion, in cases where there is a gradient of temperature or of scattering centers.

R. Kubo: In order to supplement Professor van Kampen's comment, I would like to add that the Fokker-Planck equation used here comes from the assumption that the random force is Gaussian with a white noise. One can assume this as a mathematical model. The problem is then whether or not this assumption is compatible with the nonlinear nature of the physical system. Similar questions may be raised to the mode-mode coupling theory which was discussed by Professor Zwanzig yesterday and by Professor Kawasaki just this morning. I wonder if Professor Van Kampen feels uneasy with this kind of treatment?

N. van Kampen: I do feel uneasy, because I have experienced in other fields how an inconsiderate application of the Fokker-Planck approximation to nonlinear systems may lead to erroneous results. However, in order to see whether the same objections apply to the present case, I would have to sit down and check the argument step by step. Any rash remark made now might only add to the confusion.

17

MACROSCOPIC WAVE FUNCTIONS

F. W. Cummings

1. INTRODUCTION

Attention in this article will be focused on the use of the property of "off-diagonal long-range order" (ODLRO), so called by Yang in a fundamental paper [1], in conjunction with the equations of motion for the reduced density matrices for systems of identical particles. After some preliminary remarks in this section, we will turn in the second section to a brief review of some relevant previous work, largely that of Yang and Fröhlich. Subsequent sections will be concerned with an illustration of the method discussed in section 2 as applied to superfluid ^{4}He II. In particular, in section 3, a relatively simple method for determination of the condensate fraction will be discussed, which gives this determination as a function of temperature and involves only measurements of the pair distribution function.

A central feature of quantum mechanics is that the state vectors of two or more states can be linearly superposed to form a combined state, and the absolute square of the combined state will depend on the difference in the phases of the component states. In very large systems such phase differences will most often average out due to the chaotic thermal motion, and will thus lead to no observable effect. Very interesting exceptions do occur, however, in systems of large extent, and these exceptions seem always to be related to a very strong excitation of a single mode of motion, in which case their phases then show correlations over macroscopic regions. The existence of phase correlations coincides with the existence of ODLRO in the reduced density matrices; the onset of ODLRO will signal the appearance of a new thermodynamic phase of the system, and the appearance of a type of order more subtle than the more usual spatial crystalline order, for example, and it is often referred to as "phase order."

Systems believed to be characterized by ODLRO are superconductors, superfluid helium, and laser states; biological systems are also a possibility. For systems in thermodynamic equilibrium ODLRO comes about in the low-temperature region for energetic reasons; ^{4}He II and superconductors are examples. Laser states are an example of such highly excited states being metastable, a strongly excited electromagnetic field imposing the long-range phase correlations on the thermal motion [2]. The very interesting possibility of the strong excitation of a single longitudinal "plasma-like" mode (the lowest in a narrow band) in biological systems at elevated (ambient) temperatures has recently been discussed [3].

The most natural mathematical expression of ODLRO is given in terms of the reduced density matrices [1]. Frohlich [4,5,6] has stressed the usefulness of the reduced density matrices in obtaining rigorous, general derivations of macroscopic properties of many-body systems from microscopic considerations, without requiring detailed solution of the N-body problem. For example, the form of the Euler and Navier-Stokes equations of hydrodynamics can be derived [4] as rigorous consequences of the microscopic equation of motion for a system of identical point particles interacting via two-body potentials; no continuum assumptions are necessary, nor are intermediate "master equations" required. Only the expansion of the diagonal second-order reduced density matrix about equilibrium by making use of its symmetry is required to obtain the Navier-Stokes equation from the Euler equation. Thus as one consequence of this fact we see that the Navier-Stokes equation must be valid even in the presence of ODLRO and thus must constitute one of the (exact) equations of the two-fluid model describing ^{4}He II.

If ODLRO is characteristic of and constitutes the most general statement of the presence of the thermodynamic phases of ^{4}He II and superconductors, it would seem most natural that a calculation of their properties would incorporate ODLRO from the outset; the introduction of explicit approximate models should ideally occur at some later point after it is understood to what extent general properties may be deduced in a rigorous, or nearly rigorous, way. The next section will be primarily devoted to reviewing some of Fröhlich's work (scattered through the literature) along this direction as well as portions of Yang's work on ODLRO relevant to the present discussion.

2. REDUCED DENSITY MATRICES AND ODLRO

The following is restricted to identical particles--fermions or bosons--described by nonrelativistic field operators $\psi(\underset{\sim}{x})$ satisfying appropriate commutation or anticommutation rules. The n(th) reduced density matrix, which depends on 2n space points and on time t, is defined as the trace

$$\Omega_n(\underset{\sim}{x}_1{}' \ldots \underset{\sim}{x}_n{}'; \underset{\sim}{x}_1{}'' \ldots \underset{\sim}{x}_n{}'') = \text{trace}\,[\Omega(t)\,\psi^{\dagger}(\underset{\sim}{x}_1{}'') \ldots \psi^{\dagger}(\underset{\sim}{x}_n{}'')\psi(\underset{\sim}{x}_n{}') \ldots \psi(\underset{\sim}{x}_1{}')]. \tag{2.1}$$

The density operator Ω satisfies

$$i\hbar \frac{\partial}{\partial t}\,\Omega(t) = [\mathcal{H}, \Omega(t)]\ , \tag{2.2}$$

where $\mathcal{H}$ is the Hamiltonian operator of the system

$$\mathcal{H} = T + V\ , \tag{2.3}$$

$$T = \hbar^2/2m \int \nabla\psi^{\dagger}(\underset{\sim}{x}) \cdot \nabla\psi(\underset{\sim}{x})\,d^3x, \tag{2.4}$$

$$V = \tfrac{1}{2} \int\int V(|\underset{\sim}{x} - \underset{\sim}{y}|)\psi^{\dagger}(\underset{\sim}{x})\psi^{\dagger}(\underset{\sim}{y})\psi(\underset{\sim}{y})\psi(\underset{\sim}{x})\, d\underset{\sim}{x}\, d\underset{\sim}{y}. \tag{2.5}$$

By using equations (2.1) and (2.2), the equations of motion follow at once:

$$i\hbar \frac{\partial}{\partial t} \Omega_n(\underset{\sim}{x}_1{}' \ldots \underset{\sim}{x}_n{}'') = \Big[- \frac{\hbar^2}{2m} \sum_{r=1}^{n} (\nabla_r'^{\,2} - \nabla_r''^{\,2}) + \tfrac{1}{2} \sum_{r,s} V(\underset{\sim}{x}_r{}' - \underset{\sim}{x}_s{}')$$

$$- V(\underset{\sim}{x}_r{}'' - \underset{\sim}{x}_s{}'') \Big] \Omega_n + \int \sum_r \Big[V(\underset{\sim}{x}_r' - \underset{\sim}{y}) - V(\underset{\sim}{x}_r{}'' - \underset{\sim}{y}) \Big] \Omega_{n+1} (\underset{\sim}{x}_1{}' \ldots \underset{\sim}{x}_n{}', \underset{\sim}{y};$$

$$\underset{\sim}{x}_1{}'' \ldots \underset{\sim}{x}_n{}'', \underset{\sim}{y}) d\underset{\sim}{y}\,. \tag{2.6}$$

Ω_{N+1} vanishes, and the equation of motion for Ω_N is equivalent to Schrödinger's equation, thus illustrating the connection between the second quantized and the Schrödinger formalism. From equation (2.1) follow the symmetry relations

$$\Omega_n(\underset{\sim}{x}_1{}' \ldots \underset{\sim}{x}_{r-1}{}', \underset{\sim}{x}_r{}' \ldots \underset{\sim}{x}_n{}'; \underset{\sim}{x}_1{}'' \ldots \underset{\sim}{x}_r{}'', \underset{\sim}{x}_{r-1}{}'' \ldots \underset{\sim}{x}_n{}'')$$

$$= \Omega_n(\underset{\sim}{x}_1{}' \ldots \underset{\sim}{x}_r{}', \underset{\sim}{x}_{r-1}{}' \ldots \underset{\sim}{x}_n{}'; \underset{\sim}{x}_1{}'' \ldots \underset{\sim}{x}_r{}'', \underset{\sim}{x}_{r-1}{}'' \ldots \underset{\sim}{x}_n{}'')$$

$$= \pm\, \Omega_n(\underset{\sim}{x}_1{}' \ldots \underset{\sim}{x}_r{}', \underset{\sim}{x}_{r-1}{}' \ldots \underset{\sim}{x}_n{}'; \underset{\sim}{x}_1{}'' \ldots \underset{\sim}{x}_{r-1}{}'', \underset{\sim}{x}_r{}'' \ldots \underset{\sim}{x}_n{}'')$$

$$= \Omega_n^{*}(\underset{\sim}{x}_1{}'' \ldots \underset{\sim}{x}_{r-1}{}'', \underset{\sim}{x}_r{}'' \ldots \underset{\sim}{x}_n{}''; \underset{\sim}{x}_1{}', \ldots \underset{\sim}{x}_{r-1}{}', \underset{\sim}{x}_r{}' \ldots \underset{\sim}{x}_n{}')\,, \tag{2.7}$$

where $\pm$ refers to bosons (+) or fermions (-).

Introducing

$$\underset{\sim}{x}_r = \frac{\underset{\sim}{x}_r{}' + \underset{\sim}{x}_r{}''}{2}\,, \quad \underset{\sim}{\xi}_r = \underset{\sim}{x}_r{}' - \underset{\sim}{x}_r{}''\,, \tag{2.8}$$

Ω_n can then be written in terms of two real functions

$$\Omega_n = \sigma_n(\underset{\sim}{x}_1 \ldots \underset{\sim}{x}_n; \underset{\sim}{\xi}_1 \ldots \underset{\sim}{\xi}_n) \exp[i\chi_n(\underset{\sim}{x}_1 \ldots \underset{\sim}{x}_n; \underset{\sim}{\xi}_1 \ldots \underset{\sim}{\xi}_n)] \tag{2.9}$$

where

$$\sigma_n(\underset{\sim}{x}_1 \ldots \underset{\sim}{x}_n; \underset{\sim}{\xi}_1 \ldots \underset{\sim}{\xi}_n) = \sigma_n(\underset{\sim}{x}_1 \ldots \underset{\sim}{x}_n; -\underset{\sim}{\xi}_1 \ldots -\underset{\sim}{\xi}_n)\,, \tag{2.10}$$

$$\chi_n(\underset{\sim}{x}_1 \ldots \underset{\sim}{x}_n; \underset{\sim}{\xi}_1 \ldots \underset{\sim}{\xi}_n) = -\chi_n(\underset{\sim}{x}_1 \ldots \underset{\sim}{x}_n; -\underset{\sim}{\xi}_1 \ldots -\underset{\sim}{\xi}_n). \tag{2.11}$$

Understand by "lim" the limit in which all ξ_r become zero; the lim $\sigma_n = \sigma(\underset{\sim}{x}_1 \ldots \underset{\sim}{x}_n)$ represents an n-point density, and

$$\underset{\sim}{v}_r(\underset{\sim}{x}_1 \ldots \underset{\sim}{x}_n) = \frac{\hbar}{m} \lim \nabla_{\underset{\sim}{\xi}_r} \chi_n \tag{2.12}$$

an n-point velocity, and

$$\underset{\sim}{J}_r(\underset{\sim}{x}_1 \ldots \underset{\sim}{x}_n) = \frac{\hbar}{im} \lim \nabla_{\underset{\sim}{\xi}_r} \Omega_n = \sigma(\underset{\sim}{x}_1 \ldots \underset{\sim}{x}_n)\, \underset{\sim}{v}_r(\underset{\sim}{x}_1 \ldots \underset{\sim}{x}_n) \tag{2.13}$$

a corresponding current density. From the "lim" of the equation of motion (2.16) a generalized law of continuity follows:

$$\frac{\partial}{\partial t} \sigma(\underset{\sim}{x}_1 \ldots \underset{\sim}{x}_n) + \sum_{r=1}^{n} \nabla_r \underset{\sim}{J}_r(\underset{\sim}{x}_1 \ldots \underset{\sim}{x}_n) = 0. \tag{2.14}$$

The quantities $\sigma(\underset{\sim}{x})$ and $\underset{\sim}{v}_r(\underset{\sim}{x})$ are the macroscopic density and velocity fields. Equation (2.14) is then the usual continuity equation of hydrodynamics. The Euler equation of hydrodynamics is obtained as the local limit of the derivative $\partial/\partial \xi$ of the equation of motion of $\Omega_1(\underline{x}';\underline{x}")$ given by equation (2.6). The Navier-Stokes equation is then obtained [4] by development of the force term by expansion of $\Omega_2(\underline{x},\underline{y};\underline{x},\underline{y})$ about its equilibrium bulk value, where Ω_2 is considered as a function of the density, velocity, and the difference vector $\underline{r} = \underline{x} - \underline{y}$. It is to be stressed that the Euler and the Navier-Stokes equations follow as exact results. Even more generally the viscosity coefficients may be space dependent when Ω_2 is considered as expanded about a <u>local</u> equilibrium value [7].

The occurrence of ODLRO in Ω_1 is expressed by the appearance of a factorized term

$$\Omega_1(\underline{x}';\underline{x}") = \phi^*(\underline{x}")\phi(\underline{x}') + \Lambda_1(\underline{x}';\underline{x}") , \tag{2.15}$$

where Λ_1 contains only "normal" correlations,

$$\Lambda_1(\underline{x}';\underline{x}") \longrightarrow 0 \quad \text{as} \quad |\underline{x}' - \underline{x}"| \longrightarrow \infty . \tag{2.16}$$

The function ϕ is referred to as the macroscopic wave function of the system, $\phi^*(\underline{x})\phi(\underline{x}) \equiv \rho_0(\underline{x})$ is called the condensate density and $\Lambda_1(\underline{x};\underline{x}) \equiv \rho_d(\underline{x})$ the depletion density. Since Ω_1 is Hermitian (as are all Ω_n), $\Omega_1(\underline{x}';\underline{x}") = \Omega_1^*(\underline{x}";\underline{x}')$, it possesses a diagonal spectral resolution in terms of the complete orthogonal set $\phi_i(\underline{x})$,

$$\Omega_1(\underline{x}';\underline{x}") = \sum_i \lambda_i^{(1)} \phi_i^*(\underline{x}")\phi_i(\underline{x}') . \tag{2.17}$$

ODLRO is equivalent to the presence of a single large eigenvalue, λ_o say, of order N, such that equation (2.17) is written

$$\Omega_1(\underline{x}';\underline{x}") = (\lambda_o^{(1)})^{\frac{1}{2}} \phi_o^*(\underline{x}") (\lambda_o^{(1)})^{\frac{1}{2}} \phi_o(\underline{x}') + \sum_{i \neq 0} \lambda_i^{(1)} \phi_i^*(\underline{x}") \phi_i(\underline{x}') ,$$

$$\lambda_o^{(1)} = O(N) , \tag{2.18}$$

where the last term vanishes for $\underline{x}'$ far from $\underline{x}"$; compare equation (2.15).

ODLRO can occur in Ω_1 only for a system of bosons, but not fermions because of the exclusion principle. The equation of motion for Ω_1 is from equation (2.6),

$$i\hbar \frac{\partial}{\partial t} \Omega_1(\underline{x}';\underline{x}") = -\frac{\hbar^2}{2m} (\nabla_{\underline{x}'}^2 - \nabla_{\underline{x}"}^2) \Omega_1 + \int [V(\underline{x}' - \underline{y}) - V(\underline{x}" - \underline{y})]$$

$$\Omega_2(\underline{x}',\underline{y};\underline{x}",\underline{y})d\underline{y} , \tag{2.19}$$

and use of equation (2.15), ODLRO, in equation (2.19) in the limit $|\underline{x}' - \underline{x}"| \longrightarrow \infty$ leads to an equation separable in $\underline{x}'$ and $\underline{x}"$ [5]. This results in the exact equation for the macroscopic wave function:

$$i\hbar \frac{\partial}{\partial t} \phi(\underline{x}') = -\frac{\hbar^2}{2m} \nabla_{\underline{x}'}^2 \phi(\underline{x}') + F(\underline{x}') , \tag{2.20}$$

where

$$F(\underline{x}') = \lim_{|\underline{x}' - \underline{x}"| \to \infty} \int V(\underline{x}' - \underline{y}) \frac{\Omega_2(\underline{x}',\underline{y};\underline{x}",\underline{y})}{\phi^*(\underline{x}")} d\underline{y} \tag{2.21}$$

is a function of $\underline{x}'$ alone.

There are several conditions which must be satisfied by any system of identical particles, viz.,

$$\Omega_2(\underline{x}',\underline{y}';\underline{x}",\underline{y}") \longrightarrow \Omega_1(\underline{x}';\underline{x}") \Omega_1(\underline{y}';\underline{y}") \quad \text{(liquid only)} \tag{2.22}$$

for $\underset{\sim}{x}' \approx \underset{\sim}{x}''$, $\underset{\sim}{y}' \approx \underset{\sim}{y}''$ and $|\underset{\sim}{x}' - \underset{\sim}{y}'| \to \infty$,

and $$\int \Omega_2(\underset{\sim}{x}', \underset{\sim}{y}; \underset{\sim}{x}'', \underset{\sim}{y})\, d\underset{\sim}{y} = (N - 1)\Omega_1(\underset{\sim}{x}'; \underset{\sim}{x}''), \tag{2.23}$$

as well as the symmetry conditions expressed in equation (2.7). Also, for any system interacting with a repulsive ("hard core") potential at short distances, the pair correlation function $\Omega_2(\underset{\sim}{x}, \underset{\sim}{y}; \underset{\sim}{x}, \underset{\sim}{y})$ must be required to vanish as $|\underset{\sim}{x} - \underset{\sim}{y}| \to 0$; for a realistic description of the system $\Omega_2(\underset{\sim}{x}, \underset{\sim}{y}; \underset{\sim}{x}, \underset{\sim}{y})$ should be capable of describing the structure as reflected by the pair distribution function, g, defined through

$$\Omega_2(\underset{\sim}{x}, \underset{\sim}{y}; \underset{\sim}{x}, \underset{\sim}{y}) = \rho^2 g(\underset{\sim}{x}, \underset{\sim}{y}) = \rho^2 g(|\underset{\sim}{x} - \underset{\sim}{y}|) \text{ (bulk) } (\rho = \text{density}). \tag{2.24}$$

Beyond these general conditions which apply to any system, ODLRO clearly places further restrictions, via equations (2.20) and (2.21), as well as

$$\Omega_2(\underset{\sim}{x}', \underset{\sim}{y}'; \underset{\sim}{x}'', \underset{\sim}{y}'') = \phi^*(\underset{\sim}{x}'')\phi^*(\underset{\sim}{y}'')\phi(\underset{\sim}{x}')\phi(\underset{\sim}{y}') \text{ (if ODLRO in } \Omega_1), \tag{2.25}$$

via equation (2.15), for all points far apart [1].

These conditions taken together are quite restrictive. The simplest form for Ω_2 (first suggested by Frohlich [6] in a footnote) which satisfies all of the above conditions is

$$\begin{aligned}\Omega_2 = {} & S(|\underset{\sim}{x}' - \underset{\sim}{y}'|)S(|\underset{\sim}{x}'' - \underset{\sim}{y}''|) \{\phi^*(\underset{\sim}{x}'')\phi^*(\underset{\sim}{y}'')\phi(\underset{\sim}{x}')\phi(\underset{\sim}{y}') + \phi^*(\underset{\sim}{x}'')\phi(\underset{\sim}{x}')\Lambda_1(\underset{\sim}{y}'; \underset{\sim}{y}'') \\ & + \phi^*(\underset{\sim}{y}'')\phi(\underset{\sim}{y}')\Lambda_1(\underset{\sim}{x}'; \underset{\sim}{x}'') + \phi^*(\underset{\sim}{x}'')\phi(\underset{\sim}{y}')\Lambda_1(\underset{\sim}{x}'; \underset{\sim}{y}'') + \phi^*(\underset{\sim}{y}'')\phi(\underset{\sim}{x}')\Lambda_1(\underset{\sim}{y}'; \underset{\sim}{x}'')\} \\ & + \Lambda_2(\underset{\sim}{x}', \underset{\sim}{y}'; \underset{\sim}{x}'', \underset{\sim}{y}'').\end{aligned} \tag{2.26}$$

The function Λ_2 is required to satisfy conditions analogous to the normal conditions imposed on Ω_2, e.g.,

$$\Lambda_2(\underset{\sim}{x}', \underset{\sim}{y}'; \underset{\sim}{x}'', \underset{\sim}{y}'') \longrightarrow \Lambda_1(\underset{\sim}{x}'; \underset{\sim}{x}'')\Lambda_1(\underset{\sim}{y}'; \underset{\sim}{y}'') \tag{2.27}$$

for $(\underset{\sim}{x}', \underset{\sim}{x}'')$ far from $(\underset{\sim}{y}', \underset{\sim}{y}'')$; and

$$\int \Lambda_2(\underset{\sim}{x}', \underset{\sim}{y}; \underset{\sim}{x}'', \underset{\sim}{y})\, d\underset{\sim}{y} = (N_d - 1)\, \Lambda_1(\underset{\sim}{x}'; \underset{\sim}{x}''), \tag{2.28}$$

where

$$N_d = \int \Lambda_1(\underset{\sim}{x}; \underset{\sim}{x})\, d\underset{\sim}{x} = \int \rho_d(\underset{\sim}{x})\, d\underset{\sim}{x}.$$

The factor S(r) is a rapidly varying function of its argument which goes to zero for $r < r_c$, and approaches unity for $r > r_c$. [It will be seen in section 3 that S(r) has been calculated by McMillan [8] on the basis of a rather realistic model for ^{4}He II.]

More general forms than equation (2.26) are certainly possible which satisfy the conditions mentioned below equation (2.22). For example, the S factors can be generalized to three-point functions [9]. Consistency of the form (2.26) in this case has been shown [10] by examination of its implications for the hierarchy (2.6) for $n \geq 2$.

As a passing remark, it would appear useful to apply the conditions below equation (2.21) to test an Ω_2 resulting from any trial, or approximate, solution of the many-body problem. Two examples suffice: The ideal Bose gas in the grand canonical ensemble as usually treated leads to the exact Ω_2 in the form

$$\Omega_2(\underset{\sim}{x}', \underset{\sim}{y}'; \underset{\sim}{x}'', \underset{\sim}{y}'') = \Omega_1(\underset{\sim}{x}'; \underset{\sim}{x}'')\Omega_1(\underset{\sim}{y}'; \underset{\sim}{y}'') + \Omega_1(\underset{\sim}{x}'; \underset{\sim}{y}'')\Omega_1(\underset{\sim}{y}'; \underset{\sim}{x}''). \tag{2.29}$$

This is satisfactory as long as no single-particle state is macroscopically occupied (i.e., ODLRO present in Ω_1); but when ODLRO is present in Ω_1, equation (2.29) no longer can satisfy the integration condition, equation (2.23). The grand canonical ensemble is inappropriate when condensation is present. Certain "pair models," for example that of Valatin and Butler [11], suffer from the same difficulty when ODLRO is present in Ω_1, for much the same reason--that is, the condition (2.23) cannot be met. This model has also previously been shown [12] to lead to an increasing $\Omega_2(\underset{\sim}{x}, \underset{\sim}{y}; \underset{\sim}{x}, \underset{\sim}{y})$ as $|\underset{\sim}{x} - \underset{\sim}{y}| \to 0$, in violation of the "core" condition mentioned below (2.23).

Yang [1] has shown that a system of identical fermions will exhibit ODLRO first in Ω_2, with a maximum possible eigenvalue equal to N, the fermion number. (This is to be contrasted to the boson case where the largest eigenvalue in Ω_2 is of order N^2.) The BCS pairing ansatz of superconductivity leads to the large eigenvalue being the maximum possible, namely, N; flux quantization is implied by ODLRO as well [1].

Ω_2 possesses the diagonal eigenvalue expansion

$$\Omega_2(\underset{\sim}{x}', \underset{\sim}{y}'; \underset{\sim}{x}'', \underset{\sim}{y}'') = \sum_i \lambda_i^{(2)} \phi_i^*(\underset{\sim}{x}'', \underset{\sim}{y}'')\phi_i(\underset{\sim}{x}', \underset{\sim}{y}') , \tag{2.30}$$

and ODLRO for a system of identical fermions may be expressed by

$$\Omega_2 = [(\lambda_o^{(2)})^{\frac{1}{2}} \phi_o^*(\underset{\sim}{x}'', \underset{\sim}{y}'')] [(\lambda_o^{(2)})^{\frac{1}{2}} \phi_o(\underset{\sim}{x}', \underset{\sim}{y}')] + \tilde{\Omega}_2(\underset{\sim}{x}', \underset{\sim}{y}'; \underset{\sim}{x}'', \underset{\sim}{y}''), \tag{2.31}$$

where $\lambda_o^{(2)} = O(N)$ and $\tilde{\Omega}_2$ vanishes when the pair $(\underset{\sim}{x}', \underset{\sim}{y}')$ is far from the pair $(\underset{\sim}{x}'', \underset{\sim}{y}'')$. The macroscopic wave function

$$\phi(\underset{\sim}{x}', \underset{\sim}{y}') = (\lambda_o^{(2)})^{\frac{1}{2}} \phi_o(\underset{\sim}{x}', \underset{\sim}{y}') \tag{2.32}$$

will vanish when $|\underset{\sim}{x}' - \underset{\sim}{y}'|$ is greater than the "coherence length," or the distance beyond which an electron pair is uncorrelated. Use of equation (2.31) in the equation of motion for Ω_2, (2.6), in the limit that $(\underset{\sim}{x}', \underset{\sim}{y}')$ is very far from $(\underset{\sim}{x}'', \underset{\sim}{y}'')$, leads at once to an exact equation for $\phi(\underset{\sim}{x}', \underset{\sim}{y}')$ [6]. The resulting equation is satisfied in equilibrium by the conventional microscopic (BCS) theory when certain reasonable assumptions are made concerning the three-body correlations [13]. Further apparently reasonable assumptions are required to obtain the Landau-Ginzberg [14] equation when a magnetic field is present.

Having briefly reviewed the basic concepts involved, attention will now be turned to application of equation (2.26) to the problem of superfluid helium (^{4}He II).

3. CONDENSATE FRACTION IN ^{4}He II

It is widely believed that the ratio of the (bulk) condensate to the total density $\equiv \rho_o/\rho$ (see below equation 2.16) is in the range of 0.05 [15] to 0.15 [8] at T = 0°K for ^{4}He II, based on theoretical estimates. An experimental confirmation of this fact would be very desirable. In the present section a method [16, 17] will be presented for determining the condensate density fraction for all temperatures from knowledge of only g(r, T), the pair distribution function.

It must first be understood that ODLRO in Ω_1 is a statement that each helium atom must be regarded as being partly in the condensate, and partly localized within a distance given by the range of $\Lambda_1(|\underset{\sim}{x}' - \underset{\sim}{x}''|)$. The situation may be pictured as follows as the temperature is varied: above the lambda transition, $T_\lambda = 2.18°$ K, the helium atoms are localized to within an angstrom or so, and as the temperature is lowered in this region, the diminishing thermal motion, and increasing localization, is reflected experimentally in a heightening of the maxima of $g(r, T > T_\lambda)$. As the temperature is lowered below T_λ, each helium atom will begin to contribute to the (structureless) uniform condensate density. Scattering of X-rays or neutrons will take place only from the localized "lumps," the depletion part of the density associated with Λ_1. Since total density is nearly constant, as the temperature is lowered in the region below T_λ, the "melting" of the "lumps" into the condensate will be reflected as a diminishing scattering intensity from the depletion, and a resultant lowering of the maxima of $g(r,T)$ with decreasing temperature. Thus, from this picture, a reversal in trend of maxima height of $g(r,T)$ is predicted as T is lowered through T_λ.

It may further be expected that the relative localization, or structure, of the lumps will remain relatively unchanged as T decreases below T_λ. Above T_λ the system can lower both average kinetic and potential energy by decreasing thermal motion, since the average potential energy per particle is

$$\langle V\rangle = \tfrac{1}{2}\rho \int V(r)g(r)d\underset{\sim}{r} \tag{3.1}$$

(fig. 1), and the largest maxima of $g(r)$ multiplies the negative part of $V(r)$. The lambda transition is presumably the point below which the system cannot decrease energy in this manner any longer, for increasing localization with decreasing temperature would lead to an increasing kinetic energy via the uncertainty principle. Below this point, $T = T_\lambda$, energy can be lowered further by having each helium particle go partly into the condensate, which results in a lowering of both kinetic and potential energy.

To make the above intuitive picture quantitative, we write equation (2.26) for Ω_2 in the bulk ($\equiv$ translational-rotational invariance) situation as

$$\begin{aligned}\Omega_2(\underset{\sim}{x}, \underset{\sim}{y}; \underset{\sim}{x}, \underset{\sim}{y}) &= \Omega_2(|\underset{\sim}{x} - \underset{\sim}{y}|) = \Omega_2(r) \equiv \rho^2 g(r,T) \\ &= S^2(r)(\rho_o^{\,2} + 2\rho_o\rho_d + 2\rho_o\Lambda_1(r)) + \Lambda_2(r) \, .\end{aligned} \tag{3.2}$$

The function $\Lambda_1(0) = \rho_d$ and $\Lambda_1(r) = 0$ for $r > \ell_1$, where ℓ_1 is about 4 Å [8] (fig. 1). Also, $\Lambda_2(0) = 0$, and $\Lambda_2(r) = \rho_d^{\,2}$ for $r > \ell_2$, where ℓ_2 is several times larger than ℓ_1. Thus define

$$\Lambda_1(r) = \rho_d h(r) \, , \tag{3.3}$$

and

$$\Lambda_2(r) = \rho_d^{\,2}\,\tilde{g}(r,T) \, . \tag{3.4}$$

It will be assumed in what follows that we are at zero pressure, and that $h(r)$ and $\tilde{g}(r,T)$ are functions of $|\underset{\sim}{r}|$ and temperature only. (A theoretical determination of $h(r)$ at $T = 0°$ K has been given by McMillan [8].) Above T_λ where $\rho_o = 0$, $\tilde{g}(r) = g(r)$.

Consider the expression (3.2) in the region $\ell_1 < r < \ell_2$, where

$$\Omega_2 = \rho^2 g(r,T) = \rho_o^2 + 2\rho_o\rho_d + \rho_d^2 \tilde{g}(r,T) ,$$

or

$$\rho^2[g(r,T) - 1] = \rho_d^2[\tilde{g}(r,T) - 1] \quad (\ell_1 < r < \ell_2) . \tag{3.5}$$

Equation (3.5) is an exact consequence of equation (2.26) for $\ell_1 < r < \ell_2$.

In keeping with the discussion above equation (3.2), $\tilde{g} - 1$, which represents the structure of the depletion lumps, may be expected to be given by its value at $T \simeq 2.2°\,K \simeq T_\lambda$, which is the same as g − 1, which is measurable. Then with this assumption, equation (3.5) becomes

$$\rho_d(T)/\rho = \left[\frac{g(r,T) - 1}{g(r,T \simeq T_\lambda) - 1}\right]^{\frac{1}{2}} \quad (\ell_1 < r < \ell_2;\ T < T_\lambda) . \tag{3.6}$$

Thus equation (3.6) gives ρ_d/ρ as a function of the measurable g(r,T) − 1.

Since the largest experimental error in the liquid structure factor $S(\underset{\sim}{k})$ occurs for very small momentum transfer $\underset{\sim}{k}$, we expect the maxima of g(r) to be increasingly in error as r is increased, since g is related to S(k) by a Fourier transform, and thus the most reliable data for ρ_d/ρ is expected to come from the second maximum or minimum of g, even though ideally equation (3.6) should be independent of r for $\ell_1 < r < \ell_2$.

Existing experimental data for g are available at only two temperatures above T_λ [18] and unfortunately one of them is the boiling point $T_B \simeq 4.2°$ K. The other is 2.4° K. By using g(r, T = 2.4° K) − 1 from Gordon et al. [18] in the denominator of equation (3.6), it can be estimated from data of (g − 1) at T = 1.4° K (second maximum of g)

$$\rho_d/\rho \simeq 0.9 \quad (T = 1.4°\,K) . \tag{3.7}$$

The data of Achter and Meyer [19] at 0.79° K lead to a similar result.

If the assumption of temperature independence of maxima of $\tilde{g}$ below T_λ is not made in equation (3.5) to obtain equation (3.6) but instead a linear extrapolation of the second maximum of g is made from the two points at T = 2.4° and 4.2° K (boiling point), it is found that $\rho_d/\rho \simeq 0.8$ at T = 1.4° K, which apparently represents a lower limit. By using the scant data available, it can be estimated that the ratio in equation (3.5) is independent of r roughly to within 10 per cent. The zeros of g − 1 remain constant to within about 1 per cent.

More data are clearly necessary to determine ρ_d/ρ as a function of temperature, and it will be interesting to see how this compares with the superfluid fraction $\rho_s(T)/\rho$. The reversal in trend of peak heights of g as T decreases through T_λ (fig. 1) seems to clearly reflect the existence of a single-particle condensate, and thus the presence of ODLRO in Ω_1.

One question which arises is whether S(r) of equation (3.2) can be truly said to be unity for $r > \ell_1$, as has been assumed. This cannot be answered definitively,[1] but it is very pertinent to realize that S(r) has been calculated at a temperature of absolute zero from first principles by McMillan [8]; and, at least in his calculation, such is the case. McMillan calculates a

[1] Equation (2.21) implies that S(r) = 1 for $r > \ell_1$, since $F(\underset{\sim}{x}')$ is independent of $\underset{\sim}{x}''$ for $|\underset{\sim}{x}' - \underset{\sim}{x}''| > \ell_1$.

quantity, called the pairing function, given by

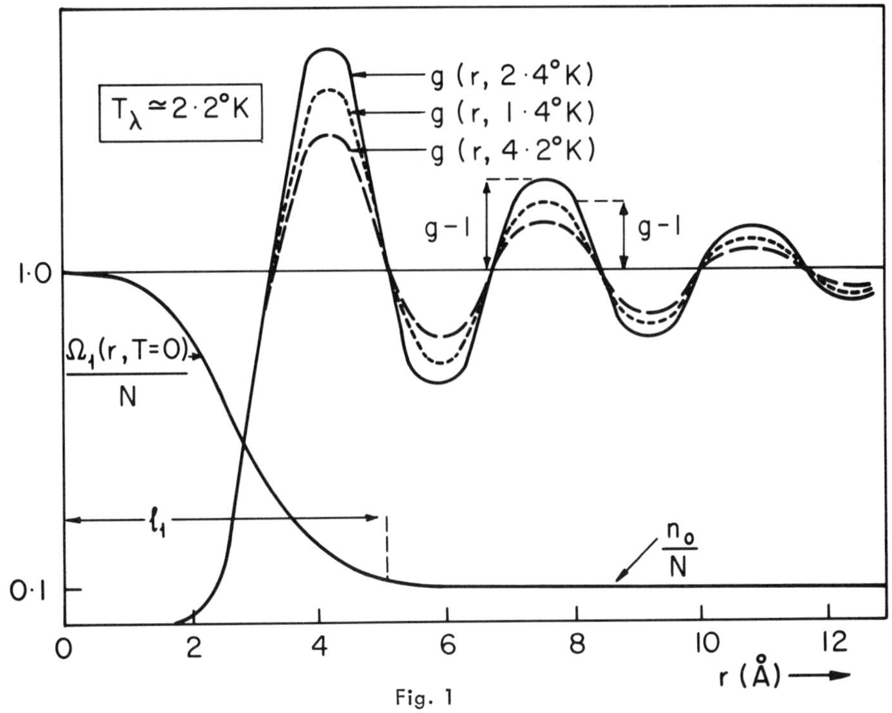

Fig. 1

$$m(\underset{\sim}{x}_N - \underset{\sim}{x}_{N+1}) = \int \Psi(\underset{\sim}{x}_1, \underset{\sim}{x}_2, \ldots, \underset{\sim}{x}_{N-1}) \Psi(\underset{\sim}{x}_1, \underset{\sim}{x}_2, \ldots \underset{\sim}{x}_{N-1}, \underset{\sim}{x}_N, \underset{\sim}{x}_{N+1}) d\underset{\sim}{x}_1 \ldots d\underset{\sim}{x}_{N-1}, \tag{3.8}$$

where Ψ is the (real) ground-state wave function, and m can also be expressed in terms of the particle annihilation operator $\psi(\underset{\sim}{x})$ and the ground states for N and N + 2 particles as

$$m(\underset{\sim}{x}_N - \underset{\sim}{x}_{N+1}) = \langle N | \psi(\underset{\sim}{x}_N)\psi(\underset{\sim}{x}_{N+1}) | N+2 \rangle . \tag{3.9}$$

Reference to the definition of Ω_2, equation (2.1), as well as to the form equation (2.26) in the limit as $(\underset{\sim}{x}', \underset{\sim}{y}')$ is taken far from the pair $(\underset{\sim}{x}'', \underset{\sim}{y}'')$, leads to the identification

$$\Omega_2 \longrightarrow m(\underset{\sim}{x}'' - \underset{\sim}{y}'')m(\underset{\sim}{x}' - \underset{\sim}{y}') \tag{3.10}$$

in the bulk, so that we have

$$\rho_o S(r) = m(r) . \tag{3.11}$$

With the approximation that Ψ is a Jastrow product of functions $f(\underset{\sim}{x}_i - \underset{\sim}{x}_j)$,

$$m(r) = \Omega_1(r)f(r) \tag{3.12}$$

(equation [19] of reference [8]), and since f(r) approaches unity for large r, m(r) approaches ρ_o, the condensate density, in this same limit. Thus S(r) can be computed, since both Ω_1 and f are given by reference [8]; S has the same general shape as g(r) for r below about 4 Å, after which it quickly becomes unity, whereas g(r) continues to oscillate.

It is possibility worth noting that McMillan's g(r) shows maxima noticeably lower than the 1.4° K experimental data of reference [19], which fact seems to have been accepted as a lack of agreement; from the previous discussion of this section, however, such is the expected behavior for g(r).

It is interesting to compute the ground-state energy per particle at zero pressure by use of equation (2.20) with equations (2.26), (3.8), and (3.9). Since in the bulk ground state we may suppose that the macroscopic wave function ϕ may be expressed as

$$\phi(\underset{\sim}{x},t) = \langle N|\psi(\underset{\sim}{x},t)|N+1\rangle = \langle N|e^{i\mathcal{H}t}\psi(\underset{\sim}{x})e^{-i\mathcal{H}t}|N+1\rangle$$
$$= \rho_o^{\frac{1}{2}} \exp[-i(E_{N+1} - E_N)] ,$$

then

$$\phi(\underset{\sim}{x},t) = \rho_o^{\frac{1}{2}} e^{-i\mu t} , \tag{3.13}$$

where μ is the chemical potential. Use of equations (3.13) and (2.26) in equation (2.20) gives

$$\hbar\mu = \int V(r)\, S(r)[\rho + \rho_d h(r)]\, d\underset{\sim}{r} \tag{3.14}$$

for the ground-state energy per particle at zero pressure. All functions under the integral, including the constant depletion density ρ_d, are consistently given by reference [8]. The numerical value of the integral is then -7.1° K [20]. The close agreement with the experimental value of -7.2° K is very possibly fortuitous; McMillan's value for the ground-state energy per particle is computed from an exact expression for the energy and is about 18% higher than the experimental result, but his calculation involved both the curvature of f as well as f itself.

4. EQUATIONS OF VELOCITY FIELDS AT T = 0° K

In this section an equation will be derived for the velocity of mass transport at T = 0° K for a system of high density bosons in the presence of ODLRO. This velocity, which we denote by $\underset{\sim}{v}_s$, is not curl-free, as is the velocity associated with the gradient of the phase of the condensate wave function. The mass current is given by

$$\underset{\sim}{j} = \rho\underset{\sim}{v} = \frac{\hbar}{2im} \lim_{\underset{\sim}{x}\to\underset{\sim}{x}'} (\nabla_{\underset{\sim}{x}} - \nabla_{\underset{\sim}{x}'})\, \Omega_1(\underset{\sim}{x};\underset{\sim}{x}') , \tag{4.1}$$

and use of the ODLRO condition for Ω_1, equation (2.15), gives

$$\underset{\sim}{j} = \rho_o\underset{\sim}{v}_o + \rho_d\underset{\sim}{v}_d , \quad \nabla\times\underset{\sim}{v}_o = 0 , \tag{4.2}$$

where

$$\underset{\sim}{v}_o = \frac{\hbar}{m}\nabla\theta , \tag{4.3}$$

θ is the phase of the condensate wave function ϕ, and

$$\underset{\sim}{v}_d = \frac{\hbar}{2m} \lim_{\underset{\sim}{x}\to\underset{\sim}{x}'} (\nabla_{\underset{\sim}{x}} - \nabla_{\underset{\sim}{x}'})\, \chi(\underset{\sim}{x};\underset{\sim}{x}') , \tag{4.4}$$

where χ is the phase of $\Lambda_1(\underset{\sim}{x};\underset{\sim}{x}')$.

The energy of the system will now be expressed as a function of $\underset{\sim}{v}_o$ and $\underset{\sim}{v}_d$, and it will be assumed that gradients of densities, $\nabla\rho_o$ and $\nabla\rho_d$, can be neglected, which is a high-density limit. We will require the mass-flow velocity field, called $\underset{\sim}{v}_s$, which minimizes the energy, subject to the condition (4.2). The energy E is given by [4]

$$E = \int \mathcal{E}(\underset{\sim}{R})d\underset{\sim}{R} = \int -\frac{\hbar^2}{2m} \lim_{\underset{\sim}{r}\to 0} \nabla_{\underset{\sim}{r}}^2 \Omega_1(\underset{\sim}{r}, \underset{\sim}{R})d\underset{\sim}{R} + \tfrac{1}{2}\int\int V(\underset{\sim}{x}-\underset{\sim}{y})\,\Omega_2(\underset{\sim}{x}, \underset{\sim}{y}; \underset{\sim}{x}, \underset{\sim}{y})d\underset{\sim}{x}\,d\underset{\sim}{y}\,, \tag{4.5}$$

where $\underset{\sim}{r} = \underset{\sim}{x} - \underset{\sim}{x}'$, and $\underset{\sim}{R} = (\underset{\sim}{x} + \underset{\sim}{x}')/2$.

Using the forms for Ω_1 and Ω_2 from equations (2.15) and (2.26) gives

$$\mathcal{E}(\underset{\sim}{R}) = \tfrac{1}{2}\rho_o v_o^2 + \tfrac{1}{2}\rho_d v_d^2 - \frac{\rho_o\rho_d}{2\rho}(\underset{\sim}{v}_d - \underset{\sim}{v}_o)^2 - \frac{\hbar^2}{2m}\lim_{\underset{\sim}{r}\to 0}\nabla_{\underset{\sim}{r}}^2 |\Omega_1(\underset{\sim}{x}; \underset{\sim}{x}')|$$

$$+ \tfrac{1}{2}\int V(r)S^2(r)\Lambda_1(\underset{\sim}{r}, \underset{\sim}{R})\phi^*(\underset{\sim}{R}+\underset{\sim}{r}/2)\phi(\underset{\sim}{R}-\underset{\sim}{r}/2)d\underset{\sim}{r} + \text{(complex conjugate)}$$

$$+ \tfrac{1}{2}\int V(r)\Lambda_2(\underset{\sim}{r}, \underset{\sim}{v}_d(\underset{\sim}{R}), \rho_d)d\underset{\sim}{r} + (\text{terms independent of } \underset{\sim}{v}_o \text{ and } \underset{\sim}{v}_d)\,. \tag{4.6}$$

The first three terms of equation (4.6) are simply $\tfrac{1}{2}\rho v^2$, and the fourth term is the "internal" kinetic energy (independent of the macroscopic flow velocities). The exponential function in the fifth term is now expanded [6], and only terms up to quadratic in the velocity $\underset{\sim}{v}_d - \underset{\sim}{v}_o$ are retained; the amplitude of $\Lambda_1(\underset{\sim}{r}, \underset{\sim}{R})$ is given by $\rho_d\, h(r)$. The term involving Λ_2, the "normal" part of Ω_2, is considered to be only a function of $\underset{\sim}{r}$, $\underset{\sim}{v}_d(\underset{\sim}{R})$, and ρ_d; symmetry considerations lead to a contribution proportional to $(\nabla \times \underset{\sim}{v}_d)^2$, $(\nabla \cdot \underset{\sim}{v}_d = 0)$.

Thus we find to this order

$$\mathcal{E}(\underset{\sim}{R}) = \tfrac{1}{2}\rho_o v_o^2 + \tfrac{1}{2}\rho_d v_d^2 - \tfrac{1}{2}\frac{\rho_o\rho_d}{\rho}(1+a)(\underset{\sim}{v}_d - \underset{\sim}{v}_o)^2$$

$$+ \tfrac{1}{2}\zeta(\nabla \times \underset{\sim}{v}_d)^2 + \epsilon(\rho_o, \rho_d)\,. \tag{4.7}$$

Here the dimensionless constant a is defined by

$$a = \frac{4\pi}{3}\frac{m\rho}{\hbar^2}\int V(r)S^2(r)h(r)r^4\,dr\,, \tag{4.8}$$

and ϵ contains all contributions to $\mathcal{E}$ which are independent of $\underset{\sim}{v}_d$ and $\underset{\sim}{v}_o$.

Now we will find the configuration of $\underset{\sim}{v}_d(R)$ which gives a minimum to E subject to equation (4.2). Thus we set the variation with respect to $\underset{\sim}{v}_d$ and $\underset{\sim}{v}_o$ of E' to zero, where

$$E' = \int [\mathcal{E} - \underset{\sim}{A}\cdot\underset{\sim}{j}]d\underset{\sim}{R}. \tag{4.9}$$

This leads at once to the equations

$$\underset{\sim}{A} = \underset{\sim}{v}_o + (1+a)\frac{\rho_d}{\rho}(\underset{\sim}{v}_d - \underset{\sim}{v}_o)\,, \tag{4.10}$$

$$\rho_d\underset{\sim}{v}_d - \underset{\sim}{A}\rho_d - (1+a)\frac{\rho_d\rho_o}{\rho}(\underset{\sim}{v}_d - \underset{\sim}{v}_o) + \zeta\nabla\times(\nabla\times\underset{\sim}{v}_d) = 0\,. \tag{4.11}$$

Elimination of the Lagrange multiplier function $\underset{\sim}{A}$ gives

$$\zeta\nabla\times(\nabla\times\underset{\sim}{v}_d) = a\,\rho_d(\underset{\sim}{v}_d - \underset{\sim}{v}_o)\,. \tag{4.12}$$

Since $\nabla\times\underset{\sim}{v}_o = \nabla\cdot\underset{\sim}{v}_o = \nabla\cdot\underset{\sim}{v}_d = 0$, by use of equation (4.2) we also have the equation for the mass flow velocity $\underset{\sim}{v}_s$,

$$\lambda^2\nabla\times(\nabla\times\underset{\sim}{v}_s) = \underset{\sim}{v}_s - \underset{\sim}{v}_o \quad (T = 0^\circ K)\,, \tag{4.13}$$

where

$$\lambda^2 = \zeta / a\rho_d \ . \tag{4.14}$$

Equation (4.13) has been mentioned previously by Sarfatt [21], and later discussed by Thouless [22] who gave a quasi-phenomenological derivation of it. This equation along with $\nabla^2 \underset{\sim}{v}_o = 0$ shows that everywhere except in a region of thickness λ (presumably of order of an angstrom) the solution coincides with the two-fluid result, $\underset{\sim}{v}_s = \underset{\sim}{v}_o$ and $\underset{\sim}{j} = \rho \underset{\sim}{v}_s$; the entire fluid gets carried along at the curl-free condensate velocity except near the boundaries where presumably $\underset{\sim}{v}_s \simeq 0$. It may also be amusing to note the formal correspondence between equation (4.13), written as

$$\lambda^2 \nabla^2 (\underset{\sim}{v}_o - \underset{\sim}{v}_s) = (\underset{\sim}{v}_o - \underset{\sim}{v}_s) \ , \tag{4.15}$$

and an equation for the vector potential in the simplest theory of superconductivity.

Retention of terms in the energy which will lead to terms in equation (4.13) nonlinear in $(\underset{\sim}{v}_s - \underset{\sim}{v}_o)$ will lead to a "critical velocity" phenomenon, so that $\underset{\sim}{v}_o$ cannot be larger than a certain value. The upper limit on $\underset{\sim}{v}_o$ may perhaps be expected to be given by the slope $v_c = \epsilon(p)/p$, where $\epsilon(p)$ is the excitation spectrum given by neutron-scattering experiments, since we have in section 4 given a high-density, high-pressure theory, where vortices are suppressed as a low-lying excitation.

REFERENCES

1. C. N. Yang, Rev. Mod. Phys. 34, 694 (1962); O. Penrose, Phil. Mag. 42, 1373 (1951) first generalized the concept of condensation in an ideal Bose gas to interacting systems. Cf. also G. V. Chester, in International School of Physics "Enrico Fermi," course 21, "Liquid Helium" (New York: Academic Press, 1963).

2. R. J. Glauber, Phys. Rev. 130, 2529 (1963).

3. H. Fröhlich, Int. J. Quant. Chem. 2, 641 (1968).

4. H. Fröhlich, Physica, 37, 215 (1967). Equation (3.3) of this reference should have $\frac{1}{2}[\rho(\underset{\sim}{y})\rho(\underset{\sim}{x}',\underset{\sim}{x}'') + \rho(\underset{\sim}{x}',\underset{\sim}{x}'')\rho(\underset{\sim}{y})]$ under the integral.

5. H. Fröhlich, J. Phys. Soc. Japan, 26 (Suppl.), 189 (1969).

6. H. Fröhlich, Phys. Kondens. Materie, 9, 350 (1969).

7. G. J. Hyland and G. Rowlands, Phys. Kondens. Materie (to be published).

8. W. L. McMillan, Phys. Rev. 138, A442 (1965).

9. H. Fröhlich, four lectures delivered at the Summer Semester of 1969 at the Institute for Theoretical Physics, University of Stuttgart, ed. R. Hubner.

10. P. Lal and C. Terreaux, Phys. Kondens. Materie (to be published).

11. J. G. Valatin and D. Butler, Nuovo Cimento, 10, 37 (1958).

12. M. Girardeau and R. Arnowitt, Phys. Rev. 113, 755 (1959).

13. A. W. B. Taylor, J. Phys. C (Solid State) Letters, 3, L37 (1970).

14. G. V. Ginzberg and L. Landau, JETP (USSR), 20, 1064 (1950).

15. O. Penrose and L. Onsager, Phys. Rev. 104, 577 (1956).

16. G. Hyland, G. Rowlands, and F. W. Cummings, Phys. Letters, 31A, 465 (1970).

17. F. W. Cummings, G. J. Hyland, and G. Rowlands, Phys. Kondens. Materie, 12, 90 (1970). For a high-momentum-transfer neutron-scattering method, see W. C. Kerr, K. N. Pathak, and K. S. Singwi, Phys. Rev. 2A, 2416 (1970).

18. W. L. Gordon, C. H. Shaw, and J. C. Daunt, Phys. Chem. Solids, 5, 117 (1958).

19. E. K. Achter and L. Meyer, Phys. Rev. 188, 291 (1969).

20. K. L. Bengtson, private communication.

21. J. Sarfatt, Phys. Letters, 24A, 287 (1967); 24A, 399 (1967).

22. D. J. Thouless, Annals Phys. 52, 403 (1969).

DISCUSSION

K. Matsuma: Can you get a somewhat singular behavior in the pair correlation function near the λ point within your mean field theory ?

F. W. Cummings: I don't see why not. For example, $\tilde{g}(r)$ or ρ_o can have discontinuous derivatives at $T = T_\lambda$ (if I understand the question).

H. Matsuda: Can there be a state of the system of atoms in which ODLRO and On-Diagonal Long-Range Order coexist? I think the discussion of Onsager and Penrose's paper that there is no Bose condensation in solid ^4He is not satisfactory.

F. W. Cummings: I think that Penrose and Onsager's treatment is only for translationally invariant systems. Chester has recently discussed this matter, and pointed out that one can have both diagonal (spatial) long-range order in Ω_2 and Bose condensation (ODLRO in Ω_1) in finite systems.

M. E. Girardeau: I didn't quite understand what you meant by your statement that the condensate should not scatter. One can calculate the neutron-scattering function $S(q, \omega)$ of the ground state of the ideal Bose gas (which is 100% condensate) and one finds that it is proportional to $\delta(\omega - 1/2q^2)$. This is certainly not zero, though it is rather trivial.

F. W. Cummings: Perhaps I should rephrase my comment. The difference between scattering from the ground state of an ideal Bose gas and the true liquid-helium ground state is that the scattering from the ground state of an ideal Bose gas (uniform condensate) is a sharp line, whereas that from real liquid helium is broadened. The condensate "scattering" does not affect the diffraction peaks of g(r).

M. Cole: With regard to the question of whether the condensate contributes to neutron scattering, the latter measures the dynamic structure factor $S(k, w)$ which is determined by matrix elements of the density fluctuation operator $S(k)$. This does include a term due to quasi-particle excitation from the condensate. Miller, Noyieres, and Pines showed in fact that in the long-wavelength limit, the scattering originates exclusively from this source. Incidentally, Hallock now recently published results for $S(k)$ at $T = .4°$ K which can further test your model.

F. W. Cummings: Yes, I agree that one gets a contribution to $S(k, \omega)$, but I would not expect it to contribute to the "structure" $[g(r) - 1]$ in the region considered. In the ideal gas at $T = 0°$ K, $S(k, \omega) = N\delta(\omega - 1/2\, k^2)$, but there is no contribution to $[g(r) - 1]$. I probably should amend my statement "no scattering" to mean in the above sense only.

J. Hansen: Could the variation in $S(r)$ peak heights not be explained simply by a competition between thermal and zero-point kinetic energies? Is the precision of $S(k)$ measurements sufficient to detect the $g(r)$ variations?

M. Nelkin: This question could be settled by looking at liquid ^{3}He.

F. W. Cummings: The experiments on $S(k)$ seem sufficiently precise to detect differences. I do not see that a reversal in trend for the peak heights of $g(r)$ could be explained as such a competition. I would think of the zero-point energy rather as establishing an upper limit to the peak heights.

G. Wannier: What is the relationship between the condensate discussed here, and the superfluid component discussed in the early two-fluid theories? Numerically, your condensate appears to be a much smaller fraction.

F. W. Cummings: At $T = 0°$ K, the superfluid fraction is 100 per cent, whereas the condensate fraction is much less. There is no simple relationship between these two densities. Gavoret and Nozieres (Ann. Phys., 1964) have shown that the entire mass of the fluid is uninfluenced by a small impressed rotational velocity; they use infinite-order perturbation summations, but their work is essentially rigorous. One might hope for a more intuitive connection. For what this is worth, in terms of the lumps and condensate picture, the lumps get "stuck" in the condensate at $T = 0°$ K, a finite energy is needed to dislodge a lump, so that everything moves in unison, i.e., $\rho = \rho_s$, and $\underset{\sim}{j} = \rho \underset{\sim}{v}_s$.

V. Liquids

DISTRIBUTION FUNCTIONS IN CLASSICAL AND QUANTUM FLUIDS

J. K. Percus

1. INTRODUCTION

I would like to describe a relatively new approach [1] to the analysis of stationary many-body systems, classical or quantal, pure state or thermal state. For several years, it has tantalizingly held just beyond reach the possibility of simple and direct, rigorous yet intuitively motivated, computation of observable quantities. It is in a way complementary to the standard battery of variational principles which remains the prime tool of so many effective investigations, with the innocuous substitution almost everywhere of the expression "lower bound" for "upper bound." But here the similarity ends, for the detailed N-body structure of the system at hand, or of the reference or model systems that in some fashion are always required, is not needed-- the low-order distributions for the former, a few expectation values for the latter, suffice. Another novelty is that interactions in unavailable regions of configuration space can often be eliminated at once, yielding a notably nonsingular structure. The net effect, when the technique is applicable, is to reduce computations to the back-of-the-postage-stamp variety. These are the cases I will emphasize, of course, leaving as starred exercises for the reader the corresponding analyses for the numerous identifiable many-body problems with which nature abounds.

2. PROTOTYPE: N $\longrightarrow$ 1 BODY REDUCTION [2]

It is intuitively transparent, but otherwise opaque, that the study of a particle system represented by a symmetric Hamiltonian can be reduced to that of the sub-Hamiltonian which when symmetrized yields the whole. To see how this might be done, let us consider a Hamiltonian which is the sum of identical one-body parts and ask for the quantum-mechanical ground-state energy (see fig. 1)

$$-\left(\frac{\partial^2}{\partial x_1^2} + \frac{\partial^2}{\partial x_2^2} + \ldots + \frac{\partial^2}{\partial x_N^2}\right)\psi_o(x_1 \ldots x_N) = \omega_o^{\,2}\,\psi_o(x_1 \ldots x_N) \;. \quad (1)$$

Supported in part by the U.S. Atomic Energy Commission, contract AT(30-1)-1480.

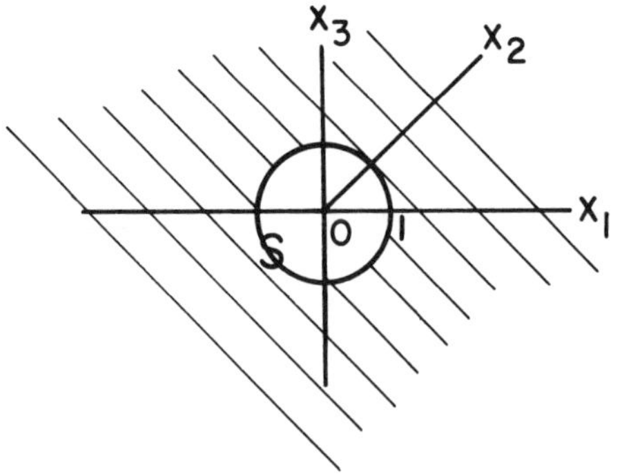

Fig. 1

To avoid triviality, we assume a hyperspherical boundary S:

$$\psi_o(x_1 \ldots x_N) = 0 \text{ when } x_1^2 + \ldots x_N^2 = 1 . \tag{2}$$

Thus we are really solving the scalar wave equation inside an N-dimensional spherical cavity.

Although the Cartesian symmetry of equation (1) has been broken, spherical symmetry remains and thus an exact solution can be written down, but let us make believe we do not know this. Standard procedure would then call for setting up the Rayleigh-Ritz variational principle

$$\omega_o^2 = \mathrm{Min}_\varphi \int \ldots \int_{\Sigma x_i^2 \le 1} \sum \left(\frac{\partial \varphi}{\partial x_i}\right)^2 dx_1 \ldots dx_N , \tag{3}$$

where

$$\int \ldots \int_{\Sigma x_i^2 \le 1} \varphi^2 \, dx_1 \ldots dx_N = 1 \text{ and } \varphi = 0 \text{ on } S.$$

One always has a ground-state wave function with the symmetry of the Hamiltonian, so the correct ψ_o can be assumed symmetric in $x_1, \ldots, x_N$. Thus equation (3) can be rewritten as

$$\omega_o^2 = \mathrm{Min}_\varphi \, N \int \ldots \int_{\underset{\sim}{x} \text{ inside } S} \frac{\partial}{\partial x} \frac{\partial}{\partial y} \varphi(x\, x_2 \ldots x_N) \times \varphi(y\, x_2 \ldots x_N) \Big|_{y=x} dx\, dx_2 \ldots dx_N \tag{4}$$

where

$$N \int \ldots \int_{\underset{\sim}{x} \text{ inside } S} \varphi(x_1 \ldots x_N)^2 \, dx_1 \ldots dx_N = N , \quad \varphi \text{ is symmetric, and } \varphi = 0 \text{ on } S.$$

Coordinates $x_2, \ldots, x_N$ are apparently coming along for the ride. To bury them completely, we need only define the one-body density matrix

$$\gamma(x,y) \equiv N \int \ldots \int \varphi(x\, x_2 \ldots x_N) \varphi(y\, x_2 \ldots x_N) \, dx_2 \ldots dx_N , \quad x^2 + x_2^2 + \ldots + x_N^2 = 1 , \quad y^2 + x_2^2 + \ldots + x_N^2 = 1 , \tag{5}$$

in which case equation (4) reduces to

$$\omega_o^2 = \mathrm{Min}_\gamma \int_{-1}^{1} \frac{\partial^2}{\partial x \partial y} \gamma(x,y) \Big|_{y=x} dx \quad . \tag{6}$$

If γ is regarded as the kernel of a one-body operator, we have thus reduced the "N-body" problem to a "one-body" problem.

The principle (6) is of course a bit incomplete, since we have not specified directly what class of γ to vary over, except via the definition of equation (5). Some of the characteristics of one-body density matrices are easily written down by inspection of equation (5). These are

$$\gamma(x,y) = \gamma(y,x) \quad , \tag{7a}$$

$$\int_{-1}^{1} \gamma(x,x)\, dx = N \, , \tag{7b}$$

$\gamma(x,y)$ is positive semidefinite as a matrix, (7c)

$$\gamma(\pm 1, x) = 0 \, , \tag{7d}$$

$$\left[\int_{-1}^{1} x^2 \gamma(x,x) dx \le 1 , \ . \ . \ . \right] . \tag{7e}$$

Equation (7c) means that

$$\int_{-1}^{1} \int f(x) f(y) \gamma(x,y) dx\, dy \ge 0 \, ,$$

following from writing the integral as

$$\int \cdot \cdot \cdot \int_{x_2^2 + \ . \ . \ . \ + x_N^2 \le 1} \left(\int_{x^2 \le 1 - \sum_2^N x_i^2} f(x) \varphi (x\, x_2 \ . \ . \ . \ x_N) dx \right)^2 dx_2 \ . \ . \ . \ dx_N \, ,$$

while equation (7e) is representative of a number of possible inequalities, and itself is a consequence of the obvious

$$\int \cdot \cdot \cdot \int \sum x_i^2 \varphi^2 dx_1 \ . \ . \ . \ dx_N \le \int \cdot \cdot \cdot \int \varphi^2 dx_1 \ . \ . \ . \ dx_N = 1 \quad .$$

If our catalog of conditions is not large enough to guarantee that γ be realizable in the form of equation (5), we are varying over a class "γ" including the legitimate one-body density matrices, and more. Thus the minimum attained is too low, and equation (6) must be replaced by

$$\omega_o^2 \ge \mathrm{Min}_{"\gamma"} \int_{-1}^{1} \frac{\partial^2}{\partial x \partial y} \gamma(x,y) \Big|_{y=x} dx \, , \tag{8}$$

a lower-bound principle.

Now let us evaluate expression (8) using only the nonspecific universal conditions (7a), (7b), (7c), and boundary condition (7d). The real matrix γ is positive semidefinite Hermitian, with Tr $\gamma = N$. Hence if we diagonalize by an orthonormal sequence, the diagonal elements will be nonnegative and add up to N:

$$\gamma(x,y) = \sum_\alpha{}' \, C_\alpha \varphi_\alpha(x)\, \varphi_\alpha(y) \, ,$$

where

$$C_\alpha \geq 0, \ \sum_\alpha C_\alpha = N, \text{ and } \varphi_\alpha(\pm 1) = 0 . \tag{9}$$

Inserting into equation (8), we obtain

$$\omega_o^2 \geq N \ \mathrm{Min}_{\{\varphi_\alpha, C_\alpha\}} \sum_\alpha \frac{C_\alpha}{N} \int_{-1}^{1} \left(\frac{\partial \varphi_\alpha(x)}{\partial x}\right)^2 dx . \tag{10}$$

But since the average over $\alpha\left(\sum C_\alpha/N = 1\right)$ can only exceed the smallest member, then (barring degeneracy) equation (10) reduces to a single term:

$$\omega_o^2 \geq N \ \mathrm{Min}_\varphi \int_{-1}^{1} \left(\frac{\partial \varphi}{\partial x}\right)^2 dx , \tag{11}$$

where

$$\int_{-1}^{1} \varphi^2 \, dx = 1 , \quad \varphi(\pm 1) = 0 .$$

This is the problem of the one-dimensional box, $\varphi'' + \Omega^2 \varphi = 0$, $\varphi(\pm 1) = 0$, with ground-state solution

$$\varphi(x) = \cos \tfrac{1}{2}\pi x, \quad \Omega = \tfrac{1}{2}\pi. \tag{12}$$

We conclude that

$$\omega_o \geq \tfrac{1}{2}\pi \ N^{\frac{1}{2}} . \tag{13}$$

How good is the result of equation (13)? Upon solving equations (1) and (2) exactly in ultraspherical coordinates, one readily finds

$$\psi_o(r) = r^{\frac{1}{2}N - 1} \ J_{\frac{1}{2}N - 1}(\omega_o r) , \qquad r \equiv \left(\sum x_i^2\right)^{\frac{1}{2}} , \tag{14}$$

where $J_{\frac{1}{2}N - 1}(\omega_o) = 0$. The resulting ω_o differs from equation (13) by less than 10% for $N = 2$, departs more and more as N increases, and for large N becomes asymptotically

$$\omega_o \cong \tfrac{1}{2} N + 1.856 \left(\tfrac{1}{2}N\right)^{1/3} + \ldots \ . \tag{15}$$

Thus equation (13) is certainly true for large N, but is a pretty miserable lower bound. The major trouble is clear enough. Equation (11) restricts each coordinate to $|x_i| \leq 1$, and consequently the configuration point $(x_1, \ldots, x_N)$ to a hypercube which for large N is enormously larger than the inscribed hypersphere--the true permitted volume. Direct evidence comes from computing

$$\int_{-1}^{1} x^2 \, \gamma(x,x) dx = N \int_{-1}^{1} x^2 \, \varphi(x)^2 \, dx$$

for $\varphi(x)$ of equation (12); we find

$$N \int_{-1}^{1} x^2 \, \varphi(x)^2 \, dx = N/7.7 , \tag{16}$$

so that equation (7e) fails for $N \geq 8$.

Now our path is clear. We must append condition (7e) when carrying out the minimization in equation (11). (Actually, two C_α in equation (10) can now be nonvanishing, but this does not affect the result.) The weakest form for equation (7e) is as an equality--when $N > 7$--and is most conveniently applied via a Lagrange parameter. Thus,

$$\omega_o^2 + \lambda^2 \geq N \operatorname{Min}_\varphi \int_{-1}^{1} \left[\left(\frac{\partial \varphi}{\partial x}\right)^2 + \lambda^2 x^2 \varphi^2 \right] dx , \tag{17}$$

where

$$\int_{-1}^{1} \varphi^2 dx = 1, \quad \int_{-1}^{1} x^2 \varphi^2 dx = \frac{1}{N}, \quad \varphi(\pm 1) = 0 .$$

For large N, the boundary condition plays no role, and we have only to solve the harmonic oscillator $-\varphi'' + \lambda^2 x^2 \varphi = \Omega^2 \varphi$. Hence $\varphi = \exp(-\frac{1}{2}\lambda x^2)$, $\Omega^2 = \lambda$, and $\int \varphi^2 dx = \lambda^2 \int x^2 \varphi^2 dx = \frac{1}{2}\lambda$, requiring $\lambda = N/2$. We therefore conclude that

$$\omega_o \geq N/2 , \tag{18}$$

which is a superb result for large N. With only the single special side condition (7e), the "N→1-body" reduction has performed as demanded.

3. CLASSICAL GROUND STATE: N ⟶ 2 BODY REDUCTION

For a symmetric sum of one-body terms in the Hamiltonian, the reduced-density-matrix technique permitted us to end up with a suitably restricted one-body problem. In nature, however, we typically meet two-body interaction potentials

$$V = \frac{1}{2} \sum_{i \neq j} V(\underset{\sim}{x}_i - \underset{\sim}{x}_j) , \tag{19}$$

and so presumably are faced with an effective two-body problem. This is really an extension by many orders of magnitude, and so let us concentrate upon the essential points by first restricting our attention to the classical ground state. That is, we seek the point minimum of equation (19) over all N-particle positions, or for that matter over all ensembles thereof, which certainly cannot lower the minimum.

Here the density matrix is superfluous. We need only consider the particle and particle-pair densities

$$n_1(\underset{\sim}{x}) \equiv \langle \sum_{i=1}^{N} \delta(\underset{\sim}{x} - \underset{\sim}{x}_i) \rangle$$

$$= n \qquad \text{for a nonuniform system;}$$

$$n_2(\underset{\sim}{x}, \underset{\sim}{y}) \equiv \langle \sum_{i \neq j} \delta(\underset{\sim}{x} - \underset{\sim}{x}_i)\delta(\underset{\sim}{y} - \underset{\sim}{x}_j) \rangle$$

$$= n^2 g(\underset{\sim}{x} - \underset{\sim}{y}) \qquad \text{for a uniform system.} \tag{20}$$

By "uniform" we refer to an ensemble which is translationally invariant in space. The densities n_1 and n_2 allow all one- and two-body expectations to be formed; thus

$$\langle \sum_i f(x_i) \rangle = \int n_1(\underset{\sim}{x}) f(\underset{\sim}{x})\, d\underset{\sim}{x},$$

and

$$V_{min} = \tfrac{1}{2}\, \text{Min}_{n_2} \int\int n_2(\underset{\sim}{x}, \underset{\sim}{y}) v(\underset{\sim}{x} - \underset{\sim}{y})\, d\underset{\sim}{x}\, d\underset{\sim}{y}$$

$$= N \frac{n}{2}\, \text{Min}_g \int g(\underset{\sim}{z}) v(\underset{\sim}{z})\, d\underset{\sim}{z} \tag{21}$$

for a uniform system.

Which brings us to the perennial question: What set of conditions allows us to fully characterize the class of radial distribution functions g?

Clearly g is nonnegative and normalized:

$$g(z) \geq 0\ , \tag{22a}$$

$$\int g(\underset{\sim}{z})\, d\underset{\sim}{z} = \frac{N-1}{N}\, \Omega\ , \tag{22b}$$

where Ω is the volume of the system. To this we shall add [3]

$$\tilde{h}(\underset{\sim}{k}) + \frac{1}{n} \geq 0\ , \tag{22c}$$

where $h(\underset{\sim}{z}) \equiv g(\underset{\sim}{z}) - 1$ is the pair correlation function, asymptotically tending to zero as $|z| \to \infty$ for a large disordered system, and $\tilde{h}(\underset{\sim}{k}) \equiv \int \exp(i\underset{\sim}{k} \cdot \underset{\sim}{z})\, h(\underset{\sim}{z})\, d\underset{\sim}{z}$ denotes a Fourier transform. The $\underset{\sim}{k} \to 0$ case of equation (22c) expresses the nonnegativity of the compressibility in a thermodynamic state. To prove equation (22c), we observe that $\langle |\sum_i' \exp(i\underset{\sim}{k} \cdot \underset{\sim}{x}_i)|^2 \rangle - |\langle \sum_i \exp(i\underset{\sim}{k} \cdot \underset{\sim}{x}_i) \rangle|^2 \geq 0$ becomes $\int\int \exp[i\underset{\sim}{k} \cdot (\underset{\sim}{x} - \underset{\sim}{y})]\, (n_2(\underset{\sim}{x}, \underset{\sim}{y}) + n_1(\underset{\sim}{x})\, \delta(\underset{\sim}{x} - \underset{\sim}{y}) - n_1(\underset{\sim}{x}) n_1(\underset{\sim}{y}))\, d\underset{\sim}{x}\, d\underset{\sim}{y} \geq 0$, or $\int \exp(i\underset{\sim}{k} \cdot \underset{\sim}{z})\, (g(\underset{\sim}{z}) + n^{-1}\, \delta(\underset{\sim}{z}) - 1)\, d\underset{\sim}{z} \geq 0$. Are these conditions sufficient? Of course not, and therefore using only some of the restrictions, we must replace equation (21) by

$$V_{min} \geq N \frac{n}{2}\, \text{Min}_{\text{"}g\text{"}} \int g(\underset{\sim}{z}) v(\underset{\sim}{z})\, d\underset{\sim}{z}, \tag{23}$$

another lower-bound principle.

Now let us see how effective expressions (22) and (23) can be. If we apply only equations (22a), (22b), positivity and normalization, equation (23) of course reduces to

$$V_{min} \geq \tfrac{1}{2} N(N-1)\, v_{min}\ , \tag{24}$$

where v_{min} is the point minimum of the pair potential (see fig. 2). For the dotted line pair potential shown, this indeed is fine: all particles can be at the same point in space, so that each of the $\frac{1}{2}N(N-1)$ pairs contributes v_{min}. Here we have an example of a nonsaturating potential, one for which V_{min}/N is not bounded from below as $N \to \infty$. But consider the solid-line potential shown for convenience in a one-dimensional box of length L. Here the minimum of $\int v(z) g(z)\, dz$ is achieved at the g shown, so that if $a > L/N$ we have the amusing but impossible result that the minimum separation between particles [where g(z) first exceeds zero] is greater than the average separation L/N. Of course equation (24) still does not saturate.

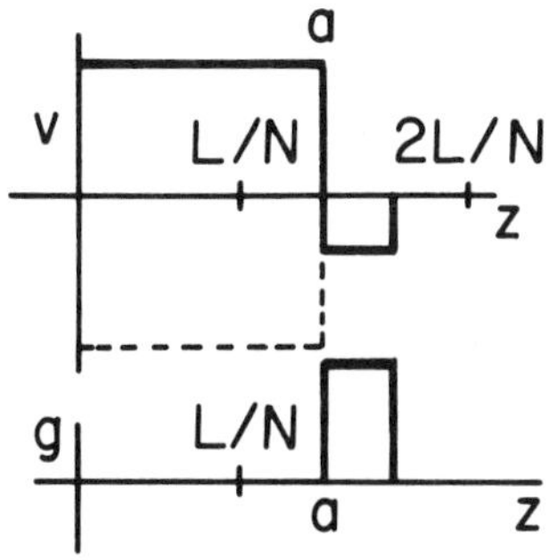

Fig. 2

What is wrong with the above example is that the correlation is too concentrated and its fluctuation is too large. But this is precisely what equation (22c) can test, and it fails dramatically for the above g, as can readily be shown. Can we then simultaneously make use of the positivity of g and of the "Ursell positivity" equation (22c)? Suppose in fact that we can write [4]

$$v(\underset{\sim}{z}) = u(\underset{\sim}{z}) + w(\underset{\sim}{z}) , \tag{25}$$

where $\tilde{u}(\underset{\sim}{k}) \geq 0, \; w(\underset{\sim}{z}) \geq 0,$

the sum of a "positive semidefinite" and a nonnegative potential. Then

$$\begin{aligned} V_{min} &\geq \tfrac{1}{2} Nn \int g(\underset{\sim}{z}) v(\underset{\sim}{z}) d\underset{\sim}{z} \\ &\geq \tfrac{1}{2} Nn \int g(\underset{\sim}{z}) u(\underset{\sim}{z}) d\underset{\sim}{z} \\ &= \tfrac{1}{2} Nn \left[\int h(\underset{\sim}{z}) u(\underset{\sim}{z}) d\underset{\sim}{z} + \int u(\underset{\sim}{z}) d\underset{\sim}{z} \right] \\ &= \tfrac{1}{2} Nn \left[\frac{1}{\Omega} \sum \tilde{h}(\underset{\sim}{k}) \tilde{u}(\underset{\sim}{k}) + \tilde{u}(0) \right] \\ &\geq \tfrac{1}{2} N \left[- \frac{1}{\Omega} \sum \tilde{u}(\underset{\sim}{k}) + n\tilde{u}(0) \right] \end{aligned} \tag{26}$$

and we conclude that

$$V_{min}/N \geq \tfrac{1}{2} (n\tilde{u}(0) - u(0)) , \tag{27}$$

which does at least saturate.

How effective then is equation (27)? It does have the interesting and unique property previously referred to that a highly singular repulsive portion of the potential can (and in fact, must) be dropped prior to the computation. As an example, we consider (see fig. 3) the "one-dimensional Lennard-Jones" potential

$$v(x) = \frac{1}{x^8} - \frac{1}{x^4} , \tag{28}$$

with a minimum of $-1/4$ at $x^4 = 2$. As an off-the-cuff approximation, we use a nonsingular repulsion and asymptotic x^{-4} attraction,

$$u(x) = \frac{4B}{(x^2 + \gamma^2)^3} - \frac{A}{(x^2 + \gamma^2)^2} , \tag{29}$$

which is positive semidefinite if $\gamma^2 A \leq 3B$. Then $u(x)$ and $v(x)$ will be tangent at $x^4 = 2$ with

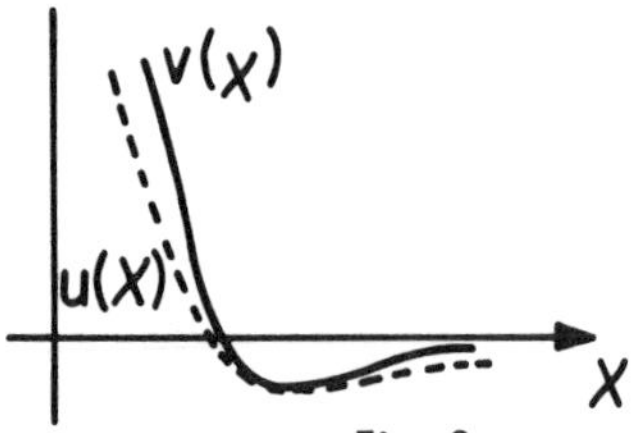

Fig. 3

u(x) always lying below (indeed much below) if

$$A = \frac{3}{4}\,(\sqrt{2} + \gamma^2)^2 \;, \quad B = \frac{1}{8}\,(\sqrt{2} + \gamma^2)^3 \,. \tag{30}$$

From

$$u(0) = \frac{4B}{\gamma^6} - \frac{A}{\gamma^4} \;, \quad \tilde{u}(0) = \frac{\pi}{2\gamma^5}(3B - \gamma^2 A) \;, \tag{31}$$

equation (27) can be maximized with respect to the available parameter γ. It follows that

$$V_{min}/N \geq \begin{cases} -\frac{1}{2} & \text{for small } n \\ +\frac{1}{5\sqrt{2}}\left(\frac{5\,n\,\pi}{16}\right)^6 & \text{for large } n, \end{cases} \tag{32}$$

qualitatively quite good. [On a regular lattice, $V = N \sum_{1}^{\infty} v(t/m) \sim N(n^8 - n^4)$].

At this point, one might well pause to observe that everybody knows--although it can rarely be proved--that most many-body potentials reach their minima at regular lattice configurations. How far away is the result of equation (27)? For this purpose, suppose that the interparticle lattice vectors are $\{\underset{\sim}{x}_s^{\,o}\}$. Then at a lattice configuration

$$\begin{aligned} V_L &= \frac{N}{2} \sum_{s \neq 0} [u(\underset{\sim}{x}_s^{\,o}) + w(\underset{\sim}{x}_s^{\,o})] \\ &= \frac{N}{2}\Big[\sum_{s \neq 0} w(\underset{\sim}{x}_s^{\,o}) + \sum u(\underset{\sim}{x}_s^{\,o}) - u(0)\Big] \\ &= \frac{N}{2}\Big[\sum_{s \neq 0} w(\underset{\sim}{x}_s^{\,o}) + n \sum \tilde{u}(\underset{\sim}{K}) - u(0)\Big] \;, \end{aligned} \tag{33}$$

where $\underset{\sim}{K}$ denotes a Bragg vector of the reciprocal lattice. Hence

$$\begin{aligned} V_L/N &= \tfrac{1}{2}\,[n\,\tilde{u}(0) - u(0)] \\ &\quad + \tfrac{1}{2}\Big[\sum_{s \neq 0} w(\underset{\sim}{x}_s^{\,o}) + n \sum_{\underset{\sim}{K} \neq 0} \tilde{u}(\underset{\sim}{K})\Big] \,. \end{aligned} \tag{34}$$

On comparing with equation (27), we see that the minimum is certainly achieved at a lattice configuration if $\tilde{u}$ vanishes at all $\underset{\sim}{K} \neq 0$ and w vanishes at all $\underset{\sim}{x}_s^{\,o} \neq 0$.

4. BOSON FLUID: N ⟶ 3 BODY REDUCTION

We have now arrived at a stage of sophistication where we can handle the reduction of a full quantum-mechanical ground state with some degree of confidence. In this section, we will have in mind an N-particle spinless boson fluid, with Hamiltonian

$$H = \sum_{1}^{N} p_i^{\,2}/2m + \tfrac{1}{2} \sum_{i \neq j} v(\underset{\sim}{x}_i - \underset{\sim}{x}_j) \tag{35}$$

bounded by a periodic cube of volume Ω. The form equation (35) clearly calls for an effective two-body representation, although we shall see later that relaxation of this restriction may be helpful. At any rate, we start again by writing the ground-state energy in standard variational form

$$E_o = \mathrm{Min}_{\Phi} \langle \Phi | H | \Phi \rangle \tag{36}$$

where $\langle \Phi | \Phi \rangle = 1$, and rewrite this as

$$E_o = \mathrm{Min}_{\Phi} \mathrm{Tr}\, H | \Phi \rangle \langle \Phi | \tag{37}$$

where $\mathrm{Tr}\, | \Phi \rangle \langle \Phi | = 1$. The quantity $\Gamma^N = | \Phi \rangle \langle \Phi |$ is simply the pure-state N-body density matrix. Now spelling out the variables over which the trace is taken,

$$\begin{aligned} &\mathrm{Tr}_{1\,.\,.\,.\,N} [p_1^{\,2} \,| \Phi(1\,.\,.\,.\,N)\rangle \langle \Phi(1\,.\,.\,.\,N)|] \\ &\quad = \mathrm{Tr}_1 [p_1^{\,2}\, \mathrm{Tr}_{2\,.\,.\,.\,N} | \Phi(1\,.\,.\,.\,N)\rangle \langle \Phi(1\,.\,.\,.\,N)|] \;, \end{aligned} \tag{38}$$

and similarly

$$\begin{aligned} &\mathrm{Tr}_{1\,.\,.\,.\,N} [v(1,2) | \Phi(1\,.\,.\,.\,N)\rangle \langle \Phi(1\,.\,.\,.\,N)|] \\ &\quad = \mathrm{Tr}_{12} [v(1,2)\, \mathrm{Tr}_{3\,.\,.\,.\,N} | \Phi(1\,.\,.\,.\,N)\rangle \langle \Phi(1\,.\,.\,.\,N)|] \;. \end{aligned} \tag{39}$$

Thus, most of the integrations in equation (37) can be done before the Hamiltonian is applied. Since both H and Φ are symmetric, we have indeed

$$E_o = \mathrm{Min}_{\gamma,\Gamma} \left\{ \mathrm{Tr}\, [\tfrac{1}{2m}\, p_1^{\,2}\, \gamma(1)] + \mathrm{Tr}_{12} [v(1,2) \Gamma(1,2)] \right\} \tag{40}$$

where $\gamma(1) \equiv N\, \mathrm{Tr}_{2\,.\,.\,.\,N} | \Phi(1\,.\,.\,.\,N)\rangle \langle \Phi(1\,.\,.\,.\,N)|$ or ,

$\langle 1 | \gamma | 1' \rangle = \int \Phi^*(12\,.\,.\,.\,N)\, \Phi(1'2\,.\,.\,.\,N) d2\,.\,.\,.\,dN$, and

$$\Gamma(1,2) \equiv N(N-1)\, \mathrm{Tr}_{3\,.\,.\,.\,N} | \Phi(1\,.\,.\,.\,N)\rangle \langle \Phi(1\,.\,.\,.\,N)| \text{ or} \tag{41a}$$

$$\langle 12 | \Gamma | 1'2' \rangle = N(N-1) \int \Phi^*(123\,.\,.\,.\,N)\, \Phi(1'2'3\,.\,.\,.\,N) d3\,.\,.\,.\,dN. \tag{41b}$$

The objects $\gamma(1)$, $\Gamma(1,2)$, are the boson one- and two-body reduced density matrices. The problem is now to characterize them sufficiently sharply that the minimum, equation (40), taken over the larger class of restricted one- and two-body operators, not all obtainable from equation (41), will not be absurdly low. We start with the obvious normalization and symmetry

$$\mathrm{Tr}_2\ \Gamma(1,2) = (N-1)\gamma(1)\,,$$

$$\mathrm{Tr}_1\,\gamma(1) = N\ ,$$

$$\Gamma(1,2) = \Gamma(2,1)\ , \tag{42}$$

while positivity is replaced by positive-semidefiniteness:

$$\int f^*(12)\ \langle 12|\Gamma|1'2'\rangle\ f(1'2')d1d1'd2d2' =$$

$$N(N-1)\int\left|\int f(1'2')\ \Phi(1'2'3\ .\ .\ .\ N)d1'd2'\right|^2 d3\ .\ .\ .\ dN \geq 0$$

$$\langle 1|\gamma|1'\rangle \text{ and } \langle 12|\Gamma|1'2'\rangle \tag{43}$$

are positive-semidefinite matrices.

What next? Here is where the art enters. Three more or less distinct classes of restrictions have been studied, which we shall enumerate without great elaboration. The first class comprises what may be termed kinematic or system-independent restrictions and is based on the reasoning which led to equation (22c), i.e., if A is any operator, then [5]

$$\langle A^\dagger A\rangle \geq |\langle A\rangle|^2\ . \tag{44}$$

One must choose A to give information as to γ and Γ. Simplest is the form $A = \sum_1^N f(i)$ for some f; then equation (44) reads

$$\mathrm{Tr}\ f^\dagger(1)[\Gamma(1,2) - \gamma(1)\gamma(2) + \gamma(1)\delta(1-2)]\ f(2) \geq 0\ . \tag{45}$$

Other clever choices for A are possible within the framework of second quantization. If $\sum f(i)$ corresponds to a symmetry of the system, e.g., total momentum, one can of course do better. If

$$\sum f(i)\ \Phi = \lambda\Phi,$$

$$\mathrm{Tr}_2\ [\Gamma(1,2) - \gamma(1)\gamma(2) + \gamma(1)\ \delta(1-2)]\ f(2) = 0\ . \tag{46}$$

Kinematic restrictions can be regarded as emanating from ground-state information on certain model Hamiltonians [6]. Thus equation (44) avers that the lowest eigenvalue of $(A - \langle A\rangle)^\dagger\ (A - \langle A\rangle)$ is not less than zero. More generally, then, if E_o^{mod} is known for a model Hamiltonian of the form equation (35), i.e., if E_o^{mod} is known for

$$H^{mod} = \sum{}' T^{mod}(i) + \tfrac{1}{2}\sum{}'_{i\neq j} V^{mod}(i,j)\ , \tag{47}$$

then

$$\mathrm{Tr}[T^{mod}(1)\gamma(1)] + \tfrac{1}{2}\ \mathrm{Tr}\,[V^{mod}(1,2)\Gamma(1,2)] \geq E_o^{mod}$$

for any valid γ, Γ. When available, model restrictions can be exceedingly effective. The problem is to find solvable models, and we will return to this in a moment.

Finally, there exists a limited class of restrictions in which knowledge of the system being considered is used explicitly, instead of quoting conditions which apply to anything in the universe. We may refer to these as dynamic restrictions [7]. Those thus far uncovered arise from the general equality

$$\langle [H, A]\rangle = 0 \tag{48}$$

for expectations in any eigenstate, and the general inequality

$$\langle A^\dagger [H, A] \rangle = \tfrac{1}{2} \langle [A^\dagger, [H, A]] \rangle \geq 0 \tag{49}$$

for expectations in the ground state (observe that $[H, A]|\Phi\rangle = [H - E_o, A]|\Phi\rangle = (H - E_o)A|\Phi\rangle$). As an example of the former, we have

$$[p_1^2/2m, \gamma(1)] + \mathrm{Tr}_2[v(1,2), \Gamma(1,2)] = 0 , \tag{50}$$

just the second of the quantum-mechanical BBGKY hierarchy. As an example of the latter (preferably in the form [86])

$$\frac{p_1^2}{2m} + \mathrm{Tr}_2\, v(1,2)\gamma(2) \geq \mu + \mu' \gamma(1) \tag{51}$$

in the sense of positive operators, where $\mu = E_o(N+1) - E_o(N)$, $\mu' = E_o(N+1) - 2E_o(N) + E_o(N-1)$.

The variety of available restrictions is enormous, the applicable mathematical tools very few indeed. In practice, the technique of choice appears to be the discovery of a "nearby" model system whose ground state energy is exactly known. Let us see how to construct such boson fluid models. If Φ is a real positive wave function, it trivially satisfies the equation

$$(\underset{\sim}{p}_i - \frac{\hbar}{i} \nabla_i \ln \Phi)\, \Phi = 0 . \tag{52}$$

Now the operator $(\underset{\sim}{p}_i - (\hbar/i)\nabla_i \ln \Phi)^\dagger \cdot (\underset{\sim}{p}_i - (\hbar/i)\nabla_i \ln \Phi)$ is positive semidefinite and has Φ as an eigenfunction. Hence Φ is its ground-state wave function, with eigenvalue 0. It follows that

$$\begin{aligned} H^{mod} &= \sum \frac{1}{2m} (\underset{\sim}{p}_i - \frac{\hbar}{i} \nabla_i \ln \Phi)^\dagger \cdot (\underset{\sim}{p}_i - \frac{\hbar}{i} \nabla_i \ln \Phi) \\ &= \sum \frac{p_i^2}{2m} + \frac{\hbar^2}{2m} \sum \nabla_i^2 \ln \Phi + \frac{\hbar^2}{2m} \sum |\nabla_i \ln \Phi|^2 \end{aligned} \tag{53}$$

has $E_o^{mod} = 0$ (and incidentally that the ground state is nondegenerate).

Except in very special cases, we cannot force H^{mod} of equation (53) into the mold of of equation (47). But we can do almost as well. Choose Φ in the Bijl-Dingle-Jastrow [8] form:

$$\Phi = \exp \tfrac{1}{2} \sum_{i \neq j} \varphi(\underset{\sim}{x}_i - \underset{\sim}{x}_j) . \tag{54}$$

We then have from equation (53)

$$\begin{aligned} \langle H^{mod} \rangle = \langle \sum \frac{p_i^2}{2m} &+ \frac{\hbar^2}{2m} \sum{}' \nabla^2 \varphi(\underset{\sim}{x}_i - \underset{\sim}{x}_j) \\ &+ \frac{\hbar^2}{2m} \sum_i \Big| \sum_j{}' \nabla \varphi(\underset{\sim}{x}_i - \underset{\sim}{x}_j) \Big|^2 \rangle \geq 0 , \end{aligned} \tag{55}$$

a relation among the one-, two-, and three-body densities. The simplest way to apply equation (55) is to use it to eliminate the momentum-dependent one-body terms from $E_o = \langle H \rangle$, leading at once to

$$\begin{aligned} E_o \geq \langle \tfrac{1}{2} \sum{}' (v(\underset{\sim}{x}_i - \underset{\sim}{x}_j) &- \frac{\hbar^2}{m} \nabla^2 \varphi(\underset{\sim}{x}_i - \underset{\sim}{x}_j)) \\ &- \frac{\hbar^2}{2m} \sum_i \Big| \sum_j{}' \nabla \varphi(\underset{\sim}{x}_i - \underset{\sim}{x}_j) \Big|^2 \rangle . \end{aligned} \tag{56}$$

Through equation (56), we have returned to a classical minimization and thus want the point minimum of the potential energy which is exhibited. One possible approach is numerical. We use the alternative representation of equation (56)

$$E_o \geq \mathrm{Min}_{\{\ldots x_i\, p_i \ldots\}} \Big\{ \sum \frac{p_i^2}{2m} - \sum \frac{p_i - p_j}{2m} \cdot \hbar \nabla\varphi(x_i - x_j) + \tfrac{1}{2} \sum{}' [v(x_i - x_j) - \frac{\hbar^2}{m} \nabla^2 \varphi(x_i - x_j)] \Big\} \tag{57}$$

and do a straight numerical minimization of this classical pair-interaction Hamiltonian; the computation is being carried out for Lennard-Jones potential. Another approach is to apply restrictions on both two- and three-body densities to bound equation (56); this is a bit delicate and is being studied. Quickest, however, is to assume that the minimum of equation (56) occurs at a regular lattice configuration if V has been chosen so that the lattice is locally stable, and compute on this basis. Referring to the full potential of equation (56) as V, a few steps of algebra yield

$$\frac{1}{N} \sum_{i,j} \exp(i p \cdot x_i^o) \exp(-i q \cdot x_j^o)\, \partial^2 V / \partial x_{i\alpha}^o\, \partial x_{j\beta}^o$$

$$= \Big[\hat{v}_{\alpha\beta}(p) - \frac{\hbar^2}{m} (\nabla^2 \varphi_{\alpha\beta}(p) + \sum_\gamma \hat{\varphi}_{\alpha\gamma}(p)\, \hat{\varphi}_{\beta\gamma}(p)) \Big] \delta_{p,q} \tag{58}$$

where $\hat{f}_{\alpha\beta}(p) \equiv \sum_i f''_{\alpha\beta}(x_i^o)(1 - \cos p \cdot x_i^o)$.

We therefore want to choose φ to maximize

$$E_o \geq \tfrac{1}{2} N \sum_{s \neq 0} [v(x_s^o) - \frac{\hbar^2}{m} \nabla^2 \varphi(x_s^o)] \tag{59}$$

subject to the positive semidefiniteness of the diadic, equation (58), for each configuration.

Let us simplify. Suppose that $v = u + w$, where $w \geq 0$ and u has Fourier components only in the first Brillouin zone. We can assume that φ has the same property as u. Now for a function such as u or φ,

$$\hat{f}_{\alpha\beta}(p) = \sum_i f_{\alpha\beta}''(x_i^o)(1 - \cos p \cdot x_i^o) = n \int f_{\alpha\beta}''(x)(1 - \cos p \cdot x)\, dx$$

$$= n\, p_\alpha p_\beta \tilde{f}(p) \,, \tag{60}$$

so that stability for equation (58) reduces to

$$\tilde{u}(p) + \frac{\hbar^2 p^2}{m} (\tilde{\varphi}(p) - n \tilde{\varphi}(p)^2) \geq 0 \tag{61}$$

and equation (59) reduces to

$$E_o \geq \tfrac{1}{2} N\, [n\tilde{u}(0) - u(0) + \frac{\hbar^2}{m} \nabla^2 \varphi(0)]$$

$$= \tfrac{1}{2} N\, [n\tilde{u}(0) - u(0) - \frac{1}{\Omega} \sum \frac{\hbar^2 p^2}{m} \tilde{\varphi}(p)] \,. \tag{62}$$

The minimum $\tilde{\varphi}(p)$ satisfying equation (61) is

$$\tilde{\varphi}(p) = \frac{1}{2n} [1 - (1 + \frac{4nm}{\hbar p^2} \tilde{u}(p))^{\frac{1}{2}}] \,, \tag{63}$$

and so the standard result of Bogoliubov [9] is validated as a lower bound under the above restrictions. The motivated reader may wish to carry out the fairly simple applications to a charged boson fluid.

5. FERMION GROUND STATE: N $\longrightarrow$ 2 BODY REDUCTION

For a fluid of fermions, the indirect influence of the remainder of the system on a chosen pair of particles is exceedingly strong. Thus the list of proven theorems is very long, but the list of solid successes rather meager. It is particularly helpful to use the notation of second quantization when dealing with fermions. Our problem then is to determine the ground-state energy of

$$H = \int \psi^\dagger(x) \langle x| T |x'\rangle \psi(x')dx'\,dx$$
$$+ \tfrac{1}{2} \int \psi^\dagger(y)\psi^\dagger(x) \langle xy| v |x'y'\rangle \psi(x')\psi(y')dx'dxdy'dy \ , \qquad (64)$$

where x denotes both position and spin component (± 1). Here, the field operators satisfy the usual anticommutation relations

$$[\psi^\dagger(x), \psi(y)]_+ = \delta(x-y)\,, \quad [\psi(x), \psi(y)]_+ = 0 = [\psi^\dagger(x), \psi^\dagger(y)]_+\,, \qquad (65)$$

and the two-body nature of the interaction is built into the form of equation (64). Since one- and two-body sums can be written as

$$\sum A(i) = \int \psi^\dagger(x) \ \langle x| A |x'\rangle \psi(x')dxdx'$$

$$\sum_{i \neq j}' Q(i\,j) = \int \psi^\dagger(y)\psi^\dagger(x) \langle xy| Q |x'y'\rangle \psi(x')\psi(y')dxdx'dydy'\,, \qquad (66)$$

one- and two-body reduced density matrices can be introduced via

$$\langle \sum A(i)\rangle = \mathrm{Tr}\, A\gamma \ , \qquad (67)$$

where $\langle x'|\gamma|x\rangle = \langle \psi^\dagger(x)\psi(x')\rangle$, and

$$\langle \sum_{i \neq j} Q(i\,j)\rangle = \mathrm{Tr}\ Q\,\Gamma\,,$$

where $\langle x'y'|\Gamma|xy\rangle = \langle \psi^\dagger(y)\psi^\dagger(x)\psi(x')\psi(y')\rangle$, and the basic variational principle becomes

$$E_o = \mathrm{Min}_{\gamma,\,\Gamma}(\mathrm{Tr}\ T\gamma + \tfrac{1}{2}\,\mathrm{Tr}\ v\,\Gamma) \ . \qquad (68)$$

Having mentally disposed of the obvious normalization, antisymmetry, and positive semidefiniteness conditions on γ and Γ, how shall we organize the remaining conditions? Here it is even significant to ask for the additional properties of the one-body matrix. There is really just one, namely, that just as

$$\int f(x) \ \langle x|\gamma|x'\rangle f^*(x')dxdx' = \langle (\int f(x)\psi(x)dx)^\dagger (\int f(x)\psi(x)dx)\rangle \geq 0$$

establishes that for all eigenvalues $\lambda(\gamma) \geq 0$, the relation

$$0 \leq \langle (\int f(x)\psi^\dagger(x)dx)^\dagger \ (\int f(x)\psi^\dagger(x)dx)\rangle = \langle \int\int f^*(x')\psi(x')\psi^\dagger(x)\, f(x)dxdx'\rangle =$$

$$\int f^*(x)f(x)dx \ - \ \langle \int\int f^*(x')\psi^\dagger(x)\psi(x')\, f(x)dxdx' \ =$$

$$\int f^*(x)\, f(x) dx - \int f^*(x')\, \langle x' | \gamma | x \rangle\, f(x)\, dx dx'$$

extends this to

$$0 \le \lambda(\gamma) \le 1 \,. \tag{69}$$

This is the Pauli principle, that each one-body occupation lies between 0 and 1.

By the exercise of ingenuity [10], expression (69) already yields information on E_o. Return for a moment to the first quantized form of equation (68), and use symmetry to write it as

$$E_o = \text{Min}\, \langle \sum T(i) + \tfrac{1}{2} \sum{}' v(i\, j) \rangle$$
$$= \text{Min}\, \frac{N}{N-1} \langle \sum_2^N T(i) + \frac{N-1}{2} \sum_2^N v(i1) \rangle \,. \tag{70}$$

The minimum is over fully antisymmetric wave functions. Hence if particle 1 is singled out and not required to share the antisymmetry, the resulting minimum must decrease. But for a relative coordinate potential $v(i\, j) = v(\underset{\sim}{x}_i - \underset{\sim}{x}_j)$, the transformation $\underset{\sim}{x}_i \to \underset{\sim}{x}_i + \underset{\sim}{x}_i$ eliminates particle 1 in equation (70), and we have

$$E_o \ge \frac{N}{N-1} \text{ Min}\, \langle \sum_2^N [T(i) + \frac{N-1}{2} v(i)] \rangle$$
$$= \frac{N}{N-1} \text{ Min}_\gamma \text{ Tr } (T + \frac{N-1}{2} v)\, \gamma \,. \tag{71}$$

Thus, the potential felt by each particle has been altered by bunching all of its interacting neighbors at the origin, a reasonable effective field for an attractive interaction causing the particles to bunch and a long-range one making particle location less crucial. Now if the eigenfunctions and eigenvalues in ascending order are

$$(T + \frac{N-1}{2} v)\, \varphi_\alpha = e_\alpha \varphi_\alpha \,, \tag{72}$$

equation (71) becomes

$$E_o \ge \frac{N}{N-1} \sum_{\alpha=1}^{\infty} e_\alpha \langle \varphi_\alpha | \gamma | \varphi_\alpha \rangle \,. \tag{73}$$

By normalization and the Pauli condition,

$$\sum_{\alpha=1}^{\infty} \langle \varphi_\alpha | \gamma | \varphi_\alpha \rangle = \text{Tr}\, \gamma = N - 1 \,, \quad 0 \le \langle \varphi_\alpha | \gamma | \varphi_\alpha \rangle \le 1 \,, \tag{74}$$

so that equation (73) reads simply

$$E_o \ge \frac{N}{N-1} \sum_{\alpha=1}^{N-1} e_\alpha \,. \tag{75}$$

Example: gravitational attraction $v(r) = -\frac{1}{2} G m^2/r$; then $e_{n\ell m} = -(N-1)^2 m^5 G^2/8\hbar^2 n^2$. For spin $\frac{1}{2}$ and large N, we have $N = 2 \sum_o^{n_{max}} n^2 = 2 n_{max}^3/3$, so $2 \sum_o^{n_{max}} n^2/n^2 = 2\,(\frac{3}{2} N)^{1/3}$ and equation (75) becomes

$$E_o \ge -\frac{m^5 G^2}{2\hbar^2} \left(\frac{3}{2}\right)^{1/3} N^{7/3} \,. \tag{76}$$

The Pauli principle is also responsible for a strengthening of the uncertainty principle in a fermion system. Suppose we fix $\langle x_1^2\rangle = N^{-1}\langle\sum x_i^2\rangle$ and ask for a minimum value of $\langle p_1^2\rangle = N^{-1}\langle\sum p_i^2\rangle$. Using a Lagrange parameter, we must minimize

$$\langle H'\rangle = \langle \tfrac{1}{2}\sum p_i^2 + \frac{\omega^2}{2}\sum x_i^2\rangle . \qquad (77)$$

The eigenvalues of equation (77) are $(\frac{3}{2} + n_1 + n_2 + n_3)\hbar\omega$, so for the ground-state energy $N = 2\sum_{n_1+n_2+n_3 \le n_{max}} 1 = 2\sum_0^{n_{max}} n^2/2 = n_{max}^3/3$, and then $\langle H'\rangle = 2\sum_0^{n_{max}} (\frac{3}{2}+n)\frac{n^2}{2}\hbar\omega = \hbar\omega n_{max}^4/4 = (3N)^{4/3}\hbar\omega/4$. Hence $\langle N^{-1}\sum x_i^2\rangle = (2/N\omega^2)\langle\frac{1}{2}\omega^2\sum x_i^2\rangle = (3\hbar/2\omega)(3N)^{1/3}$, and

$$\langle \frac{1}{N}\sum p_i^2\rangle = \frac{2}{N}\langle\tfrac{1}{2}\sum p_i^2\rangle \ge (3\hbar\omega/2)(3N)^{1/3}$$

or [11]

$$\langle p_1^2\rangle^{\frac{1}{2}}\langle x_1^2\rangle^{\frac{1}{2}} \ge \frac{3}{2}(3N)^{1/3}\hbar . \qquad (78)$$

Equation (28), in the form

$$\langle\sum p_i^2\rangle \ge \left(\frac{\hbar}{2}\right)^2 (3N)^{8/3}/\langle\sum x_i^2\rangle , \qquad (79)$$

can be used as well to "eliminate" the kinetic energy directly from a fermion system, converting it to a classical lower bound

$$\begin{aligned} &E_o(\sum p_i^2/2m + \sum U(\underset{\sim}{x}_i) + \tfrac{1}{2}\sum' v(\underset{\sim}{x}_i - \underset{\sim}{x}_j)) \\ &\quad \ge \frac{\hbar^2}{8m}(3N)^{8/3}\frac{1}{\langle\sum x_i^2\rangle} + \sum\langle U(\underset{\sim}{x}_i)\rangle + \tfrac{1}{2}\sum'\langle v(\underset{\sim}{x}_i - \underset{\sim}{x}_j)\rangle . \end{aligned} \qquad (80)$$

This bound is not superb, but can be greatly improved by asking instead for the minimum of $\langle\sum p_i^2\rangle$ at fixed $\langle\sum u(\underset{\sim}{x}_i)\rangle$; reader please note.

Having extracted nearly the last drop of relevant information from γ, we proceed to the problem of effective restrictions on the two-body matrix Γ. The division into kinematic, model, and dynamic restrictions made in the boson case remains appropriate. The complete set of kinematic restrictions in second quantized form now derives from the single [12]

$$\begin{aligned} &\langle A^\dagger A\rangle \ge 0, \text{ where} \\ &A = c + \int f(x)\psi(x)\,dx + \int g(x)\psi^\dagger(x)\,dx \\ &\quad + \iint w(x,y)\psi(x)\psi(y)\,dx\,dy \\ &\quad + \iint g(x,y)\psi^\dagger(x)\psi(y)\,dx\,dy + \iint r(x,y)\psi^\dagger(x)\psi^\dagger(y)\,dx\,dy. \end{aligned} \qquad (81)$$

For a state of fixed particle number, equation (81) then neatly decomposes into

$\langle A_i^\dagger A_i\rangle \ge 0$, $i = 1, \ldots, 5$, where

$$
\begin{aligned}
A_1 &= \int f(x)\psi(x)\,dx\ , \qquad A_2 = \int g(x)\psi^\dagger(x)\,dx\ , \\
A_3 &= \iint w(x,y)\psi(x)\psi(y)\,dx\,dy\ , \\
A_4 &= \iint r(x,y)\ \psi^\dagger(x)\psi^\dagger(y)\,dx\,dy\ , \\
A_5 &= c + \iint g(x,y)\ \psi^\dagger(x)\psi(y)\,dx\,dy\ .
\end{aligned} \tag{82}
$$

Now A_1 gives the positive semidefiniteness of the 1-matrix γ, A_3 that of the 2-matrix Γ, A_2 the Pauli restriction $\lambda(\gamma) \le 1$. The remaining restrictions have more meat. According to A_4, the "Q-matrix"

$$
\begin{aligned}
\langle xy|\,Q\,|x'y'\rangle &\equiv \langle \psi(y)\psi(x)\psi^\dagger(x')\psi^\dagger(y')\rangle \\
&= \delta(x-x')\delta(y-y') - \delta(y-x')\delta(x-y') - \delta(x-x')\,\langle \psi^\dagger(y')\psi(y)\rangle \\
&\quad - \delta(y-y')\langle\psi^\dagger(x')\psi(x)\rangle + \delta(x-y')\langle\psi^\dagger(x')\psi(y)\rangle \\
&\quad + \delta(x'-y)\,\langle\psi^\dagger(y')\psi(x)\rangle + \langle\psi^\dagger(x')\psi^\dagger(y')\psi(y)\psi(x)\rangle \\
&= \langle xy|\Gamma|x'y'\rangle + \delta(x-y')\langle y|\gamma|x'\rangle + \delta(y-x')\langle x|\gamma|y'\rangle \\
&\quad - \delta(x-x')\langle y|\gamma|y'\rangle - \delta(y-y')\langle x|\gamma|x'\rangle + \delta(x-x')\delta(y-y') - \delta(y-x')\delta(x-y')
\end{aligned} \tag{83}
$$

is positive semidefinite. For A_5, we first minimize over c, obtaining

$$
\langle A^\dagger A\rangle \ge \langle A^\dagger\rangle\,\langle A\rangle\ , \tag{84}
$$

where $A = \iint g(x,y)\psi^\dagger(x)\psi(y)\,dx\,dy$; and expanding as in equation (83), then find that the "G-matrix"

$$
\begin{aligned}
\langle xy|\,G\,|x'y'\rangle &= \langle\psi^\dagger(y)\psi(x)\psi(x')\psi(y')\rangle - \langle\psi^\dagger(y)\psi(x)\rangle\,\langle\psi^\dagger(x')\psi(y')\rangle \\
&= \delta(x-x')\langle y'|\gamma|y\rangle - \langle x|\gamma|y\rangle\,\langle y'|\gamma|x'\rangle - \langle xy'|\Gamma|x'y\rangle
\end{aligned} \tag{85}
$$

is positive semidefinite.

As for model restrictions and dynamic restrictions, the discussion, equations (47)-(51), can be taken over essentially unchanged to the domain of fermion systems. Model systems are again not found in profusion, although a fair amount of mileage can be obtained from the generalized BCS model. Rather, taking advantage of the second quantization formalism, one can construct a variety of triplet-interaction models, further reducible to mixed fluids with pair interactions. Dynamic equalities and inequalities are available based upon each of the A_i of equations (82), but since several of the A_i can change particle number, formula (49) should be written instead as

$$
\begin{aligned}
\langle A^\dagger[H - E_o(N), A]\rangle &= \tfrac{1}{2}\langle[A^\dagger, [H - E_o(N), A]]\rangle \\
&= \langle[A^\dagger, H - E_o(N)]\,A\rangle \ge 0\ .
\end{aligned} \tag{86}
$$

Now, how does one deal with this embarrassment of riches? The technical job of minimizing subject to a few inequalities can be formidable, not to speak of the uncountable number at our disposal. In the few cases where effective computations have been carried out, the G-matrix condition has served as principal tool. In general, this is not surprising since the

G-matrix is the quantum-mechanical generalization of the Ursell positivity, equation (22c), in classical systems which is so effective in limiting the fluctuations in pair correlation. On a more detailed level, it is also not surprising, for if the ground state can be reasonably well characterized by its one-body near-symmetries

$$\left(\int \Lambda(x,x')\psi^{\dagger}(x)\psi(x')\,dx\,dx' - \lambda\right)\Psi \sim 0 \tag{87}$$

and its approximate quasi-particle annihilators

$$\int C(x,x')\psi^{\dagger}(x)\psi(x')\,dx\,dx' \quad \Psi \sim 0 \tag{88}$$

each of equation (87) and equation (88) yields a G-matrix expectation which nearly vanishes. Thus the inequalities corresponding to equation (85) are so strong that they are almost equalities, which is a fine criterion for effectiveness.

As an example, let us briefly consider the basic BCS fermion model for superconductivity, for convenience at a given chemical potential

$$H - \mu N = \sum_{k>0} (T_k - \mu)(a_k^{\dagger} a_k + a_{-k}^{\dagger} a_{-k}) - \frac{2}{\Omega} \sum_{k,\ell>0} w_k^* w_\ell a_{-k} a_k a_\ell a_{-\ell}\,. \tag{89}$$

Via the "G-matrix" condition

$$\left\langle \left(\frac{w_\ell}{\alpha_{k\ell}} a_k^{\dagger} a_{-\ell}^{\dagger} + w_k \alpha_{k\ell}^* a_\ell^{\dagger} a_{-k}\right)^{\dagger} \left(\frac{w_\ell}{\alpha_{k\ell}} a_k^{\dagger} a_{-\ell} + w_k \alpha_{k\ell}^* a_\ell^{\dagger} a_{-k}\right)\right\rangle \geq 0, \tag{90}$$

we can eliminate all momentum transfer terms, resulting in

$$\begin{aligned} \langle H\rangle - \mu\langle N\rangle \geq & \sum_{k>0} \left(T_k - \mu - \frac{|w_k|^2}{\Omega} \sum_{\ell\neq 0} |\alpha_{\ell k}|^2\right) \langle \hat{N}_k + \hat{N}_{-k}\rangle \\ & + \frac{1}{\Omega} \sum_{k,\ell>0} |w_k|^2 |\alpha_{\ell k}|^2 \langle \hat{N}_k \hat{N}_{-\ell} + \hat{N}_{-k}\hat{N}_\ell\rangle, \end{aligned} \tag{91}$$

a function of the occupation number $N_k = \langle \hat{N}_k\rangle = \langle a_k^{\dagger} a_k\rangle$ above. Since only the commuting $\hat{N}_k$, each of eigenvalue 0 or 1, enter into formula (91), a classical minimization over the $\hat{N}_k$ and maximization over the $\alpha_{k\ell}$ can be carried out. The result is highly accurate only for a highly depleted Fermi sea.

One more set of correlation inequalities has, however, been patiently waiting in the wings--one whose ancestral form is the second Griffiths inequality for a ferromagnetic spin system [13]. It can be shown that for the Hamiltonian of equation (89) in the thermodynamic limit,

$$\langle \hat{N}_k \hat{N}_\ell\rangle \geq \langle \hat{N}_k\rangle\langle \hat{N}_\ell\rangle, \tag{92}$$

and indeed at any temperature. Hence expression (91) reduces further to a function of the c-number expectations $N_k = \langle \hat{N}_k\rangle$ alone. Now maximizing over the $\alpha_{k\ell}$ and minimizing over the N_k recovers the standard BCS result as a lower bound. Extension of equation (92) to a wider variety of fermion systems is now under intensive study.

6. CLASSICAL STATISTICAL MECHANICS: N ⟶ 2 REDUCTION

It is time to arise from the ground state. In thermal equilibrium we may focus attention upon the Helmholtz free energy, taking advantage of the fact that it too obeys a minimum principle. The obvious difficulty is that the entropy is not a function of any few-body reduced density, and the art consists of somehow circumventing this difficulty. To avoid problems which are not specifically germane to this question, we will restrict the discussion to classical systems, but only in the case of the Bogoliubov inequality does this require more than a trivial modification in the quantum domain.

The free energy minimum principle is simply that

$$F = \mathrm{Min}_{\rho} \int \rho(x_1 \ldots p_N)\, H(x_1 \ldots p_N)\, dx^N\, dp^N - \beta^{-1} S[\rho] , \tag{93}$$

where

$$S[\rho] \equiv - \int \rho \ln (\rho / N!)\, dx^N\, dp^N$$

and

$$\int \rho\, dx^N\, dp^N = 1 .$$

Since the momentum dependence of $\rho(x_1 \ldots p_N)$ is Maxwell-Boltzmann, we can rewrite equation (93) at once as

$$F = F_c + F_M \tag{94}$$

where

$$F_M = \beta^{-1} [\ln N! - \frac{3N}{2} \ln (2\pi m/\beta)] ,$$

where, of course, β denotes $1/kT$, and now determine the configurational free energy by

$$F_c = \mathrm{Min}_{\rho} \int \rho\, V\, dx^N - \beta^{-1} S_c[\rho] \tag{95}$$

where

$$S_c[\rho] = - \int \rho \ln \rho\, dx^N$$

and

$$\int \rho(x_1 \ldots x_N)\, dx^N = 1.$$

To write equation (95) in two-body form, we must obtain both n_2 and S_c from some legitimate p which we may assume is symmetric and nonnegative. But why not do just that? In other words, write equation (95) as

$$F_c = \mathrm{Min}_{n_2, S_c} \tfrac{1}{2} \int n_2(\underset{\sim}{x}, \underset{\sim}{y})\, v(\underset{\sim}{x}, \underset{\sim}{y})\, d\underset{\sim}{x}\, d\underset{\sim}{y} - \beta^{-1} S_c \tag{96}$$

over all realizable pairs of two-body function n_2 and scalar S_c. Equation (96) is indeed the desired principle, and our task is to give it meaning in practice.

We now interpret equation (96) as a lower bound and seek necessary conditions on (n_2, S_c), or assuming uniformity, on (g, S_c):

$$F_c \geq \mathrm{Min}_{g, S_c} \frac{N}{2} n \int g(\underset{\sim}{z})\, v(\underset{\sim}{z})\, d\underset{\sim}{z} - \beta^{-1} S_c . \tag{97}$$

The obvious class of kinematic conditions consists of equations (22a, b, c) and leaves us no wiser as to the form of the entropy. But model conditions are a different story. If $V^M = \frac{1}{2} \sum' v(\underset{\sim}{x}_i - \underset{\sim}{x}_j)$ is a model potential for which the free energy F_c^M is known at a model reciprocal temperature

β^M, we must of course have for any realizable pair n_2, S_c

$$F_c^M \leq \tfrac{1}{2}\int n_2(\underset{\sim}{x}, \underset{\sim}{y})\, v^M(\underset{\sim}{x}, \underset{\sim}{y})\, d\underset{\sim}{x}\, d\underset{\sim}{y} - \frac{1}{\beta^M} S_c . \tag{98}$$

Equation (98) allows us to eliminate the entropy and hence

$$B F_c \geq \text{Min}_{n_2} [\tfrac{1}{2}\int (\beta v(\underset{\sim}{x}, \underset{\sim}{y}) - \beta^M v^M(\underset{\sim}{x}, \underset{\sim}{y}))\, n_2(\underset{\sim}{x}, \underset{\sim}{y})\, d\underset{\sim}{x} d\underset{\sim}{y} + \beta^M F_c^M] \tag{99}$$

which is now a practical lower bound principle and may of course be derived in many other ways [14]. We can regard equation (99) as a generalized interpolation method. A growing collection of solvable models allows us to maximize equation (99) over models and hence restrict the free energy of a given system more and more tightly. Indeed, if one of the models coincides with $V : V^{M_o} = V$, then

$$\text{Min}_{n_2}\ \text{MAX}_M\ \tfrac{1}{2}\iint (\beta v(\underset{\sim}{x}, \underset{\sim}{y}) - \beta^M v^M(\underset{\sim}{x}, \underset{\sim}{y}))\, n_2(\underset{\sim}{x}, \underset{\sim}{y}) d\underset{\sim}{x}\, d\underset{\sim}{y}$$

$$+ \beta^M F_c^M \geq \text{Min}_{n_2}\ \beta^{M_o} F_c^{M_o} = \beta F_c ,$$

so that the result of equation (99) is precisely βF_c. However, one should observe that since only linear inequalities are employed, a model which is "near" the true system will not produce a free energy error quadratic in some error parameter; rather [14] it will be linear in the absolute deviation.

There is one domain in which a reference model is by definition available; this is when one wishes to perturb an already solved system. As an example, let us apply a magnetic field $B > 0$ to the two-dimensional Ising model [14]:

$$H = \tfrac{1}{2} J \sum_{\langle ij \rangle} \sigma_i \sigma_j - B \sum_i \sigma_i \tag{100}$$

where $\langle ij \rangle$ indicates a nearest-neighbor sum and $\sigma_i = \pm 1$. As model system, we choose the field-free case at the same temperature but different coupling (or the reverse):

$$H^M = -\tfrac{1}{2} J' \sum_{\langle ij \rangle} \sigma_i \sigma_j . \tag{101}$$

The model system has for its free energy

$$\beta F^M(J')/N = f(J') = -\ln(2 \cosh \beta J')$$

$$- \frac{1}{2\pi} \int_0^{\pi} \tfrac{1}{2}[1 + (1 - \kappa^2 \sin^2\varphi)^{\frac{1}{2}}]\, d\varphi , \tag{102}$$

where $\kappa(J') = 2 \tanh 2\beta J'/\cosh 2\beta J'$. Now if $g(\sigma, \sigma')$ is the nearest-neighbor distribution, the energy difference of equation (99) becomes

$$\frac{1}{N} \langle H - H^M \rangle = 2 \sum_{\sigma, \sigma'} (\tfrac{B}{4}\sigma + \tfrac{B}{4}\sigma' - \Delta \sigma \sigma')\, g(\sigma, \sigma') \tag{103}$$

where $\Delta = J - J'$. If one inspects the point minimum of $\frac{1}{4} B\sigma + B\sigma' - \Delta\sigma\sigma'$, it is clear that

$$\frac{1}{N}\langle H - H^M \rangle \geq \begin{cases} -2\Delta - B & \text{for } \Delta \geq -B/4 \\ 2\Delta & \text{for } \Delta \leq -B/4 \end{cases} , \tag{104}$$

so that according to equation (99)

$$f \geq \begin{cases} f(J-\Delta) - 2\Delta - B & \text{for } \Delta \geq -B/4 \\ f(J-\Delta) + 2\Delta & \text{for } \Delta \leq -B/4 \end{cases} . \tag{105}$$

The optimal bound now results from maximizing equation (105) over Δ. Choosing $\Delta = -B/4$, we have

$$f \geq f(J + \tfrac{1}{4}B) - \tfrac{1}{2}B , \tag{106}$$

thereby giving a bound on the spontaneous magnetization per site

$$I = -\frac{\partial f}{\partial B}\Big|_{B=0} \leq \frac{1}{2} - \frac{1}{4}\frac{\partial f(J)}{\partial J} . \tag{107}$$

The low temperature value is found to be $1 - 2e^{-8\beta J}$, corresponding exactly to the well-known result of Yang [15].

Finally, we proceed to the matter of dynamic restrictions. The classical analog of equation (48) is the Poisson bracket relation $\langle [H, A] \rangle = 0$, which does become useful when the special form $\underset{\sim}{A} = \sum A_i(x) \quad \underset{\sim}{p}_i$ is chosen, coupled with the observation that

$$\langle \underset{\sim}{p}_i \cdot \underset{\sim}{p}_j \rangle = \frac{m}{\beta} \underset{\sim}{I} \delta_{ij} . \tag{108}$$

We then obtain at once the dynamic equality

$$\beta \langle \sum \underset{\sim}{A}_i \nabla_i V \rangle = \langle \sum \nabla_i \underset{\sim}{A}_i \rangle . \tag{109}$$

This should occasion no surprise since, by choosing arbitrary $\underset{\sim}{A}_i(\underset{\sim}{x}_i)$, equation (109) is seen to imply

$$\nabla n_1(\underset{\sim}{x}) = -\beta \int n_2(\underset{\sim}{x}, \underset{\sim}{y}) \nabla v(\underset{\sim}{x} - \underset{\sim}{y}) d\underset{\sim}{y} , \tag{110}$$

the first of the BBGKY hierarchy. Similarly, the classical analog of equation (49) is the Poisson bracket relation

$$\langle [A^\dagger, [H, A]] \rangle \leq 0 \tag{111}$$

which may be strengthened at finite temperature by using equation (109) to eliminate $\nabla_i V$ from the trivial inequality

$$\langle | \sum (A_i \nabla_i V - \frac{1}{\beta}\nabla_i A_i - \frac{1}{\beta}\underset{\sim}{c}_i)|^2 \rangle \geq 0 . \tag{112}$$

This yields

$$\begin{aligned} &\langle \sum A_i^* A_j \nabla_i \cdot \nabla_j V + \frac{1}{\beta}\sum \nabla_i A_j \cdot \nabla_j A_i^* \rangle \\ &\geq \frac{1}{\beta}\langle \sum (A_i^* \nabla_i \cdot \underset{\sim}{c}_j + A_i \nabla_i \cdot \underset{\sim}{c}_j^* - \underset{\sim}{c}_i^* \cdot c_j) \rangle , \end{aligned} \tag{113}$$

the case $\underset{\sim}{c} = 0$ corresponding to equation (111).

For a uniform system, the full content of equation (113) is expressed by choosing $A_i = \exp(i\underset{\sim}{k}\cdot\underset{\sim}{x}_i)$, $\underset{\sim}{c}_i = i\underset{\sim}{\gamma}_k \exp(i\underset{\sim}{k}\cdot\underset{\sim}{x}_i)$, and maximizing the right-hand side over $\underset{\sim}{\gamma}_k$. In the case of a grand ensemble, we thereby obtain

$$\tilde{h}(k) = c_k/(1 - n\ c_k) \tag{114}$$

with

$$c_k + \beta v_k \geq \frac{1}{k^2}\,[\widetilde{(h\,\nabla^2\beta v)}(k) - \widetilde{(h\,\nabla^2\beta v)}(0)]\,.$$

An immediate consequence [16] is the bounding of $\tilde{h}(k)$ by the linearized Debye-Huckel approximation for Coulomb forces. However, the consequences of all three classes of restrictions remain to be delineated.

REFERENCES

1. J. E. Mayer, Phys. Rev. 100, 1579 (1957); F. Bopp, Z. Physik, 156, 1421 (1957); C. N. Yang, Rev. Mod. Phys. 34, 694 (1962); A. J. Coleman, Rev. Mod. Phys. 35, 668 (1963); J. K. Percus, Trans. N.Y. Acad. Sci. 26, 1062 (1964); C. Garrod and J. K. Percus, J. Math. Phys. 5, 1756 (1964); L. J. Kijewski and J. K. Percus, Phys. Rev. 164, 228 (1967).

2. J. K. Percus, 1968 Lectures on Reduced Density Matrices, Courant Institute 1971.

3. See, e.g., J. K. Percus, reference 1.

4. For further discussion, see D. Ruelle, Statistical Mechanics (New York: W. A. Benjamin).

5. See, e.g., C. Garrod and J. K. Percus, reference 1.

6. L. J. Kijewski and J. K. Percus, J. Math. Phys. 8, 2184 (1967).

7. M. Rosina, J. K. Percus, L. J. Kijewski, and C. Garrod, J. Math. Phys. 10, 1761 (1969).

8. See, e.g., R. Jastrow in "Many-Body Problem," ed. J. K. Percus (New York: Interscience, 1963).

9. N. N. Bogoliubov, J. Phys. (USSR), 11, 23 (1947).

10. J. M. Levy-Leblond, Phys. Letters, 26A, 540 (1968).

11. E. R. Davidson, Phys. Rev. A1, 30 (1970).

12. C. Garrod and J. K. Percus, reference 1.

13. R. B. Griffiths, J. Math. Phys. 8, 478 (1967).

14. L. J. Kijewski and J. K. Percus, reference 1.

15. C. N. Yang, Phys. Rev. 85, 809 (1952).

16. N. D. Mermin, Phys. Rev. 171, 272 (1968).

DISCUSSION

G. Wannier: In the case of bosons, the forces involving the wave functions themselves seemed to have some resemblance to the hidden forces of Bohm which convert a quantum system into a classical system. Is there anything correct in this impression?

J. K. Percus: The form of this potential (57), and the origin of its form, are certainly very similar to that involved in Bohm's work. It was not my intention at this time to discuss any deeper meaning, and indeed an approximation to the wave function is quite sufficient for the energy bounds desired. However, I have been intrigued by the fact that on using the exact wave function as input, the equivalent classical system is characterized by having the absolute minimum energy, subject to the (quantum?) uncertainty due to an infinite degeneracy of the point medium.

B. Simon: Do you think there is a hope of your methods providing a "back of a postage stamp" derivation of the Dyson-Lenard result on saturation of Coulomb forces?

J. K. Percus: Yes. I think there are probably enough inequalities on the pair correlation even now, but I wouldn't want to make a bet. Notice, I took the example of purely attractive Coulomb forces as one that does not saturate.

T. Tanaka: In the case of the BCS Hamiltonian it is known that the Weiss (molecular field) approximation gives rigorous results. Is this fact an obvious consequence of your formulation?

J. K. Percus: Not quite. In fact, to obtain an exact lower bound, we would have to graft on the zero-correlation approximation or its weaker form, nonnegative correlation. This is certainly a defect of the restrictions now available, or at least of the fashion of their use.

M. Green: If you had used the fundamental maximum principle of statistical mechanics without eliminating the entropy functional, would you have obtained a Rayleigh-type upper bound rather than a lower bound?

J. K. Percus: Yes, if the system were then solved exactly. If the entropy functional is exact, reduced density matrix methods still give a lower bound (unles the functional is constructed to exist only for realizable pair distributions). But the entropy functional is a difficult object, and any mere approximation of it would vitiate either bound.

M. E. Fisher: As a further response to Dr. Karl Freed it may be worth emphasizing the point on which you touched, namely, that one can obtain bounds on an expectation $\langle A \rangle$ by adding a term $-\lambda A$ to the Hamiltonian. The energy $E_o(\lambda)$ or free energy $F(\lambda)$ must then be convex in λ. Hence if one has, say, a pointwise upper bound at $\lambda = 0$ and a lower bound as a function of λ, one can obtain upper and lower bounds on the derivative of $E_o(\lambda)$ or $F(\lambda)$ at $\lambda = 0$. These bounds are bounds on $\langle A \rangle$.

J. K. Percus: In a way, I would regard the existence of this principle as the major implicit justification for paying so much attention to lower bounds. I believe that it has been used by Parr for some model systems, and I know that I have done the same. Its relation to the extensive work of Weinhold on bounds on observables is not quite so clear.

H. Falk: Have you compared your free energy lower bound for the spin system with the chord lower bound which follows from the convexity?

J. K. Percus: No, we have not, but it would be easy to do so. I doubt that it would alter the result (107). In general, of course, the "unperturbed" free energy is not known either, and the class of bounds mentioned by Fisher becomes mandatory.

COMMENTS ON THE THEORY OF QUANTUM FLUIDS

Eugene Feenberg

I want to bring to your attention certain problems, questions, and results connected closely with the microscopic theory of the helium liquids under realistic conditions. By realistic I mean simply the experimental density range and use of adequate interatomic potentials. One of the problems concerns the accuracy of an approximation which has been used in developing the method of correlated basis functions [1, 2]. However, several topics appear in a more general context, and these are sketched first.

The first question concerns the liquid structure function S(k) associated with the ground state of the boson system (liquid ^{4}He or the hypothetical boson-type ^{3}He liquid):

$$S(k) = \frac{1}{N} \int \psi_o^2 |\rho_{\underset{\sim}{k}}|^2 d\underset{\sim}{r}_{12...N} = 1 + \rho \int \exp(i\underset{\sim}{k} \cdot \underset{\sim}{r}) [g(r) - 1] d\underset{\sim}{r},$$

$$\rho_{\underset{\sim}{k}} = \sum_1^N \exp(i\underset{\sim}{k} \cdot \underset{\sim}{r}_i) . \qquad (1)$$

POLYNOMIAL APPROXIMATIONS FOR STRUCTURE FUNCTIONS

For small k($k \leq 1$ Å^{-1}, $\hbar k/ms \leq 0.664$)

$$S(k) = \frac{\hbar k}{2ms} [1 + S_2 \frac{\hbar k}{ms} + S_3 (\frac{\hbar k}{ms})^2 + S_4 (\frac{\hbar k}{ms})^3 + \ldots] . \qquad (2)$$

What is the character of the square bracket factor? Does it contain both odd and even powers? If both odd and even powers occur, can one show at least that $S_2 = 0$? Some insight into these questions can be gained by examining the Bogolubov formalism for the low-density weakly interacting boson system. Under the stated conditions the energy $\epsilon(k)$ of an elementary excitation is given by the formula

---

Supported in part by NSF grant GP-22564.

$$\epsilon(k) = \left[\left(\frac{\hbar^2 k^2}{2m}\right)^2 + 2\rho\, v_k \frac{\hbar^2 k^2}{2m} \right]^{\frac{1}{2}}$$

$$= \hbar k s \left[\frac{v_k}{v_o} + \left(\frac{\hbar k}{2ms}\right)^2 \right]^{\frac{1}{2}} , \qquad (3)$$

$$v_k = \int v(r) \exp(i\underset{\sim}{k} \cdot \underset{\sim}{r}) d\underset{\sim}{r} ,$$

$$ms^2 = \rho v_o , \quad v_k > -\left(\frac{\hbar k}{2ms}\right)^2 v_o .$$

Here s is the phase velocity of density fluctuations (sound waves). The formalism also gives

$$S(k) = \frac{\hbar^2 k^2}{2m\, \epsilon(k)}$$

$$= \frac{\hbar k / 2ms}{[v_k/v_o + (\hbar k/2ms)^2]^{\frac{1}{2}}} . \qquad (4)$$

If now $r^n v(r)$ is integrable for all values of $n > -2$ (as is the case for finite linear combinations of Yukawa and Gaussian functions), then the power series for v_k contains only even powers and the power series for $\epsilon(k)$ and $S(k)$ contain only odd powers. Consider however a more physical situation with $v(r)$ falling off asymptotically as r^{-6}:

$$\lim_{r \to \infty} r^6 v(r) = W, \qquad (5)$$

W a constant, positive or negative. In the physical problem W is negative. It is then possible to show that

$$v_k = v_o - \frac{2}{3}\pi \int_0^\infty v(r)\, r^4 dr\, k^2 + \frac{1}{12}\pi^2 W k^3 + O(k^4) . \qquad (6)$$

Polynomial approximations for $S(k)$ and $\epsilon(k)$ now contain both odd and even powers. In particular, $S_2 = 0$ and

$$S_3 = -\frac{1}{8} + \frac{\pi m \rho}{\hbar^2} \int_0^\infty v r^4 dr ,$$

$$S_4 = -\frac{\pi^2}{24} \frac{m\rho}{\hbar^2} \frac{ms}{\hbar} W . \qquad (7)$$

In applications to the real superfluid these results have only suggestive force. However, it is clear that no a priori basis exists for postulating that $S(k)$ and $\epsilon(k)$ must be represented by polynomials containing only odd powers. In this connection it is interesting that Pines and Woo [3] have obtained an estimate of $S_4 \neq 0$. A second interesting point is that Woods and Cowley [4] find that only first and fifth power terms are needed to represent $\epsilon(k)$ quite well in the range $k \leq 1$ Å^{-1}. The apparent absence or smallness of second, third, and fourth power terms in $\epsilon(k)$ appears to call for a theoretical explanation. Maris and Massey [5] have suggested that a small positive-valued third power term in $\epsilon(k)$, within the small upper limit determined by Woods and Cowley [4], is helpful in accounting for the absorption of sound in superfluid liquid ^{4}He.

The questions raised here can be extended to other structure functions, in particular to the distribution function n(k) for particles in momentum space, the density-density fluctuation function χ_k, the intensity function Z(k) for inelastic neutron scattering with production of an elementary excitation, and the kinetic structure function D(k). Lacking are theorems on the general character of polynomial approximations for small values of k and rigorous examples exhibiting admissible forms under extreme conditions.

PHONONS IN LIQUID ^{4}He AT T = 0

In the range of small k, the theoretical formula for the energy $\epsilon(k)$ of an elementary excitation has the form

$$\epsilon(k) = \epsilon_o(k)\,[1 - \gamma'\,(\frac{\hbar k}{ms})^2 + \ldots] \tag{8}$$

in which

$$\epsilon_o(k) = \frac{\hbar^2 k^2}{2m\,S(k)} \tag{9}$$

is the Bijl-Feynman formula. The experimental relation

$$\epsilon(k) \simeq \hbar ks\ , \quad k \leq 0.6\ \text{Å}^{-1}\ , \tag{10}$$

requires

$$|S_2/S_3| << 1 \quad \text{and} \quad S_3 \simeq -\gamma'\ . \tag{11}$$

Equation (11) is an empirical relation with, as yet, no theoretical justification. Does it mean something, or is it simply accidental?

Calculations based on standard perturbation formulae yield the estimate $\gamma' \sim 1.46$ [6,7]. An alternative estimate of γ' is generated by a procedure involving second and third moments of $h = H - E_o - \epsilon_o(k)$ with respect to the Bijl-Feynman trial function for an elementary excitation [1,8]. Let

$$h = H - E_o - \epsilon(k)\ ,$$

$$\psi_a = \rho_k\,\psi_o/[NS(k)]^{\frac{1}{2}}\ ,$$

$$\psi_b = h\ \psi_a/\langle a|h^2|a\rangle^{\frac{1}{2}}\ ,$$

$$\Delta_n = \epsilon_o^{-n}\ \langle a|h^n|a\rangle\ . \tag{12}$$

These definitions yield

$$h_{aa} = 0\ ,$$

$$h_{ab} = \epsilon_o\sqrt{\Delta_2} \sim O(k^{3/2})\ ,$$

$$h_{bb} = \epsilon_o\ \Delta_3/\Delta_2 \sim O(k^0)\ . \tag{13}$$

The optimum linear combination of ψ_a and ψ_b gives an improved formula for the energy of an elementary excitation:

$$\epsilon(k) < \epsilon_x(k) \equiv \left\{1 + \frac{\Delta_3}{2\Delta_3} - \frac{1}{2}\left[\left(\frac{\Delta_3}{\Delta_2}\right)^2 + 4\Delta_2\right]^{\frac{1}{2}}\right\}$$

$$\simeq \epsilon_o\left[1 - \Delta_2^{\,2}/\Delta_3 + \ldots\right], \quad \frac{\hbar k}{ms} << 1 . \tag{14}$$

Comparison with equation (8) yields

$$\gamma' \geq \lim_{k \to 0} \frac{1}{\Delta_3(0)} \left[\frac{ms}{\hbar k} \Delta_2(k)\right]^2$$

$$= \lim_{k \to 0} \left[\frac{(ms^2)^2}{\epsilon_o^{\,3}} \frac{h_{ab}^{\,2}}{h_{bb}}\right] . \tag{15}$$

The inequality in equation (15) reflects the incompleteness of the basis system ψ_a, ψ_b. It is possible that a more complete basis system will produce an additional positive contribution to γ'. This does indeed appear to be the case. A third basis function ψ_c, normalized and orthogonal to ψ_a and ψ_b, is available in the function space ψ_a, $h\psi_a$, $h^2\psi_a$. The optimum linear combination of ψ_a, ψ_b, ψ_c now yields a modified version of equation (15); h_{bb} is replaced by $h_{bb} - h_{bc}^{\,2}/h_{cc}$. The new quantities h_{bc} and h_{cc} can be expressed in terms of second, third, fourth, and fifth moments of h. No information is available on the magnitudes of the fourth and fifth moments; however, it appears reasonable to assume $h_{cc} >> h_{bb}$ and $h_{bb}\, h_{cc} >> h_{cb}^{\,2}$. In practice these assumptions amount to assuming the equality in applications of equations (15).

ANALYSIS OF THE KINETIC STRUCTURE FUNCTION

The second moment of $h = H - E_o - \epsilon_o$ mentioned earlier involves the kinetic structure function

$$D(k) = \frac{N-1}{k^2} \int (\cos \underset{\sim}{k} \cdot \underset{\sim}{r}_{12} - 1)(\underset{\sim}{k} \cdot \underset{\sim}{\nabla}_1 \psi_o)(\underset{\sim}{k} \cdot \underset{\sim}{\nabla}_2 \psi_o)\, d\underset{\sim}{r}_{12\,.\,.\,N} . \tag{16}$$

Feynman introduced the working hypothesis that the function $\rho_{\underset{\sim}{k}}\, \psi_o$ approximates closely to an eigenstate of the Hamiltonian as $k \to 0$. This hypothesis finds a precise mathematical formulation in the equivalent statements:

$$\langle a|h^2|a\rangle \sim O(k^3) , \tag{17a}$$

$$\Delta_2(k) \equiv \frac{\langle a|h^2|a\rangle}{\epsilon_o^{\,2}} \sim O(k) , \tag{17b}$$

$$D(k) = \frac{ms}{2\hbar} k + O(k^2) . \tag{17c}$$

The same conclusion is reached by direct evaluation of D(k) with the ground-state eigenfunction ψ_o replaced by a suitable two-particle product-type trial function. "Suitable" here refers to correct asymptotic behavior of the two-particle factors in the trial function. To discuss these statements I introduce the product-type trial function

$$\psi = \prod_{i<j} \exp[\tfrac{1}{2} U(r_{ij})] / \left[\int \prod_{i<j} \exp[U(r_{ij})] d\underline{r}_{12 \ldots N}\right]^{\frac{1}{2}} \tag{18}$$

subject to the asymptotic condition [9]

$$U(r) = - \frac{ms}{\pi^2 \hbar \rho r^2} + O(1/r^4) , \; r >> \rho^{-1/3} . \tag{19}$$

The defining equation for D(k) reduces to the explicit form [10]

$$D(k) = \frac{1}{4} k^2 (S(k) - 1)$$

$$- \frac{\pi\rho}{2k} \left[\int_{\infty}^{\infty} (\frac{1}{3} + \frac{d^2}{dx^2} \frac{\sin x}{}) g(|x|/k) (x^2 \frac{d^2}{dx^2} - x \frac{d}{dx}) U(|x|/k) dx \right.$$

$$\left. + \int_{-\infty}^{\infty} (1 - \frac{\sin x}{x}) g(|x|/k) \, x \frac{d}{dx} U(|x|/k) \, dx \right] , \tag{20}$$

in which g(r) is the radial distribution function generated by the trial function.

In the evaluation of $\Delta_2(k)$ near the origin, equation (19) tells us only that $\Delta_2 \sim O(k)$. To connect the behavior of U(r) in the near region ($r \sim \rho^{-1/3}$) with the slope of Δ_2 at the origin, let

$$U(r) = U_o(r) + \delta U(r) ,$$

$$\lim_{r \to \infty} r^4 U_o(r) = 0 ,$$

$$\delta U(r) = - \frac{ms}{\pi^2 \hbar \rho} \sum_p \frac{Z_p}{r^2 + 1/k_p^{\,2}} ,$$

$$\sum Z_p = 1 . \tag{21}$$

Now equation (19) is satisfied without imposing any constraint on $U_o(r)$ or on the magnitude and sign of $\delta U(r)$ in the near region.

For $k << 2\pi\rho^{1/3}$, the substitution δU for U in the right-hand member of equation (20) is permitted. The error introduced by the replacement of g(r) by 1 has been estimated and found to be small. The integrals can be evaluated with the result [8, 10]

$$D(k) = \frac{ms}{2\hbar} k - \left[\frac{1}{4} + \frac{ms}{2\hbar} \sum_p \frac{Z_p}{k_p}\right] k^2 + \ldots \tag{22}$$

and

$$\Delta_2(k) = k\left[\frac{\hbar}{2ms} - \sum \frac{Z_p}{k_p}\right] + O(k^2) . \tag{23}$$

Equations (23) and (15) require

$$\frac{\hbar}{ms} [\gamma' \Delta_3(0)]^{\frac{1}{2}} \geq \frac{\hbar}{2ms} - \sum \frac{Z_p}{k_p} . \tag{24}$$

Inserting the theoretical estimates $\gamma' \simeq 1.46$ and $\Delta_3(0) \simeq 7.0$, equation (24) becomes

$$\sum (Z_p/k_p) \geq -1.80\,\text{Å} \ . \tag{25}$$

Subject to the plausible assumption that the equality in equations (15), (24), and (25) is nearly satisfied, equation (25) tells us that at least one of the amplitude coefficients Z_p must be negative. It is likely then that the normalization condition of equation (21) can be met only if the series representing δU contains more than one element and at least one coefficient Z_p is negative.

CONSEQUENCES OF SUM RULES IN THE LONG-WAVELENGTH LIMIT

Difficult and careful experimental studies of X-ray [11-12] and slow neutron scattering [4, 13-18] by liquid ^{4}He have produced a large body of information related to the dynamic form factor $S(k, \omega)$. The definition

$$S(k, \omega) = \frac{1}{N} \sum_n |\langle n|\rho_{\underset{\sim}{k}}|0\rangle|^2 \, \delta(E_n - E_o - \omega) \tag{26}$$

involves the ground state and excited states with momentum $\hbar \underset{\sim}{k}$. Sum rules in which $S(k, \omega)$ appears as a weight factor are highly useful in analyzing measurements and developing and testing theoretical ideas [19-22]. Five sum rules are stated below, a first group of three characterized by explicit and simple structure and a second group of two involving second and third moments of $H - E_o - \epsilon_o(k)$ with respect to the Bijl-Feynman trial function [23, 24]:

$$S(k) = \int_0^\infty S(k, \omega) d\omega \ , \tag{27a}$$

$$\frac{\hbar^2 k^2}{2m} = \int_0^\infty S(k, \omega)\, \omega d\omega \ , \tag{27b}$$

$$\frac{1}{2ms^2} = \lim_{k \to 0} \int_0^\infty S(k, \omega) \frac{d\omega}{\omega} \ , \tag{27c}$$

$$N^{-1} \langle 0|\rho_{-\underset{\sim}{k}} [H, [H, \rho_{\underset{\sim}{k}}]]|0\rangle = \int_0^\infty S(k, \omega)\, \omega^2 d\omega$$

$$= \epsilon_o(k)^2 S(k) (\Delta_2(k) + 1) \ , \tag{27d}$$

$$- N^{-1} \langle 0|[H, \rho_{-\underset{\sim}{k}}] [H, [H, \rho_{\underset{\sim}{k}}]]|0\rangle = \int_0^\infty S(k, \omega)\, \omega^3 d\omega$$

$$= \epsilon_o(k)^3 S(k) \left[\Delta_3(k) + 3\Delta_2(k) + 1 \right] . \tag{27e}$$

The left-hand member of equation (27d) involves the kinetic structure function $D(k)$ mentioned earlier which has been estimated by substituting a nearly optimum Jastrow-type trial function for

ψ_o. In equation (27e) the left-hand member involves the mean kinetic energy per particle in the ground state and one-dimensional radial integrals. It is a remarkable circumstance that the three-particle distribution function generated by ψ_o does not occur in these evaluations.

Equation (27c) can be presented as a boundary condition on the density-density fluctuation function χ_k defined by

$$-\frac{\chi_k}{2ms^2} = \int_0^\infty S(k, \omega) \frac{d\omega}{\omega} . \qquad (28)$$

The boundary condition is $\chi_o = -1$.

Let n = 1 denote the excited state describing a single elementary excitation. For comparison with intensity measurements the dynamic form factor is expressed as a linear combination of elementary and multiexcitation summands [25]:

$$S(k, \omega) = Z(k)\, \delta(\omega - \epsilon(k)) + S^{(1)}(k, \omega) , \qquad (29)$$

in which

$$Z(k) = \frac{1}{N} |\langle 1|\rho_{\underset{\sim}{k}}|0\rangle|^2 ,$$

$$S^{(1)}(k, \omega) = \frac{1}{N} \sum_{n>1} |\langle n|\rho_{\underset{\sim}{k}}|0\rangle^2\, \delta(E_n - E_o - \omega) . \qquad (30)$$

A slightly more detailed notation is useful [8]:

$$S(k, \omega) = S(k) \left[a(k)^2\, \delta(\omega - \epsilon(k) + b(k)^2) X(k, \omega)\right] ,$$

$$a(k)^2 + b(k)^2 = 1, \quad \int_0^\infty X(k, \omega)\, d\omega = 1 ,$$

$$Z(k) = S(k)\, a(k)^2 = S(k)\, [1 - b(k)^2] . \qquad (31)$$

The mean energy of the multiexcitation component is defined by

$$\epsilon'(k) = \int_0^\infty X(k, \omega)\, \omega d\omega . \qquad (32)$$

The second sum rule yields

$$b(k)^2 = \frac{\epsilon_o - \epsilon}{\epsilon' - \epsilon} , \qquad (33)$$

expressing the statistical weight of the multiexcitation component in terms of energy quantities which can be measured, computed, or estimated.

I have presented this formidable array of definitions and sum rules to provide background for some useful inequalities and limiting forms. First the density-density fluctuation function χ_k is bounded above and below by simple functions of S(k), ϵ(k), and ϵ'(k):

$$\frac{S(k)}{\epsilon(k)} \geq -\frac{\chi_k}{2ms^2} \geq \frac{S(k)}{\epsilon(k)}\left[1 - \frac{\epsilon_o(k) - \epsilon(k)}{\epsilon'(k)}\right]$$

$$> \frac{S(k)}{\epsilon(k)}\left[1 - b(k)^2\right] = \frac{Z(k)}{\epsilon(k)} . \tag{34}$$

The upper limit follows from the fact that $S(k, \omega)$ vanishes for $\omega < \epsilon(k)$; the first lower limit, from the theorem that the mean value of $1/\omega$ is greater than the reciprocal of the mean value of ω. The first lower limit is thought to be rather close, but uncertainty in the estimation of $\epsilon'(k)$ from the observations prevents a careful test.

Some implications of equations (31) and (34) are made explicit by introducing polynomial approximations for the structure functions involved. Let $x = \hbar k/ms$ and write

$$S(k) = \frac{1}{2}x\,[1 + S_3x^2 + S_4x^3 + \ldots],$$

$$\epsilon(k) = ms^2x[1 + \epsilon_3x^2 + \epsilon_4x^3 + \ldots],$$

$$-\chi(k) = 1 + \chi_2x^2 + \chi_3x^3 + \ldots,$$

$$Z(k) = \frac{1}{2}x\,[1 + Z_3x^2 + Z_4x^3 + \ldots]. \tag{35}$$

Equation (11) translates into $|\epsilon_3| << 1$; equations (31) yield

$$Z_3 = S_3 \cong -\gamma' , \tag{36a}$$

$$Z_4 = S_4 - \frac{ms^2}{\epsilon'(0)}\gamma' . \tag{36b}$$

Equation (34) reduces to

$$0 \geq (\epsilon_3 + \chi_2 - S_3)x^2 + (\epsilon_4 + \chi_3 - S_4)x^3 + \ldots$$

$$\geq -\frac{ms^2\gamma'}{\epsilon'(0)}x^3 + \ldots . \tag{37}$$

It is clear that the inequality will fail on one side or the other for sufficiently small x unless the coefficient of x^2 vanishes; consequently,

$$\chi_2 = S_3 - \epsilon_3 \simeq S_3 \simeq -\gamma' . \tag{38}$$

Now terms in x^3 dominate for sufficiently small x and the two sided inequality requires

$$\frac{ms^2\gamma'}{\epsilon'(0)} \geq \epsilon_4 + \chi_3 - S_4 + \frac{ms^2\gamma'}{\epsilon'(0)}$$

$$= \epsilon_4 + \chi_3 - Z_4 \geq 0 . \tag{39}$$

Equations (36a) and (38) are known relations, as is also equation (36b) with $Z_4 = 0$ [3]. Equation (39) is new and untested.

Terms in S_2, ϵ_2, z_2, and χ_1 have not been considered, but they may be included in the analysis with the result that $S_2 = z_2 = -\epsilon_2$ and $\chi_1 = S_2 - \epsilon_2 = -2\epsilon_2$. Thus if $\epsilon_2 = 0$, all these coefficients vanish.

Equation (27d) can be used to generate an upper limit on $\epsilon'(k)$ and a lower limit on $b(k)^2$:

$$\epsilon'(k) - \epsilon_o(k) \le \frac{\epsilon_o^2(k)\ \Delta_2(k)}{\epsilon_o(k) - \epsilon(k)} ,$$

$$b(k)^2 \ge \frac{(\epsilon_o - \epsilon)^2}{\epsilon_o^2 \Delta_2 + (\epsilon_o - \epsilon)^2} . \tag{40}$$

For $k << 2\pi\rho^{1/3}$ these inequalities are strengthened by introducing equation (15) for $\Delta_2(0)$ with the results

$$\epsilon'(0) \le ms^2 \left(\frac{\Delta_3(0)}{\gamma'} \right)^{\frac{1}{2}} ,$$

$$b(k)^2 \ge \left[1 + \left(\frac{ms}{\hbar k}\right)^3 \left(\frac{\Delta_3(0)}{\gamma'^3} \right)^{\frac{1}{2}} \right]^{-1} . \tag{41}$$

The equality in equation (40) holds only if $X(k, \omega)$ reduces to a delta function. Let us consider the possibility $X(0, \omega) = \delta[\omega - \epsilon'(0)]$. The measured and computed values $\epsilon'(0) \simeq 20°$ K, $ms^2 \simeq 27°$ K, $\Delta_3(0) \simeq 7.0$ in equation (41) yield $\gamma' \sim 13$, a value exceeding the theoretical microscopic estimate of Lai, Sim, and Woo [7] by a factor of nine. This discrepancy appears to exclude the possibility that the multiple excitation phenomena observed by Woods and Cowley [18] define a sharp discrete state at $k = 0$.

The ω^2 and ω^3 sum rules yield information on the spread of the multiple excitation distribution about $\epsilon'(k)$. In particular, at $k = 0$ [8]:

$$\int_0^\infty X(0, \omega)\,(\omega - \epsilon'(0))^2 d\omega = \epsilon'(0) \left[\frac{ms^2}{\gamma'} \left(\frac{\Delta_2(k)}{\hbar k/ms} \right)_{k=0} - \epsilon'(0) \right]$$

$$\le \epsilon'(0)^2 \left[\frac{ms^2}{\epsilon'(0)} \left(\frac{\Delta_3(0)}{\gamma'} \right)^{\frac{1}{2}} - 1 \right] , \tag{42a}$$

$$\int_0^\infty X(0, \omega)\,[\omega - \epsilon'(0)]^3 d\omega = \epsilon'(0)^3 \left[2 + \left(\frac{ms^2}{\epsilon'(0)} \right)^2 \frac{\Delta_3(0)}{\gamma'} - 3\, \frac{ms^2}{\epsilon'(0)} \left(\frac{\Delta_2}{\gamma'\,\hbar k/ms} \right)_0 \right]$$

$$\ge \epsilon'(0)^3 \left[\frac{ms^2}{\epsilon'(0)} \left(\frac{\Delta_3(0)}{\gamma'} \right)^{\frac{1}{2}} - 1 \right] \left[\frac{ms^2}{\epsilon'(0)} \left(\frac{\Delta_3(0)}{\gamma'} \right)^{\frac{1}{2}} - 2 \right] . \tag{42b}$$

These forms give $X(0, \omega)$ a root mean square spread about $\epsilon'(0)$ comparable with $\epsilon'(0)$ and a positive-valued third moment about $\epsilon'(0)$.

THE BINARY BOSON [26] SYSTEM

The binary boson system is of interest in connection with the microscopic theory of ^{3}He dissolved in liquid ^{4}He. The Schrodinger equation

$$\left[- \frac{\hbar^2}{2m_4} \sum_1^{N_4} \Delta_i - \frac{\hbar^2}{2m_3} \sum_{N_4+1}^{N} \Delta_i + \sum_{i<j} v(r_{ij}) \right] \psi = E\psi \tag{43}$$

possesses a ground-state solution $\psi_B(r_1, r_2, \ldots r_{N_4}; r_{N_4+1} \ldots r_N)$ with symmetry appropriate to N_4 bosons of mass m_4 and N_3 bosons of mass m_3 (symmetrical in $r_1, r_2, \ldots, r_{N_4}$ and also symmetrical in $r_{N_4+1}, \ldots, r_N$). This function serves as the correlation factor in the microscopic theory of the ^{3}He, ^{4}He system developed by Woo, Tan and Massey [26]. The information in ψ_B needed for the evaluation of physical matrix elements is largely contained in the liquid structure functions

$$S^{\alpha\beta}(k) = (N_\alpha N_\beta)^{-\frac{1}{2}} \int \psi_B^2 \rho_{\underset{\sim}{k}}^{(\alpha)} \rho_{-\underset{\sim}{k}}^{(\beta)} d\underset{\sim}{r}_{12\ldots N} \tag{44}$$

in which

$$\rho_{\underset{\sim}{k}}^{(3)} = \sum_{N_4+1}^{N = N_3 + N_4}{}' \exp(i\underset{\sim}{k} \cdot \underset{\sim}{r}_i) \, ,$$

$$\rho_{\underset{\sim}{k}}^{(4)} = \sum_{1}^{N_4}{}' \exp(i\underset{\sim}{k} \cdot \underset{\sim}{r}_i) \, . \tag{45}$$

Information on the behavior of $S^{\alpha\beta}(k)$ near the origin can be derived from sum rules involving the dynamical structure functions

$$S^{\alpha\beta}(k, \omega) = \tfrac{1}{2} (N_\alpha N_\beta)^{-\frac{1}{2}} \sum_n{}' \Big[\langle B|\rho_{\underset{\sim}{k}}^{(\alpha)}|n\rangle \langle n|\rho_{\underset{\sim}{k}}^{(\beta)}|B\rangle$$

$$+ \langle B|\rho_{\underset{\sim}{k}}^{(\beta)}|n\rangle \langle n|\rho_{\underset{\sim}{k}}^{(\alpha)}|B\rangle \Big] \, \delta(E_n - E_B - \omega) \, . \tag{46}$$

These functions serve as weight factors in the sum-rule statements derived by Woo, Tan and Wu [27] and used by then to characterize the behavior of $S^{\alpha\beta}(k)$ near the origin:

$$\int_0^\infty S^{\alpha\beta}(k, \omega) d\omega = S^{\alpha\beta}(k) \, ,$$

$$\int_0^\infty S^{\alpha\beta}(k, \omega) \omega d\omega = \frac{\hbar^2 k^2}{2m_\alpha} \delta_{\alpha\beta} \, ,$$

$$\lim_{k \to 0} \int_0^\infty S^{\alpha\beta}(k, \omega) \frac{d\omega}{\omega} = A^{\alpha\beta} \, ; \tag{47}$$

which, in turn, yield

$$\int_0^\infty S^{\alpha\alpha}(k, \omega) \left(\sqrt{x\omega} - \frac{1}{\sqrt{x\omega}}\right)^2 d\omega$$

$$= x \frac{\hbar^2 k^2}{2m_\alpha} + \frac{1}{x} A^{\alpha\alpha} - 2S^{\alpha\alpha}(k) \geq 0 \tag{48}$$

for $k << 2\pi\rho^{1/3}$. Minimum value of the last integral occurs at $[A^{\alpha\alpha}/(\hbar^2 k^2/2m_\alpha)]^{\frac{1}{2}}$ resulting in the inequality

$$\lim_{k \to 0} \frac{S^{\alpha\alpha}(k)}{k} \leq \left\{ \frac{\hbar^2}{2m_\alpha} A^{\alpha\alpha} \right\}^{\frac{1}{2}} . \tag{49}$$

These considerations and related statements involving $S^{34}(k)$ suggest the plausible working hypothesis that $S^{\alpha\beta}(k)$ vanishes at the origin with finite slope $[S^{\alpha\beta}(k) = B^{\alpha\beta}k]$. All this constitutes a natural extension to the binary boson system of the well-substantiated working hypothesis introduced by Feynman in the theory of liquid ^{4}He.

Campbell [28] completed the statement of the $1/\omega$ sum rule by deriving explicit formulae for the parameters $A^{\alpha\beta}$:

$$A^{33} = \frac{e_{33}}{e_{33}e_{44} - e_{34}^2} ,$$

$$A^{44} = \frac{e_{44}}{e_{33}e_{44} - e_{34}^2} ,$$

$$A^{34} = \frac{-e_{34}}{e_{33}e_{44} - e_{34}^2} . \tag{50}$$

Here $e(n_3, n_4) = E_B/\Omega$ is the energy density and the subscripts denote partial derivatives with respect to the partial number densities n_3 and n_4.

At this point an apparent contradiction becomes visible. Consider the limiting condition in which $m_3 = m_4$ (pure liquid ^{4}He). The partial liquid structure functions $S^{\alpha\beta}(k)$ can all be expressed in terms of $S(k)$, the liquid structure function of the single component system. We find

$$S^{33}(k) = x_4 + x_3 \; S(k) ,$$

$$S^{44}(k) = x_3 + x_4 \; S(k) ,$$

$$S^{34}(k) = (x_3 x_4)^{\frac{1}{2}} [S(k) - 1] , \tag{51}$$

in which $x_\alpha = N_\alpha/N$. Clearly these functions do not vanish at $k = 0$.

The apparent contradiction is resolved by noting, as pointed out by Campbell, that the energy density depends on n_3 and n_4 only through the sum $n_3 + n_4$ when $m_3 = m_4$. Then $e_{33}e_{44} - e_{34}^2$ vanishes and the coefficients $A^{\alpha\beta}$ are infinite. Thus the slope of $S^{\alpha\beta}(k)$ at the origin becomes infinite as $m_3 \to m_4$. This translates into a discontinuity in $S^{\alpha\beta}(0)$ as a function of m_3 at $m_3 = m_4$.

APPROXIMATE EVALUATION OF THE THREE PHONON STRUCTURE FUNCTION

The three-phonon structure function,

$$S^{(3)}(\underset{\sim}{k}, \underset{\sim}{\ell}, -\underset{\sim}{k} - \underset{\sim}{\ell}) = \frac{1}{N} \int \psi_0^2 \, \rho_{\underset{\sim}{k}} \, \rho_{\underset{\sim}{\ell}} \, \rho_{-\underset{\sim}{k} - \underset{\sim}{\ell}} \, d\underset{\sim}{r}_{12 \ldots N} , \tag{52}$$

occurs in a number of physical problems. For orientation I list three [1,2]:

1. Perturbation calculations on the energy $\epsilon(k)$ of an elementary excitation in liquid

^{4}He. Here ψ_o is either the ground-state eigenfunction or the optimum two-particle product-type trial function.

2. Improvements in the description of the ground state of liquid ^{4}He beyond the optimum two-particle product type.

3. Cluster expansions for the diagonal and off-diagonal matrix elements of the identity and the Hamiltonian in the theory of liquid ^{3}He. In this case ψ_o is the boson-type ground-state solution of the physical Schrodinger equation.

The first point to be made in the discussion of $S^{(3)}$ is that it is a real function. This follows from the invariance of ψ_o^2 under reflection through the origin (replacement of $\underset{\sim}{r}_1, \underset{\sim}{r}_2, \ldots, \underset{\sim}{r}_N$ by $-\underset{\sim}{r}_1, -\underset{\sim}{r}_2, \ldots, -\underset{\sim}{r}_N$).

The evaluation of $S^{(3)}$ requires a working approximation for the three-particle distribution function generated by ψ_o. To see what is involved let $h(r) = g(r) - 1$ and write

$$p^{(3)}(1,2,3) = \rho^3\Big[1 + h(r_{12}) + h(r_{23}) + h(r_{31}) + h(r_{12})h(r_{23}) + h(r_{23})h(r_{31}) + h(r_{31})h(r_{12})\Big] + \delta p^{(3)}(1,2,3)\ , \tag{53}$$

a partial cluster representation good when at least one particle is far from the other two plus a remainder $\delta p^{(3)}$ which is large only when all three particles are close (within distances of the order $\rho^{-1/3}$). Since $p^{(3)}$ vanishes when two particles coincide, we get

$$\delta p^{(3)}(1,1,3) = -\rho^3 h^2(r_{13})\ . \tag{54}$$

The sequential relation connecting the two- and three-particle distribution functions requires

$$\int \delta p^{(3)}(1,2,3)d\underset{\sim}{r}_1 = -\rho^3 \int h(r_{21})h(r_{13})d\underset{\sim}{r}_1\ . \tag{55}$$

The integral condition on $\delta p^{(3)}$ is satisfied by the function

$$\delta p_c^{(3)}(1,2,3) \equiv \rho^4 \int h(r_{14})h(r_{24})h(r_{34})d\underset{\sim}{r}_4\ , \tag{56}$$

known as the convolution remainder form. These definitions imply

$$\int \Big[\delta p^{(3)}(1,2,3) - \delta p_c^{(3)}(1,2,3)\Big] f(\underset{\sim}{r}_1, \underset{\sim}{r}_2) d\underset{\sim}{r}_{123} = 0 \tag{57}$$

for arbitrary integrable function $f(\underset{\sim}{r}_1, \underset{\sim}{r}_2)$. By using $\delta p_c^{(3)}(1,2,3)$ in evaluating $S^{(3)}$, the three-phonon structure function can be expressed as a simple explicit approximation formula plus a remainder:

$$S^{(3)}(\underset{\sim}{k}, \underset{\sim}{\ell}, -\underset{\sim}{k}-\underset{\sim}{\ell}) = S(k)\, S(\ell)\, S(|\underset{\sim}{k}+\underset{\sim}{\ell}|) + \delta S^{(3)}(\underset{\sim}{k}, \underset{\sim}{\ell}, -\underset{\sim}{k}-\underset{\sim}{\ell})$$

$$\delta S^{(3)}(\underset{\sim}{k}, \underset{\sim}{\ell}, -\underset{\sim}{k}-\underset{\sim}{\ell}) = \frac{1}{N}\int \Big[\delta p^{(3)}(1,2,3) - \delta p_c^{(3)}(1,2,3)\Big]\ . \tag{58}$$

$$\exp\Big\{i\,[\underset{\sim}{k}\cdot\underset{\sim}{r}_1 + \underset{\sim}{\ell}\cdot\underset{\sim}{r}_2 - (\underset{\sim}{k}+\underset{\sim}{\ell})\cdot\underset{\sim}{r}_3]\Big\}\, d\underset{\sim}{r}_{123}\ .$$

The remainder function $\delta S^{(3)}$ vanishes if $k\ell|\underline{k}+\underline{\ell}| = 0$, and it surely approaches zero rapidly as $k\ell|\underline{k}+\underline{\ell}|$ increases without limit. To develop a more explicit estimate for $\delta S^{(3)}$ I introduce a set of continuous, linearly independent, short-range functions $\{h_m(r)\}$ with the essential property

$$\int h_m(r)d\underline{r} = 0 \ . \tag{59}$$

These functions are used to generate the approximate form

$$\delta p_{AP}^{(3)}(1,2,3) = \delta p_c^{(3)}(1,2,3) + \sum_{mnp} \Lambda_{mnp}\,\rho^4 \int h_m(r_{14}) h_n(r_{24})\, h_p(r_{34}) \ldots d\underline{r}_4 \ , \tag{60}$$

which automatically satisfies the sequential condition imposed by equation (55). The coefficients Λ_{mnp} are symmetrical functions of the indices.

The sequential relation and the positive-definite condition on $p^{(3)}$ can be used to determine the expansion coefficients. These conditions require

$$\rho^3 h(r_{13})^2 + \delta p_c^{(3)}(1,1,3) + \sum_{mnp} \Lambda_{mnp}\,\rho^4 \int h_m(r_{14}) h_n(r_{14})\, h_p(r_{34}) dr_4 \geq 0. \tag{61}$$

The left-hand member of equation (61) can be made to vanish for all values of r_{13} only if the set of functions $\{h_m(r)\}$ is (a) infinite and (b) properly chosen. Just what condition (b) implies is not yet clear. To proceed in a practical direction, I replace equation (61) by the least-squares condition

$$\delta \int \Big[\rho h(r_{13})^2 + \delta p_c^{(3)}(1,1,3) + \sum_{mnp} \Lambda_{mnp}\, h_m(r_{14}) h_n(r_{14})\, h_p(r_{34}) dr_4\Big]^2 dr_3 = 0 \ , \tag{62}$$

and propose that Λ_{mnp} should be determined by solving the system of linear inhomogeneous equations generated by equation (62). In this way we arrive at an extended approximate form for $\delta p^{(3)}$, but must recognize that the problem is only half solved because criteria for testing the uniqueness and accuracy of such forms are difficult to apply [30].

A simple example will serve to illustrate the possibilities and limitations of the procedure. Suppose the sum in equation (60) is limited to a single term (with m = n = p = 1) in which

$$h_1(r) = \Delta h(\lambda r) \ . \tag{63}$$

Let

$$A(r_{13}) = h(r_{13})^2 + \rho \int h(r_{14})^2 h(r_{34})\, d\underline{r}_4 \ ,$$

$$B_\lambda(r_{13}) = \frac{1}{\rho} \int (\Delta_1 h(\lambda r_{14}))^2\ \Delta_3 h(\lambda r_{34}) d\underline{r}_4 \ . \tag{64}$$

Equation (61) reduces to

$$A(r) + \Lambda B_{\lambda}(r) \geq 0 . \tag{65}$$

Minimum value of the quadratic form

$$F(\Lambda, \lambda) = \rho \int [A(r) + \Lambda B_{\lambda}(r)]^2 d\underline{r} \tag{66}$$

occurs at

$$\Lambda(\lambda) = \frac{\int A(r) B_{\lambda}(r) d\underline{r}}{\int B_{\lambda}(r)^2 d\underline{r}} . \tag{67}$$

Finally the scale parameter λ is chosen to yield the minimum value of the function

$$F(\Lambda(\lambda), \lambda) = \rho \int A(r)^2 d\underline{r} - \frac{[\int A(r) B_{\lambda}(r) d\underline{r}]^2}{\int B_{\lambda}(r)^2 d\underline{r}} . \tag{68}$$

The remainder term in $S^{(3)}$ is now

$$\delta S_{AP}^{(3)}(\underline{k}, \underline{\ell}, -\underline{k}-\underline{\ell}) = \frac{1}{N} \int \left[\delta \rho_{AP}^{(3)}(1,2,3) - \delta \rho_c^{(3)}(1,2,3) \right] \exp\left\{ i[\underline{k} \cdot \underline{r}_1 + \underline{\ell} \cdot \underline{r}_2 - (k + \ell) \cdot r_3] \right\} d\underline{r}_{123}$$

$$= \frac{\Lambda}{\rho^2 N} \rho^4 \int \Delta_1 h(\lambda r_{14}) \Delta_2 h(\lambda r_{24}) \Delta_3 h(\lambda r_{34}) \exp\{i[\underline{k} \cdot \underline{r}_1 + \underline{\ell} \cdot \underline{r}_2 - (\underline{k} + \underline{\ell}) \cdot \underline{r}_3]\} d\underline{r}_{1234}$$

$$= - \frac{\Lambda}{\lambda^9 \rho^2} (k \ell | \underline{k} + \underline{\ell} |)^2 [S(k) - 1] [S(\ell) - 1] [S(|\underline{k} + \underline{\ell}|) - 1] . \tag{69}$$

The evaluation of Λ and λ for liquid ^{4}He using available experimental and theoretical information on the radial distribution function is feasible, but has not yet been carried out. I report numerical results based on the Gaussian form

$$h(r) = - e^{-(r/a)^2} , \quad a^3 \rho \pi^{3/2} = 1 , \tag{70}$$

which is a reasonable approximation for the charged-boson system at metallic densities. Now $F(\Lambda(\lambda), \lambda)$ attains a minimum value at $\lambda^2 \simeq 1.631$ with

$$\Lambda(\lambda) \simeq 6.3 \times 10^{-4} ,$$

$$F(\Lambda(\lambda), \lambda) \simeq 10^{-4} . \tag{71}$$

For comparison observe that $F(0, \lambda) = 0.0606$.

The left-hand member of equation (65) can be expressed as a power series in r^2. The condition that the constant term and the coefficient of r^2 vanish determines values of Λ and λ^2 differing slightly from the values which minimize $F(\Lambda, \lambda)$. This procedure yields

$$\Lambda = -6.64 \times 10^{-4} , \quad \lambda^2 = 1.738 , \tag{72}$$

and a small positive coefficient (0.136) for the fourth power term.

The ratio $\delta S_{AB}^{(3)}/S_c^{(3)}$ gives information on the accuracy of $S_c^{(3)}$. Consider the equilateral triangle situation $k = \ell = |k + \ell|$. For small k ($\frac{1}{2}$ ka < 1)

$$\frac{\delta S_{AB}^{(3)}}{S_c^{(3)}} \sim 0.11 \exp[-0.226 (\tfrac{1}{2} ka)^2] . \tag{73}$$

For large k ($\frac{1}{2}$ ka > 1)

$$\frac{\delta S_{AP}^{(3)}}{S_c^{(3)}} \sim 0.575 \left(\frac{ka}{2\lambda}\right)^6 \exp[-3(ka/2\lambda)^2]$$

$$\leq 0.575e^{-3} = 0.0286 . \tag{74}$$

For what they are worth these estimates provide support for the view that $S_c^{(3)}$ is a fairly good approximation.

For a final remark on this topic let me mention that D. K. Lee has found a four-particle distribution function which yields $p_c^{(3)}$ in the sequential relation connecting $p^{(4)}$ and $p^{(3)}$ [6]; F. Y. Wu and M. K. Chien [29] have found the general formula for arbitrary order in the same context. Very little is known about the accuracy of these forms.

REFERENCES

1. E. Feenberg, Theory of Quantum Fluids (New York: Academic Press, 1969).
2. E. Feenberg, Am. J. Phys. **38**, 684 (1970).
3. D. Pines and C-W. Woo, Phys. Rev. Letters, **24**, 1044 (1970); **25**, 197 (1970).
4. A. D. B. Woods and R. A. Cowley, Phys. Rev. Letters, **24**, 646 (1970).
5. H. J. Maris and W. E. Massey, Phys. Rev. Letters, **25**, 220 (1970).
6. Deok Kyo Lee, Phys. Rev. **162**, 134 (1967).
7. H-W. Lai, H-K. Sim, and C-W. Woo, Phys. Rev. **A1**, 1536 (1970).
8. D. Hall and E. Feenberg, Ann. Phys. **63**, 335 (1971).
9. J. E. Enderby, T. Gaskell, and N. H. March, Proc. Roy. Soc. (London), **85**, 217 (1965).
10. T. Davison and E. Feenberg, Phys. Rev. **171**, 221 (1968).
11. E. K. Achter and L. Meyer, Phys. Rev. **188**, 291 (1969).
12. R. B. Hallock, Phys. Rev. Letters, **23**, 830 (1969).
13. D. G. Hurst and D. H. Henshaw, Phys. Rev. **100**, 994 (1955).
14. D. G. Henshaw, Phys. Rev. **119**, 14 (1960).
15. H. Palevsky, K. Otnes, and K. E. Larsson, Phys. Rev. **112**, 11 (1959).
16. J. L. Yarnell, G. P. Arnold, P. J. Bendt, and E. C. Kerr, Phys. Rev. **113**, 1379 (1959).
17. D. G. Henshaw and A. D. B. Woods, Phys. Rev. **121**, 1266 (1961).
18. R. A. Cowley and A. D. B. Woods, Can. J. Phys. **49**, 177 (1971).
19. H. W. Jackson, Phys. Rev. **185**, 186 (1969).

20. V. F. Sears, Phys. Rev. **185**, 200 (1969); Phys. Rev. **A1**, 1699 (1970).

21. R. D. Puff, Phys. Rev. **A1**, 125 (1970).

22. W. C. Kerr, P. N. Pathak, and K. S. Singwi, Phys. Rev. **A2**, 2416 (1970).

23. K. Huang and A. Klein, Ann. Phys. (N.Y.), **30**, 203 (1964).

24. N. Nihara and R. D. Puff, Phys. Rev. **171**, 221 (1968).

25. A. Miller, D. Pines, and P. Nozieres, Phys. Rev. **127**, 1452 (1962).

26. C-W. Woo, H-T. Tan, and W. E. Massey, Phys. Rev. Letters, **22**, 278 (1969); Phys. Rev. **185**, 287 (1969).

27. C-W. Woo, H-T. Tan, and F. Y. Wu, preprint (1970).

28. C. E. Campbell, J. Low Temp. Phys. **4**, 433 (1971).

29. F. Y. Wu and M. K. Chien, J. Math. Phys. **11**, 1912 (1970).

30. H. W. Jackson, doctoral dissertation, Washington University, 1962; D. K. Lee, H. W. Jackson, and E. Feenberg, Ann. Phys. (N.Y.) **44**, 84 (1967).

DISCUSSION

H. Matsuda: In what range of density is your approach valid? Is it also valid in the solid helium?

E. Feenberg: The approach is designed for the physical density range of the helium liquids. If I interpret the second question correctly, the answer is yes. A suitable product of two-particle functions as a trial wave function can represent a crystalline solid.

T. Nishiyama: May I ask if you could find any evidence of the double-branch structure found by Woods and Cowley in your theory?

E. Feenberg: The upper branch resulting from multiple phonon excitation processes in the inelastic scattering of slow neutrons is implicit in the correlated basis functions since they form a complete set. H. W. Jackson has developed the theory (with numerical results) in the range of large k ($k > 3\ \text{Å}^{-1}$).

J. P. Hausen: Does a similar theoretical analysis for the liquid ^{3}He structure factor, based on a Jastrow and Slater determinant trial wave function, show the surprising qualitative agreement between ^{3}He and ^{4}He S(k) observed experimentally?

E. Feenberg: Such an analysis has actually been carried out and confirms the surprising similarity.

M. Cole: 1. Chester and Reatto pointed out several years ago that the exponent of the Jastro-type wave function should manifest long-wavelength correlations arising from the zero-point motion of the phonons. Does your wave function include this?

2. Professor Cummings proposed yesterday that the pair distribution function, and hence S(k), should explicitly include effects due to the presence of the condensate. Is it at all clear how this would show up in your calculations?

E. Feenberg: 1. Yes, the asymptotic condition on U(r) mentioned in the talk.

2. Not at all clear, but the point made by Cummings is valuable in directing attention to the possibility of extracting significant physical information from the detailed behavior of the radial distribution function.

H. Gould: Does your method imply that γ' is necessarily positive? The heat capacity measurements of Phillips et al. seem to imply that γ' is negative at vapor pressure and then changes sign with increasing pressure. Recently Victor Wong and myself have shown that for a weakly interacting Bose gas, S(k) cannot be expanded in a given series; that is, there appears a $k^5 \ln k$ term. Of course, this term does not affect your lower order results.

E. Feenberg: In the formula

$$\epsilon(k) = \epsilon_o(k)[1 - \gamma'(\frac{\hbar k}{ms})^2 + \ldots] ,$$

γ' is necessarily positive. The actual third power term in the dispersion relation for $\epsilon(k)$ is not γ'. There is a contribution from $\epsilon_o(k)$.

EQUILIBRIUM THEORY OF CLASSICAL LIQUIDS

Loup Verlet

I. INTRODUCTION

For a long time, the main line of attack in the theory of liquids has been the extension of the range of validity of the low-density expansions: the virial series has been pushed up to the fifth virial coefficient [1]. Simple integral equations, PY [2] and HNC [3] have been solved [4] in great detail for the case of the Lennard-Jones potential

$$v_{LJ}(r) = 4\epsilon \left[\left(\frac{\sigma}{r}\right)^{12} - \left(\frac{\sigma}{r}\right)^{6} \right] , \qquad (1)$$

which will be considered in the present paper as the prototype potential for the study of simple liquids. Various improvements of those integral equations have been proposed and studied [5,6,7]. Comparison with the results obtained from computer experiments [8,9,10] show that this type of approach completely fails[1] when one gets near to the fluid-solid transition line.

It has become clear in recent years that the structure of a dense liquid is mainly determined by the repulsive part of the interaction. Evidence for this is provided by the remarkable success of the hard-sphere model [11] in representing the structure factor of simple liquids as obtained from real [11] or computer [12] experiments. Zwanzig [13] has shown how one can incorporate that idea in an expansion scheme in which the hard sphere system is used as a basis. This scheme, used first for potentials composed of a hard core with an attractive square well [14,15], is the cornerstone of the perturbation theories reviewed in the present paper. Let us divide the two-body potential v(r) into two parts: a part $v_o(r)$ containing mainly the short-range repulsive part of the potential, for which we suppose the solution of the N-body problem to be known, and the remaining part of the interaction w(r), which will be considered as a

---

[1]When making "moral" judgments of a theory, one should state clearly what one is aiming at. The ideal aim would be to build a consistent theory, able to reproduce the "exact" computer results within the error in the latter, that is, with an error in $p/\rho kT$ of the order of 0.01 at low density and of 0.05 in the highest density region where the isotherms are very steep. Being forced to compromise and accept, up to 2 or 3 times that error, we may still call the theory "good."

perturbation. We thus multiply **w(r)** by some coupling parameter λ, expand in λ, and ultimately make λ equal to 1.

It is shown very easily that the free energy per particle f is given exactly by

$$f = f_o + 2\pi\rho \int_0^\infty w(r)\; g_o(r)\; r^2\; dr + 2\pi\rho \int_0^1 d\lambda \int_0^\infty w(r)\; r^2\; dr\; [g_\lambda(r) - g_o(r)] \;. \tag{2}$$

The first term is the free energy for the reference system [i.e., for the system of particles interacting through $v_o(r)$ only]; $g_\lambda(r)$ is the radial distribution for particles interacting through $v_o(r) + \lambda\, w(r)$; ρ is the particle density. If the last term is expanded in powers of λ, the leading contribution is the mean square fluctuation of the perturbation w(r) averaged over the reference system. At high density, where the repulsive cores prevent large local density fluctuations, this term is small, especially if the division of the potential is such that w(r) is smooth.

II. HIGH-TEMPERATURE THEORY

A "classical" separation of the potential v(r), specific to the Lennard-Jones interaction, has been considered in references [16] and [17]. It consists in choosing $v_o(r) = 4\epsilon(\sigma/r)^{12}$ as the reference system potential and $w(r) = -4\epsilon\,(\sigma/r)^6$ as the perturbation. It is seen that the reference system potential can be scaled: the length σ enters only through the combination $4\epsilon\beta\sigma^{12}$. A change of temperature can be obtained by changing the length scale (and thus the number density) so as to keep this combination constant. It is therefore only necessary to consider one isotherm. The same can be said for the first-order term in (2); it is sufficient to calculate the average of $1/r^6$ over the reference system for one value of the temperature and to scale the result to reach other temperatures. Likewise, the second-order term can be computed and scaled. Such computations have been performed, by the Monte Carlo method, both for the reference system [18, 19] and for the first- and second-order terms [19]. They indicate that at high density a 1% error in the excess free energy is obtained if $kT/\epsilon \gtrsim 5$. Near the triple point the convergence is not so good [20]. There the error in the excess free energy reaches 1/15.

In the following we shall introduce other expressions for the potential of the reference system which will behave much more like hard spheres than the $1/r^{12}$ potential, and will be better suited as a zero-order approximation to the low-temperature situation. At high temperature, on the other hand, the role of the repulsive interaction is apparent and the above separation of the potential is appropriate.

III. LOW-TEMPERATURE THEORY

Barker and Henderson [21] have considered the following separation of the potential

$$
\begin{aligned}
v_o(r) &= v_{LJ}(r) && (r < \sigma) \\
&= 0 && (\text{otherwise}) ; \\
w(r) &= v_{LJ}(r) && (r > \sigma) \\
&= 0 && (\text{otherwise}) . \qquad (3)
\end{aligned}
$$

The drawback of this separation is that it yields a $w(r)$ which is not smooth, so that the remainder of the perturbation series, although small, is not negligible. Near the triple point the ratio of the remainder to the total excess free energy reaches 1/10. If quantitative answers are desired it is necessary to evaluate [21, 22, 23] the next-order term in the λ expansion. This leads to calculations which are neither simple nor exact. Barker and Henderson identify the reference system with a system of hard spheres in the following way: the potential $v_o(r)$ is replaced by

$$v_o(d + \frac{r - d}{\alpha}) .$$

This representation is chosen so that when α is equal to 1, $v_o(r)$ is recovered. When α tends to zero, a hard-sphere potential with diameter d is obtained. The free energy of the reference system is expanded in a series in α, and d is chosen so as to ensure the vanishing of the first-order term. One then obtains

$$d = d_B = \int_0^\infty \{1 - \exp[-\beta v_o(r)]\} \, dr \quad . \qquad (4)$$

The pressure of the reference system is then the pressure of a system of hard spheres with a packing fraction $\eta = \pi \rho \, d_B^3/6$.

The error in $(p/\rho kT)_o$ entailed by the Barker-Henderson approximation is quite small [24], of the order of 0.1. In order to calculate the second term in (2), Barker and Henderson use for $g_o(r)$ the Wertheim-Thiele [25] solution of the Percus-Yevick equation for the hard-sphere problem. This leads to an error which is not negligible (1/30 of the total excess free energy near the triple point). This error is due partly to the replacement of $g_o(r)$ by the hard sphere radial distribution function and partly to the use of the PY approximation.

The more recent theories of Mansoori and Canfield [26] and Rasaiah and Stell [27] are elegant variants of the Barker-Henderson theory. They suffer from the same defects.

The above-mentioned difficulties can be overcome by an appropriate use of very interesting and fruitful new ideas very recently proposed by Weeks, Chandler, and Andersen [28]. These authors first propose to use for $v_o(r)$ that part of the potential which gives rise to the repulsive forces and vanishes when the forces vanish, i.e., for $r_m = 2^{1/6} \sigma$:

$$
\begin{aligned}
v_o(r) &= v_{LJ}(r) + \epsilon && \text{for } r \le r_m , \\
&= 0 && \text{otherwise}. \qquad (5)
\end{aligned}
$$

We are thus left with

$$
\begin{aligned}
w(r) &= -\epsilon && (r \le r_m) \\
&= v_{LJ}(r) && (r \ge r_m) . \qquad (6)
\end{aligned}
$$

It is clear that w(r) is now much smoother than before. In fact, a computation made in the neighborhood of the triple point [20] shows that the remainder of the series in equation (2) is of the order of 1/120 of the total excess free energy. The error made in keeping only the first order term will be acceptable in the high-density region.

IV. EXPRESSING THE REFERENCE SYSTEM IN TERMS OF HARD SPHERES

Another proposal of Weeks, Chandler, and Andersen concerns the identification of the reference system with a hard sphere system. They write for the radial distribution function of the reference system the approximation already considered by Kim [29]:

$$\hat{g}_o(r) = \exp[-\beta v_o(r)]\, y_{HS}(r/d,\ \eta), \tag{7}$$

where

$$y_{HS}(r/d,\ \eta)$$

is the radial distribution function of a hard sphere system of diameter d, of packing fraction $\eta = \pi\rho d^3/6$, suitably extrapolated for $r \leq d$ (we may remain vague on this last point, as we shall show later on that the extrapolation of the hard-sphere radial distribution function inside the hard core is not really needed). According to Weeks, Chandler, and Andersen, d should be determined by requiring that

$$\hat{S}_o(0) = S_{HS}(0, \eta), \tag{8}$$

where $\hat{S}_o(k)$ is the approximate structure factor for the reference system, as defined by

$$\hat{S}_o(k) = 1 + \rho \int e^{i\underset{\sim}{k}\underset{\sim}{r}}\, [\hat{g}_o(r) - 1]\, d\underset{\sim}{r}, \tag{9}$$

and $S_{HS}(kd,\ \eta)$ is likewise the structure factor for the hard-sphere gas. At first, condition (8) may seem rather puzzling and arbitrary. It should be realized, however, that the statement that $\hat{g}_o(r)$ is a radial distribution function appropriate to a system with repulsive forces is a very strong statement, one which is practically sufficient to determine d: the hard-sphere model [11] can be applied and d can be determined by somehow identifying $\hat{S}_o(k)$ with $S_{HS}(kd,\ \eta)$, for example, by requiring that

$$\int_0^{\infty} |\hat{S}_o(k) - S_{HS}(kd,\ \eta)|\, dk \tag{10}$$

be a minimum when d varies.

When it is realized that $\hat{S}_o(0)$ is a very sensitive function of d, the simple requirement of equation (8) is seen to mean essentially that, for a dense system with repulsive interactions, the inverse compressibility is small and positive. It leads to values of d very similar to those obtained with the help of expression (10). This shows the consistency of the approximation.

In order to proceed further, accurate knowledge of the properties of the hard sphere system is clearly required. The computer data [30] are outstandingly well summarized by the Carnahan-Starling [31] equation of state

$$z_{HS}(\eta) = \frac{1 + \eta + \eta^2 - \eta^3}{(1 - \eta)^3} \tag{11}$$

where we use the symbol z for the compressibility factor $p/\rho kT$. Use of the virial theorem yields for the hard-sphere radial distribution function at the core the value

$$y_{HS}(1, \eta) = \frac{1 - \eta/2}{(1-\eta)^3} \quad . \tag{12}$$

At low density (for values of the packing fraction less than 0.2) the radial distribution function is excellently approximated by the PY equation and we can use the analytical solution due to Wertheim and Thiele [25]. Nearer to the transition density ($\eta_\epsilon = 0.49$), it is necessary to build an improved expression [20] for the radial distribution function in order to reproduce the computer data [32, 33]. Except near the core, the exact hard-sphere radial distribution function is very well fitted by the Wertheim-Thiele radial distribution function $y_W(r/d, \eta)$ if one makes a slight renormalization of the diameter (and of the packing fraction), so that for $r \gtrsim 1.3\ d$,

$$y_{HS}(r/d, \eta) \approx y_W(r/d_W, \eta_W) \ , \tag{13}$$

and $\eta_W = \pi \rho d_W^3/6$ is found empirically to obey the relation

$$\eta_W = \eta - \frac{1}{16} \eta^2 \ . \tag{14}$$

A simple and satisfactory adjustment of the radial distribution function near the core can then be made, using the requirements that the correct value at the core in equation (12) is obtained and that the inverse compressibility remains very small. This leads to simple semiempirical expressions [20] for the values of the first derivatives of the HS radial distribution function at the core. As we shall see, this and equation (8) are the only ingredients needed to deal with the reference system.

Let us expand $y_{HS}(r/d, \eta)\, r^2/d^2$ around the core:

$$x^2\ y_{HS}(x, \eta) = \sigma_o + (x-1)\,\sigma_1 + \ldots \ , \tag{15}$$

where σ_o is given by equation (12) and σ_1 can be calculated as just mentioned. This can be used in conjunction with the condition (8) which can be written in coordinate space as

$$\int_0^d dr\, r^2\, y_{HS}(r/d, \eta)\ \exp[-\beta v_o(r)] = \int_0^d dr\ r^2\ y_{HS}(r/d, \eta)\ \{1 - \exp[-\beta v_o(r)]\} \ . \tag{16}$$

A straightforward integration by parts leads to the following condition for d:

$$d = d_B\ \left(1 + \frac{\sigma_1}{2\sigma_o}\,\delta\right) \ , \tag{17}$$

where d_B is defined by equation (4), and where δ, which is very small (0.2% of d_B/σ), is given by

$$\delta = \frac{1}{d_B^2} \int_0^\infty (r - d_B)^2\ \frac{d}{dr}\ \exp[-\beta v_o(r)]\ dr. \tag{18}$$

We can neglect higher-order terms in equation (15) due to the fact that $\exp[-\beta v_o(r)]$ is very steep [29], so that

$$\frac{d}{dr}\ \exp[-\beta v_o(r)] \simeq \delta(r - d_B) \ . \tag{19}$$

The compressibility factor of the reference system is calculated by using the virial theorem

$$z_o = 1 - \frac{\rho}{6\,kT} \int \underset{\sim}{dr}\; r \frac{\partial v_o}{\partial r} \exp[-\beta v_o(r)]\; g_o(r). \tag{20}$$

Using the approximation equation (7) for the radial distribution function, one obtains

$$\hat{z}_o = 1 + 4\,\eta \int_0^\infty dr\, \frac{r^3}{d^3}\, y_{HS}(r/d,\, \eta)\, \frac{d}{dr} \exp[-\beta v_o(r)]\,. \tag{21}$$

We expand as above $r^3/d^3\; y_{HS}$ around the core, and obtain, by the use of equations (4), (17) and (18),

$$\hat{z}_o = z_{HS}(\eta) + O(\delta)\,, \tag{22}$$

where the small correction linear in δ can be calculated [20] in terms of the two first derivatives of the hard-sphere radial distribution function at core.

We give an example showing that the reference system has been correctly treated when the above approximations are made: for $\rho\sigma^3 = 0.84$ and $kT/\epsilon = 0.75$, an "exact" computation gives $z_o = 10.23$. From equation (22), one obtains $\hat{z}_o = 10.37$.

Once the correct equation of state is obtained, a straightforward integration on the density leads to the free energy of the reference system, which, within a small term in δ, is that of a hard-sphere gas of packing fraction η.

V. FIRST-ORDER CORRECTION TO THE FREE ENERGY

The first-order term in equation (2) is given with the help of the approximation (7) by

$$\hat{f}_1 = 2\,\pi\,\rho \int_0^\infty r^2\, dr\, y_{HS}(\tfrac{r}{d},\, \eta)\; w(r) \exp[-\beta v_o(r)]\,. \tag{23}$$

By using equation (16), this can be transformed to

$$\hat{f}_1 = 2\,\pi\,\rho \int_d^\infty r^2\, dr\, y_{HS}(r/d,\, \eta)\, w(r)\,. \tag{24}$$

By using the above-mentioned improvement of the hard-sphere radial distribution function, it can be shown [20] that, with an error of about 0.1%, one can write

$$\hat{f}_1 = 12\,\eta_W \int_1^\infty x^2\, dx\, y_W(x,\, \eta_W)\, w(x\, d_W)\,, \tag{25}$$

where η_W is given by equation (14).

We can separate this integral into two parts, by writing

$$w(x\, d_W) = v_{LJ}(x\, d_W) - v_o(x\, d_W). \tag{26}$$

The first part involves integrals of the form

$$I_n^{\,1}(\eta_W) = \int_1^\infty x^2\, dx\, \frac{y_W(x,\, \eta_W)}{x^n}\,, \tag{27}$$

with n equal to 6 and 12, for which simple approximants are known [34, 20]. The short-range part involving $v_o(x\, d_W)$ is given in terms of integrals of the form

$$I_n^2(\eta_W, d_W) = \int_1^{r_m/d_W} x^2\, dx\, \frac{y_W(x, \eta_W)}{x^n}\, dx\ , \tag{28}$$

which are accurately and simply calculated by expanding $y_W(x, \eta_W)$ around x = 1. The error introduced by this series of approximations is quite small. For the state already considered, around the triple point, one gets directly $\beta f_1 = -8.82$ and approximately $\beta \hat{f}_1 = -8.84$.

VI. RESULTS AND EXTENSION TO LOW DENSITY

The procedure we have just sketched leads to simple expressions for the thermodynamic functions at high density. From these expressions, one gets quite accurate results at high density. For instance, for the states already considered, a Monte Carlo computation yields z = 0.37, while with the approximation made in sections IV and V, one finds $\hat{z} = 0.32$. In the same way, for the excess part of the internal energy, one gets $U_i = -6.04\ \epsilon$ "exactly" and $\hat{U}_i = 6.02\ \epsilon$ approximately. The results are not so good at lower density. For instance, the critical temperature obtained with the above equation of state is found to be 1.45 ϵ/k. The value 1.36 ϵ/k should be obtained for a fluid with Lennard-Jones interactions when the large-scale density fluctuations are suppressed [24]. It is therefore necessary to evaluate somehow the last term in equation (2). A straightforward procedure consists of adding to the first two terms in equation (2), which are the beginning of the λ expansion, the remainder treated in a density expansion: this is a mixed perturbation expansion of a type considered previously on several occasions [34, 35, 36].

In this simple form, this procedure does not work, because while the equation of state becomes quite good (obviously) at low density, the high-density results are completely spoiled. A more elaborate but successful procedure consists in obtaining $g(r, \lambda)$ in the Percus-Yevick approximation and computing the λ integral in equation (2). This leads to excellent answers. For instance, for $kT/\epsilon = 1.35$ and $\rho\sigma^3 = 0.4$, the "exact" value of the excess free energy is - 1.00, the sum of the zeroth- and first-order terms is - 0.91 (a 10% error!). The remainder term in the Percus-Yevick (PY) approximation is - 0.09. It should be noted that at high density this remainder term in the PY equation gets very small, so that the excellent fit previously obtained is maintained. We can therefore say that we have obtained a successful theory of liquids valid for all densities. It should be emphasized, however, that the price for the extension of the range of validity of the theory to low densities is high. While the high-density results are written in an analytically well-defined form and can be easily obtained with the help of a slide rule or desk computer, the lower-density correction requires, for the solution of the PY equation for several values of λ, large-scale computing equipment and a numerical differentiation to obtain the equation of state. In its present state, the theory thus suffers from a great lack of homogeneity in the methods it uses.

VII. CONCLUDING REMARKS AND PROSPECTS FOR FUTURE DEVELOPMENTS

1. The theory can be partly extended to describe the solid-fluid transition. It can first be shown [37] that the λ-expansion holds as well for solids as for liquids near the transition. The reduction of the Lennard-Jones system to hard spheres is more delicate: the high-density limits of these two systems are quite different, and only near the transition can we hope to make that reduction. Being unable to generalize equation (7) to solids, we shall use the Barker and Henderson minimization procedure, which is still valid in the solid case: the reference system is thus replaced by a solid of hard spheres with the diameter d_B and the density ρ. The pressure and free energy of that system can be calculated by using the known hard-sphere results [38]. In order to obtain the first-order correction, it is first necessary to construct a simple expression for the correlation function of the hard-sphere solid. Once the free energy is known, the transition characteristics are determined by the double tangent method.

2. A weak point of perturbation theories based on the λ-expansion is that they do not provide any simple approximations for the radial distribution function. The zeroth-order approximation for this quantity is not precise enough. For instance, for $\rho\sigma^3 = 0.84$, $kT/\epsilon = 0.75$, the "exact" value for the structure factor at its first maximum is $S(k_o) = 2.75$. The zeroth-order result is $S_o(k_o) = 2.43$. The next-order terms in the λ-expansion involve three- and four-body correlation functions of the reference system and are not simple, even in the superposition approximation. In order to calculate the correction to the zeroth-order term, one might use the Lebowitz-Stell-Baer [39] expansion scheme in which the perturbation parameter is the inverse range of the perturbation. This leads to simpler expressions than with the λ-expansion, but in a Lennard-Jones fluid the range of the attraction is not large as compared to the interparticle distance so that the theory is not clearly applicable.

Weeks, Chandler, and Andersen have proposed, implicitly, to use as an approximation for the total radial distribution function g(r) the approximation (7), where $y_{HS}(r/d, \eta)$ is treated in the Percus-Yevick approximation. The theory so obtained is seen to be very satisfactory. This leads to an equation of state which turns out to be remarkably good [28], and which, by using the techniques of sections III and IV, can be cast in a very simple form [20]. The diameter calculated by equation (17) (in the PY approximation) is the diameter of the equivalent hard-sphere gas [11], the structure factor of which is written in terms of the Wertheim-Thiele solution. This approximation for the structure factor turns out to be good at the triple-point density, and for densities lower by 10%, but rather poor otherwise.

3. The theory can be easily extended to potentials giving rise to an attractive force which can be written as a sum of inverse powers. The preceding perturbation theory can (and will) be used to obtain an effective two-body potential valid for dense liquids.

4. Some simplification of the remainder of the expansion equation (4) is necessary if we want to obtain, over the whole density range a theory possessing the same simplicity as for dense systems. It is plausible that if the PY theory, or even simpler ones such as that based on the

mean spherical model [40], or the linearized perturbed PY theory [41], could be solved in simple terms for a short-range perturbing potential (such as square well or Yukawa potentials) in addition to a reference hard-sphere interaction, one could use this solution to express the remainder in equation (2) in an approximate fashion.

The author acknowledges fruitful collaboration with Jean-Jacques Weis and stimulating discussions with George Stell.

REFERENCES

1. J. A. Barker and J. J. Monaghan, J. Chem. Phys. 36, 2564 (1962); J. A. Barker, D. J. Leonard and A. Pompe, J. Chem. Phys. 44, 426 (1966).

2. J. K. Percus and G. J. Yevick, Phys. Rev. 110, 1 (1958).

3. E. Meeron, J. Math. Phys. 1, 192 (1960); T. Morita and K. Hiroike, Progr. Theor. Phys. 23, 385, 1003 (1960); M. S. Green, J. Chem. Phys. 33, 1403 (1960); J. M. J. Van Leeuwen, J. Groeneveld, and J. De Boer, Physica, 25, 792 (1959); L. Verlet, Nuovo Cimento, 18, 77 (1960); G. S. Rushbrooke, Physica, 26, 259 (1960).

4. D. Levesque, Physica, 32, 1985 (1966) and references therein.

5. S. A. Rice and J. Lekner, J. Chem. Phys. 42, 3559 (1965).

6. L. Verlet, Physica, 30, 95 (1964); L. Verlet and D. Levesque, Physica, 36, 254 (1967).

7. M. S. Wertheim, J. Math. Phys. 8, 927 (1967).

8. W. W. Wood and F. R. Parker, J. Chem. Phys. 27, 720 (1957).

9. J. McDonald and K. Singer, J. Chem. Phys. 47, 4766 (1967).

10. L. Verlet, Phys. Rev. 159, 98 (1967).

11. N. W. Ashcroft and J. Lekner, Phys. Rev. 145, 83 (1966).

12. L. Verlet, Phys. Rev. 165, 201 (1968).

13. R. W. Zwanzig, J. Chem. Phys. 22, 1420 (1954).

14. E. B. Smith and B. J. Alder, J. Chem. Phys. 30, 1190 (1959).

15. J. A. Barker and D. Henderson, J. Chem. Phys. 47, 2856 (1967).

16. D. A. McQuarrie and J. L. Katz, J. Chem. Phys. 44, 2393 (1966).

17. J. L. Lebowitz, Phys. Letters, 28A, 596 (1969).

18. W. G. Hoover, M. Ross, K. W. Johnson, D. Henderson, J. A. Barker, and B. C. Brown, J. Chem. Phys. 52, 4931 (1970).

19. J. P. Hansen, Phys. Rev. A2, 221 (1970).

20. L. Verlet and J. J. Weis (to be published).

21. J. A. Barker and D. Henderson, J. Chem. Phys. 47, 4714 (1967).

22. E. Praestgaard and S. Toxvoerd, J. Chem. Phys. 51, 1895 (1969); S. Toxvoerd and E. Praestgaard, J. Chem. Phys. 53, 2389 (1970).

23. D. Henderson and J. A. Barker, J. Chem. Phys. 52, 2315 (1970).

24. D. Levesque and L. Verlet, Phys. Rev. **182**, 307 (1969).

25. M. S. Wertheim, Phys. Rev. Letters, **10**, 321 (1963); E. Thiele, J. Chem. Phys. **38**, 1959 (1963).

26. G. A. Mansoori and F. B. Canfield, J. Chem. Phys. **51**, 4958 (1969).

27. J. Rasaiah and G. Stell, Mol. Phys. **18**, 249 (1970).

28. J. D. Weeks, D. Chandler, and H. C. Andersen (to be published). A summary of the results is given in D. Chandler and J. D. Weeks, Phys. Rev. Letters, **25**, 149 (1970).

29. S. Kim, Phys. Fluids, **12**, 2046 (1969).

30. B. J. Alder and T. E. Wainwright, J. Chem. Phys. **33**, 1439 (1960).

31. N. F. Carnahan and K. E. Starling, J. Chem. Phys. **51**, 635 (1969).

32. B. J. Alder and C. E. Hecht, J. Chem. Phys. **50**, 2032 (1969).

33. D. Schiff and L. Verlet (unpublished).

34. J. J. Kozac and S. A. Rice, J. Chem. Phys. **48**, 1226 (1968).

35. M. Chen, D. Henderson, and J. A. Barker, Canadian J. Phys. **47**, 2009 (1969).

36. G. Stell (to be published).

37. J. P. Hansen and L. Verlet, Phys. Rev. **184**, 151 (1969).

38. W. G. Hoover and F. H. Ree, J. Chem. Phys. **49**, 3609 (1968).

39. J. L. Lebowitz, G. Stell, and S. Baer, J. Math. Phys. **6**, 1282 (1965).

40. J. L. Lebowitz and J. K. Percus, Phys. Rev. **144**, 251 (1966).

41. J. K. Percus and S. Shanack, New York University report, NYO-1480-151 (1970).

DISCUSSION

H. Andersen: Chandler, Weeks, and I have been able to show that the same effective hard-core diameter can be used to calculate the free energy and the pair correlation function for a system of molecules with repulsive forces. The result follows from a systematic expansion for the Helmholtz free energy and Y(r). With our choice of the density and temperature dependent diameter, the free energies of hard- and soft-core systems differ by an amount that is fourth order in the softness of the repulsive potential. The Y functions differ by an amount of second order.

R. P. Futrelle: When you refer to $g_w(r)$, the Wertheim hard-sphere radial distribution function, do you in fact use it for r less than the hard-core diameter? That is, do you in fact use $g_w(r) \exp[+\beta v_{HS}(r)]$ which is continuous? This would seem to be helpful in evaluating $g_{LJ}(r)$ for r less than d.

L. Verlet: In our approximation scheme the extrapolation of $g_{HS}(r) \exp[\beta v_{HS}(r)]$ inside the core is not used. It can be defined (and even computed) as a proper statistical average.

C. H. Bennett: How well, and to how low a temperature, will this method work for the crystal phase?

L. Verlet: I believe that the perturbation theory is also valid in the crystal phase at relatively high temperatures, let us say above that of the triple point, where one can neglect quantum corrections and where the harmonic approximation is certainly poor.

S. Larsen: Unless you have a counter argument, I would be most skeptical of using the Axilrod-Teller form for the three-body forces. The Axilrod-Teller form is valid at very large distances and results from the use of third-order perturbation theory. However, as was shown by

McGiven and Jansen in the middle 1950's, if one does the calculation more carefully, antisymmetrizing at least some of the electron wave functions, then these forces come in second-order perturbation theory just as do the Van der Waals forces, and are much stronger than the Axilrod-Teller forces at intermediate distances. Further, as Sherwood and DeRoco showed, if one considers the third virial coefficient and Axilrod forces at large distances together with a different form at small distances, then this contribution to the virial from near distances is not trivial and is of opposite sign and tends to cancel the contribution from the Axilrod terms at large distances.

21

KINETIC THEORY OF DENSE FLUIDS

H. Ted Davis

I. INTRODUCTION

The first attempts to develop a kinetic theory of dense fluids go back to Enskog's extension [1] of Boltzmann's equation to dense hard spheres and to Kirkwood's derivation [2] of a Fokker-Planck equation using the concepts of the theory of Brownian motion. Enskog's theory has the shortcoming of applying only to nonattracting infinitely rigid particles and Kirkwood's theory depends on the assumption of only weakly deflecting collisions among the particles of the fluid.

A more realistic theory was developed by Rice and Allnatt [3] (RA), who assumed that the force of interaction of molecules could be divided into a short-range strongly repulsive part, which they approximated as a hard-sphere interaction, and a weaker longer-range part, which could be approximated, for example, by a Lennard-Jones model. They then derived a kinetic equation for the singlet distribution function in which the collision operator was the sum of Enskog's collision operator accounting for the dissipative action of the repulsive part of the intermolecular interaction and a Fokker-Planck operator accounting for the weaker part of the interaction.

More recently, Prigogine, Nicolis and Misguich [4,5,6] (PNM), using the general theory of Prigogine and co-workers, have presented a local equilibrium theory of transport processes in a dense fluid in which they assumed that (i) the singlet distribution function was adequately approximated by its local Maxwellian equilibrium value and (ii) the nonequilibrium doublet distribution function could be evaluated by essentially a linear trajectory approximation. As did RA, PNM also divided the interaction potential into a hard-sphere part and a weaker longer ranged part, the contribution to the transport coefficients of the former being treated in a manner similar to the collisional transport term in Enskog's theory, and the latter being accounted for by their nonequilibrium doublet distribution function.

Finally, Allen and Cole [7,8] have recently tried to deduce from the general theory of

Prigogine and co-workers [9, 10] kinetic equations for dense fluids by summing appropriate classes of diagrams. They sum the class of binary collision diagrams giving an Enskog-like collision term and the class of ring diagrams known to dominate in plasmas. Thus, the collision operator in their singlet kinetic equation is similar to the one in the RA theory in that it is a sum of an Enskog collision operator and a Fokker-Planck term. However, for reasons discussed in section IV, it seems that Allen and Cole's choice of the ring diagrams as a dominant class is not justifiable.

Transport coefficients of simple fluids have been computed with encouraging results from the RA and the PNM theories. However, in the RA theory (and the Allen-Cole theory) the effects of the spatial displacement and the finite duration of a collisional process have been neglected in the Fokker-Planck operator. That these effects should not be neglected is illustrated in the next section by a study of the inhomogeneous weak-coupling kinetic equation. Neglect of these terms is shown to affect the predicted hydrodynamic equations and the transport coefficients. On the other hand, a shortcoming of the PNM theory is that transport coefficients depending solely on the nonequilibrium singlet distribution function--e.g., self-diffusion, mutual diffusion, and thermal diffusion--cannot be estimated (also the kinetic parts of the coefficients of viscosity and thermal conductivity are neglected, but these become less important as a fluid becomes dense and at low temperature).

In section III, we present an approximate transport theory that incorporates the desirable features of the RA theory and the PNM theory and is, hopefully, an improvement on them.

Section IV is devoted to a numerical study of certain of the Prigogine-Balescu diagrams, and the Allen-Cole theory is discussed somewhat.

In section V a brief discussion is presented on proofs of existence of the Chapman-Enskog solutions to kinetic equations of the Rice-Allnatt type.

II. WEAK-COUPLING KINETIC EQUATION

To second order in the interaction potential, the long-time partial differential equation obeyed by the singlet distribution function is [11]

$$\partial_t f(\underset{\sim}{x}_1, \underset{\sim}{v}_1, t) + \underset{\sim}{v}_1 \cdot \nabla_1 f(\underset{\sim}{x}_1, \underset{\sim}{v}_1, t) + m^{-1} \underset{\sim}{F}(\underset{\sim}{x}_1, t) \cdot \underset{\sim}{\partial}_1 f(\underset{\sim}{x}_1, \underset{\sim}{v}_1, t) =$$

$$- \frac{1}{m^2} \int_0^\infty d\tau \int d\underset{\sim}{v}_2 \int d\underset{\sim}{r} \, \frac{\partial V(\underset{\sim}{r})}{\partial \underset{\sim}{r}} \cdot \underset{\sim}{\partial}_1 \left[\frac{\partial}{\partial r} V(\underset{\sim}{r} - \underset{\sim}{v}_{12}\tau) \right] \cdot \left(\underset{\sim}{\partial}_{12} + \tau \nabla_1 + 2\tau \frac{\partial}{\partial \underset{\sim}{r}} \right)$$

$$f(\underset{\sim}{x}_1 - \underset{\sim}{v}_1 \tau, \underset{\sim}{v}_1, t - \tau) f(\underset{\sim}{x}_1 - \underset{\sim}{r} - \underset{\sim}{v}_2 \tau, \underset{\sim}{v}_2, t - \tau), \tag{2.1}$$

where

$$\underset{\sim}{F}(\underset{\sim}{x}_1, t) \equiv - \nabla_1 \int d\underset{\sim}{x}_2 \, d\underset{\sim}{v}_2 \, V(|\underset{\sim}{x}_1 - \underset{\sim}{x}_2|) f(\underset{\sim}{x}_2, \underset{\sim}{v}_2, t) , \tag{2.2}$$

$$\partial_t \equiv \frac{\partial}{\partial t} , \tag{2.3}$$

$$\underset{\sim}{\partial}_i \equiv \frac{\partial}{\partial \underset{\sim}{v}_i} , \tag{2.4}$$

$$\nabla_1 \equiv \frac{\partial}{\partial \underset{\sim}{x}_i} , \tag{2.5}$$

$$\underset{\sim}{\partial}_{12} \equiv \underset{\sim}{\partial}_1 - \underset{\sim}{\partial}_2 , \tag{2.6}$$

$$\underset{\sim}{v}_{12} \equiv \underset{\sim}{v}_1 - \underset{\sim}{v}_2 , \tag{2.7}$$

V is the pair potential energy, and $\underset{\sim}{x}_i$ and $\underset{\sim}{v}_i$ are the position and velocity, respectively, of the i(th) molecule.

As is clear by inspection of equation (2.1), the collision operator is non-Markovian, since the history of the distribution function over the duration of a collision is involved. Moreover, the operator depends on the spatial delocalization of the interacting molecule through the dependence of f on $\underset{\sim}{r}$. In the Boltzmann operator, the Fokker-Planck operator used by RA, and the plasma operator arising from the sum of ring diagrams [10], the spatial and time delocalization of f are neglected.

What we shall do here is solve equation (2.1) consistently to the first Chapman-Enskog approximation and determine the effect on the hydrodynamic equations and on the coefficient of viscosity of neglecting the spatial and time displacements in the collision operator. First we expand the f's in the integrand of equation (2.1), keeping terms through first order in gradients and time derivatives, i.e.,

$$f(\underset{\sim}{x}_1 - \underset{\sim}{v}_1\tau, \underset{\sim}{v}_1, t - \tau) = f(1) - \underset{\sim}{v}_1\tau \cdot \nabla_1 f(1) - \tau\, \partial_t f(1) \tag{2.8}$$

and

$$f(\underset{\sim}{x}_1 - \underset{\sim}{v}_2\tau - \underset{\sim}{r}, \underset{\sim}{v}_2, t - \tau) = f(2) - (\underset{\sim}{v}_2\tau + \underset{\sim}{r}) \cdot \nabla_1 f(2) - \tau\, \partial_t f(2) , \tag{2.9}$$

where we have introduced the abbreviation

$$f(i) = f(\underset{\sim}{x}_1, \underset{\sim}{v}_i, t) . \tag{2.10}$$

We now write

$$f(i) = f(i)^{\circ} \; (1 + \phi(i)) , \tag{2.11}$$

where f° is the local Maxwellian distribution

$$f(i)^{\circ} = n(\underset{\sim}{x}_1, t) \left(\frac{m}{2\pi kT(\underset{\sim}{x}_1, t)} \right)^{3/2} \exp\left[\frac{-m}{2\,kT(\underset{\sim}{x}_1, t)} \; (\underset{\sim}{v}_1 - \underset{\sim}{u}(\underset{\sim}{x}_1, t))^2 \right] , \tag{2.12}$$

with n, $\underset{\sim}{u}$, and T the local density, hydrodynamic velocity, and temperature defined, respectively, by the relations

$$n = \int f(\underset{\sim}{x}_1, \underset{\sim}{v}, t) d\underset{\sim}{v} , \tag{2.13}$$

$$\underset{\sim}{u} = n^{-1} \int f(\underset{\sim}{x}_1, \underset{\sim}{v}, t) \underset{\sim}{v} \, d\underset{\sim}{v} , \tag{2.14}$$

$$T = \frac{2}{3nk} \int f(\underset{\sim}{x}_1, \underset{\sim}{v}, t) \tfrac{1}{2} m(\underset{\sim}{v} - \underset{\sim}{u})^2 \, d\underset{\sim}{v} . \tag{2.15}$$

To obtain the first approximation of f(i), we insert equation (2.11) into equation (2.1), use expansions (2.8) and (2.9), and keep first-order terms in gradients, time derivatives, and ϕ.

The resulting linear equation for ϕ is

$$\partial_t^{\,o} f(1)^o + \underset{\sim}{v}_1 \cdot \nabla_1 f(1)^o - \frac{1}{m} \underset{\sim}{\partial}_1 \left(\frac{f(1)^o}{n} \right) \cdot \nabla_1 P_v^{\,o} - J(f^o) - R(f^o) - M(f^o) = \mathcal{C}_2(\phi) \quad (2.16)$$

where $P_v^{\,o}$ is the first-order potential-energy contribution to the local pressure,

$$P_v^{\,o} = -\frac{1}{6} [n(\underset{\sim}{x}_1, t)]^2 \int d\underset{\sim}{r}\, r \frac{dV(r)}{dr} \, . \quad (2.17)$$

The quantities J, R, and M arise from the expansions of f in time and displacement and have been Fourier transformed to the forms

$$J(f^o) = \frac{1}{m^2} \partial_t^{\,o} \int d\underset{\sim}{v}_2 \, 8\pi^3 \int d\underset{\sim}{\ell} \, V_\ell^{\,2} \underset{\sim}{\ell} \cdot \partial_1 \frac{1}{(\underset{\sim}{\ell} \cdot \underset{\sim}{v}_{12} - i0)^2} \underset{\sim}{\ell} \cdot \underset{\sim}{\partial}_{12} f(1)^o f(2)^o \quad (2.18)$$

$$R(f^o) = \frac{1}{m^2} \int d\underset{\sim}{v}_2 \, d\underset{\sim}{x}_2 \, \delta(\underset{\sim}{x}_1 - \underset{\sim}{x}_2) \nabla_2 \cdot 8\pi^3 \int d\underset{\sim}{\ell} \frac{\partial}{\partial \underset{\sim}{\ell}} (V_\ell \underset{\sim}{\ell}) \cdot \underset{\sim}{\partial}_1$$

$$\times \frac{1}{\underset{\sim}{\ell} \cdot \underset{\sim}{v}_{12} - i0} V_\ell \underset{\sim}{\ell} \cdot \underset{\sim}{\partial}_{12} f^o(\underset{\sim}{x}_1, \underset{\sim}{v}_1, t) f^o(\underset{\sim}{x}_2, \underset{\sim}{v}_2, t) \, , \quad (2.19)$$

$$M(f^o) = \frac{1}{m^2} \int d\underset{\sim}{v}_2 \, d\underset{\sim}{x}_2 \, \delta(\underset{\sim}{x}_1 - \underset{\sim}{x}_2) 8\pi^3 \int d\underset{\sim}{\ell} \, V_\ell^{\,2} \underset{\sim}{\ell} \cdot \underset{\sim}{\partial}_1 \frac{1}{(\underset{\sim}{\ell} \cdot \underset{\sim}{v}_{12} - i0)^2}$$

$$\times (\underset{\sim}{v}_1 \cdot \nabla_1 + \underset{\sim}{v}_2 \cdot \nabla_2) \underset{\sim}{\ell} \cdot \underset{\sim}{\partial}_{12} \, f^o(\underset{\sim}{x}_1, \underset{\sim}{v}_1, t) f^o(\underset{\sim}{x}_2, \underset{\sim}{v}_2, t) \, , \quad (2.20)$$

where V_ℓ is the Fourier transform of the potential defined as

$$V_\ell = \frac{1}{8\pi^3} \int d\underset{\sim}{r} \, e^{-i\underset{\sim}{\ell} \cdot \underset{\sim}{r}} \, V(\underset{\sim}{r}) \, . \quad (2.21)$$

We have used the notation [10]

$$\frac{1}{\underset{\sim}{\ell} \cdot \underset{\sim}{v}_{12} - i0} = i\pi \, \delta(\underset{\sim}{\ell} \cdot \underset{\sim}{v}_{12}) + \mathcal{P} \frac{1}{\underset{\sim}{\ell} \cdot \underset{\sim}{v}_{12}} \, , \quad (2.22)$$

where $\delta(x)$ is the Dirac delta function and $\mathcal{P} x^{-1}$ is the principal part operator. Similarly Fourier transformed, the linearized collision operator $\mathcal{C}_2(\phi)$ is of the form

$$\mathcal{C}_2(\phi) = \frac{1}{m^2} \int d\underset{\sim}{v}_2 \, 8\pi^3 \int d\underset{\sim}{\ell} \, V_\ell^{\,2} \underset{\sim}{\ell} \cdot \underset{\sim}{\partial}_1 \frac{-i}{\underset{\sim}{\ell} \cdot \underset{\sim}{v}_{12} - i0} \underset{\sim}{\ell} \cdot \underset{\sim}{\partial}_{12} f(1)^o f(2)^o [\phi(1) + \phi(2)] .$$
$$(2.23)$$

The notation $\partial_t^{\,o}$ indicates that the linearized equation will generate the lowest-order time derivatives of the hydrodynamic quantities and that the form of the lowest-order hydrodynamic equations will be determined by the conditions of solubility of equation (2.16). These conditions are that the left-hand side of equation (2.16) be orthogonal to 1, $\underset{\sim}{v}_1$, and $\frac{1}{2} m v_1^{\,2}$, which are the solutions to the homogeneous equation

$$\mathcal{C}_2(\phi_h) = 0 \, . \quad (2.24)$$

Multiplying equation (2.16) respectively by $d\underset{\sim}{v}_1$, $\underset{\sim}{v}_1 d\underset{\sim}{v}_1$, and $\frac{1}{2} m v_1^{\,2} d\underset{\sim}{v}_1$ and integrating, we obtain (after eliminating some terms in the kinetic energy equation with aid of the momentum equation) the following hydrodynamic equations:

$$\partial_t^o n + \nabla_1 \cdot (n\underset{\sim}{u}) = 0 \ , \tag{2.25}$$

$$n \partial_t^o n + n\underset{\sim}{u} \cdot \nabla_1 \underset{\sim}{u} + \frac{1}{m} \nabla_1 p^{(2)} = 0, \tag{2.26}$$

$$C_v^{(2)} \partial_t^o T + C_v^{(2)} \underset{\sim}{u} \cdot \nabla_1 T + T \left(\frac{\partial p^{(2)}}{\partial T}\right)_n \nabla_1 \cdot \underset{\sim}{u} = 0 \ , \tag{2.27}$$

where $C_v^{(2)}$ and $P^{(2)}$ are the local specific heat and the pressure to second order in the potential, namely,

$$C_v^{(2)} = \frac{3}{2} n(\underset{\sim}{x}_1,t)k + \frac{[n(\underset{\sim}{x}_1,t)]^2}{2\,kT^2(\underset{\sim}{x}_1,t)} \int d\underset{\sim}{r}\ V^2(r) \ , \tag{2.28}$$

$$p^{(2)} = n(\underset{\sim}{x}_1,t)\,kT(\underset{\sim}{x}_1,t) \left[1 - \frac{n(\underset{\sim}{x}_1,t)}{6} \int d\underset{\sim}{r}\ r\ \frac{\partial V(r)}{\partial r} + \frac{n(\underset{\sim}{x}_1,t)}{6\,kT(\underset{\sim}{x}_1,t)} \int d\underset{\sim}{r}\ r\ V(r) \frac{\partial V(r)}{\partial r}\right]. \tag{2.29}$$

Equations (2.25)-(2.27) are the exact local equilibrium hydrodynamic equations to second order in the interaction potential. The effect of neglecting the spatial and time delocalization terms, J, R, and M is to replace $C_v^{(2)}$ by $\frac{3}{2}$ nk and $P^{(2)}$ by $nkT - \frac{1}{6} n^2 \int d\underset{\sim}{r}\ r\ dV/dr$ and thus to generate hydrodynamic equations incorrect to the order of potential included in the kinetic equation.

To obtain an equation for ϕ in terms of gradients of n, $\underset{\sim}{u}$, and T, we eliminate most of the time derivatives in equation (2.16) with the aid of equations (2.25)-(2.27). The resulting equation is

$$\frac{1}{T} f^o \left(\frac{mc_1^2}{2\,kT} - \frac{5}{2}\right) \underset{\sim}{c}_1 \cdot \nabla_1 T + f^o \frac{m}{kT} (\underset{\sim}{c}_1 \underset{\sim}{c}_1 - \frac{1}{3} c_1^2 \mathfrak{I}) : \nabla_1 \underset{\sim}{u} + \frac{f^o}{C_v^{(2)}} \left(\frac{n^2 a}{3\,kT^2} + \frac{n^2 b}{kT^2}\right)$$

$$\left(\frac{mc_1^2}{2kT} - \frac{3}{2}\right) \nabla_1 \cdot \underset{\sim}{u} - f^o \frac{nm\underset{\sim}{c}_1}{kT} \cdot \nabla_1 \left(\frac{b}{kT}\right) + 2mn^2 J_1(\hat{f}) \nabla_1 \cdot \underset{\sim}{u} - n^2 \partial_t^o J_1(\hat{f})$$

$$- n^2 R(\hat{f}) - \underset{\sim}{M}_1(\hat{f}) \cdot \nabla_1 n^2 - n^2 M(\hat{f}) = \mathfrak{a}_2(\phi) \ , \tag{2.30}$$

where

$$\underset{\sim}{c}_1 = \underset{\sim}{v}_1 - \underset{\sim}{u} \ , \tag{2.31}$$

$$a = \int d\underset{\sim}{r}\ V^2 \ , \tag{2.32}$$

$$b = \frac{1}{6} \int d\underset{\sim}{r}\ r\ V \frac{dV}{dr} \ , \tag{2.33}$$

$$\hat{f}(1) = \frac{f(1)^o}{n} \ , \tag{2.34}$$

$$J_1(\hat{f}) = - \frac{a}{mkT} \underset{\sim}{\partial}_1 \cdot \int d\underset{\sim}{v}_2 \frac{\underset{\sim}{v}_{12}}{v_{12}^2} \hat{f}(1)\hat{f}(2) \ , \tag{2.35}$$

$$\underset{\sim}{M}_1(\hat{f}) = - \frac{a}{2mkT} \underset{\sim}{\partial}_1 \cdot \int d\underset{\sim}{v}_2 \frac{\underset{\sim}{v}_{12}}{v_{12}^2} \underset{\sim}{u}_{12}\ \hat{f}(1)\hat{f}(2) \ , \tag{2.36}$$

and

$$\underset{\sim}{u}_{12} = \underset{\sim}{c}_1 + \underset{\sim}{c}_2 = (\underset{\sim}{v}_1 - \underset{\sim}{u}) + (\underset{\sim}{v}_2 - \underset{\sim}{u}) \; . \tag{2.37}$$

By eliminating the time derivatives in $\partial_t^{\,o} J_1(\hat{f})$, we obtain the desired equation for ϕ. The function ϕ may be written in the form

$$\phi = \underset{\sim}{A} \cdot \nabla_1 T + \overset{o}{\mathfrak{B}} : \nabla_1 \underset{\sim}{u} + C \, \nabla_1 \cdot \underset{\sim}{u} , \tag{2.38}$$

where the vector $\underset{\sim}{A}$, traceless tensor $\overset{o}{\mathfrak{B}}$ and scalar C are functions to be determined by solving equation (2.30) with the auxiliary conditions

$$\int d\underset{\sim}{v} \, \phi \, \phi_h = 0 \tag{2.39}$$

for $\phi_h = 1$, $\underset{\sim}{v}_1$, and v^2. These conditions, which we are free to require since ϕ_h solves $\mathfrak{a}(\phi_h) = 0$, establish our identifications (2.13)-(2.15) of the quantities n, $\underset{\sim}{u}$, and T appearing in f^o.

We give here the determination only of $\overset{o}{\mathfrak{B}}$. By the standard procedure an adequate estimate of $\overset{o}{\mathfrak{B}}$ is

$$\overset{o}{\mathfrak{B}} = \frac{5m^2}{6\,kT\,\zeta^{(2)}}\left[1 - \frac{3\,na}{10(kT)^2}\right] (\underset{\sim}{c}_1 \underset{\sim}{c}_1 - \frac{1}{3}\, c_1^{\,2}\, \mathfrak{J}) , \tag{2.40}$$

where

$$\zeta^{(2)} = \frac{16\pi^5 n}{3\,kT} \left(\frac{m}{\pi kT}\right)^{\frac{1}{2}} \int_0^\infty d\ell \, \ell^3 V_\ell^{\,2} \; . \tag{2.41}$$

This result for $\overset{o}{\mathfrak{B}}$ leads to the following expression for the kinetic part of the shear viscosity:

$$\eta_k = \frac{5nkTm}{6\,\zeta^{(2)}}\left[1 - \frac{3\,na}{10(kT)^2}\right] . \tag{2.42}$$

Neglect of the spatial and time delocalization in the collision operator would neglect the second term in brackets in equation (2.42). In the case of a dense fluid at liquid temperatures this term would certainly not be negligible. Of course, η_k is a negligible part of the viscosity of a liquid. Our point, however, is to show how important these terms are and that they must be kept in a theory of mutual diffusion and thermal diffusion of liquids.

Incidentally, we can verify the validity of Balescu's [10], Guernsey's [12], and Lenard's [13] neglect of the delocalization effects for a plasma. Taking V(r) to be the screened Coulomb potential

$$V(r) = \frac{e^2}{r} \, e^{-\kappa r} , \tag{2.43}$$

$$\kappa = \left(\frac{4\pi e^2 n}{kT}\right)^{\frac{1}{2}} , \tag{2.44}$$

we obtain for equation (2.42) the form

$$\eta_k = \frac{5nmkT}{6\,\zeta}\left[1 - \frac{3}{20}\left(\frac{e^2\kappa}{kT}\right)\right] \tag{2.45}$$

Computation shows that the second term in the brackets of equation (2.45) is less than 10^{-3} for temperatures greater than $10^{4\,o}$ K and for densities less than 10^{18} unit charges per cubic centimeter. Thus, the neglect of delocalization effects is valid for the sum of the ring diagrams for plasmas but is not justified in neutral liquids, in contradiction to what was assumed by Allen and Cole.

III. A GENERALIZATION OF THE RICE-ALLNATT THEORY

In this section we present an approximate theory of transport processes in dense fluids which incorporates the desirable features of both the RA theory and the PNM theory. In particular, deviations of the singlet distribution function from local equilibrium are retained as in the RA theory while the spatial delocalization and non-Markovian nature of the collision process are retained as in the PNM theory. For simplicity we shall ignore Enskog-like contributions initially and correct for this in a later part of this section.

Our starting point is the exact long-time kinetic equation for the singlet distribution derived by Severne [11] based on the general theory developed by Prigogine and co-workers [9, 10, 14]. The basic equation for the singlet distribution function may be expressed in the form

$$d_t f(\underset{\sim}{x}_1, \underset{\sim}{v}_1, t) = \sum_{s=2}^{N} i \int d\Gamma^{s-1}\, \delta^{s-1}(\underset{\sim}{x}_1 - \underset{\sim}{x}_i)\, \psi^s(\{\nabla\}; i0 + i\partial_t) \prod_{i=1}^{s} f(\underset{\sim}{x}_i, \underset{\sim}{v}_i, t), \tag{3.1}$$

where

$$d_t = \partial_t + \underset{\sim}{v}_1 \cdot \nabla_1 + m^{-1} \underset{\sim}{F}(\underset{\sim}{x}_1, t) \cdot \underset{\sim}{\partial}_1 , \tag{3.2}$$

$$\underset{\sim}{F}(\underset{\sim}{x}_1, t) = -\nabla_1 \int V(|\underset{\sim}{x}_1 - \underset{\sim}{x}_2|)\, f(\underset{\sim}{x}_2, \underset{\sim}{v}_2, t)\, d\underset{\sim}{x}_2\, d\underset{\sim}{v}_2, \tag{3.3}$$

$$d\Gamma^{s-1} = d\underset{\sim}{x}_2 \cdots d\underset{\sim}{x}_{s-1}\, d\underset{\sim}{v}_2 \cdots d\underset{\sim}{v}_{s-1} , \tag{3.4}$$

$$\delta^{s-1}(\underset{\sim}{x}_1 - \underset{\sim}{x}_i) = \prod_{i=2}^{s-1} \delta(\underset{\sim}{x}_1 - \underset{\sim}{x}_i) , \tag{3.5}$$

$$\psi^s(\{\nabla\}; i0 + i\partial_t) = C^{-s+1} \left\{ \sum_{n=1}^{\infty} \langle \underset{\sim}{k}_1 + \underset{\sim}{k}_2 + \cdots + \underset{\sim}{k}_s | (-\delta L)\left[(L_o - z)^{-1} (-\delta L) \right]^n | \underset{\sim}{k}_1, \underset{\sim}{k}_2, \ldots \underset{\sim}{k}_s \rangle_s^{ir} \right\}_{\substack{z \to i0 + i\partial_t \\ \underset{\sim}{k}_i \to -i\nabla_i}} . \tag{3.6}$$

C represents the average density of the system.

The right-hand side of equation (3.1) is the exact long-time non-Markovian collision operator. The superscript(s) on ψ denotes the number of different particles involved in that contribution to the collision operator. The quantities L_o and δL are the free particle and interaction parts of the Liouville operator defined, respectively, as follows:

$$L_o = -i \sum_{j=1}^{N} \underset{\sim}{v}_j \cdot \nabla_j \tag{3.7}$$

and

$$\delta L = \sum_{i,j} \delta L_{ij} = \sum_{i,j}^{N} m^{-1}\, i\, \nabla_i V(|\underset{\sim}{x}_i - \underset{\sim}{x}_j|) \cdot \underset{\sim}{\partial}_{ij} . \tag{3.8}$$

It has been assumed, of course, that the particles are monatomic and structureless. The angular brackets in equation (3.6) denote the Fourier representation of the enclosed operator, i.e.,

$$\langle \underset{\sim}{k}_1', \underset{\sim}{k}_2', \ldots \underset{\sim}{k}_s' | \alpha | \underset{\sim}{k}_1, \underset{\sim}{k}_2, \ldots \underset{\sim}{k}_s \rangle = \langle \{\underset{\sim}{k}'\} | \alpha | \{\underset{\sim}{k}\} \rangle$$

$$= \frac{1}{\Omega^s} \int d\underset{\sim}{x}_1 \ldots d\underset{\sim}{x}_s \exp(i \sum_{j=1}^{s} \underset{\sim}{k}_j' \cdot \underset{\sim}{x}_j)\, \alpha \exp(-i \sum_{j=1}^{s} \underset{\sim}{k}_j \cdot \underset{\sim}{x}_j) . \quad (3.9)$$

In ψ^s the primed state has only one nonzero vector, namely

$$\underset{\sim}{k}_1' = \sum_{j=1}^{s} \underset{\sim}{k}_j, \quad \underset{\sim}{k}_j' = 0, \quad j > 1 .$$

In the Fourier representation, the operator $(L_o - z)^{-1}$ is diagonal and of the form

$$\langle \{\underset{\sim}{k}'\} | (L_o - z)^{-1} | \{\underset{\sim}{k}\} \rangle = (\sum_j \underset{\sim}{k}_j \cdot \underset{\sim}{v}_j - z)^{-1} \delta^{Kr}_{\{\underset{\sim}{k}'\}, \{\underset{\sim}{k}\}} , \quad (3.10)$$

while under the assumed pairwise additivity of the interaction potential the operator δL_{ij} has the rather simple form

$$\langle \{\underset{\sim}{k}'\} | \delta L_{ij} | \{\underset{\sim}{k}\} \rangle = \frac{8\pi^3}{\Omega} \theta^{ij} (\underset{\sim}{k}_i' - \underset{\sim}{k}_i) \delta^{Kr}_{\underset{\sim}{k}_i' + \underset{\sim}{k}_j', \underset{\sim}{k}_i + \underset{\sim}{k}_j} \prod_{d \neq i,j}^{N} \delta^{Kr}_{\underset{\sim}{k}_\alpha' \underset{\sim}{k}_\alpha} , \quad (3.11)$$

where

$$\theta^{ij}(\underset{\sim}{k}) = m^{-1} V_k \underset{\sim}{k} \cdot \underset{\sim}{\partial}_{ij} \quad (3.12)$$

with

$$V_k = \frac{1}{(2\pi)^3} \int d\underset{\sim}{r}\, e^{-i\underset{\sim}{k} \cdot \underset{\sim}{x}} V(x) . \quad (3.13)$$

The superscript ir on the angular bracket in equation (3.6) means that, after all the Fourier matrix elements have been expressed, no intermediate "state" $\{\underset{\sim}{k}''\}$ can be expressed entirely in terms of the vectors $\underset{\sim}{k}_1, \underset{\sim}{k}_2, \ldots, \underset{\sim}{k}_s$. The subscript s on the same angular bracket means that ψ^s involves exactly s distinct particles.

The final expression for ψ^s is obtained by fully expressing the Fourier matrix elements and then replacing the vectors $\underset{\sim}{k}_i$ with the operators $-i\nabla_i$ and the variable z by $i0 + i\partial_t$. For example, the term

$$N\langle \underset{\sim}{k}_1 + \underset{\sim}{k}_2 | (-\delta L_{12})(L_o - z)^{-1}(-\delta L_{12}) | \underset{\sim}{k}_1, \underset{\sim}{k}_2 \rangle$$

$$= -8\pi^3 \int d\underset{\sim}{\ell}\, \theta^{12}(\underset{\sim}{\ell} - \underset{\sim}{k}_2) \frac{1}{\underset{\sim}{\ell} \cdot \underset{\sim}{v}_{12} + \underset{\sim}{k}_1 \cdot \underset{\sim}{v}_1 + \underset{\sim}{k}_2 \cdot \underset{\sim}{v}_2 - z} \theta^{12}(\underset{\sim}{\ell}) \quad (3.14)$$

gives the following contribution to $d_t f$:

$$-i 8\pi^3 \int d\underset{\sim}{v}_2 d\underset{\sim}{x}_2\, \delta(\underset{\sim}{x}_1 - \underset{\sim}{x}_2) \int d\underset{\sim}{\ell}\, \theta^{12}(\underset{\sim}{\ell} + i\nabla_2)$$

$$\times \frac{1}{\underset{\sim}{\ell} \cdot \underset{\sim}{v}_{12} - i\underset{\sim}{v}_1 \cdot \nabla_1 - i\underset{\sim}{v}_2 \cdot \nabla_2 - i0 - i\partial_t} \theta^{12}(\underset{\sim}{\ell}) f(\underset{\sim}{x}_1, \underset{\sim}{v}_1, t) f(\underset{\sim}{x}_2, \underset{\sim}{v}_2, t). \quad (3.15)$$

The weak coupling approximation studied in section II is in fact obtained immediately by approximating the right-hand side of equation (3.1) by the term equation (3.15).

The first Chapman-Enskog approximation to the singlet distribution function is obtained by substituting

$$f(\mathbf{x}_i, \mathbf{v}_i, t) = f^{o}(\mathbf{x}_i, \mathbf{v}_i, t) + f^{o}(\mathbf{x}_i, \mathbf{v}_i, t)\, \phi(\mathbf{x}_i, \mathbf{v}_i, t) \tag{3.16}$$

in equation (3.1), expanding ψ^{s} in a power series in ∇_i and ∂_t, and keeping linear terms in = ∂_t, and ϕ. To this order, equation (3.1) becomes

$$d_t f^{o} + \mathcal{B}(f^{o}) = \mathcal{A}(\phi)\ , \tag{3.17}$$

where

$$\mathcal{B}(f^{o}) = \mathcal{L} \sum_{s=2}^{N} i \int d\Gamma^{s-1}\, \delta^{s-1}(\mathbf{x}_1 - \mathbf{x}_i)\, \psi^{s}(\{\nabla\}; i0 + i\partial_t) \prod_{i=1}^{s} f^{o}(\mathbf{x}_i, \mathbf{v}_i, t) \tag{3.18}$$

and

$$\mathcal{A}(\phi) = \sum_{s=2} i \int d\Gamma^{s-1}\, \delta^{s-1}(\mathbf{x}_1 - \mathbf{x}_i)\, \psi^{s}(\{0\}; i0) \prod_{i=1}^{s} f^{o}(\mathbf{x}_i, \mathbf{v}_i, t) \sum_{\alpha=1} \phi(\mathbf{x}_\alpha, \mathbf{v}_\alpha, t). \tag{3.19}$$

The operator $\mathcal{L}$ in equation (3.18) denotes linearization with respect to ∇_i and ∂_t. Equation (3.17) is the exact linear long-time transport equation. We shall now introduce approximations to render the equation solvable.

Our first assumption is that the PNM approximation may be used in evaluating $\mathcal{B}(f^{o})$. Thus, as has been shown by Misguich [5,6], to this approximation

$$\mathcal{B}(f^{o}) = i\, \frac{8\pi^{3}}{m^{2}} \int d\mathbf{v}_2 \int d\boldsymbol{\ell}\; \theta^{12}(\boldsymbol{\ell})\; g_2^{PNM}(\mathbf{x}_1, \mathbf{v}_1, \mathbf{v}_2, \boldsymbol{\ell})\ , \tag{3.20}$$

where [15]

$$g_2^{PNM} = g_{2s}^{1} + g_{2p}^{1}\ , \tag{3.21}$$

with

$$g_{2s}^{1} = -\frac{mi}{kT} \left\{ \frac{\partial}{\partial \boldsymbol{\ell}}\ \frac{kT}{\boldsymbol{\ell}\cdot\mathbf{v}_{12} - i0} \cdot \left[\nabla_1\, \mathbf{u}\cdot\boldsymbol{\ell} + \frac{\boldsymbol{\ell}\cdot\mathbf{c}_2}{T} \nabla_1 T \right] + \frac{\partial}{\partial\boldsymbol{\ell}} \cdot \left[\frac{m\mathbf{c}_2}{kT} \cdot (\nabla_1 \mathbf{u})^{\dagger} + \frac{1}{T}\left(\frac{mc_2^{2}}{2kT} - \frac{3}{2} \right) \nabla_1 T \right] \right\} G_\ell f(1)^{o} f(2)^{o} - \frac{mi}{kT}\, \frac{f(1)^{o} f(2)^{o}}{n^{2}}\, \frac{\partial}{\partial\boldsymbol{\ell}} \cdot \nabla_1 \left(\frac{kT}{2} n^{2} G_\ell \right) \tag{3.22}$$

and

$$g_{2p}^{1} = \frac{2i\, G_\ell f(1)^{o} f(2)^{o}}{\boldsymbol{\ell}\cdot\mathbf{v}_{12} - i0} \left\{ \frac{m}{4kT} (\mathbf{u}_{12}\mathbf{u}_{12} + \mathbf{v}_{12}\mathbf{v}_{12} + 2\mathbf{u}_{12}\mathbf{u}) : \nabla\mathbf{u} + \left[\frac{m}{4kT} [(\frac{1}{4} u_{12}^{2} + \frac{1}{4} v_{12}^{2})\, \mathfrak{I} + \frac{1}{2}\mathbf{v}_{12}\mathbf{v}_{12}] + \frac{1}{4}[T (\frac{\partial \ln G_\ell}{\partial T})_n - \frac{3}{2}]\, \mathfrak{I} \right] : \mathbf{u}_{12} \nabla_1 \ln T + \frac{1}{2} \left[\frac{m}{4kT} (u_{12}^{2} + v_{12}^{2}) - 3 + T (\frac{\partial \ln G_\ell}{\partial T})_n \right] \mathbf{u} \cdot \nabla_1 \ln T + (\frac{1}{2}\mathbf{u}_{12} + \mathbf{u}) \cdot \nabla_1 \ln n \right\} + \partial_t \left\{ \frac{i G_\ell f(1)^{o} f(2)^{o}}{\boldsymbol{\ell}\cdot\mathbf{v}_{12} - i0} \right\}. \tag{3.23}$$

As in section II, $\mathbf{u}_{12} = \mathbf{v}_1 + \mathbf{v}_2 - 2\mathbf{u}$, $\mathbf{v}_{12} = \mathbf{v}_1 - \mathbf{v}_2$, and $f(j)^{o} = f^{o}(\mathbf{x}_1, \mathbf{v}_j, t)$. The quantity G_ℓ in equations (3.22) and (3.23) is the Fourier transform of the equilibrium radial distribution function, which we shall denote as $g(r) - 1$, evaluated at the local conditions at $\mathbf{x}_1$. The formula for G_ℓ is

$$G_{\ell} = \frac{1}{(2\pi)^3} \int [g(r) - 1] e^{-i\underset{\sim}{\ell}\cdot\underset{\sim}{r}} d\underset{\sim}{r} . \qquad (3.24)$$

Let us now consider the term $\mathcal{A}(\phi)$ = Since $\psi^s(\{0\}; i0)$ does not depend on $\{\underset{\sim}{x}_i\}$, the integration over $d\underset{\sim}{x}^{s-1}$ can be carried out immediately to obtain

$$\mathcal{A}(\phi) = \sum_{s=2}^{N} i \int d\underset{\sim}{v}^{s-1}\, \psi^s(\{0\}; i0) \prod_{j=1}^{s} f(j)^o \sum_{\alpha=1}^{s} \phi(\alpha) , \qquad (3.25)$$

where

$$\phi(\alpha) \equiv \phi(\underset{\sim}{x}_1, \underset{\sim}{v}_\alpha, t) . \qquad (3.26)$$

Thus, we see that the collision operator is completely localized at $\underset{\sim}{x}_\alpha$ the variation of f due to the delocalization and duration of collisions having been absorbed in $\mathcal{B}(f)^o$. The collision operator $\psi^s(\{0\}; i0)$, given by the expression

$$\psi^s(\{0\}; i0) = C^{-s-1} \sum_{n=2}^{\infty} \langle 0|(-\delta L)\left[(L_o - i0)^{-1}(-\delta L)\right]^{n-1}|0\rangle_s^{ir} , \qquad (3.27)$$

is now simply the diagonal fragment operator well known from the theory of homogeneous systems. By explicitly expressing the first two interaction matrix elements $\mathcal{A}(\phi)$ can be rewritten in the form

$$\mathcal{A}(\phi) = -i\,8\pi^3 \int d\underset{\sim}{v}_2\, d\underset{\sim}{\ell}\; \theta^{12}(\underset{\sim}{\ell}) \frac{1}{\underset{\sim}{\ell}\cdot\underset{\sim}{v}_{12} - i0} \Big\{\theta^{12}(\ell)\, f(1)^o f(2)^o [\phi(1) + \phi(2)]$$

$$+ \frac{(2\pi)^3}{\Omega} \sum_{\underset{\sim}{\ell}'} \theta^{12}(\underset{\sim}{\ell}' - \underset{\sim}{\ell}) \sum_{s=2}^{N} \sum_{n=1}^{\infty} C^{-s+2} \int d\underset{\sim}{v}^{s-2} \langle 1_{\underset{\sim}{\ell}'}, 2_{-\underset{\sim}{\ell}'}|[(L_o - i0)^{-1}(-\delta L)]^n|0\rangle_s^{ir}$$

$$\prod_{j=1}^{s} f(j)^o \sum_{\alpha=1}^{s} \phi(\alpha) + \int d\underset{\sim}{v}_3 \Big[\frac{(2\pi)^3}{\Omega} \sum_{\underset{\sim}{\ell}'} \theta^{13}(\underset{\sim}{\ell} - \underset{\sim}{\ell}') \sum_{s=3}^{N} \sum_{n=1}^{\infty} N C^{-s+2} \int d\underset{\sim}{v}^{s-3}$$

$$\times \langle 1_{\underset{\sim}{\ell}'}, 2_{-\underset{\sim}{\ell}}, 3_{\underset{\sim}{\ell}-\underset{\sim}{\ell}'}|[(L_o - i0)^{-1}(-\delta L)]^n|0\rangle_s^{ir} \prod_{j=1}^{s} f(j)^o \sum_{\alpha=1}^{s} \phi(\alpha) + (1 \to 2)\Big]$$

$$+ \int d\underset{\sim}{v}_3 \Big[\theta^{13}(\underset{\sim}{\ell})(2\pi)^3 \sum_{s=3}^{N} \sum_{n=1}^{\infty} C^{-s+3} \int d\underset{\sim}{v}^{s-3}$$

$$\times \langle 2_{-\underset{\sim}{\ell}}, 3_{\underset{\sim}{\ell}}|[(L_o - i0)^{-1}(-\delta L)]^n|0\rangle_s^{ir} \prod_{j=1}^{s} f(j)^o \sum_{\alpha=1}^{s} \phi(\alpha) + (1 \to 2)\Big]\Big\}. \qquad (3.28)$$

The notation $(1 \to 2)$ means to repeat the first term in the brackets but with an interchange of the indices 1 and 2 [5].

Equation (3.28) is still the exact linearized collision operator. We shall now introduce an approximation which is equivalent to Rice and Allnatt's approximation. The approximation is to neglect all the terms in equation (3.28) except those for which the quantity $\sum_\alpha \phi(\alpha)$ may be taken to the left of the matrix elements denoted by the pointed brackets. With this assumption equation (3.28) reduces to the form

$$\mathcal{C}(\phi) = -i8\pi^3 \int d\underset{\sim}{v}_2 d\underset{\sim}{\ell}\, \theta^{12}(\underset{\sim}{\ell}) \frac{1}{\underset{\sim}{\ell}\cdot\underset{\sim}{v}_{12} - i0} \Big\{ \theta^{12}(\underset{\sim}{\ell}) f(1)^\circ f(2)^\circ$$

$$+ \frac{(2\pi)^3}{\Omega} \sum_{\underset{\sim}{\ell}'} \theta^{12}(\underset{\sim}{\ell}-\underset{\sim}{\ell}')\, g_2^\circ(1_{\underset{\sim}{\ell}'}, 2_{-\underset{\sim}{\ell}'}; \underset{\sim}{x}_1) + 8\pi^3 \int d\underset{\sim}{v}_3 \frac{(2\pi)^3}{\Omega} \sum_{\underset{\sim}{\ell}'} \Big[\theta^{13}(\underset{\sim}{\ell}-\underset{\sim}{\ell}')$$

$$g_3^\circ(1_{\underset{\sim}{\ell}'}, 2_{-\underset{\sim}{\ell}'}, 3_{\underset{\sim}{\ell}-\underset{\sim}{\ell}'}; \underset{\sim}{x}_1) + \theta^{23}(\underset{\sim}{\ell}'-\underset{\sim}{\ell}) g_3^\circ(1_{\underset{\sim}{\ell}}, 2_{-\underset{\sim}{\ell}'}, 3_{\underset{\sim}{\ell}'-\underset{\sim}{\ell}}; \underset{\sim}{x}_1)\Big] + 8\pi^3 \int d\underset{\sim}{v}_3$$

$$\Big[\theta^{13}(\underset{\sim}{\ell})\, g_2^\circ(2_{-\underset{\sim}{\ell}}, 3_{\underset{\sim}{\ell}}; \underset{\sim}{x}_1) f(1)^\circ + \theta^{23}(\underset{\sim}{\ell})\, g_2^\circ(1_{\underset{\sim}{\ell}}, 3_{-\underset{\sim}{\ell}}; \underset{\sim}{x}_1) f(2)^\circ\Big]\Big\} \Big[\phi(1)+\phi(2)\Big], \quad (3.29)$$

where $g_2^\circ(i_{\underset{\sim}{\ell}}, j_{-\underset{\sim}{\ell}}; \underset{\sim}{x}_1)$ and $g_3^\circ(i_{\underset{\sim}{\ell}'}, j_{-\underset{\sim}{\ell}}, n_{\underset{\sim}{\ell}-\underset{\sim}{\ell}'}; \underset{\sim}{x}_1)$ denote the Fourier transforms of the equilibrium correlation functions g_2 and g_3 evaluated for the thermodynamic conditions existing at point $\underset{\sim}{x}_1$ and time t. To see how equation (3.28) reduces to equation (3.29) let us consider the second term inside the braces of equation (3.28), namely,

$$I = \sum_{s=3}^{N} \sum_{n=1}^{\infty} C^{-s+2} \int d\underset{\sim}{v}^{s-2} \Big(\sum_{\alpha=1}^{s} \phi(\alpha)\Big) \langle 1_{\underset{\sim}{\ell}'}, 2_{-\underset{\sim}{\ell}'} | [(L_o - i0)^{-1}(-\delta L)]^n | 0\rangle_s^{ir} \prod_{j=1}^{s} f(j)^\circ. \quad (3.30)$$

We examine first the part of I corresponding to $\alpha = 1$ and 2. The contribution from these two terms is

$$\Big[\phi(1)+\phi(2)\Big] \sum_{s=3}^{N} \sum_{n=1}^{\infty} C^{-s+2} \int d\underset{\sim}{v}^{s-2} \langle 1_{\underset{\sim}{\ell}'}, 2_{-\underset{\sim}{\ell}'} | [(L_o - i0)^{-1}(\delta L)]^n | 0\rangle_s^{ir} \prod_{j=1}^{s} f(j)^\circ$$

$$= \Big[\phi(1)+\phi(2)\Big]\, g_2^\circ(1_{\underset{\sim}{\ell}'}, 2_{-\underset{\sim}{\ell}'}; \underset{\sim}{x}_1), \quad (3.31)$$

where the identification of g_2° and the summation in equation (3.31) follows from Severne's theory [11, 16]. Consider next the part of I for which $\alpha \neq 1, 2$, say $\alpha = 3$. For this case, the contribution to I is

$$\int d\underset{\sim}{v}_3\, \phi(3) \sum_{s=3}^{N} \sum_{n=1}^{\infty} C^{-s+2} \int d\underset{\sim}{v}^{s-3} \langle 1_{\underset{\sim}{\ell}'}, 2_{-\underset{\sim}{\ell}'} | [(L_o - i0)^{-1}(-\delta L)]^n | 0\rangle_s^{ir} \prod_{j=1}^{s} f(j)^\circ$$

$$= \int d\underset{\sim}{v}_3\, \phi(3)\, C^{-1} g_2^\circ(1_{\underset{\sim}{\ell}'}, 2_{-\underset{\sim}{\ell}'}; \underset{\sim}{x}_1) f(3)^\circ = 0. \quad (3.32)$$

To set this term equal to zero we have anticipated the orthogonality condition

$$\int d\underset{\sim}{v}_3\, \phi(3) f^\circ(3) = 0 \quad (3.33)$$

which may be imposed on ϕ because $\phi = 1$ is a solution to the homogeneous equation $\mathcal{C}(\phi) = 0$.

Thus, we see that I_1 is the total contribution to I. By similar analysis the other terms in equation (3.29) may be verified.

To further simplify equation (3.29) let us introduce the spatial correlation functions h_s°

$$g^\circ(1_{\underset{\sim}{\ell}}, 1_{-\underset{\sim}{\ell}}; \underset{\sim}{x}_1) = f(1)^\circ f(2)^\circ h_2^\circ(1_{\underset{\sim}{\ell}}, 2_{-\underset{\sim}{\ell}}; \underset{\sim}{x}_1),$$

$$g_3^\circ(1_{\underset{\sim}{\ell}}, 2_{-\underset{\sim}{\ell}}, 3_{\underset{\sim}{\ell}-\underset{\sim}{\ell}'}; \underset{\sim}{x}_1) = f(1)^\circ f(2)^\circ f(3)^\circ h_3^\circ(1_{\underset{\sim}{\ell}'}, 2_{-\underset{\sim}{\ell}}, 3_{\underset{\sim}{\ell}-\underset{\sim}{\ell}'}; \underset{\sim}{x}_1). \quad (3.34)$$

Here h_2° and h_3° are Fourier transforms of the corresponding equilibrium spatial correlation functions evaluated for the thermodynamic conditions at $\underset{\sim}{x}_1$ at time t. The quantities h_2° and h_3° are independent of velocity. In terms of these quantities equation (3.29) may be rearranged to obtain

$$\mathcal{A}(\phi) = -\frac{8\pi^3}{m^2}\int d\underset{\sim}{v}_2 d\underset{\sim}{\ell}\, V_\ell\, \underset{\sim}{\ell}\cdot\underset{\sim}{\partial}_{12}\frac{1}{\underset{\sim}{\ell}\cdot v_{12}-i0}\Big\{\int d\underset{\sim}{\ell}'[\delta(\underset{\sim}{\ell}')+$$

$$h_2^{\circ}(1_{\underset{\sim}{\ell}'},2_{-\underset{\sim}{\ell}'};\underset{\sim}{x}_1)]V_{|\underset{\sim}{\ell}-\underset{\sim}{\ell}'|}(\underset{\sim}{\ell}-\underset{\sim}{\ell}')\cdot\underset{\sim}{\partial}_{12}+\int d\underset{\sim}{\ell}'\, V_{|\underset{\sim}{\ell}-\underset{\sim}{\ell}'|}(\underset{\sim}{\ell}-\underset{\sim}{\ell}')\cdot 8\pi^3 n(\underset{\sim}{x}_1,t)$$

$$\Big[h_3^{\circ}(1_{\underset{\sim}{\ell}'},2_{-\underset{\sim}{\ell}},3_{\underset{\sim}{\ell}-\underset{\sim}{\ell}'};\underset{\sim}{x}_1)\underset{\sim}{\partial}_1-h_3^{\circ}(1_{\underset{\sim}{\ell}},2_{-\ell'},3_{\underset{\sim}{\ell}-\underset{\sim}{\ell}};\underset{\sim}{x}_1)\underset{\sim}{\partial}_2\Big]$$

$$+V_\ell\underset{\sim}{\ell}\cdot 8\pi^3\Big[h_2^{\circ}(2_{-\underset{\sim}{\ell}},3_{\underset{\sim}{\ell}};\underset{\sim}{x}_1)\underset{\sim}{\partial}_1-h_2^{\circ}(1_{\underset{\sim}{\ell}},3_{-\underset{\sim}{\ell}};\underset{\sim}{x}_1)\underset{\sim}{\partial}_2\Big]\Big\}f(1)^{\circ}f(2)^{\circ}\Big[\phi(1)+\phi(2)\Big]. \tag{3.35}$$

By somewhat lengthy manipulations, making use of the BBGKY hierarchy of equations for the h_s°, the three-body correlation function h_3° can be eliminated from equation (3.35) to obtain the following rather simple result for $\mathcal{A}(\phi)$:

$$\mathcal{A}(\phi) = -\frac{8\pi^3}{m^2}\int d\underset{\sim}{v}_2 d\underset{\sim}{\ell}\, V_\ell\,\underset{\sim}{\ell}\cdot\underset{\sim}{\partial}_1\frac{-ikT}{\underset{\sim}{\ell}\cdot\underset{\sim}{v}_{12}-i0}\,G_\ell\,\underset{\sim}{\ell}\cdot\underset{\sim}{\partial}_{12}f(1)^{\circ}f(2)^{\circ}(\phi(1)+\phi(2)), \tag{3.36}$$

where G_ℓ was defined in equation (3.24). This is precisely the form of the Fokker-Planck operator appearing in the Rice-Allnatt kinetic equation.

Alternatively, in place of the commutation approximation used to obtain equation (3.36), we could approximate $\mathcal{A}(\phi)$ by the ring diagrams of plasma theory to obtain

$$\mathcal{A}_R(\phi) = \frac{8\pi^4}{m^2}\,\mathcal{L}\int d\underset{\sim}{v}_2 d\underset{\sim}{\ell}\,\frac{V_\ell^2}{|\epsilon(\underset{\sim}{x}_1,\underset{\sim}{v}_1)|^2}\,\delta(\underset{\sim}{\ell}\cdot\underset{\sim}{v}_{12})\underset{\sim}{\ell}\cdot\underset{\sim}{\partial}_{12}\,f(1)\,f(2)\,, \tag{3.37}$$

where

$$\varepsilon(\underset{\sim}{x}_1,\underset{\sim}{v}_1) = 1+\frac{8\pi^4}{m}\int d\underset{\sim}{v}_3\Big(-\frac{i}{\underset{\sim}{\ell}\cdot\underset{\sim}{v}_{13}-i0}\Big)\,V_\ell\,\underset{\sim}{\ell}\cdot\underset{\sim}{\partial}_3\,f(3)\,. \tag{3.38}$$

The operator $\mathcal{L}$ means that only linear terms in ϕ are kept in equation (3.37). As we shall see in section IV, it is doubtful that equation (3.37) is as good an approximation for dense fluids as is equation (3.36).

We shall now modify our results to include Enskog-like contributions to the collision integral. Following Rice and Allnatt we assume that the pair potential energy V is composed of the sum of a short-range strongly repulsive part V^H and a longer range smaller part V^s. Inserting this decomposition in the left most interaction term in the collision operator, we can rewrite equation (3.1) in the form

$$d_t f(1) = \sum_s^N i \int d\Gamma^{s-1}\, \delta^{s-1}(\underset{\sim}{x}_1 - \underset{\sim}{x}_i)\, \psi_H^{\,s}(\{\nabla\}; i0 + i\partial_t) \prod_{i=1}^{s} f(\underset{\sim}{x}_i, \underset{\sim}{v}_i, t)$$

$$+ \sum_s^N i \int d\Gamma^{s-1}\, \delta^{s-1}(\underset{\sim}{x}_1 - \underset{\sim}{x}_i)\, \psi_s^{\,s}(\{\nabla\}; i0 + i\partial_t) \prod_{i=1}^{s} f(\underset{\sim}{x}_i, \underset{\sim}{v}_i, t), \tag{3.39}$$

where $\psi_H^{\,s}$ and $\psi_s^{\,s}$ have the same form as ψ^s given by equation (3.6) except that the leftmost interaction terms are, respectively,

$$\delta L_{12}^{\ H} = i\nabla_1 V^H(|\underset{\sim}{x}_1 - \underset{\sim}{x}_2|)\partial_i \text{ and } \delta L_{12}^{\ s} = i\nabla_1 V^s(|\underset{\sim}{x}_1 - \underset{\sim}{x}_2|)\partial_i \ .$$

We now approximate the part of equation (3.29) involving $\psi_H^{\,s}$ by the Enskog collision operator for hard spheres of diameter σ and treat the part involving $\psi_s^{\,s}$ to the approximation yielding the terms (3.20) and (3.26). Thus, our approximate linearized kinetic equation becomes

$$d_t f(1)^o + \mathcal{B}_s(f^o) = \mathcal{A}_s(\phi) + J^E, \tag{3.40}$$

where

$$J^E = \mathcal{L} \iint_{\underset{\sim}{v}_{21}\cdot\hat{k} > 0} \Big[f(\underset{\sim}{x}_1, \underset{\sim}{v}_1', t) f(\underset{\sim}{x}_1 + \sigma\hat{k}, v_2', t)\, g(\underset{\sim}{x}_1, \underset{\sim}{x}_1 + \sigma\hat{k})$$

$$- f(\underset{\sim}{x}_1, \underset{\sim}{v}_1, t) f(\underset{\sim}{x}_1 - \sigma\hat{k}, v_2, t)\, g(\underset{\sim}{x}_1, \underset{\sim}{x}_1 - \sigma\hat{k}) \Big] \sigma^2\, \underset{\sim}{v}_{21}\cdot\hat{\underset{\sim}{k}}\, d\hat{\underset{\sim}{k}}\, d\underset{\sim}{v}_2 \tag{3.41}$$

and $\mathcal{B}_s$ and $\mathcal{A}_s$ have the forms given by equations (3.20) and (3.26) except that the term $\theta^{12}(\underset{\sim}{\ell})$ in equation (3.20) is replaced by $m^{-1} V_\ell^{\ s}\, \underset{\sim}{\ell}\cdot\underset{\sim}{\partial}_{12}$ and the term V_ℓ in equation (3.36) is replaced by $V_\ell^{\ s}$. As before, the operator $\mathcal{L}$ means that only the terms linear in ϕ and in gradients of $n(\underset{\sim}{x}_1, t)$, $\underset{\sim}{u}(\underset{\sim}{x}_1, t)$ and $T(\underset{\sim}{x}_1, t)$ are kept. The quantity $\hat{\underset{\sim}{k}}$ in equation (3.41) denotes a unit vector directed along the line of centers of the pair of colliding molecules, $d\hat{\underset{\sim}{k}}$ the corresponding unit solid angle, $g(\underset{\sim}{x}_1, \underset{\sim}{x}_1 \pm \sigma\hat{\underset{\sim}{k}})$ the contact value of the local equilibrium radial distribution function, and $\underset{\sim}{v}_1'$, $\underset{\sim}{v}_2'$ the velocities of a pair of molecules for which a binary hard-sphere collision occurring at $\underset{\sim}{x}_1$, $\underset{\sim}{x}_1 + \sigma\hat{\underset{\sim}{k}}$ would result in the velocities $\underset{\sim}{v}_1, \underset{\sim}{v}_2$.

The lowest-order hydrodynamic equations can be obtained from equation (3.40) by multiplying by $d\underset{\sim}{v}_1$, $\underset{\sim}{v}_1 d\underset{\sim}{v}_1$, and $\frac{1}{2} m v_1^{\ 2} d\underset{\sim}{v}_1$, respectively, and integrating the resulting equations with respect to $\underset{\sim}{v}_1$. The results may be summarized as follows:

$$\partial_t^o n + \nabla_1 \cdot (n\underset{\sim}{u}) = 0 \ , \tag{3.42}$$

$$n\partial_t^o \underset{\sim}{u} + n\underset{\sim}{u}\cdot\nabla_1\underset{\sim}{u} + \frac{1}{m}\nabla P = 0 \ , \tag{3.43}$$

$$C_v \partial_t^o T + C_v \underset{\sim}{u}\cdot\nabla_1 T + T\left(\frac{\partial\tilde{P}}{\partial T}\right)_n \nabla\cdot\underset{\sim}{u} = 0 \ , \tag{3.44}$$

where P and C_v are the local pressure and specific heat, respectively,

$$P = nkT\left(1 + \frac{2\pi}{3} n\sigma^3 g(\sigma+)\right) - \frac{n^2}{6}\int d\underset{\sim}{r}\; r\, \frac{dV^s}{dr}\, g(r) \ , \tag{3.45}$$

$$C_v = \frac{3}{2} nk + \left[\frac{\partial}{\partial T}\, \frac{n^2}{2}\int V^s(r) g(r) d\underset{\sim}{r}\right]_n \ , \tag{3.46}$$

and

$$T\left(\frac{\partial \tilde{P}}{\partial T}\right)_n = nkT\left(1 + \frac{2\pi}{3} n\sigma^3 g(\sigma+)\right) + P_v^s - \frac{2}{3} n \tilde{E}_v^s \tag{3.47}$$

The quantities P_v^s and $\tilde{E}_v^s$ are excess pressure and energies defined as follows:

$$P_v^s = -\frac{1}{6} n^2 \int d\underline{r}\, r \frac{dV(r)^s}{dr} [g(r) - 1] , \tag{3.48}$$

$$n\tilde{E}_v^s = \frac{1}{2} n^2 \int d\underline{r}\, V(r)^s [g(r) - 1] , \tag{3.49}$$

where g(r) is the equilibrium radial distribution function evaluated for thermodynamic conditions existing at $\underset{\sim}{x}_1$ at time t, and $g(\sigma+)$ is the value of g(r) for spheres in contact. It should be noted that the theory gives the correct continuity and momentum balance equations, gives the correct heat capacity terms in the kinetic energy balance equation, but gives $(\partial P/\partial T)_n$ only approximately. It is in Misguich's approximation of the part of the doublet correlation function denoted g_{2P}^1 that the error in $(\partial P/\partial T)_n$ is introduced. We are currently trying to improve this approximation. In the RA approximation the potential-energy part of the heat capacity is omitted and $T(\partial P/\partial T)_n$ is approximated by the hard-sphere result $nkT[1 + \frac{2\pi}{3} n\sigma^3 g(\sigma+)]$.

We have solved equation (3.40), by the usual method of moments [1], for the terms giving the kinetic parts of the shear viscosity and the thermal conductivity. The lowest approximation, of the form

$$\phi = -A\left(\frac{5}{2} - \frac{m\underset{\sim}{c}_1^2}{2kT}\right) \underset{\sim}{c}_1 \cdot \nabla_1 T - B\left(\underset{\sim}{c}_1\underset{\sim}{c}_1 - \frac{1}{3} c_1^2 \mathfrak{I}\right) : \nabla_1 \underset{\sim}{u} , \tag{3.50}$$

has been shown to give a good approximation to the transport coefficients in dilute gas studies and will be assumed to suffice here. We have obtained the following results for the coefficients A and B:

$$A = -\frac{5m}{4\zeta T}\left[1 + \frac{2\pi}{5} n\sigma^3 g(\sigma+) + \frac{8}{25} X_s\right] \tag{3.51}$$

$$B = \frac{5m^2}{6\zeta kT}\left[1 + \frac{4\pi}{15} n\sigma^3 g(\sigma+) + \frac{2P_v^s + \frac{5}{3} n\tilde{E}_v^s}{5nkT}\right] , \tag{3.52}$$

where

$$\zeta = -\frac{16\pi^5 n}{3}\left(\frac{m}{\pi kT}\right)^{\frac{1}{2}} \int_0^\infty \ell^3 V_\ell^s G_\ell \, d\ell + \frac{8}{3} n\sigma^2 g(\sigma+)(\pi m kT)^{\frac{1}{2}} \tag{3.53}$$

and

$$X_s = \frac{25}{4nk}\left(\frac{\partial P_v^s}{\partial T}\right)_n - \frac{25}{6nk^2T^2}\left(\frac{\partial P}{\partial T}\right)_n \tilde{E}_v^s + \frac{15}{8}\frac{P_v^s}{nkT} + \frac{175\tilde{E}_v^s}{48kT} + \frac{25}{24k}\left(\frac{\partial \tilde{E}_v^s}{\partial T}\right)_n . \tag{3.54}$$

The terms $\frac{8}{25} X_s$ in A and $(2P_v^s + \frac{5}{3} n\tilde{E}_v^s)/5nkT$ in B are neglected in Rice and Allnatt's and Cole and Allen's approximations.

The expression (3.50) yields the following results for the kinetic parts of the thermal conductivity and viscosity coefficients, respectively,

$$\kappa_k = -\frac{5nk^2T^2A}{2m}, \tag{3.55}$$

$$\eta_k = \frac{n}{m} k^2T^2B . \tag{3.56}$$

Ignoring the hard-sphere contribution to equation (3.56) and reducing B to the weak-coupling limit, we obtain the weak-coupling limit of equation (3.56)

$$\eta_k = \frac{5\,mnkT}{6\zeta^{(2)}}\left[1 - \frac{8na}{30(kT)^2}\right], \tag{3.57}$$

which is to be compared with the exact weak-coupling result of equation (2.42). The difference is that the factor 8/30 in the second term in equation (3.57) should be 9/30 to agree with the exact result. The difference between the two is only about 3%.

The nonequilibrium doublet distribution function may be obtained directly by comparing the exact BBGKY equation

$$d_t f = -\frac{i8\pi^3}{m}\int d\underset{\sim}{v}_2 d\underset{\sim}{\ell}\, V_\ell^{\,s}\, \underset{\sim}{\ell}\cdot\underset{\sim}{\partial}_1\, g_2(\underset{\sim}{v}_1, \underset{\sim}{v}_2, \underset{\sim}{\ell}, \underset{\sim}{x}_1, t)$$

$$-\frac{i8\pi^3}{m}\int d\underset{\sim}{v}_2 d\underset{\sim}{\ell}\, V_\ell^{\,H}\, \underset{\sim}{\ell}\cdot\underset{\sim}{\partial}_1\, g_2(\underset{\sim}{v}_1, \underset{\sim}{v}_2, \underset{\sim}{\ell}, \underset{\sim}{x}_1, t) \tag{3.58}$$

with our approximation equation (3.50). Equating the corresponding integrands of the terms in equations (3.50) and (3.58) involving the soft part $V_\ell^{\,s}$ of the potential, we obtain

$$g_2(\underset{\sim}{v}_1, \underset{\sim}{v}_2, \underset{\sim}{\ell}, \underset{\sim}{v}_1, t) = g_2^{PNM}(\underset{\sim}{v}_1, \underset{\sim}{v}_2, \underset{\sim}{\ell}, \underset{\sim}{x}_1, t)$$

$$-\frac{(kT/m)}{\underset{\sim}{\ell}\cdot\underset{\sim}{v}_{12} - i0}\, G_\ell\, \underset{\sim}{\ell}\cdot\underset{\sim}{\partial}_{12}\, f(1)^0 f(2)\left[1 + \phi(1) + \phi(2)\right]. \tag{3.59}$$

The potential parts of the momentum and energy fluxes, $\underset{\approx}{J}_M^{\,V}$ and $\underset{\sim}{J}_E^{\,V}$, may be written in the form

$$\underset{\approx}{J}_M^{\,V} = +\frac{m\sigma^3}{2}\int_{\underset{\sim}{v}_{12}\cdot\underset{\sim}{k}<0} d\underset{\sim}{v}_1 d\underset{\sim}{v}_2 d\underset{\sim}{k}\, (\underset{\sim}{v}_{12}\cdot\underset{\sim}{k})^2\, \underset{\sim}{k}\,\underset{\sim}{k}\, f_2(\underset{\sim}{x}_1, \underset{\sim}{x}_1 - \sigma\underset{\sim}{k}, \underset{\sim}{v}_1, \underset{\sim}{v}_2, t)$$

$$-\frac{1}{2}\int d\underset{\sim}{v}_1 d\underset{\sim}{v}_2 d\underset{\sim}{r}\, \underset{\sim}{r}\, \frac{\partial V^s(r)}{\partial\underset{\sim}{r}}\, f_2(\underset{\sim}{x}_1, \underset{\sim}{x}_1 + \underset{\sim}{r}, \underset{\sim}{v}_1, \underset{\sim}{v}_2, t) \tag{3.60}$$

and

$$\underset{\sim}{J}_E^{\,V} = -\frac{m\sigma^3}{4}\int_{\underset{\sim}{v}_{12}\cdot k<0} d\underset{\sim}{v}_1 d\underset{\sim}{v}_2 d\underset{\sim}{k} (\underset{\sim}{v}_{12}\cdot\underset{\sim}{k})^2\, \underset{\sim}{k}\,\underset{\sim}{k}\cdot\underset{\sim}{u}_{12}\, f_2(\underset{\sim}{x}_1, \underset{\sim}{x}_1 - \sigma\underset{\sim}{k}, \underset{\sim}{v}_1, \underset{\sim}{v}_2, t)$$

$$+\frac{1}{4}\int d\underset{\sim}{v}_1 d\underset{\sim}{v}_2 d\underset{\sim}{r}\, \underset{\sim}{u}_{12}\cdot\left[\underset{\approx}{\mathfrak{I}}\, V^s(r) - \underset{\sim}{r}\, \frac{\partial V^s(r)}{\partial\underset{\sim}{r}}\right] f_2(\underset{\sim}{x}_1, \underset{\sim}{x}_1 + \underset{\sim}{r}, \underset{\sim}{v}_1, \underset{\sim}{v}_2, t). \tag{3.61}$$

The first members of the right-hand sides of equations (3.60) and (3.61) are the Enskog hard-sphere collisional contribution to the fluxes while the second members are the contributions arising from the soft part of the interaction. The vector $\underset{\sim}{k}$ is a unit vector directed along the line of centers of a pair of molecules undergoing a hard core collision. Integration over the corresponding solid angle is denoted by $\int d\underset{\sim}{k}$.

Evaluating the expressions in equations (3.60) and (3.61) with the aid of the results equations (3.50) and (3.59), and comparing the predicted energy and momentum fluxes with Fourier's law of heat conduction and Newton's stress tensor, respectively, we obtain the following expressions for the potential parts of the potential contribution to the coefficients of thermal conductivity and viscosity:

$$\varkappa_V = \varkappa^{PNM} + \varkappa_K \left[\frac{2}{5} \pi n \sigma^3 g(\sigma+) + \frac{\pi}{15} n\sigma^3 (g(\sigma+) - 1)(1 + 2\chi_3) - \frac{9 P_V^S + n\tilde{E}_V^S}{15\, n\, kT} \right) \quad (3.62)$$

and

$$\eta_V = \eta^{PNM} + \eta_K \left[\frac{4\pi}{15} n\sigma^3 g(\sigma+) + \frac{2\pi}{15} n\sigma^3 (g(\sigma+) - 1)(1 - 3\chi_2) + \frac{2(n\tilde{E}_V^S - P_V^S)}{5\, n\, kT} \right], \quad (3.63)$$

where

$$\chi_2 = \frac{2}{\pi^{\frac{1}{2}}} + \frac{1}{4\pi^{\frac{1}{2}}[g(\sigma+) - 1]} \int_1^\infty dx\, [g(x) - 1]\, x \frac{d}{dx} [x P_2(x)]$$

$$\chi_3 = 1 - \frac{1}{16[g(\sigma+) - 1]} \int_1^\infty dx\, [g(x) - 1]\, x \frac{d}{dx} [x P_3(x)]$$

The quantities P_n are defined by Eq. (3.67). K^{PNM} and η^{PNM} denote the results obtained from the PNM theory. These results are

$$\kappa^{PNM} = \frac{2}{3m} k n^2 \sigma^4 (\pi m kT)^{\frac{1}{2}} \Big\{ g(\sigma+) + \frac{1}{16} \Big[-8[g(\sigma+) - 1] - \int_1^\infty dx\, [g(x) - 1][P_2'(x) - P_4'(x)] \Big] + \frac{1}{32} \int_1^\infty dx\, x\, [g(x) - 1][8P_4(x) - P_2(x)] + \frac{1}{16} \int_1^\infty dx\, x\, P_2(x) T \left(\frac{\partial g(x)}{\partial T} \right)_n \Big\} + \frac{4\pi^4}{3} n^2 \left(\frac{\pi k}{mT} \right)^{\frac{1}{2}} \Big\{ T \Big[\frac{\partial}{\partial T} (6W_1 + W_2) \Big]_n - 5W_1 + \frac{7}{2} W_2 + W_3 \Big\} \quad (3.64)$$

and

$$\eta^{PNM} = \frac{4}{15} n^2 \sigma^4 (\pi m kT)^{\frac{1}{2}} \Big\{ \frac{1}{2}[g(\sigma+) + 1] - \frac{1}{16} \int_0^\infty dx\, [g(x) - 1][P_2'(x) - P_4'(x)] + \frac{1}{8} \int_1^\infty dx\, x\, [g(x) - 1][3P_4(x) - P_2(x)] \Big\} + \frac{8\pi^4}{15} n^2 \left(\frac{\pi m}{kT} \right)^{\frac{1}{2}} (3W_2 + W_3), \quad (3.65)$$

where

$$W_n = \int_0^\infty d\ell\, G_\ell\, \ell^n \frac{\partial^{n-1}}{\partial \ell^{n-1}} V_\ell^S \qquad (3.66)$$

and

$$P_n(x) = 16 \int_{-1}^{0} dr\, r^n (x^2 - 1 + r^2)^{-\frac{1}{2}} \ . \qquad (3.67)$$

$P_n'(x)$ denotes the derivative of $P_n(x)$.

In table 1 the shear viscosity predicted for argon by the theory developed here is compared with experiment and with the predictions of the RA and PNM theories. Comparison of rows 2 and 3 of table 1 shows that the terms neglected by the PNM theory but retained in our theory amount to 3 to 20% of the PNM contributions. Thus, even for the pure-fluid transport coefficients neglect of the deviation of the singlet distribution function from local equilibrium involves substantial error. On the other hand, the actual agreement between theory and experiment in our results is no better than in the RA or PNM results. This can be taken to mean anything, since in all the calculations in table 1 use has been made of Kirkwood, Lewinson, and Alder's pair correlation calculations based on the superposition approximation. As mentioned in the beginning of this section, the advantages of the present theory over the original RA theory are that (a) some terms they discard are retained with the consequence that the predicted hydrodynamic equations are improved and (b) the doublet distribution function is determined directly from the singlet equation, thus avoiding having to solve a Smoluchowski equation. Moreover, unlike RA, we do not have to treat a reference particle differently from the other particles of the system in obtaining the Fokker-Planck operator. And, of course, unlike the PNM theory, the present theory will yield diffusion and thermal diffusion coefficients.

Table 1

Shear Viscosity of Liquid Argon

State	90° K, 1.3 atm.	128° K, 50 atm.	133.5° K, 100 atm.	185.5° K, 500 atm.
η (RA)	1.41	0.69	0.70	0.86
η (PNM)	1.31	0.82	0.83	0.86
η (This work)	1.34	0.92	0.93	1.03
η (exp)	2.39	0.84	0.84	0.87

Note: Units are in millipoise. Experimental values are taken from Lowry, Rice and Gray (see ref. 3).

In closing this section it is interesting to note that the RA and PNM approximations are equivalent in the sense that they both retain only the contributions arising from the "inhomogeneous" effects of ∇_i, ∂_t, and ϕ in the "latest" or left most position in the matrix elements of the collision operator. This point is discussed more fully in Appendix II.

IV. COMMENTS ON THE ALLEN-COLE THEORY

In a series of recent papers Allen and Cole [7,8] have presented a kinetic theory of dense fluids in which they use the Prigogine-Balescu [9,10] diagram theory to try to pick out dominate classes of terms in the collision operator, equation (3.6) of section III, and in the similar creation operator which gives the doublet distribution function as a functional of the singlet. We shall discuss briefly some computations Dowling and the author [16] have done recently which are relevant to Allen and Cole's theory.

A few words on the diagram representation are perhaps useful at this point. In the diagram theory, a matrix element of the free-particle resolvent $(L_o - i0)^{-1}$ is represented by a number of superposed lines equal to the number of nonzero Fourier vectors $\{\underset{\sim}{k}\}$ involved in the matrix element, and each line is labeled with the index of the particle associated with that nonzero Fourier vector. The matrix element of an interaction term, say δL_{ij}, is represented as a number of superposed lines equal to the number of nonzero vectors associated with particles other than i and j (and therefore conserved in the interaction) and a vertex representing the possible transitions $\underset{\sim}{k}_i, \underset{\sim}{k}_j \rightarrow \underset{\sim}{k}_i', \underset{\sim}{k}_j'$ under δL_{ij} consistent with the conservation law, equation (3.11), imposed by the assumption of pairwise additive forces. A few examples, typical [9] of those we shall discuss in this section, are

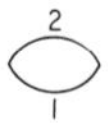

$$= \frac{8\pi^3}{m^2} C \int d\underset{\sim}{\ell}\, V_\ell\, \underset{\sim}{\ell} \cdot \underset{\sim}{\partial}_{12} \frac{1}{\underset{\sim}{\ell} \cdot \underset{\sim}{v}_{12} - i0} V_\ell\, \underset{\sim}{\ell} \cdot \underset{\sim}{\partial}_{12} \, ,$$

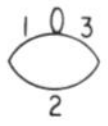

$$= \frac{(8\pi^3 C)^2}{m^3} \int d\underset{\sim}{\ell}\, V_\ell\, \underset{\sim}{\ell} \cdot \underset{\sim}{\partial}_{12} \frac{1}{\underset{\sim}{\ell} \cdot \underset{\sim}{v}_{12} - i0} V_\ell\, \underset{\sim}{\ell} \cdot \underset{\sim}{\partial}_{13} \frac{1}{\underset{\sim}{\ell} \cdot \underset{\sim}{v}_{12} - i0} V_\ell\, \underset{\sim}{\ell} \cdot \underset{\sim}{\partial}_{23}$$

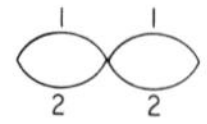

$$= \frac{8\pi^3 C}{m^3} \int d\underset{\sim}{\ell}\, d\underset{\sim}{\ell}'\, V_\ell\, \underset{\sim}{\ell} \cdot \underset{\sim}{\partial}_{12} \frac{1}{\underset{\sim}{\ell} \cdot \underset{\sim}{v}_{12} - i0} V_{|\underset{\sim}{\ell} - \underset{\sim}{\ell}'|} (\underset{\sim}{\ell} - \underset{\sim}{\ell}') \cdot \underset{\sim}{\partial}_{12}$$

$$\frac{1}{\ell' \cdot v_{12} - i0} V_{\ell'}\, \ell' \cdot \underset{\sim}{\partial}_{12} \quad . \tag{4.1}$$

These diagrams represent the first few terms in the homogeneous collision operator, $\psi(\{0\}; i0)$. As was shown in section III, equation (3.19), it is the homogeneous operator that acts on the unknown ϕ after the exact equation (3.1) has been linearized. In table 2, the diagrams contributing to the singlet kinetic equation are depicted through fourth order in the interaction.

In order to estimate the order of magnitude of the diagrams shown in table 1 we shall replace the velocity and velocity operators by thermal velocity, i.e., $\underset{\sim}{v} = (kT/m)^{\frac{1}{2}}$ and $\underset{\sim}{\partial}_i = (m/kT)^{\frac{1}{2}}$, and introduce the reduced variables

$$\underset{\sim}{\ell}_r = \sigma \underset{\sim}{\ell} \, , \tag{4.2}$$

Table 2

Diagrams for Gaussian Potential*

Diagram	Lower Bound	Exact	Upper Bound
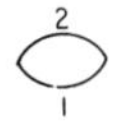		1.63×10^5	
		9.20×10^9	
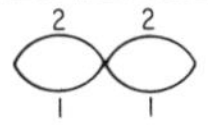	5.36×10^9	5.74×10^{10}	5.16×10^{11}
		1.06×10^{14}	
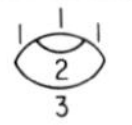	3.68×10^{14}		
	3.11×10^{14}		
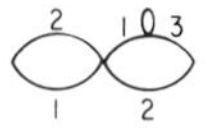	6.85×10^{13}		
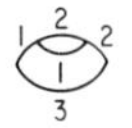	4.02×10^{13}		6.43×10^{16}
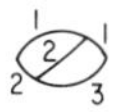	2.83×10^{13}		
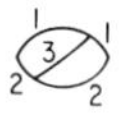	9.00×10^{12}		5.10×10^{17}
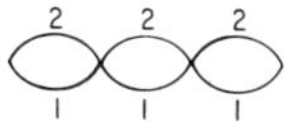	1.04×10^{14}		

*Diagrams reduced by $(8\pi^3)(4\pi)[b/8(\gamma\pi)^{3/2}]^s(kT/m)^{1/2}\sigma^{-1}$ and $c\sigma^3$ and ϵ/kT have been set to unity. s denotes the order of interaction of a given diagram.

$$V_r(\ell_r) = V_\ell/\epsilon\sigma^3 , \tag{4.3}$$

where σ and ϵ are the characteristic molecular diameter and interaction energy, respectively. With this reduction the diagrams in equation (4.1) become

 $\sim (C\sigma^3)\,\frac{(kT/m)^{\frac{1}{2}}}{\sigma}\,(\frac{\epsilon}{kT})^2 8\pi^3 \int d\underset{\sim}{\ell}_r\, V_r^{\,2}(\ell_r)\ell_r$,

 $\sim (C\sigma^3)^2\frac{(kT/m)^{\frac{1}{2}}}{\sigma}\,(\frac{\epsilon}{kT})^3\,(8\pi^3)^2 \int d\underset{\sim}{\ell}_r\, V_r^{\,3}(\ell_r)\ell_r$,

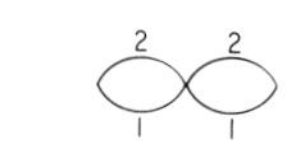 $$\sim (C\sigma^3)\,\frac{(kT/m)^{\frac{1}{2}}}{\sigma}\,8\pi^3 \int d\underset{\sim}{\ell}_r\, d\underset{\sim}{\ell}_r{}'\, V_r(\ell_r) V_r(|\underset{\sim}{\ell}_r - \underset{\sim}{\ell}_r{}'|)\,|\underset{\sim}{\ell}_r - \underset{\sim}{\ell}_r{}'|\, V(\ell_r{}'). \quad (4.4)$$

Similar reductions may be performed for the other diagrams depicted in table 1.

Allen and Cole argue that the collision operator is dominated by the chain diagrams

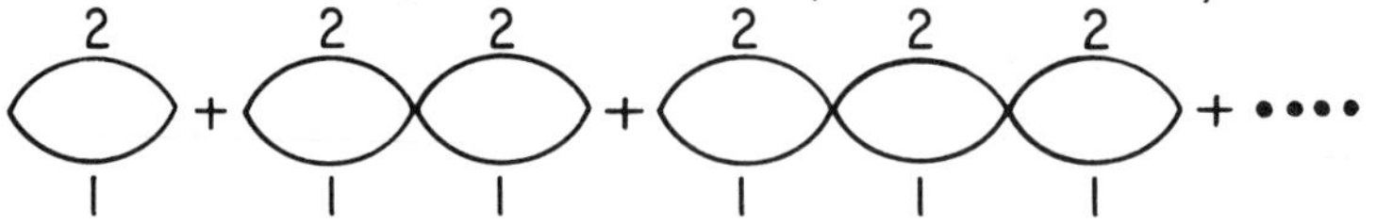 (4.5)

known to lead to Boltzmann's binary collision operator [9, 17], and the ring diagrams

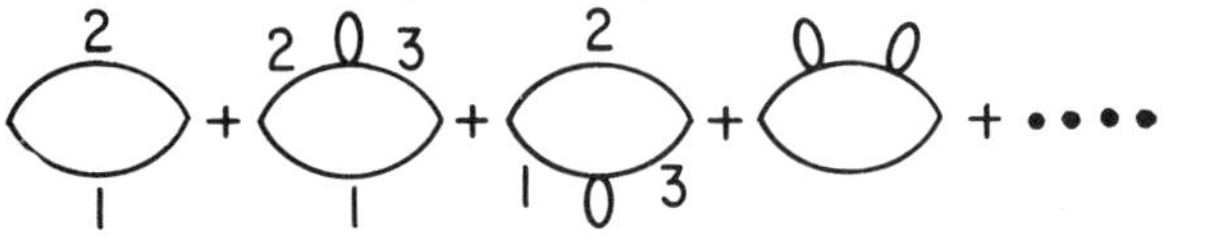 (4.6)

known to dominate for low-density plasmas. Summing these two classes of diagrams, they obtain for the collision operator Boltzmann's collision operator plus the plasma operator given in equation (3.37) minus the cycle operator  since it was counted in both equations (4.5) and (4.6). Moreover, unlike Rice and Allnatt, they argue that it is appropriate to use the total potential energy in the Boltzmann and plasma operators rather than decompose it into a strongly repulsive part contributing to the Boltzmann operator and a "soft" part contributing to the plasma operator.

In order to utilize their theory, Allen and Cole have to require that the pair potential have a Fourier transform. In their numerical work [8] they use the truncated Lennard-Jones potential

$$V(r) = 4\epsilon\left[\left(\frac{\sigma}{r}\right)^{12} - \left(\frac{\sigma}{r}\right)^6\right], \quad r \geq 0.7\sigma ;$$

$$= 0, \quad r \leq 0.7\sigma , \qquad (4.7)$$

the justification of the cutoff being that not many particles have the energy to achieve an interparticle separation $r < 0.7\sigma$. An alternative to equation (4.7), used by Dowling and the author, is to use a potential of the form

$$V(r) = \epsilon b \left\{ A \exp[-\alpha(\frac{r}{\sigma})^2] - \exp[-\gamma(\frac{r}{\sigma})^2] \right\}, \qquad (4.8)$$

where the constants ϵ and σ are the 6-12 Lennard-Jones constants and b, A, α, and γ are determined so as to ensure that equation (4.8) has the same position and depth of the minimum as equation (4.7) and that equation (4.8) be 0 at $r = \sigma$, where equation (4.7) is zero. For argon these parameters are $\epsilon/k = 121^\circ$ K, $\sigma = 3.418 \times 10^{-8}$ cm, $b = 6.07$, $A = 1400$, $\alpha = 8.54$, and $\gamma = 1.3$. In figure 1, equations (4.7) and (4.8) are compared, and in figure 2 the second virial coefficient is compared for the two potentials. Clearly, equations (4.7) and (4.8) should yield similar predicted values for equilibrium and transport properties. The motivation, of course, to choosing equation (4.8) over equation (4.7) is that its Fourier transform can be obtained explicitly with a resulting simplification of the numerical work.

Using the potential function given by equation (4.8), Dowling and the author have estimated the orders of magnitudes of the diagrams ranging through fourth order in the interaction

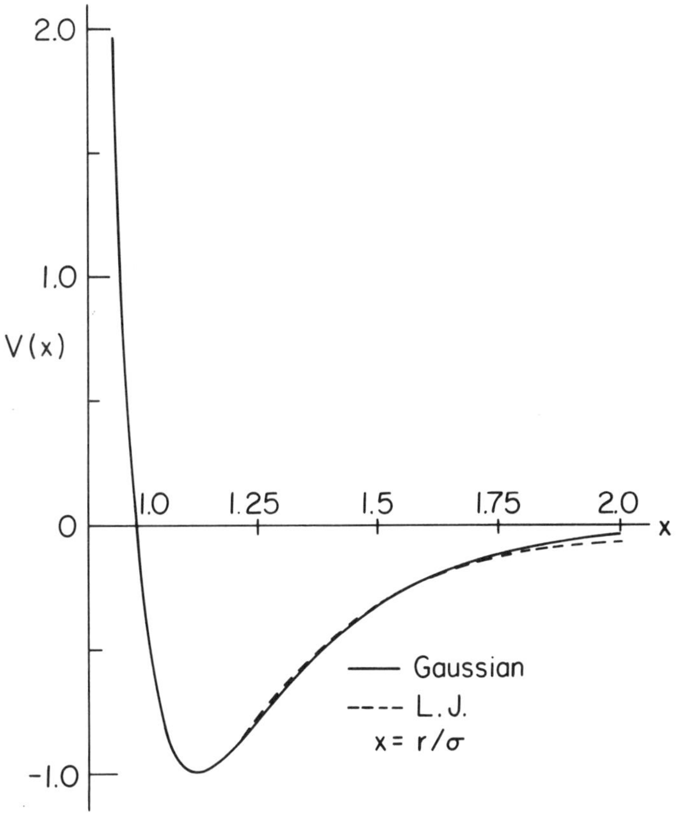

Fig. 1. Comparison of the Lennard-Jones and the Gaussian (eq. [4.8]) potential functions, respectively. The potential energies are plotted in units of ϵ versus the reduced distance r/σ. The parameters of the Gaussian potential are those for argon given in the text.

potential (see table 2). The column headed "Exact" corresponds to using the estimates in equation (4.4) and a similar estimate for  . In the other columns upper and lower bounds are placed on the terms. The details of these bounds are given elsewhere. Examination of the entries in table 2 leads to the conclusion that all of the diagrams are comparable to the ring diagram of the same order in the interaction. This conclusion is made even stronger by recognizing that $C\sigma^3 \lesssim \frac{1}{2}$ rather than the value of unity used for table 1. Thus, there seems to be no justification for keeping the ring diagrams over any others except for the appealing fact that the rings are a summable class.

As a further examination of the Allen-Cole theory, Dowling and the author computed the equilibrium radial distribution function from the expression Allen [8] used for the doublet distribution function in obtaining thermal conductivity from their theory. In the equilibrium limit the expression given is simply the ring sum which can be done explicitly. The result is

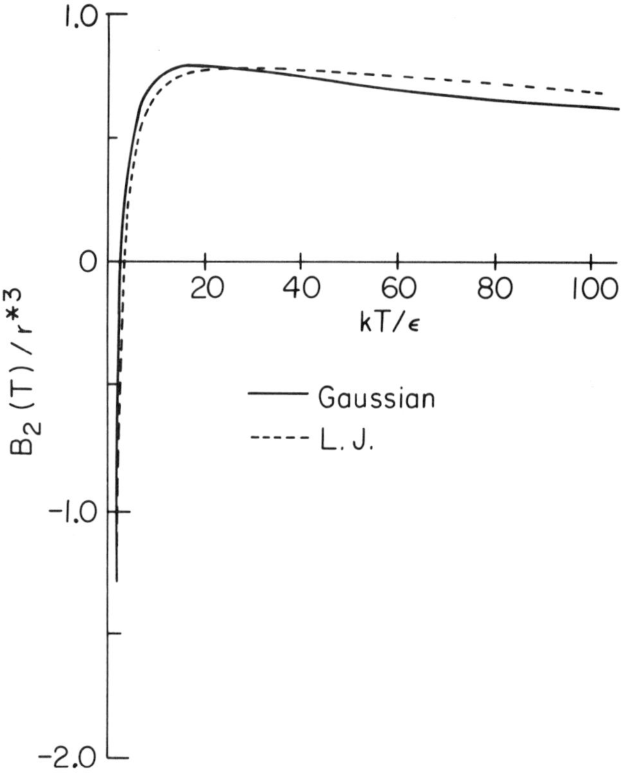

Fig. 2. Comparison of reduced second virial coefficients predicted by the Lennard-Jones and Gaussian potentials.

$$g(r) - 1 = -\beta \int d\underset{\sim}{\ell} \, \frac{V_{\ell} \exp(i \underset{\sim}{\ell} \cdot \underset{\sim}{r})}{1 + 8\pi^3 C \beta V_{\ell}} \quad . \tag{4.9}$$

As seen in figure 3, this expression drastically underestimates the first peak in the correlation function. Another test of equation (4.9) may be obtained by recalling the grand-canonical ensemble expression for the isothermal compressibility κ_T:

$$C kT \kappa_T - 1 = C \int d\underset{\sim}{r} \, [g(r) - 1] \quad . \tag{4.10}$$

Inserting equation (4.9) into the integrand of equation (4.10), we obtain

$$C kT \kappa_T - 1 = \frac{8\pi^3 C V_o}{1 + 8\pi^3 C \beta V_o} \quad , \tag{4.11}$$

where

$$V_o = \frac{1}{8\pi^3} \int d\underset{\sim}{r} \, V(r) \quad . \tag{4.12}$$

For argon at 87° K and under its vapor pressure, the Gaussian potential yields $\kappa_T = 1.82 \times 10^{-6}$ atm^{-1}

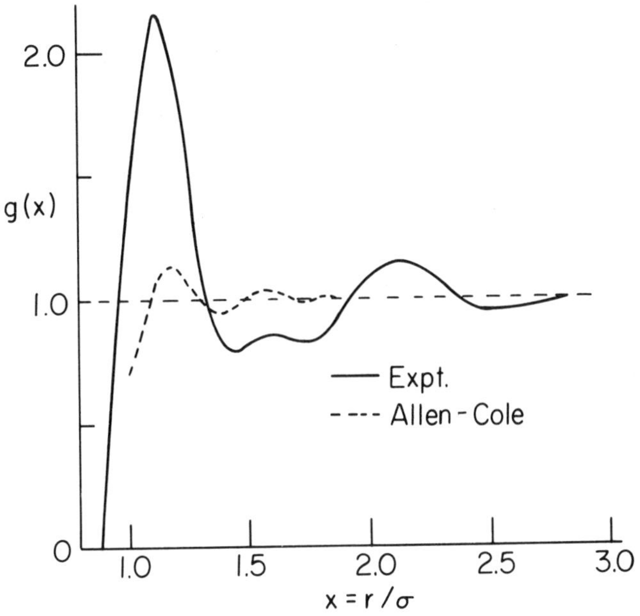

Fig. 3. Comparison of the experimental radial distribution function (C.J. Pings, private communication) with that predicted by eq. (4.9) for the Gaussian potential. The experimental data are for argon at 138.15° K and a density of 0.982 g/cm^3.

compared to the experimental value of $\kappa_T = 2.18 \times 10^{-4}$ atm^{-1}. For the truncated Lennard-Jones model equation (4.12) has the unacceptable feature of depending completely on the truncation value of r.

As another test of the theory Dowling and the author have computed self-diffusion coefficients for potential models equations (4.7) and (4.8) and obtained negative diffusion coefficients in both cases, the reason being that the contribution of the subtracted cycle diagram simply swamps the other contributions.

In view of preceding discussion, we feel that either (a) the diagrams kept in the Allen-Cole theory are not the best choice or (b) refusal to treat the hard and soft parts of the potential separately is responsible for the poor results.

V. EXISTENCE THEORY OF CHAPMAN-ENSKOG SOLUTIONS OF EQUATIONS OF THE RICE-ALLNATT TYPE

The Chapman-Enskog scheme for obtaining the so-called normal solutions to singlet kinetic equations involves solving a linear equation of the form

$$A\phi = h \ , \tag{5.1}$$

where A is the linear collision operator, h the inhomogeneous term, and $f^o\phi$ the deviation of the singlet distribution function from its local equilibrium value f^o. For example, the first Chapman-

Enskog approximation of equation (3.40) is obtained by solving an equation of the form of equation (5.1) with the operator A defined as follows:

$$A\phi = \frac{8\pi^3 kT}{f^\circ(1)m^2}\int d\underset{\sim}{v}_2 d\underset{\sim}{\ell} V_\ell \underset{\sim}{\ell}\cdot\underset{\sim}{\partial}_1 \frac{i}{\underset{\sim}{\ell}\cdot\underset{\sim}{v}_{12} - i0} G_\ell \underset{\sim}{\ell}\cdot\underset{\sim}{\partial}_{12} f^\circ(1)f^\circ(2)\left[\phi(1)+\phi(2)\right],$$

$$-\sigma^2 g(\sigma+) \int_{\underset{\sim}{v}_{21}\cdot\hat{\underset{\sim}{k}}>0} f^\circ(2)\left[\phi'(1)+\phi'(2)-\phi(1)-\phi(2)\right]\hat{\underset{\sim}{k}}\cdot\underset{\sim}{v}_{21}\, d\underset{\sim}{k}\, d\underset{\sim}{v}_2. \tag{5.2}$$

The domain $\mathfrak{D}_A$ of the operator A appropriate for kinetic theory is defined to be the set of functions $\phi\in\mathfrak{D}_A$, where

$$\mathfrak{D}_A = \left\{\phi(\underset{\sim}{v})\,|\,\phi(\underset{\sim}{v})\in C^2(R_3)\cap\mathfrak{L}^2(R_3;f^\circ),\ A\phi\in\mathfrak{L}^2(R_3;f^\circ)\right\}. \tag{5.3}$$

$C^n(R_3)$ denotes the space of functions of the three dimensional peculiar velocity vector, $\underset{\sim}{v} = \underset{\sim}{v} - \underset{\sim}{u}$, whose n(th) derivatives are continuous, and $\mathfrak{L}^2(R_3;\,f^\circ)$ is the space of functions which are Lebesque square integrable with weighting function f°. Thus, the inner product of ϕ and ψ in $\mathfrak{L}^2(R_3;\,f^\circ)$ is by definition

$$(\phi,\psi) \equiv \int d\underset{\sim}{v}\, f^\circ(\underset{\sim}{v})\,\bar{\phi}(\underset{\sim}{v})\,\psi(\underset{\sim}{v}), \tag{5.4}$$

and the norm of ϕ is

$$||\phi|| \equiv (\phi,\phi)^{\frac{1}{2}}. \tag{5.5}$$

For a certain class of operators the Fredholm alternative theorem [18] states that equation (5.1) has a solution if and only if h is orthogonal to the solutions of the homogeneous adjoint equation, i.e., if and only if

$$(\psi_h, h) = 0, \tag{5.6}$$

where ψ_h is any solution of the equation

$$A^*\psi_h = 0. \tag{5.7}$$

A^* in $\mathfrak{D}_{A^*}$ denotes the adjoint of the operator A in $\mathfrak{D}_A$. Hecke [19], Carleman [20], and Grad [21] have shown that the hard-sphere Boltzmann operator--the second part of the left-hand side of equation (5.2) aside from the constant $g(\sigma+)$--can be rewritten in the form

$$A_{Boltz}\phi = \nu(\underset{\sim}{v})\,\phi(\underset{\sim}{v}) + K[\phi], \tag{5.8}$$

where $\nu(\underset{\sim}{v})$ is a positive function greater than some constant $c_o > 0$ and K is a completely continuous operator. For operators of this form the Fredholm alternative theorem is valid, thus assuring the existence of a solution to the Chapman-Enskog sequence for the hard-sphere Boltzmann equation, and to the Enskog equation since $g(\sigma+)A_{Boltz}\phi = A_{Ens}\phi$.

Operators of the Rice-Allnatt type, however, are of the form

$$A\phi = B\phi + C\phi, \tag{5.9}$$

where C is a completely continuous symmetric operator and B is a differential operator of the form

$$B\phi = \frac{1}{f^\circ(1)}\left\{-\underset{\sim}{\partial}_1\cdot\left[\underset{\approx}{a}(\underset{\sim}{v}_1)\cdot\underset{\sim}{\partial}_1\phi\right] + q(\underset{\sim}{v})\,\phi\right\}, \tag{5.10}$$

where $\underset{\approx}{a}$ and q have the following properties:

1) $\underset{\approx}{a}(\underset{\sim}{v})$ is a real-valued symmetric tensor function;

2) $\underset{\approx}{a} : \underset{\sim}{\xi}\,\underset{\sim}{\xi} \geq \rho(\underset{\sim}{v})\, \xi^2$ for arbitrary vector $\underset{\sim}{\xi}$; $\rho(\underset{\sim}{v})$ is real valued and $\rho(\underset{\sim}{v}) > 0$ for $\underset{\sim}{v} \in R_3$;

3) $\underset{\approx}{a}(\underset{\sim}{v}) \in C'(R_3)$;

4) $q(\underset{\sim}{v})$ is a real valued positive definite scalar function, i.e., $q(\underset{\sim}{v}) \geq c_o > 0$.

Moreover, the operator A in $\mathfrak{D}_A$ is symmetric [22] and positive semidefinite; i.e.,

$$(\psi, A\phi) = (A\psi, \phi) \quad \text{and} \quad (\psi, A\psi) \geq 0 .$$

For operators of the type just described, there does not exist a Fredholm alternative theorem for the existence of a solution to equation (5.1) in the classical sense. We can, however, prove Theorem 1: If A is of the form given by equations (5.9)-(5.10) and $\mathfrak{D}_A$ is defined as in equation (5.3), then the equation

$$A\phi = h \tag{5.11}$$

has a solution in the <u>weak-$\mathfrak{L}^2$ sense</u> for $h \in \mathfrak{L}^2(R_3; f^o)$ if and only if

$$(h, \psi_h) = 0 \tag{5.12}$$

for every weak-$\mathfrak{L}^2$ solution ψ_h of the equation

$$A\psi_k = 0 \tag{5.13}$$

ϕ is said to be a weak-$\mathfrak{L}^2$ solution [23] of equation (5.11) if it satisfies the equation

$$(Au, \phi) = (u, h) \quad \text{for all} \quad u \in S \tag{5.14}$$

where S is a dense subset of $\mathfrak{L}^2(R_3; f^o)$ which is composed of infinitely differentiable functions of compact support, i.e., each function vanishes outside a compact subset of R_3 depending on the function itself. Theorem 1 states that h must be orthogonal to at least 1, $\underset{\sim}{v}_1$ and $\frac{1}{2}v_1^2$ since these ordinary solutions of equation (5.13) are certainly weak solutions. Moreover, on physical grounds one would not expect any other weak solutions to equation (5.13) since the homogeneous solutions express the conservation laws governing the collision dynamics.

To prove Theorem 1, we note that since B is a positive definite symmetric operator it possesses at least one self-adjoint extension [18] $\bar{B}$ in $\mathfrak{D}_{\bar{B}}$ such that $\bar{B}$ is also positive definite with the same lower bound as B and such that $\mathfrak{D}_B \subseteq \mathfrak{D}_{\bar{B}}$. The domain $\mathfrak{D}_B$ is identical to the domain $\mathfrak{D}_A$ since the domain $\mathfrak{D}_C = \mathfrak{L}^2(R_3; f^o)$ owing to the fact that C is a completely continuous operator [18].) Extending the operator in equation (5.11), we have the equation

$$\bar{B}w + Cw = h, \tag{5.15}$$

which can be rearranged to the form

$$w + \bar{B}^{-1} Cw = \bar{B}^{-1} h, \tag{5.16}$$

since the inverse of $\bar{B}$ exists and has the domain $\mathfrak{L}^2(R_3; f^o)$. Since $\bar{B}^{-1}$ is bounded, the product $\bar{B}^{-1}C$ is a completely continuous operator [18], and therefore, equation (5.16) is in a form for which the Fredholm alternative theorem has been proved. The orthogonality condition on h for the existence of a solution to equation (5.16) may be expressed as

$$(\theta_h, h) = 0 \ , \tag{5.17}$$

where θ_h is any solution to the homogeneous equation

$$\bar{B}\theta_h + C\theta = 0 \ . \tag{5.18}$$

Of course a solution to equation (5.16) is a solution to equation (5.15).

Assume now that w is a solution to equation (5.15). Taking the inner product of equation (5.15) with $u \in S$, we obtain the result

$$\begin{aligned}(u, h) &= (u, \bar{B}w + Cw) \\ &= (\bar{B}u + Cu, w) \\ &= (Bu + Cu, w) \quad \text{for all } u \in S \ .\end{aligned} \tag{5.19}$$

The second equality in equation (5.19) obtains because $\bar{B} + C$ is symmetric, and the third, because the subspace S is contained in $\mathfrak{D}_A$. Similarly, we find

$$(Bu + Cu, \theta_h) = 0 \quad \text{for all } u \in S \ . \tag{5.20}$$

Thus, Theorem 1 is proved since w provides a weak solution to equation (5.11) and the θ_h's are the weak solution to equation (5.13).

Although the proof of existence of a weak-$\mathfrak{L}^2$ solution is the best we have been able to do thus far for operators with the properties given above for A, Baleiko and the author [24] have shown that the weak solution to equation (5.11) is also a classical solution if for the Fokker-Planck part of the operator one used the simple "constant friction coefficient" used by Rice and Allnatt, namely,

$$\mathfrak{a}(\phi) = -\frac{\zeta_s}{f^o(1)} \, \underset{\sim}{\partial}_1 \cdot f^o(1) \, \underset{\sim}{\partial}_1 \, \phi \ . \tag{5.21}$$

For this case, Baleiko and the author showed that the condition equation (5.14) obeyed by the weak solution ϕ implies that $\phi \in C^2(R_3) \cap \mathfrak{L}^2(R_3; f^o)$ if $h \in C'(R_3) \cap \mathfrak{L}(R_3; f^o)$. This result and a similar result for ψ_h imply that ϕ and $\psi_h \in \mathfrak{D}_A$ and, therefore, that ϕ is a classical solution to equation (5.11). They also showed that the eigenfunctions and eigenpackets of the self-adjoint extension and the unextended operator are the same. The condition $h \in C'(R_3) \cap \mathfrak{L}^2(R_3; f^o)$ is obeyed by the inhomogeneous terms in the Chapman-Enskog sequence of equations so that the result obtained by Baleiko and the author provides an adequate existence theory for the Rice-Allnatt equation with the simplified Fokker-Planck operator equation (5.21). Such a theory is desirable but, unfortunately, not presently available for the more general RA collision operator, nor for the Balescu-Guernsey-Lenard plasma operator equation (3.36), for that matter. A thorough investigation into the properties of these operators is of interest not only as a mathematical nicety but also because of the role these properties--their spectra, for example--play in determining the decay to, the existence of, and stability (or instability) of, local hydrodynamic equilibria.

The author wishes to thank Gary Dowling and Marc Baleiko for helping prepare this manuscript. The author is also grateful for financial support of this research furnished by the National Science Foundation and the U.S. Army Research Office-Durham.

APPENDIX I

The PNM theory is based on four approximations. First is the neglect of deviations of the singlet distribution function from local equilibrium. The second approximation is to neglect the contributions of the inhomogeneity and nonstationarity, i.e., contributions from the operators ∇_i and ∂_t, except for the leftmost contribution in the matrix elements of the collision operator. For example, in the third-order contribution to

$$d_t f|_{\delta L^3} = -i 8\pi^3 \int d\underset{\sim}{v}_2 \int d\underset{\sim}{v}_3 \int d\underset{\sim}{x}_2 d\underset{\sim}{x}_3\, \delta(\underset{\sim}{x}_1 - \underset{\sim}{x}_2)\delta(\underset{\sim}{x}_1 - \underset{\sim}{x}_3)\, \frac{8\pi^3}{\Omega} \int d\underset{\sim}{\ell}$$

$$\theta^{12}(\underset{\sim}{\ell}) \exp(-i\nabla_2 \cdot \frac{\partial}{\partial \underset{\sim}{\ell}}) \left[\underset{\sim}{\ell} \cdot \underset{\sim}{v}_{12} - i \underset{\sim}{v}_1 \cdot (\nabla_1 + \nabla_3) - i \underset{\sim}{v}_2 \cdot \nabla_2 - i0 - i\partial_t\right]^{-1}$$

$$\theta^{13}(\underset{\sim}{\ell} - i\nabla_3) \left[\underset{\sim}{\ell} \cdot \underset{\sim}{v}_{32} - i(\underset{\sim}{v}_1 \cdot \nabla_1 + \underset{\sim}{v}_2 \cdot \nabla_2 + \underset{\sim}{v}_3 \cdot \nabla_3) - i0 - i\partial_t\right]^{-1} \theta^{23}(\underset{\sim}{\ell})$$

$$\times \left[\prod_{i=1}^{3} f^o(\underset{\sim}{x}_i, \underset{\sim}{v}_i, t)\right] / \left[n(\underset{\sim}{x}_3, t)\right] , \qquad \text{(A1.1)}$$

the operators ∇_i and ∂_t are kept only in the factor

$$\exp(-i\nabla_2 \cdot \frac{\partial}{\partial \underset{\sim}{\ell}}) \left[\underset{\sim}{\ell} \cdot \underset{\sim}{v}_{12} - i\underset{\sim}{v}_1 \cdot (\nabla_1 + \nabla_3) - i\underset{\sim}{v}_2 \cdot \nabla_2 - i0 - i\partial_t\right]^{-1} . \qquad \text{(A1.2)}$$

The third approximation is that the system has relaxed to the "fully delocalized" local equilibrium before the action of the operators ∇_i and ∂_t as prescribed by the second approximation. The fourth approximation is that for the purpose of eliminating the local equilibrium triplet correlation function in favor of the local equilibrium doublet correlation function the BBGKY hierarchy for equilibrium with conditions prevailing at $\underset{\sim}{x}_1$ and t can be used.

APPENDIX II

We demonstrate here the sense in which the PNM and the RA approximation are equivalent. Consider the contribution of $d_t f$ of order $(\delta L)^3$:

$$d_t f|_{\delta L^3} = -i 8\pi^3 \int d\underset{\sim}{v}_2 d\underset{\sim}{v}_3\, \delta(\underset{\sim}{x}_1 - \underset{\sim}{x}_2)\, \delta(\underset{\sim}{x}_1 - \underset{\sim}{x}_3) \frac{8\pi^3}{\Omega} \int d\underset{\sim}{\ell}\, \theta^{12}(\underset{\sim}{\ell}) \exp(-i\nabla_2 \cdot \frac{\partial}{\partial \underset{\sim}{\ell}})$$

$$\times \left[\underset{\sim}{\ell} \cdot \underset{\sim}{v}_{12} - i\underset{\sim}{v}_1 \cdot (\nabla_1 + \nabla_2) - i\underset{\sim}{v}_2 \cdot \nabla_2 - i0 - i\partial_t\right]^{-1} \theta^{13}(\underset{\sim}{\ell} - i\nabla_3)$$

$$\times \left[\underset{\sim}{\ell} \cdot \underset{\sim}{v}_{32} - i(\underset{\sim}{v}_1 \cdot \nabla_1 + \underset{\sim}{v}_2 \cdot \nabla_2 + \underset{\sim}{v}_3 \cdot \nabla_3) - i0 - i\partial_t\right]^{-1} \theta^{23}(\underset{\sim}{\ell})$$

$$\times \frac{1}{n(\underset{\sim}{x}_3, t)} \prod_{i=1}^{3} f^o(\underset{\sim}{x}_i, \underset{\sim}{v}_i, t) \left[1 + \sum_{i=1}^{3} \phi(\underset{\sim}{x}_i, \underset{\sim}{v}_i, t)\right] \qquad \text{(A2.1)}$$

This term can be linearized in ∇_i, ∂_t, and ϕ by expanding the propagation operators. The PNM approximation ignores the terms in ϕ and the terms in ∇_i, ∂_t other than those arising from the leftmost operator; i.e., they keep linear terms arising from the factor

$$\exp(-i\nabla_2 \cdot \frac{\partial}{\partial \underset{\sim}{\ell}}) \left[\underset{\sim}{\ell} \cdot \underset{\sim}{v}_{12} - i\underset{\sim}{v}_1 \cdot (\nabla_1 + \nabla_3) - i\underset{\sim}{v}_2 \cdot \nabla_2 - i0 - i\partial_t\right]^{-1}$$

and neglect the others.

Now in the terms linear in ϕ no ∇_i, ∂_t terms appear since they are of second order. The contribution of this part of equation (A2.1) can be written in the form

$$- i\, 8\pi^3 \int d\underset{\sim}{v}_2 d\underset{\sim}{v}_3\, d\underset{\sim}{\ell}\, \frac{8\pi^3}{\Omega}\, \theta_{12}(\underset{\sim}{\ell})\, \frac{1}{\underset{\sim}{\ell}\cdot\underset{\sim}{v}_{12} - i0}\, \theta_{12}(\underset{\sim}{\ell}) \left(\sum_i \phi(1)\right) \frac{1}{\underset{\sim}{\ell}\cdot\underset{\sim}{v}_{23} - i0}\, \theta_{32}(\underset{\sim}{\ell})$$

$$\frac{\prod_{i=1}^{3} f^{o}(i)}{n(3)} - i\, 8\pi^3 \int d\underset{\sim}{v}_2 d\underset{\sim}{v}_3\, d\underset{\sim}{\ell}\, \frac{8\pi^3}{\Omega}\, \theta_{12}(\underset{\sim}{\ell})\, \frac{1}{\underset{\sim}{\ell}\cdot\underset{\sim}{v}_{12} - i0}\, \theta_{13}(\underset{\sim}{\ell})\, \frac{1}{\underset{\sim}{\ell}\cdot\underset{\sim}{v}_{23} - i0}$$

$$\left[m^{-1} V_{\ell}\, \underset{\sim}{\ell}\cdot \underset{\sim}{\partial}_{32} \left(\sum_i \phi(i)\right)\right] \prod_{i=1}^{3} \frac{f^{o}(i)}{n(3)}\;, \qquad \text{(A2.2)}$$

where the quantity in square brackets is no longer an operator. The pattern in equation (A2.2) for higher-order interaction terms is to move the quantity $\sum_i' \phi(i)$ successively to the left, generating finally a sum of operators in which the "inhomogeneous" term $\sum_i \phi(i)$ appears to the right of two, three, etc., interaction operators. The RA approximation neglects all the contributions except those in which the inhomogeneity appears to the right of two interaction operators. The term for which the inhomogeneity appears to the right of one interaction operator is zero. Therefore, the PNM and the RA approximations are equivalent in the sense that they both retain only the contributions arising from the first leftmost appearance (actually last appearance in the time evolution) of an inhomogeneity in the collision operator.

NOTES ADDED IN PROOF

1. In the GRA theory it is unnecessary, and in fact inconsistent, to use in J^E of equation (3.41) Enskog's chaos assumption that the doublet distribution function factors into a product of singlets and the local equilibrium radial distribution function. The author has now eliminated the chaos assumption in the GRA theory and has used instead the approximation given by equation (3.59) for the doublet function in both J^E and the "soft" part of the collision operator. This work will appear in Annual Reviews of Chemical Physics. The corrections obtained are in the values of the coefficients A and B of equations (3.55) and (3.56). The forms of equations (3.62) and (3.63) are unchanged.

2. Misguich, in a work to be published in Molec. Phys., has eliminated the fourth PNM approximation mentioned in Appendix 1 with resulting corrections to the thermal conductivity. Viscosity and self-diffusion are unaffected by the corrections.

REFERENCES

1. S. Chapman and T. G. Cowling, The Mathematical Theory of Nonuniform Gases (London: Cambridge University Press, 1939).

2. J. G. Kirkwood, J. Chem. Phys. 14, 180 (1946).

3. S. A. Rice and P. Gray, The Statistical Mechanics of Simple Liquids (New York: John Wiley and Sons, 1966).

4. I. Prigogine, G. Nicolis, and J. Misguich, J. Chem. Phys. 43, 4516 (1965).

5. J. Misguich, Ph.D. thesis, Free University of Brussels, Brussels, Belgium (1968).

6. J. Misguich, J. Physique, 30, 221 (1969).

7. P. M. Allen and G. H. A. Cole, Molec. Phys. 14, 413 (1968); 15, 549 (1968); 15, 557 (1968).

8. P. M. Allen, Molec. Phys. 18, 349 (1970).

9. I. Prigogine, Nonequilibrium Statistical Mechanics (New York: Interscience, 1963).

10. R. Balescu, Statistical Mechanics of Charged Particles (New York: Interscience, 1963).

11. G. Severne, Physica, 31, 877 (1965).

12. R. L. Guernsey, The Kinetic Theory of Fully Ionized Gases, Off. Nav. Res., Contract No. Nonr. 1224 (15), July, 1960.

13. A. Lenard, Ann. Phys. 3, 390 (1966).

14. I. Prigogine and P. Resibois, Physica, 27, 629 (1961).

15. We approximated the contributing terms in g_{2s}^{1} by

$$g_{2s}^{1} = \int d\underline{x}_2\, \delta(\underline{x}_1 - \underline{x}_2) \nabla_2 \cdot \frac{\partial}{\partial \underline{\ell}} \frac{i}{\underline{\ell} \cdot \underline{v}_{12} - i0} \underline{\ell} \cdot \left[+ \frac{m \underline{v}_2}{kT(\underline{x}_2, t)} - \frac{m \underline{v}_1}{kT(\underline{x}_1, t)} \right]$$

$$kT \frac{(\underline{x}_1 + \underline{x}_2, t)}{2} \quad g_2^{o}(\underline{x}_1, \underline{x}_2, \underline{v}_1, \underline{v}_2, t) ,$$

whereas Misguich assumed that the temperature factor multiplying g_2^{o} was independent of $\underline{x}_2$. It can be shown that our form is necessary to reproduce the exact weak-coupling results.

16. G. Dowling and H. T. Davis, Can. J. Phys. 50, 317 (1972).

17. P. Resibois, in Physics of Many-Particle Systems, ed. E. Meeron (New York: Gordon and Breach, 1966).

18. F. Reisz and B. Sz. Nagy, Functional Analysis (New York: Frederick Ungar Publ. Co., 1955).

19. E. Hecke, Math. Z. 12, 274 (1922).

20. T. Carleman, Problèmes Mathématiques dans la Théorie Cinétique des Gaz (Uppsala: Almqvist and Wiksells, 1957).

21. H. Grad, Appl. Mech. Vol. 1, Suppl. 2, 26 (1963).

22. For a proof that operators of this form are symmetric see, for example, G. Helwig, Differential Operators of Mathematical Physics (Reading, Mass.: Addison-Wesley Publ. Co., 1967).

23. See, for example, A. Friedman, Partial Differential Equations (New York: Holt, Rinehart and Winston, 1969).

24. H. T. Davis and M. Baleiko, J. Stat. Phys., 3, 47 (1971).

DISCUSSION

D. Forster: I wonder to what extent calculations of transport coefficients assess the validity of the kinetic theory from which they are obtained, in the "kinetic state" where it is purported to be appropriate. Severne's equation omits the "destruction fragment." In the hydrodynamic regime, that is certainly all right. But I doubt that the equation would be suitable to discuss the kinetic behavior for nonvanishing frequency. Indeed, e.g., computer experiments make the existence of a kinetic stage in dense liquids rather unlikely. We have computed the shear viscosity of liquid argon by using generalized hydrodynamics and sum rules (Phys. Rev. **170**, 160 [1968]). The calculation is extremely simple and very crude, with an accuracy of only about 30%, say (of course, the experimental values seem to be not much better, really). What that calculation does show is how relatively insensitive the transport coefficient is to details of the dynamics in the kinetic stage.

H. T. Davis: Calculations of transport coefficients simply afford one test of a kinetic theory. Another test that I have pointed out here is provided by comparing the hydrodynamic equations generated by the approximate kinetic theory with the exactly known lowest-order or nondissipative equations.

Another reason I feel the kinetic equation approach is fruitful is that in principle one can extend models successful in treating linear problems to nonlinear problems with the same formalism. Up to now, of course, not much progress has been made along these lines except in the dilute gas limit.

Although the "destruction fragment" is neglected in the long-time limit of Severne's kinetic equation that I have used here, Severne's equation retains that term. In fact, aside from mathematical questions of convergence of operator expansions, Severne's kinetic equation is exact. To study nonvanishing frequency behavior it would be appropriate to begin with Severne's original equation rather than with the long-time limit that I have used. The condition that the destruction fragment be negligible in the long-time limit is that the correlations in the initially prepared system be of finite range. Such a condition should not affect the long-time hydrodynamic and transport properties of a fluid.

As for numerical agreement between theoretical and experimental transport coefficients, I believe that, given the state of the art in computing equilibrium correlation functions, there is no way to distinguish among theories coming to within 30 or 40% of experiment.

R. P. Futrelle: In calculating higher moments for the time correlation methods it would seem that higher-order correlation functions, other than $g_2(r)$, are needed.

H. T. Davis: Yes, it's true. This is, in fact, the major problem in attempting to fit time correlation functions or their memory functions to some functional form containing parameters to be determined from moments or sum rules.

PROFESSOR ROBERT BROUT
Faculté des Sciences
Université Libre de Bruxelles
Avenue F.D. Roosevelt, 50
Brussels, Belgium

PROFESSOR MORREL H. COHEN
The James Franck Institute
The University of Chicago
5640 Ellis Avenue
Chicago, Illinois 60637

PROFESSOR JACK D. COWAN
Chairman, Department of Theoretical Biology
Experimental Biology Building 105A
The University of Chicago
Chicago, Illinois 60637

PROFESSOR F. W. CUMMINGS
Department of Physics
University of California
Riverside, California 92502

PROFESSOR H. TED DAVIS
Department of Chemical Engineering
University of Minnesota
Minneapolis, Minnesota 55455

PROFESSOR EUGENE FEENBERG
Department of Physics
Washington University
St. Louis, Missouri 63130

PROFESSOR ROBERT B. GRIFFITHS
Physics Department
Carnegie-Mellon University
Pittsburgh, Pennsylvania 15213

PROFESSOR KYOZI KAWASAKI
Department of Physics
Temple University
Philadelphia, Pennsylvania 19122

DR. ROBERT H. KRAICHNAN
Dublin, New Hampshire 03444

DR. ROLF LANDAUER
IBM Thomas J. Watson Research Center
P. O. Box 218
Yorktown Heights, New York 10598

PROFESSOR J. L. LEBOWITZ
Belfer Graduate School of Science
Yeshiva University
Amsterdam Avenue and 185th
New York, New York 10033

PROFESSOR ELLIOTT W. MONTROLL
Department of Physics and Astronomy
The University of Rochester
River Campus Station
Rochester, New York 14627

PROFESSOR DONALD S. ORNSTEIN
Department of Mathematics
Stanford University
Stanford, California 94305

PROFESSOR J. K. PERCUS
Courant Institute of Mathematical Physics
New York University
New York, New York 10012

DR. ANEESUR RAHMAN
Argonne National Laboratory
9700 Cass Avenue
Argonne, Illinois 60439

PROFESSOR PIERRE RESIBOIS
Faculté des Sciences
Université Libre de Bruxelles
Avenue F. D. Roosevelt, 50
Brussels, Belgium

PROFESSOR ANTHONY ROBERTSON
Department of Theoretical Biology
Experimental Biology Building, 109
University of Chicago
Chicago, Illinois 60637

PROFESSOR STEVE SMALE
Department of Mathematics
University of California, Berkeley
Berkeley, California 94720

PROFESSOR RENE THOM
Institut des Hautes Etudes Scientifiques
91 Bures-sur-Yvette
Seine et Oise, France

PROFESSOR LOUP VERLET
Laboratoire de Physique Théorique
et Hautes Energies
Faculté des Sciences, D'Orsay, BP no. 12
Université de Paris
Orsay, France

DR. JAMES W. F. WOO
IBM Thomas J. Watson Research Center
P. O. Box 218
Yorktown Heights, New York 10598

PROFESSOR ROBERT ZWANZIG
Institute for Fluid Dynamics
and Applied Mathematics
University of Maryland
College Park, Maryland 20740

Institute for Molecular Physics
University of Maryland
College Park, Maryland 20740